Microbial Biotechnology Energy and Environment

Adorations and salutations to

Surya Narayana
the Sun God
(Chandogya Upanishad, 3.19.1)

The Lord of Light,
the source of all Energy,
the purifier of Environment
and the Sustainer of Life on this planet

'Surya Atma Jagatastasthushashcha'
(the Sun God is the Soul of all beings, moving and non-moving)
- from the Vedas

This book is dedicated to

*Late Professor C.M. GOVIL
my Teacher
with a deep sense of gratitude and love*

Microbial Biotechnology Energy and Environment

Editor

Rajesh Arora, PhD

Staff Officer
Office of the Distinguished Scientist and
Chief Controller Research and Development
Defence Research and Development Organization
Ministry of Defence, Government of India
New Delhi, India

CABI is a trading name of CAB International

CABI	CABI
Nosworthy Way	875 Massachusetts Avenue
Wallingford	7th Floor
Oxfordshire OX10 8DE	Cambridge, MA 02139
UK	USA
Tel: +44 (0)1491 832111	T: +1 800 552 3083 (toll free)
Fax: +44 (0)1491 833508	T: +1 (0)617 395 4051
E-mail: info@cabi.org	E-mail: cabi-nao@cabi.org
Website: www.cabi.org	

© CAB International 2012. All rights reserved. No part of this publication may be reproduced in any form or by any means, electronically, mechanically, by photocopying, recording or otherwise, without the prior permission of the copyright owners.

Disclaimer: The views expressed in this book are those of the Editor and do not necessarily reflect those of the Defence Research and Development Organization, Ministry of Defence or the Government of India.

A catalogue record for this book is available from the British Library, London, UK.

Library of Congress Cataloging-in-Publication Data

Microbial biotechnology : energy and environment / editor, Rajesh Arora.
       p. ; cm.
   Includes bibliographical references and index.
   ISBN 978-1-84593-956-4 (alk. paper)
   I. Arora, Rajesh.
   [DNLM: 1. Bioelectric Energy Sources. 2. Biodegradation, Environmental. 3. Biofuels. QT 34]

  333.72--dc23

                                                                 2012009991

ISBN-13: 978 1 84593 956 4

Commissioning editor: Rachel Cutts
Production editor: Fiona Chippendale

Printed and bound in the UK from copy supplied by the authors by CPI Group (UK) Ltd, Croydon, CR0 4YY.

Contents

Contributors	x
Foreword	xii
Acknowledgements	xiii
About the Editor	xiv

Part 1: Microbial Biotechnology: Present and Future Prospects

1. **Emerging Trends in Microbial Biotechnology: Energy and Environment** — Rajesh Arora … 1

Part 2: Harnessing Sustainable Energy Sources from Microbes

2. **The Microbiology of Microbial Electric Systems** — Sarah A. Hensley, Madeline Vargas and Ashley E. Franks … 16

3. **A Comparative Assessment of Bioelectrochemical Systems and Enzymatic Fuel Cells** — Deepak Pant, Gilbert Van Bogaert, Ludo Diels and Karolien Vanbroekhoven … 39

4. **Electrical Energy from Microorganisms** — Sheela Berchmans … 58

5. **Rumen Microbial Fuel Cells** — Chin-Tsan Wang, Che-Ming J. Yang, Yung-Chin Yang … 78

Part 3: Mechanistics of Bioenergy Production

6. **Systems Microbiology Approach to Bioenergy** — Qasim K. Beg and Ritu Sarin … 97

7. **Nanotechnology and Bioenergy: Innovations and Applications** — Mrunalini V. Pattarkine … 112

8. **Host Engineering for Biofuel-Tolerant Phenotypes** — Becky J. Rutherford and Aindrila Mukhopadhyay … 148

Part 4: Bioenergy from Wastes and Pollutant Removal

9. **Microbial Fuel Cells: Electricity Generation from Organic Wastes by Microbes** — 162
   Kun Guo, Daniel J. Hassett and Tingyue Gu

10. **Integration of Anaerobic Digestion and Oil Accumulation: Bioenergy Production and Pollutants Removal** — 190
    Mi Yan, Jianguo Zhang, Bo Hu

11. **Biohydrogen Generation Through Solid Phase Anaerobic Digestion from Organic Solid Waste** — 207
    S. Jayalakshmi

Part 5: Microalgae for Biofuels

12. **Algae – A Novel Biomass Feedstock for Biofuels** — 224
    Senthil Chinnasamy, Polur Hanumantha Rao, Sailendra Bhaskar, Ramasamy Rengasamy and Manjinder Singh

13. **Biofuel from Microalgae: Myth versus Reality** — 240
    Jubilee Purkayastha, Hemanta Kumar Gogoi, Lokendra Singh and Vijay Veer

Part 6: Bioremediation Technologies for Petroleum Hydrocarbons, PAHs and Xenobiotics

14. **Biodegradation of Petroleum Hydrocarbons in Contaminated Soils** — 250
    Aniefiok E. Ite and Kirk T. Semple

15. **Bioremediation of Polycyclic Aromatic Hydrocarbons (PAHs)** — 279
    Carl G. Johnston and Gloria P. Johnston

16. **The Role of Biological Control in the Creation of Bioremediation Technologies** — 297
    Yana Topalova

Part 7: Bioremediation of Nuclear Waste

17. **Bioremediation of Uranium, Transuranic Waste and Fission Products** — 310
    Evans M.N. Chirwa

18. **Uranium Bioremediation: Nanotechnology and Biotechnology Advances** 349
   Mrunalini V. Pattarkine

Part 8: Extremophilic Microbes: Role in Environmental Cleanup

19. **Going Extreme for Small Solutions to Big Environmental Challenges** 363
    Chris Bagwell

Index 382

Contributors

Rajesh Arora, Office of the Distinguished Scientist and Chief Controller Research and Development (Life Sciences and International Cooperation), Room No. 339, DRDO Bhawan, Rajaji Marg, New Delhi 110105, India

Christopher (Kitt) E. Bagwell, Savannah River National Laboratory, SRNL Environmental Sciences and Biotechnology, Savannah River Site Aiken, South Carolina 29808, USA

Qasim Beg, Boston University, Boston, Massachusetts, USA

Sheela Berchmans, Electronics and Electrocatalysis Division, Central Electrochemical Research Institute, Karaikudi 630006, India

Sailendra Bhaskar, Aban Informatics Private Limited, Biotechnology Division, Chennai 600008, India

Gilbert Van Bogaert, Separation and Conversion Technology, VITO – Flemish Institute for Technological Research, Boeretang 200, Mol 2400, Belgium

Senthil Chinnasamy, Aban Infrastructure Private Limited, Biotechnology Division, Chennai 600008, India

Evans Chirwa, Department of Chemical Engineering, University of Pretoria, South Africa

Ludo Diels, Separation and Conversion Technology, VITO – Flemish Institute for Technological Research, Boeretang 200, Mol 2400, Belgium

Yujie Feng, Harbin Institute of Technology, China

Ashley E. Franks, Department of Microbiology, University of Massachusetts Amherst, Amherst, Massachusetts, USA and Department of Microbiology, La Trobe University, Bundoora, Victoria, Australia

Hemanta Kumar Gogoi, Defence Research Laboratory, Tezpur 784001, Assam, India

Tingyue Gu, Department of Chemical and Biomolecular Engineering, Ohio University, Athens, Ohio 45701, USA

Kun Guo, National Key Laboratory of Biochemical Engineering, Institute of Process Engineering, Chinese Academy of Sciences, Beijing 100190, People's Republic of China

Sarah A. Hensley, Department of Microbiology, University of Massachusetts Amherst, Amherst, Massachusetts, USA

Daniel J. Hassett, Department of Molecular Genetics, Biochemistry and Microbiology, University of Cincinnati College of Medicine, Cincinnati, Ohio 45267, USA

Bo Hu, Department of Bioproducts and Biosystems Engineering, 316 Bio AgEng, 1390 Eckles Ave., University of Minnesota, St Paul, Minnesota 55108-6005, USA

Aniefiok E. Ite, Lancaster Environment Centre, Lancaster University, Lancaster LA1 4YQ, UK

S. Jayalakshmi, Institute of Remote Sensing, Department of Civil Engineering, Anna University, Chennai 25, India

Carl Johnston, Biology Department, Youngstown State University, Youngstown, Ohio 44555, USA

G. Patricia Johnston, Department of Biological Sciences, Kent State University, Kent, Ohio 44242, USA

Aindrila Mukhopadhyay, Physical Biosciences Division, Lawrence Berkeley National Laboratory, Berkeley, California 94720; Director, Fuels Transport and Toxicity, Director, Omics studies, Joint BioEnergy Institute, Emeryville, California 94608, USA

Deepak Pant, Separation and Conversion Technology, VITO – Flemish Institute for Technological Research, Boeretang 200, Mol 2400, Belgium

Mrunalini V. Pattarkine, Harrisburg University of Science and Technology, 326 Market Street, Harrisburg, Pennsylvania 17101, USA

Jubilee Purkayastha, Defence Research Laboratory, Tezpur 784001, Assam, India

Polur Hanumantha Rao, Aban Infrastructure Private Limited, Biotechnology Division, Chennai 600008, India

Ramasamy Rengasamy, Centre for Advanced Studies in Botany, University of Madras, Guindy Campus, Chennai 600025, India

Becky J. Rutherford, Lawrence Berkeley National Laboratory, Berkeley, California 94720, USA

Ritu Sarin, Yale University, New Haven, Connecticut, USA

Kirk T. Semple, Lancaster Environment Centre, Lancaster University, Lancaster LA1 4YQ, UK

Lokendra Singh, Defence Research Laboratory, Tezpur 784001, Assam, India

Manjinder Singh, Biorefining and Carbon Cycling Program, Department of Biological and Agricultural Engineering, The University of Georgia, Athens, Georgia 30602, USA

Yana Topalova, Faculty of Biology, Sofia University 'St. Kliment Ohridski', Dragan Tzankov Bld. 8, 1164 Sofia, Bulgaria

Karolien Vanbroekhoven, Separation and Conversion Technology, VITO – Flemish Institute for Technological Research, Boeretang 200, Mol 2400, Belgium

Madeline Vargas, Department of Microbiology, University of Massachusetts Amherst, Amherst, Massachusetts, and Department of Biology, College of the Holy Cross, Worcester, Massachusetts, USA

Chin-Tsan Wang, National Ilan University, 1, Sec. 1, Shen-Lung Road, I-Lan, 260, Taiwan, Republic of China

Mi Yan, Department of Bioproducts and Biosystems Engineering, 316 Bio AgEng, 1390 Eckles Ave., University of Minnesota, St Paul, Minnesota 55108-6005, USA

Che-Ming J. Yang, National Ilan University, 1, Sec. 1, Shen-Lung Road, I-Lan, 260, Taiwan, Republic of China

Yung-Chin Yang, National Ilan University, 1, Sec. 1, Shen-Lung Road, I-Lan, 260, Taiwan, Republic of China

Jianguo Zhang, Department of Bioproducts and Biosystems Engineering, 316 Bio AgEng, 1390 Eckles Ave., University of Minnesota, St Paul, Minnesota 55108-6005, USA

Foreword

Microbes were the first forms of life to appear nearly 3.8 billion years ago on this planet and have developed the ability to synthesize a plethora of molecules and also catalyse a number of reactions, while evolving hand-to-hand with the higher species. The immense potential of microbes has been harnessed for industrial and other applications.

In recent years, there has been a spurt of interest in microbial biotechnology, which has been innovatively utilized for energy generation and environmental clean-up. The major reasons responsible for the spurt of interest in alternative green fuels can be attributed to the depletion of fossil fuels, increase in vehicular population, rapid industrialization, growing energy demand due to the burgeoning population, environmental pollution and consequent global warming. Biofuels, being renewable, are the future fuels, which could meet the energy requirements of the civilian and defence sector. Microbes have also been shown to produce electricity and it is anticipated that microbial fuel cells would one day power small equipment and wearable clothing of soldiers and astronauts. However, at present, there is much that needs to be perfected in terms of microbial technology pertaining to electricity generation, synthesis of biofuels and environmental clean-up – and this forms the basis for this book.

Energy and environment in today's world are interlinked and are attracting immediate attention at global level. Endeavouring to meet the growing energy demand, man has already polluted the environment to such an extent that the planet is on the brink of an environmental catastrophe if things are not set right in real earnest. Hydrocarbons, nuclear materials, heavy metal pollution – all need to be taken care of and bioremediation by microbes is a useful and pragmatic way to clean the environment. With increasing focus on harnessing nuclear power for meeting energy needs, the problem of managing high- and low-level nuclear wastes also has to be addressed. This book focuses on the two key issues viz., energy and environment that are confronting humanity. There is a need to devise strategies for protecting the environment, at the same time adequately meeting the ever-growing energy needs of the world. Harnessing the power of microbes is one step towards finding cheap, green and sustainable solutions to the problems of energy and environment.

This unique and comprehensive book with a focus on energy and environment is a well-timed effort undertaken by Dr Rajesh Arora, who has carefully brought together the international subject experts in the area on a common platform to share their thoughts on these contemporarily relevant problems of mankind. I congratulate Dr Arora on his efforts in undertaking the task of compiling the state-of-the-art in the field and am confident that the book will be useful to all stakeholders, including researchers, environmentalists, academicians, students and the industry.

Dr W. Selvamurthy
Distinguished Scientist and Chief Controller Research and Development
Defence Research and Development Organization
(Government of India), New Delhi, India

Acknowledgements

I am greatly indebted to all those who have inspired me.
My special thanks to:
Dr V.K. Saraswat
Scientific Adviser to Defence Minister
Secretary and Director General Research and Development (DRDO)
Ministry of Defence, Government of India
New Delhi, India

Dr W. Selvamurthy
Distinguished Scientist and Chief Controller Research and Development
DRDO Headquarters
New Delhi, India

Late (Mrs) Shanti and Mr Gopal Kumar Arora
my dear parents

Mrs Preeti Arora
my dear wife

Tanmay and Geetansh
my Angelic children

My Teachers and Students

The Esteemed Contributors

Dr Nigel Farrar
Editorial Director, CABI, UK

Ms Rachel Cutts
Associate Editor, CABI, UK

Ms Fiona Chippendale nee Harrison
Former Production Editor, CABI, UK

Mr Simon Hill
Senior Production Editor, CABI, UK

Ms Alexandra Lainsbury
Editorial Assistant, CABI, UK

and all those who helped me, but I have omitted to mention...

Rajesh Arora

The Editor

Dr Rajesh Arora is a Scientist with India's Defence Research and Development Organization and currently holds the position of Staff Officer in the Office of the Distinguished Scientist and Chief Controller Research and Development (Life Sciences and International Cooperation). He is well known for his contributions to the area of novel drug design and development, particularly countermeasure agents for CBRN Defence. Dr Arora's name is included in the *Who's Who of the World*, USA, *Who's Who in Science and Engineering*, USA and *International Biography*, UK and has also been awarded several prestigious fellowships and awards. Dr Arora is a recipient of the esteemed DRDO Laboratory Scientist of the Year Award (2010). He has been a Visiting Scientist in the European Union during 2009–2010. Dr. Arora is also an alumnus of the George C. Marshall European Center for Security Studies. Dr Arora has more than 150 publications, 12 patents and 10 books to his credit. He is a life member of several professional societies and sits on the editorial board of numerous journals. Several of his students currently occupy responsible positions in the government/academia/industry. Dr Arora is a PhD in Biotechnology and also holds post-graduate diplomas in Human Resource Development and Marketing Management and an MBA degree in Human Resource Management. Dr Arora has also been actively involved in several Techno-managerial assignments, including Human Resource Development for leveraging human capital in cutting edge areas of Science and Technology.

e-mail: rajesharoradr@rediffmail.com or rajesharoradr@gmail.com

Chapter 1

Emerging Trends in Microbial Biotechnology: Energy and Environment

Rajesh Arora

Introduction

The source of most energy on Earth is the Sun.

> The sun is the great mother. All life on earth might be considered as transient materialization of the exhaustless floods of radiance which she pours on the planet's surface. This enables green plants to synthesize sugars and starches from water in the soil and form carbon dioxide gas in the atmosphere, thus making possible the emergence of all other forms of life on earth by producing the essential foods. We eat sunshine in sugar, bread, and meat, burn sunshine of millions of years ago in coal and oil, wear sunshine in wool and cotton. Sunshine makes the wind and the rain, the summers and winters of years and of ages. Inextricably interwoven are the threads of life and light.
>
> Thomas R. Henry, in his article, 'The Smithsonian Institution', *The National Geographic Magazine*, September 1948

The web of life revolves because of energy and most of it comes from the Sun. Ever since the existence of humans on this planet, attempts have been made to harness energy, including solar energy in its various forms. It is a well known fact that energy is the main driving force for sustainable development. In the last few decades, energy consumption has increased exponentially at global level and it is apparent that energy security is vital to the future economic prosperity and environmental safety. Since energy is linked to accessibility to water, agricultural growth, industrial production, better human health and overall quality of life, and infrastructure development, immense attention is being paid in this area.

Attempts have been made to predict future energy needs and simultaneously to harness various sources of energy e.g. fossil energy sources, solar, wind, hydro-wave and nuclear power etc. to meet the growing energy demand in the civil and military sector. The military world over is exceedingly dependent on energy and in view of threats from various quarters, homeland security is also increasingly becoming energy-dependent. The world has witnessed several energy upheavals in recent times, e.g. the Deepwater Horizon oil spill (the BP oil spill in the Gulf of Mexico) of 2010, the recent Fukushima nuclear reactor accident in Japan (2011), oil-price volatility and uncertainty in petroleum product supply due to colossal uprisings in the Arab world. This has further aggravated the oil situation the world over. All these and various other reasons necessitate the development of alternative sources of energy to avert future energy crises.

Energy and environment are interlinked. On the one hand, there is a need to judiciously harness the available energy sources and on the other to develop novel sustainable energy generation technologies for meeting futuristic needs. Simultaneously, there is an urgent need to protect the environment while meeting the global energy requirements – a tough challenge indeed!

Global Energy Scenario and Emerging Trends

The ever increasing energy demand, geopolitical pressures, and the search for alternative sources of energy continue to challenge the global energy landscape. While fossil fuels have been a major energy source, petroleum has been commonly used as an important 'primary energy' source. Over the last five decades, oil has become the world's most important source of energy, mainly due to its high energy density, easy transportability and relative abundance. The accelerated consumption of petroleum worldwide is causing serious climate change, impacting the global environment. The greenhouse gases, primarily CO_2 released mainly as a result of transportation, are expected to touch 2.7 billion t by 2030. The economy of most developed and developing nations is dependent to a great extent on oil and its derivatives and consequently any disruption in the oil supply whether due to geopolitical unrest or otherwise, adversely impacts not only the economy but also homeland security. Consequently, the thrust of most nations now is on economy, energy security, climate change and environment protection. With a spurt in the population in China, India and other small nations, the energy demand is increasing substantially and with the limited oil reserves, it would be hard to meet the demand in coming years.

The demand for oil and other sources of energy is growing at extremely fast pace and worldwide energy consumption is projected to increase by approximately 36% by the year 2030. It is being anticipated that maximum growth will occur in countries with emerging economies such as China and India. In addition, rising energy demand from economic output and improved standards of living in most nations will also put added pressure on energy supplies. In China alone, energy demand is expected to increase by nearly 75% by 2035 and India is not going to be far behind. To meet the growing demand, there is a need to develop conventional and new sources of oil and gas, use energy more wisely and efficiently, develop renewable sources of energy and harness the next generation energy sources. However, even if the use of renewables triples over the next 25 years, the world is still likely to depend on fossil fuels for meeting at least 50% of its energy needs and hence reducing CO_2 emissions is going to be a big challenge, especially in developing countries.

As per estimates of International Energy Outlook (IEO 2011) (Reference case), world energy consumption increases by nearly 53%, from 505 quadrillion Btu in 2008 to 770 quadrillion Btu in 2035. World energy demand increases strongly as a result of robust economic growth and expanding populations in the world's developing economies. Organization for Economic Cooperation and Development (OECD) member countries constitute the major advanced energy consensus and it is evident that energy demand in OECD economies grows slowly over the projection period, at an average annual rate of 0.6%, whereas energy consumption in the non-OECD emerging economies expands by an average of 2.3% per year. China and India continue to lead world economic and energy demand growth in the Reference case and energy consumption has substantially increased in these nations. Together these two countries accounted for about 10% of total world energy consumption in 1990 and 21% in 2008. While the US energy consumption declined by 5.3% in 2009, energy use in China is estimated to have surpassed that of the USA for the first time. In the IEO 2011 Reference case, strong economic growth continues in China and India and their combined energy use more than doubles, accounting for 31% of total world energy consumption in 2035. It is anticipated that by 2035, China's energy demand would be significantly higher (nearly 68%) than US energy demand.

Apart from petroleum, electricity is the world's fastest growing form of end-use energy consumption. As per the Reference case, net electricity generation worldwide rises by 2.3% per year on average from 2008 to 2035, while total world energy demand grows by 1.6% per year.

It is evident that the demand for oil and electricity is going to increase substantially in coming years and is going to depend on levels of economic development and political, social and demographic factors. Considering the prevailing scenario, there is a need to holistically harness all available sources of energy, including renewable sources with great zeal (Fig. 1.1).

Harnessing alternative energy sources

In view of the ever increasing energy demands and energy deficits projected, there is a need for global investments in renewable and alternative energy. In particular, renewables such as geothermal energy and solar energy, cellulosic, algal and microbial biofuels, which do not undermine the food supply, and other novel energy sources will likely provide power for the world and additionally benefit the environment. Fossil fuels (oil and natural gas) are likely to remain the world's predominant sources of energy for decades to come. Even under the most aggressive climate policy scenario presented by the International Energy Agency, fossil fuels are still expected to contribute at least 50% of the world's energy supplies until 2035. However, in order to meet the growing demand of developing economies, every energy source available, including efficiency and renewables, will need to be harnessed ingeniously. It has been felt that the energy security of any country remains vulnerable until alternative fuels are developed from renewable sources (plants, microbes etc.), which can substitute or supplement petro-based fuels.

Sustainable Energy from Microbes

Microbial fuel cells: generating electricity the green way

Renewable bioenergy is viewed as an effective means to alleviate the current energy and global warming crisis. To meet the increasing energy demand, research efforts are being made around the globe to develop alternative electricity production methods. In the present scenario, electricity production from renewable resources without a net carbon dioxide (CO_2) emission is highly desirable from environmental point of view (Lovely, 2006, 2008; Davis and Higson, 2007; see Chapters 2–5, this volume). Microbial fuel cells (MFCs) that convert the energy stored in chemical bonds in organic molecules to electrical energy have generated considerable interest in recent years. Several researchers are of the opinion that

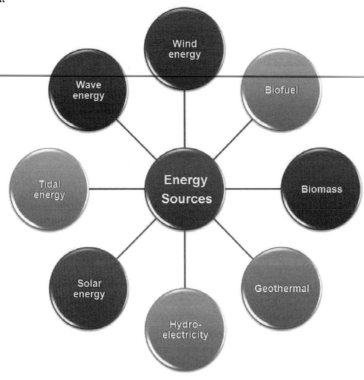

Fig. 1.1. Different sources of energy.

microbial production of electricity may one day become one of the important form of bioenergy (Allen and Bennetto, 1993; Gil *et al.*, 2003; Moon *et al.*, 2006; Du *et al.*, 2007; Ringeisen *et al.*, 2007; Lovely, 2008), particularly because they offer a recyclable, efficient and clean (green) way to generate electricity. Simultaneously, the use of MFC technology holds promise in biodegradation of organic matters and wastes (Park and Zeikus, 2003; Oh and Logan, 2005; Lovely, 2008). Miniature microbial fuel cells may one day provide emergency backup for soldiers and civilians. They could power military equipment (night vision devices, wearable military power packs, global positioning systems, personnel protective clothing, sensors, digital communication systems, helmet-mounted displays etc.).

Rapid strides have been made in MFC research in recent years and a number of groups are working in the area presently with the fond hope that they will come up with a breakthrough. The area of MFC has been reviewed in recent years by a number of researchers, who are focusing their attention on electrochemically active species, extracellular electron transport modes. Logan *et al.* (2006) have reviewed the MFC designs, characterizations and performance. The issue of microbial metabolism in MFCs has been reviewed by Rabaey and Verstraete (2005). Pham *et al.* (2006) have reviewed the advantages and disadvantages of MFC for the production of renewable energy (biogas). Bullen *et al.* (2006) reviewed several biofuel cells vis-à-vis MFC reactor design, MFC performance and optimization of operating parameters. Chang *et al.* (2006) have compiled the properties of electrochemically active bacteria used in mediator-less MFC and the rate limiting steps in electron transport. Du *et al.* (2007) have exhaustively presented a state-of-the-art review on microbial fuel cells and discussed how the technology can be used for wastewater treatment and bioenergy. Zhuwei *et al.* (2008) have demonstrated electricity generation using membrane-less MFC during wastewater treatment. Pant *et al.* (2010) recently reviewed the role of MFCs as a mode of converting organic waste, including waste waters and lignocellulosic biomass, into electricity. Osman *et al.* (2010) have reviewed the recent progress and challenges in biofuel cells, and focused on current challenges in performance, stability requirements, modularity etc. Virdis *et al.* (2011) have recently discussed the role of MFCs in energy recovery from waste streams to the production of value-added chemicals. Liu *et al.* (2011) have recently reported an MFC-based biosensor for *in situ* monitoring of anaerobic digestion process.

Zhang *et al.* (2012) have incorporated biocathodes into a three-chamber MFC to yield electricity from sewage sludge at maximum power output of 3.2 ± 1.7 W m^{-3} during polarization, much higher than those previously reported. The massively parallel sequencing technology, 454 pyrosequencing technique, was adopted to probe microbial community on anode biofilm, with the most dominant phyla belonging to Proteobacteria (45% of total bacteria), and followed by Bacteroidetes, uncultured bacteria, Actinobacteria, Firmicutes, Chloroflex. At genera level, the most abundant taxa were *Rhodoferax, Ferruginibacter, Rhodopseudomonas, Ferribacterium, Clostridium, Chlorobaculum, Rhodobacter* and *Bradyrhizobium* (relative abundances >2.0%).

Photosynthetic microbial fuel cells (PMFCs) offer a novel approach for producing electrical power in a CO_2-free self-sustainable manner in the absence of organic fuel. The ability of cyanobacteria to display electrogenic activity under illumination emphasizes the necessity to develop improved anode materials capable of harvesting electrons directly from photosynthetic cultures. Zou *et al.* (2010) have shown that nanostructured electrically conductive polymer polypyrrole substantially improved the efficiency of electron collection from photosynthetic biofilm in PMFCs. Nanostructured fibrillar polypyrrole showed better performance than granular polypyrrole. Cyclic voltammetry and impedance spectroscopy analyses revealed that better performance of nanostructured anode materials was due to the substantial improvement in electrochemical properties including higher redox current and lower interface electron-transfer resistance. At loading density of 3 mg cm^{-2}, coating of anode with fibrillar polypyrrole resulted in a 450% increase in the power density (Zou *et al.*, 2010).

Nanotechnological advances in MFCs

Nanotechnology is likely to revolutionize MFCs. Ghasemi *et al.* (2012) have fabricated and utilized self-made carbon nanofibre (CNF)/Nafion and activated carbon nanofibre (ACNF)/Nafion nanocomposite membranes tested in an MFC. The electrospinning method was used for the production of CNFs as it is simple, highly cost effective and the most prolific method for the production of CNF. Nanocomposite membranes were used because they had higher production power and coulombic efficiency (CE) than Nafion 117 and Nafion 112 in MFC systems. These workers reported that the system operated by the ACNF/Nafion membrane produces the highest voltage of 57.64 mW m^{-2}, while Nafion 112 produces the lowest power density (13.99 mW m^{-2}).

The recent trend is fast moving towards development of µl-scale MFCs (Choi and Chae, 2012; Wang *et al*., 2011) and nano-scale MFCs. Wang *et al*. (2011) have reviewed the developments in the area of micro-sized MFCs (including ml-scale and µl-scale setups), along with characteristics, fabrication methods, performances, potential applications and future trends. Both two-chambered and air-breathing cathodes are promising configurations for ml-scale MFCs. However, most of the existing µl-scale MFCs generate significantly lower volumetric powered density compared to their ml-counterparts because of the high internal resistance. Thus far, µl-scale MFCs have not provided sufficient power for operating conventional equipment, however, they exhibit immense potential in rapid screening of electrochemically active microbes and in the future such MFCs would be common.

Since the sustainable microbially induced current density can be one of the major limitations to power production in MFCs, increasing and maintaining biocatalyst density presents a prime opportunity to continue the dramatic improvements in MFC performance (Ren *et al*., 2010). Chen *et al*. (2011) have developed electrospun layered carbon fibre mats by layer-by-layer (LBL) electrospinning of polyacrylonitrile on to thin natural cellulose paper and subsequent carbonization. The layered carbon fibre mat has been proved to be a promising MFC anode for high density layered biofilm propagation and high bioelectrocatalytic anodic current density and holds promise.

It is now feasible to translate laboratory-scale designs into commercial-scale facilities by replacing low surface-area carbon-based electrodes (e.g. graphite felt, vitreous carbon and carbon cloth) with emerging nanomaterials (Gadhamshetty and Koratkar, 2011). Specific surface areas of nanomaterials, e.g. carbon nanotubes (CNTs) and graphene (1000 $m^2\ g^{-1}$), are at least 1000-fold higher than that of conventional carbon-based electrodes. Both CNTs and graphene offer high surface and high electrical-conductivity nano-carbon materials, which provides a unique opportunity to boost power densities by enhancing bioreaction kinetics, electrochemical kinetics and mass transfer characteristics in MFCs. Individual carbon nanotube and graphene sheets can be assembled into self-standing, self-supporting macroscopic paper-like electrode materials that are scalable. Freestanding nano-structured materials possess ultra-high surface area as well as high conductivity; these dual properties render nano-structured materials superior in terms of achieving acceptable levels of oxygen reduction without the need for metal catalyst and current collectors and lead to the development of affordable and sustainable MFC technologies (Gadhamshetty and Koratkar, 2011).

MFCs with nano-engineered electrode architectures provide a sustainable and environment-friendly approach for waste management and energy generation (Gadhamshetty and Koratkar, 2011; Lamp *et al*., 2011). Based on the results available, it is anticipated that replacing conventional wastewater treatment technologies (e.g. aeration tank of activated sludge process, trickling filters, or membrane bioreactors) with nano-engineered electrode-based MFCs may yield significant energy savings. Well-optimized MFC technology can offer far-reaching benefits in producing renewable energy from organic waste, minimizing fossil fuel consumption during wastewater treatment and reducing greenhouse gas emissions.

Despite the advances made in the area of MFCs, several problems associated with the performance of high delivering MFCs include: (i) long lag times before onset of electricity generation; (ii) use of right consortium of bacteria for high power densities; (iii) instability at higher voltages; (iv) selection of appropriate electron mediator; and (v) use of expensive electrodes (Reimers *et al.*, 2001; Rabaey *et al.*, 2004; Logan *et al.*, 2006). Multiwalled nanotubes and single-walled nanotubes have attracted a lot of attention in view of their unique physical and chemical properties.

A recent emerging area is the use of carbon nanostructures (CNSs) to enhance MFC anode properties. Carbon nanotubes (CNTs) have received significant attention in fuel cell research in view of their unique structural and conductive properties such as nanometer size, high accessible surface area, good electronic conductivity and high stability. Utilizing CNTs, Sharma *et al.* (2008) have shown a six-fold increase in power density as compared to graphite electrodes. Improvements in anodic power density have been reported following the addition of CNTs (Sun *et al.*, 2010) and graphene (Zhang *et al.*, 2011). Fan *et al.* (2011) have shown that nanoparticle decoration can greatly enhance the performance of microbial anodes. The anodes decorated with Au nanoparticles were shown to produce a current density up to 20-fold higher than those of the plain graphite anodes, while those decorated with Pd only produced 50–150% increase in current as compared to control, indicating that the chemical composition of the nanoparticles affects the anodic performance.

As alluded to earlier, while MFCs are likely to meet the future electricity requirements in a small way (Chapters 2–5), simultaneously, keeping the environment clean, microbes are an important source of next generation biofuels and hold immense promise as outlined below.

Biofuels

Biofuels, i.e. fuels made from living organisms, can be divided into three categories: (i) first-generation biofuels, which are produced from readily available crops containing edible sugars, starches and oil, e.g. bioethanol, biodiesel and biobutanol; (ii) second-generation biofuels that are made from non-edible difficult to hydrolyse raw materials such as lignocellulosic material; and (iii) third-generation biofuels (also called advanced biofuels), which are made from algae and other microbes. While first-generation biofuels can jeopardize the already short food supply in various countries, the second-generation biofuels are dependent mainly on lignocellulosic material, placing less burden on food supply, while the third-generation biofuels are made from algae and other microbes and, therefore, place no burden on the already constrained food supply. In addition, their sulfur content is very low as compared to conventional petroleum-based fuels and first-generation biofuels. Though the primary biofuels currently in use are biodiesel and bioethanol and occupy nearly 90% of the total biofuel market, it is anticipated that the third-generation biofuels could play an important role in diversifying the world's energy supplies and curbing greenhouse gas emissions. Biofuels offer multiple advantages as highlighted in Fig. 1.2.

Rapid progress in the field of microbial biotechnology has resulted in the development of established fermentation methods mainly based on *Saccharomyces cerevisiae*. However, *S. cerevisiae* lacks the ability to ferment many pentose sugars, resulting in reduction of efficiency of the conversion process. An ideal microbial strain for bioethanol production should produce high yields of ethanol, with few side products and have low inhibitor sensitivity and high ethanol tolerance. A plethora of microbes are capable of utilizing the pentose sugars as a carbon source during fermentation, e.g. *Aerobacter hydrophila, Bacillus polymyxa, Clostridium acetobutylicum, Clostridium thermocellum, Escherichia coli, Fusarium oxysporum, Klebsiella oxytoca, Kluyveromyces marxians, Mucor* sp.,

Neurospora crassa, *Pichia stipitis*, *Streptococcus fragilis*, *Thermoanerobacter ethanolicus*, *Zymomonas mobilis* etc.; however, they are not efficient ethanol producers and exhibit poor ethanol tolerance. Genetic engineering approaches are being developed for the production of suitable fermentative microbes for efficient bioethanol production.

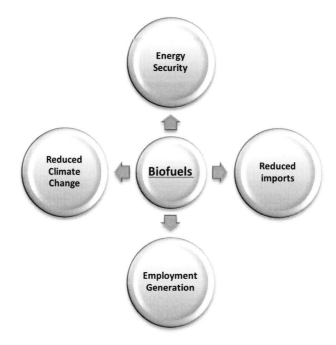

Fig. 1.2. Advantages of biofuels to economy, environment and homeland security.

Biodiesel is currently a mature technology for use in compression-ignition engines. As compared to developed nations, the consumption of diesel is much greater in underdeveloped and developing nations, and they therefore need to focus more on biodiesel than on ethanol for supplementing the petro-based fuels. In India, the government has taken up a number of initiatives and attention is being paid in this direction under the aegis of several ministerial departments. For biofuels, jatropha – which is very rich in oil (40%) – and other crops such as maize, oats, cotton, hemp, soybean, coffee, flax, euphorbia, sesame, safflower, groundnut, castor beans, jojoba, mustard seed, rice bran oil, palm oil, neem oil, Indian beech etc. have been explored for biodiesel production. Experience with jatropha oil has been successful in India for meeting the diesel fuel requirements of rural and forest communities and is also being tested for running military vehicles. Jatropha grows easily in dry, non-fertile, barren lands and is commercially viable and, consequently, villagers and farmers have benefited by way of augmented income without much investment. Since Jatropha oil is carbon-neutral, its large-scale production is likely to improve the country's carbon emissions profile. The focus of India's 'National Biofuel Policy' is to meet 20% of the demand through biofuels.

As far as third-generation biofuels are considered, microalgae are being explored as an economically viable resource apart from extremophilic microbes. Microalgae offer several advantages for the production of lipids for conversion into biodiesel: (i) they accumulate lipids and oils in large quantities; (ii) they grow rapidly (biomass doubling time is usually less than 24 h); (iii) they are able to grow in saline waters and waste waters; (iv) photobioreactors for growth of microalgae can be located in arid or semi-arid areas that are not suitable for agriculture; (v) the nutrients needed for growth can be provided from waste sources such as agricultural run-off, industrial or municipal waste water and animal feeds; (vi) they remove CO_2 emitted from burning fossil fuels; (vii) unlike crops, their growth is not seasonal; (viii) a large number of microalgae produce valuable by-products such as biopolymers, pigments and polysaccharides, which can be harvested; (ix) after lipid extraction, the algal biomass can be anaerobically converted into biogas, which can provide more energy than the energy produced from the lipids (Barnard *et al.*, 2010). Part 5 (Chapters 12 and 13) of this book focus on third-generation biofuels – their production, improvement and applications in various sectors.

Microbes for Environmental Clean-up of Hydrocarbons and Nuclear Wastes

Apart from production of energy, microbes have the unique ability of cleaning up the environment, including hazardous wastes such as petroleum waste products, a variety of toxic industrial chemicals and radioactive materials. In recent years, they have been used particularly for the bioremediation of hydrocarbons and nuclear wastes (see Parts 6 and 7, this volume).

The increasing use of nuclear energy around the globe, despite all precautions in place, poses serious risk of release of such agents in the environment as has been witnessed in the Chernobyl reactor accident and the recent Fukushima incident in Japan (Sharma and Arora, 2011). As many as 11 radionuclides, including ^{131}I, ^{137}Cs, are supposed to have been released in the sea water in the Fukushima incident. A several-fold increase in the radioactivity in the sea poses a danger that may lurk for several years. As per reports of the International Atomic Energy Agency (IAEA) in the case of Fukushima, the isotopic composition of the fallout included mainly volatile radionuclides, e.g. I, Te and Cs, however, ^{134}Cs and ^{137}Cs remain the dominant contaminants and are present in the topsoil. Remediation efforts currently on-going in the region cover approximately a 500 km^2 region having radiation dose levels above 20 mSv/year) and approximately 1300 km^2 (region with radiation dose levels between 5 mSv/year to 20 mSv/year). Bioremediation can be effectively utilized for bioremediation of such radionuclides, which have a half life of several years and have the potential of entering the food chain (Arora and Bell, 2012).

Several microbes during the course of evolution have evolved the ability to survive in hazardous environments, including radiation environments. Microbes that continuously have been exposed to radiation have usually developed immense radioresistance and consequently have potential for environmental clean-up (Kumar *et al.*, 2010; Arora and Bell, 2012). The release of radionuclides from nuclear sites and their mobility in the environment is a cause of public concern. In the USA alone there are over 120 Department of Energy (DOE) nuclear sites, and several other facilities in Europe and the former USSR. It has been anticipated that due to compromise in storage, contamination of trillions of gallons of groundwater and millions of cubic metres of soil and debris has taken place. The cost of remediation of these sites is estimated to be of the order of trillions of dollars. Microbial bioremediation is a cheap and effective way of remediating such sites. The high cost of remediating radioactive waste sites from nuclear weapons production has stimulated the development of novel, cost-effective bioremediation strategies. The 2011 Tohoku (Great East Japan) Earthquake, Fukushima nuclear power plant disaster in Japan has shaken the faith of most countries, particularly European nations, in the safety of nuclear power. Several countries have already begun to reconsider plant life-extensions or construction of new nuclear generating capacity. The German government has already announced its plans to close all nuclear reactors in the country by 2022, while the Swiss government will phase out nuclear power by 2034, and Italian voters in a country-wide referendum have rejected plans to build nuclear power reactors in Italy. In India, safety concerns have been raised regarding the upcoming 2000 mW Kudankulam nuclear plant and the nuclear plant has been embroiled in a controversy over its safety by the local population. There have been several struggles by local people against the project, since they are afraid that in case of a disaster the reactor may explode. A Public Interest Litigation (PIL) has also been filed against the government's civil nuclear programme at the apex Supreme Court. The PIL specifically asks for the 'staying of all proposed nuclear power plants till satisfactory safety measures and cost-benefit analyses are completed by independent agencies'. A central panel constituted by the Government of India, which carried out a survey of the safety features in the plant, said the Kudankulam reactors are the safest and fears of the people are not based on scientific principles. It is evident that populations in several countries have begun to question the safety of nuclear reactors during disasters and this is likely to result in delayed or aborted projects. The Chernobyl Incident of 1986 and the 2011 Fukushima/Daiichi nuclear power plant following the massive 9.0 Richter Tohoku Earthquake and Tsunami have once again brought the importance of bioremediation to the forefront. A plethora of microorganisms have been evaluated in recent years for their ability to degrade or detoxify radioactive materials such as ^{237}Np, Pu isotopes, Am, ^{3}H, ^{14}C, ^{85}Kr, ^{90}Si, ^{99}Tc, ^{129}I, ^{137}Cs, ^{235}U etc., e.g. *Deinococcus radiodurans*, *Geobacter sufurreducens*, *Methanococcus jannaschii*, *Shewanella oneidensis*, *Citrobacter* sp., *Desulfovibrio vulgaris*, *Shewanella putrefaciens* etc. In addition, genetic engineering approaches have also been employed to generate microbes that can degrade radioactive wastes. Extremophiles have also been exploited for bioremediation of radioactive wastes (Table 1.1; also see Chapter 19).

Table. 1.1. Some extremophilic microbes that have potential for bioremediation.

Extremophilic microbe	Source	Radiation tolerance range (kGy)	Reference
Thermus thermophilis	thermal vent in Mine-hot spring, Japan	0.8	Horikoshi, 2011
Kineococcus radiotolerans	high-level radioactive environment, Savannah River Site	20	Bagwell et al., 2008
Rubrobacter xylanophilus	Thermally polluted industrial runoff, Salisbury, UK	15	Horikoshi, 2011
Deinococcus radiodurans	Sterilised Meat can, Oregon, USA	2–20	Bagwell et al., 2008

The presence of toxic chemicals, heavy metals, halogenated solvents and radionuclides in many waste materials presents a challenging problem for separating different species and disposing of individual contaminants. *Deinococcus radiodurans* strains and other detoxifying microorganisms may be utilized to detoxify halogenated organics and toxic metals such as mercury, and it is anticipated that it could be used to remove these classes of compounds selectively from mixed wastes under mild conditions (Daly, 2009).

Brim *et al*. (2000) have developed a radiation-resistant bacterium for the treatment of mixed radioactive wastes containing ionic mercury using *D. radiodurans*, the most radiation-resistant organism known. *Deinococcus geothermalis* is another extremely radiation-resistant thermophilic bacterium and is well known for its ability to grow at temperatures as high as 55°C. The bacterium has been engineered for *in situ* bioremediation of radioactive wastes (Brim *et al*., 2003). These workers generated a Hg(II)-resistant *D. geothermalis* strain capable of reducing Hg(II) at elevated temperatures and in the presence of 50 Gy h^{-1}. Additionally, *D. geothermalis* is capable of reducing Fe(III)-nitrilotriacetic acid, U(VI) and Cr(VI). These characteristics support the prospective development of this thermophilic radiophile for bioremediation. Brim *et al*. (2006) engineered *D. radiodurans*, which naturally reduces Cr(VI) to the less mobile and less toxic Cr(III), for complete toluene degradation by cloned expression of *tod* and *xyl* genes of *Pseudomonas putida*. Toluene and other fuel hydrocarbons are commonly found in association with radionuclides at numerous US Department of Energy sites, frequently occurring together with Cr(VI) and other heavy metals and pose a major problem.

A process, namely, heavy metal remediation, is commonly used in the cleanup of radioactive metals such as mercury, chromium and uranium. One way to achieve heavy metal remediation is to insert an enzyme into *D. radiodurans* X2, giving *D. radiodurans* the ability to reduce and diminish the toxicity of a particular metal. merA, which converts mercury to a less volatile and dangerous oxidation state, has been expressed in *D. radiodurans* and used in such applications. PhoN, a nonspecific acid phosphatase, has also been expressed in *D. radiodurans* and has been used to precipitate radioactive uranium at nuclear weapon production sites (Daly, 2000).

It has been observed by many researchers that *D. radiodurans* is adept at detoxifying reactive oxygen species (ROS) during radiation exposure when many free radicals are generated from hydrolysis of water (Daly, 2000, 2009). High-level radioactive waste (HLW) is an anthropogenic disturbance to which only a scarce number of organisms are resistant. Usually the waste sites contain metal ions such as Fe, Al, Si, Ca, F, K, alkali cations, organic and toxic solvents, radionuclides and few other products. Organic constituents include used complexants, radiolysis products from degradation of complexants and solvents, and waste tank decontamination reagents.

Deinococcus has been genetically engineered for bioremediation to consume and digest solvents and heavy metals even in highly radioactive sites. The deinococcal mercuric reductase gene has been cloned from *E. coli* into the bacterium to detoxify the ionic mercury frequently found in radioactive waste generated from nuclear weapons manufacture. Researchers developed a strain of *Deinococcus* that could detoxify both mercury and toluene in mixed radioactive wastes. U(VI) and Tc(VII) metal ion availability can be reduced by *D. radiodurans* along with humic acids or synthetic electron shuttle agents and Cr(VI) can be reduced in the absence of humic acid (Mann, 2009; Daly, 2000). Thus, it has prospective applications for remediation at the metal- and radionuclide-contaminated sites where ionizing radiation or other DNA-damaging agents may hamper the activity of other sensitive microorganisms (Fredrickson *et al*., 2000).

Another mechanism for the external precipitation of heavy metals like U(VI) on the cell surface involves utilization of PhoN protein. This is a non-specific periplasmic acid phosphatase expressed by *Citrobacter* sp. and numerous bacteria and has been observed to hydrolyse organic and inorganic phosphates that further interact with metals in the environment and hence precipitating them on the cell surface in the form of insoluble metal phosphates (Mann, 2009). It has been reported that *Bacillus sphaericus* has evolved a crystalline S-layer that covers the outside of the cell. This layer also serves to accumulate high amounts of toxic metals such as Ur, Pb, Cu, Al and Cd and hence plays a vital role in sequestering them from the environment. This mechanism further aids in contributing to various pathways for performing bioprecipitation of these heavy metals (Merroun *et al*., 2005). A number of microbes such as *Desulfovibrio*, *Geobacter*, *Shewanella* etc. possess the ability to carry out reductive precipitation of radionuclides, but the sensitivity of these organisms to heavy metals may perhaps limit their *in situ* activities. In such cases, transfer of genes, from resistant microbes to susceptible species, which are responsible for providing resistance against increasing concentrations of radionuclides and heavy metals, can prove to be a boon for reclaiming contaminated sites (Martinez *et al*., 2006). Employing microorganisms to tackle the toxic heavy metal waste is a progressive branch, which promises tremendous potential in future.

More attention needs to be focused on the bioremediation of radioactive wastes because such compounds pose a great risk to the health of humans and the environment.

Conclusion

Energy and environment are and will continue to remain the key issues of attention in the coming decades. Man has over-utilized the fossil resources and in the process has caused severe damage to the environment. On the one hand, fossil fuel reserves are fast becoming depleted, and on the other the global environment is precariously changing. This calls for a paradigm shift in the research focus on green energy and green environmental technologies. The survival of humanity on the planet in the future will depend on judiciously harnessing renewable sources of energy, mainly solar, wind, water, geothermal and nuclear power, and developing unthought-of energy sources. Microbes could help solve some of the energy needs, as alluded to in this chapter and the ensuing chapters of this book.

Rapid advances in MFC technology, coupled with nanotechnology, will provide useful products and systems in the future for the civil and military sectors. Understanding of the intricate molecular pathways that operate in microbes at genomic and proteomic level will help achieve better electricity generation, biofuel production and bioremediation of compounds and by-products of hydrocarbons and nuclear technology. Biofuels is an important area among renewable energy that needs focus. Biofuels that are not dependent on arable land and which complement conventional transportation fuels entirely, or those that fall in the category of novel green biofuels, are likely to play an important role in meeting the world's growing energy needs of the future. There is a need to bring down the costs of cultivation, harvesting and transporting biomass so that large-scale production is economically feasible. Further, to enable rapid market acceptance the advanced biofuels must be compatible with existing infrastructure and vehicles. There is a need to harness the potential of mesophilic microbes, extremophiles and the novel enzymes they synthesize for producing efficient biofuels. As it would be inevitable to establish more nuclear power stations in developing nations to meet the on-going energy needs, we need to be prepared with bioremediation technologies for handling nuclear wastes.

Microbial biotechnology can help effectively meet the challenges of both energy and environment by maintaining a subtle balance and providing innovative solutions to problems facing mankind. The future is going to be an energy-based economy and we need to strive towards achieving this by holistically all conventional, alternative and renewable sources of energy.

References

Allen, R.M. and Bennetto, H.P. (1993) Microbial fuel-cells: electricity production from carbohydrates. *Applied Biochemistry and Biotechnology* 39, 27–40.

Arora, R. and Bell, E. (2012) Biotechnological applications of extremophiles: Promise and Prospects. In: Bell, E. (ed.) *Life at Extremes: Environments, Organisms and Strategies for Survival.* CAB International, United Kingdom, pp. 498–521.

Bagwell, C.E., Bhat, S., Hawkins, G.M., Smith, B.W. and Biswas, T. *et al.* (2008) Survival in nuclear waste, extreme resistance, and potential applications gleaned from the genome sequence of *Kineococcus radiotolerans* SRS30216. PLoS ONE 3(12), e3878.

Barnard, D., Casanueva, A., Tuffin, M. and Cowan, D. (2010) Extremophiles in biofuel synthesis. *Environmental Technology* 31(8–9), 871–888.

Brim, H., McFarlan, S.C., Fredrickson, J.K. and Kenneth, W. (2000) Engineering *Deinococcus radiodurans* for metal remediation in radioactive mixed waste environments. *Nature Biotechnology* 18, 85–90.

Brim, H., Osborne, J.P., Kostandarithes, H.M., Fredrickson, J.K., Wackett, L.P. and Daly, M.J. (2006) *Deinococcus radiodurans* engineered for complete toluene degradation facilitates Cr(VI) reduction. *Microbiology* 152, 2469–2477.

Brim, H., Venkateswaran, A., Kostandarithes, H.M., Fredrickson, J.K. and Daly, M.J. (2003) Engineering *Deinococcus geothermalis* for bioremediation of high-temperature radioactive waste environments. *Applied and Environmental Microbiology* 69(8), 4575–4582.

Bullen, R.A., Arnot, T.C., Lakeman, J.B. and Walsh, F.C. (2006) Biofuel cells and their development. *Biosensors and Bioelectronics* 21, 2015–2045.

Chang, I.S., Moon, H., Jang, J.K., Park, H.I. and Nealson, K.H. (2006) Electrochemically active bacteria (EAB) and mediator-less microbial fuel cells. *Journal of Microbiology and Biotechnology* 16, 163–177.

Chen, S., He, G., Carmona-Martinez, A.A., Agarwal, S., Greiner, A., Haoqing Hou, H. and Schröder, U. (2011) Electrospun carbon fiber mat with layered architecture for anode in microbial fuel cells. *Electrochemistry Communications* 13(10), 1026–1029.

Choi, S. and Chae, J. (2012) An array of microliter-sized microbial fuel cells generating 100 μW of power. *Sensors and Actuators A: Physical* 177, 10–15

Daly, M.J. (2000) Engineering radiation-resistant bacteria for environmental biotechnology. *Current Opinion in Biotechnology* 11, 280–285.

Daly, M.J. (2009) A new perspective on radiation resistance based on *Deinococcus radiodurans*. *Nature Reviews Microbiology* 7, 237–245.

Davis, F. and Higson, S.P.J. (2007) Biofuel cells-recent advances and applications. *Biosensors and Bioelectronics* 22, 1224–1235.

Du, Z., Li, H. and Gu, T. (2007) A state of the art review on microbial fuel cells: a promising technology for wastewater treatment and bioenergy. *Biotechnology Advances* 25, 464–482.9.

Fan, Y., Xu, S., Schaller, R., Jia, J., Chplen, F. and Liu, H. (2011) Nanoparticle decoration for enhanced current generation in microbial electrochemical cells. *Biosensors and Bioelectronics* 26(5), 1908–1912.

Fredrickson, J.K., Kostandarithes, H.M., Li, S.W., Plymale, A.E. and Daly, M.J. (2000) Reduction of Fe(III), Cr(VI), U(VI), and Tc(VII) by *Deinococcus radiodurans* R1. *Applied and Environmental Microbiology* 66(5), 2006–2011.

Gadhamshetty, V. and Koratkar, N. (2012) Nano-engineered biocatalyst-electrode structures for next generation microbial fuel cells. *Nano Energy Nano Energy* 1(1), 3–5.

Ghasemi, M., Shahgaldi, S., Ismail, M., Yaakob, Z. and Daud, W.R.W. (2012) New generation of carbon nanocomposite proton exchange membranes in microbial fuel cell systems. *Chemical Engineering Journal* (in press), Available online 9 January 2012.

Gil, G.C., Chang, I.S., Kim, B.H., Kim, M., Jang, J.Y. and Park, H.S. (2003) Operational parameters affecting the performance of a mediator-less microbial fuel cell. *Biosensors and Bioelectronics* 18, 327–334.

Horikoshi, K. (ed.) (2011) *Extremophiles Handbook*, Vols 1 and 2. Springer, Tokyo, 1243 pp.

Kumar, R., Patel, D.D., Bansal, D.D., Mishra, S., Mohammed, A., Arora, R., Sharma, A., Sharma, R.K. and Tripathi, R.P. (2010) Extremophiles: Sustainable Resource of Natural Compounds-Extremolytes. In: Singh, O.V. and Harvey, S.P. (eds) *Sustainable Biotechnology: Sources of Renewable Energy*. Springer Dordrecht, Heidelberg, pp. 279–294.

Lamp, J.L., Guest, J.S., Naha, S., Radavich, K.A., Love, N.G., Ellis, M.W. and Puri, I.K. (2011) Flame synthesis of carbon nanostructures on stainless steel anodes for use in microbial fuel cells. *Journal of Power Sources* doi 10.1016/j.powsour.2011.02.077

Liu, Z., Liu, J., Zhang, S, Xingh, X.H. and Su, Z. (2011) Microbial fuel cell based biosensor for *in situ* monitoring of anaerobic digestion process. *Bioresource Technology* 102, 10221–10229.

Logan, B.E., Hamelers, B., Rozendal, R., Schroder, U., Keller, J. and Freguia, S. (2006) Microbial fuel cells: methodology and technology. *Environmental Science and Technology* 40(17), 5181–5192.

Lovely, D.R. (2006) Microbial fuel cells: novel microbial physiologies and engineering approaches. *Current Opinion in Biotechnology* 17, 327–332.

Lovely, D.R. (2008) The microbe electric: conversion of organic matter to electricity. *Current Opinion in Biotechnology* 19, 1–8.

Mann, J.E. (2009) Recent advances in the development of *Deinococcus* spp. for use in bioremediation of mixed radioactive waste. MMG 445 *Basic Biotechnology* 5, 60–65.

Martinez, R.J., Wang, Y., Raimondo, M.A., Coombs, J.M., Barkay, T. and Sobecky, P.A. (2006) Horizontal gene transfer of PIB-type ATPases among bacteria isolated from radionuclide- and metal-contaminated subsurface soils. *Applied and Environmental Microbiology* 72(5), 3111–3118.

Merroun, M.L., Raff, J., Rossberg, A., Hennig, C., Reich, T. and Selenska-Pobell, S. (2005) Complexation of uranium by cells and S-layer sheets of *Bacillus sphaericus* JG-A12. *Applied and Environmental Microbiology* 71, 5532–5543.

Moon, H., Chang, I.S. and Kim, B.H. (2006) Continuous electricity production from artificial wastewater using a mediator-less microbial fuel cell. *Bioresource Technology* 97, 621–627.

Oh, S.E. and Logan, E. (2005) Hydrogen and electricity production from a food processing waste water using fermentation and microbial fuel cell technologies. *Water Research* 39, 4673–4682.

Osman, M.H., Shah, A.A. and Walsh, F.C. (2010) Recent progress and continuing challenges in bio-fuel cells. Part II. *Microbial Biosensors and Bioelectronics* 26(3), 953–963.

Pant, D., Bogaert, G.V., Diels, L. and Vanbroekhoven, K. (2010) A review of the substrates used in microbial fuel cells (MFCs) for sustainable energy production. *Bioresource Technology* 101(6), 1533–1543.

Park, D.H. and Zeikus, J.G. (2003) Improved fuel cell and electrode designs for producing electricity from microbial degradation. *Biotechnology and Bioengineering* 81, 348–355.

Pham, T.H., Rabaey, K., Aelterman, P., Clauwaert, P., De Schamphelaire, L., Boon, N. and Verstraete, W. (2006) Microbial fuel cells in relation to conventional anaerobic digestion technology. *Engineering in Life Sciences* 6, 285–292.

Rabaey, K. and Verstraete, W. (2005) Microbial fuel cells: novel biotechnology for energy generation. *Trends in Biotechnology* 23, 291–298.

Rabaey, K., Boon, N., Siciliano, S.D., Verhaege, M. and Verstraee, W. (2004) Biofuel cells select for microbial consortia that self mediate electron transfer. *Applied Environmental Microbiology* 70, 5373–5382.

Reimers, C.E., Tender, L.M., Fertig, S. and Wang, W. (2001) Harvesting energy from the marine sediment-water interface. *Environmental Science and Technology* 35, 192–195.

Ren, Z., Ramasamy, R.P., Cloud-Owen, S.R., Yan, H., Mench, M.M. and Regan, J.M. (2010) Time-course correlation of biofilm properties and electrochemical performance in single-chamber microbial fuel cells. *Bioresource Technology* 102, 416–421.

Ringeisen, B., Henderson, E., Wu, P.K., Pietron, J., Ray, R. and Little, B. (2006) High power density from a miniature microbial fuel cell using *Shewanella oneidensis* DSP10. *Environmental Science and Technology* 39, 4666–4671.

Sharma, R.K. and Arora, R. (2011) Fukushima, JAPAN – an Apocalypse in the making? *Journal of Pharmacy and Bioallied Sciences* 3(2), 315–316.

Sharma, T., Reddy, A.L.M., Chandra, T.S. and Ramaprabhu, S. (2008) Development of carbon nanotubes and nanofluids based microbial fuel cell. *International Journal of Hydrogen Energy* 33, 6749–6754.

Sun, J.J., Zhao, H.Z., Yang, Q.Z., Song, J. and Xue, A. (2010) A novel layer-by-layer self-assembled carbon nanotube-based anode: Preparation, characterization, and application in microbial fuel cell. *Electrochimica Acta* 55(9), 3041–3047.

Virdis, B., Freguia S., Rozendal, R.A., Rabaey, K., Yuan, Z. and Keller, J. (2011) Microbial Fuel Cells. In: Wilderer, P. (ed.) *Treatise on Water Science*. Elsevier, Oxford, UK, pp. 641–665.

Wang, H.Y., Bernarda, A., Huang, C.Y., Lee, D.J. and Chang, J.S. (2011) Micro-sized microbial fuel cell: a mini-review. *Bioresource Technology* 102(1), 235–243.

Zhang, G., Zhao, Q., Jiao, Y., Wang, K., Lee, D.-J. and Ren, N. (2012) Efficient electricity generation from sewage sludge using biocathode microbial fuel cell. *Water Research* 46(1), 43–52.

Zhang, Y., Mo, G., LI, X., Zhang, W., Zhang, J., Ye, J., Huang, X. and Yu, C. (2011) A graphene modified anode to improve the performance of microbial fuel cells. *Journal of Power Sources* 196, 5402–5407.

Zhuwei, D.U., Li, Q., Tong, M., Li, S. and Li, H. (2008) Electricity generation using membrane-less microbial fuel cell during wastewater treatment. *Chinese Journal of Chemical Engineering* 16(5), 772–777.

Zou, Y., Pisciotta, J. and Baskakov, I.V. (2010) Nanostructured polypyrrole-coated anode for sun-powered microbial fuel cells. *Bioelectrochemistry* 79(1), 50–56.

Chapter 2

The Microbiology of Microbial Electric Systems

Sarah A. Hensley, Madeline Vargas and Ashley E. Franks

Introduction

While discovery of the ability of microorganisms to utilize electricity is more than a century old (Potter, 1911), the current requirement for alternative and renewable energy sources has led to a recent explosion in research in this field. Microbial electric systems (MESs) exploit the ability of microorganisms to electrically interact with an electrode for the reduction or oxidation of a substrate. Initial studies in this field focused on microbial fuel cell (MFC) technology for the production of electricity from the reduction of organic compounds. Recent research has expanded to microbial electrosynthesis cell (MEC) technology that utilizes current flow in an opposite direction to MFCs, enabling microorganisms to oxidize an electrode in order to produce compounds (Rabaey, K. *et al.*, 2005, 2007; Thrash *et al.*, 2007; Foley *et al.*, 2010). However, large improvements are required before MESs are able to compete with existing technologies (Logan, 2009).

Various electron shuttles such as neutral red, thionine, benzylviologen, 2,6-dichlorophenolindophenol, 2-hydroxy-1,4-naphthoquinone, as well as natural electron shuttles such as phenazines, phenothiazines, phenoxoazines and iron chelators, have been examined in initial MFC research to promote the transfer of electrons between the microorganism and the electrode surface but were found to be unsuitable for a variety of reasons (Logan and Regan, 2006). The poor efficiency of transfer, cost of the electron shuttle, continuously replenishing the electron shuttle and toxicity of many of the electron shuttles made them impractical for use in real world situations (Lovley 2008a, 2008b; Lovley and Nevin, 2008; Logan, 2009).

An important breakthrough in MFC research was the discovery of a microorganism capable of directly interacting with an electrode surface, without the need for an electron shuttle (Bond *et al.*, 2002; Bond and Lovley, 2003). Subsequent microorganisms capable of direct electron transfer to an electrode have been discovered across bacterial phyla and may be ubiquitous in nature. Further microorganisms, termed electricigens (Lovley, 2006e), capable of completely reducing an organic compound to carbon dioxide (CO_2) were discovered (Bond *et al.*, 2002; Bond and Lovley, 2003). These microorganisms utilize the electrode as a final electron acceptor for anaerobic respiration, resulting in highly efficient (>90%) electricity production. The ability of these microorganisms to conserve energy for growth while using an electrode to respire makes the population potentially self sustaining as long as an electron donor is supplied (Lovley, 2010). However, although these organisms are efficient, they are not capable of high rates of electricity production. Further improvements are most likely to be made through the adaptive evolution and genetic engineering of electrogenic microorganisms.

While significant improvements in MFCs' power production were initially achieved thorough reactor designs (discussed further in other of chapters of this volume), power output due to engineering improvements has since plateaued (Lovley and Nevin, 2008; Logan, 2009; Lovley, 2010). For further improvements to MES technology, a greater understanding of the microbiology and microbial processes of these systems is required. MFC power output is still too small to be considered a viable technology for large-scale electricity production (Logan, 2009). The application of MFCs is therefore currently limited to fuelling low power marine devices in remote locations (Tender et al., 2008) or to processes where electricity production is not the major concern (Williams et al., 2010).

In comparison, the reverse process of electron transfer from an electrode to a microbial species has not currently been as extensively studied and may have broad implications in compound production and bioremediation. Recent discoveries cite organisms capable of directly accepting electrons from an electrode for metabolic processes, including the fixation of CO_2 and production of organic compounds (Lovley, 2010; Nevin et al., 2010; Rabaey and Rozendal, 2010; Lovley and Nevin, 2011). The microbes and processes associated with MFC and MEC systems will be discussed in further detail in this chapter.

Microbial Communities Capable of Electricity Production

The first major studies using mixed communities of microorganisms to produce electricity from natural inoculums without the edition of exogenous electron shuttles were conducted in marine sediments. Termed benthic MFCs, these devices utilize natural microbial communities to facilitate the oxidation of organic matter in aquatic sediments to electricity (Fig. 2.1). The first successful practical application of MFC technology is nicknamed BUG (Benthic Unattended Generator) and is capable of powering a small meteorological device (Reimers et al., 2001; Tender et al., 2002; Ryckelynck et al., 2005). Benthic MFCs are comprised of two electrodes: one buried in anaerobic marine sediment (anode) and the other in the overlaying oxygenated water column (cathode). The difference in redox states between the anaerobic sediments and the oxic sea water acts to poise the anode, enabling electrogenic microorganisms to utilize it as an electron sink. After transfer of the electrons to the anode surface, the electrons are combined with oxygen at the cathode surface creating an electrical current. Benthic MFCs produce electrical current without the need for the addition of exogenous electron mediators, utilizing just the local organic matter and bacterial community found in the sediment (Reimers et al., 2001; Bond and Lovley, 2002; Tender et al., 2002). Due to this, the field-deployed BUG has continually produced power for several years without the need for any external input (Tender et al., 2008). These devices hold great promise for powering monitoring devices in hard to reach or remote areas. Through understanding the ecology and physiology behind the microorganisms capable of catalysing the reduction of the anode, it is hoped that the application of MFCs will have wider versatility.

While an immense colonization pressure exists on all surfaces in the marine environment, the anode provides an environmental niche for those organisms capable of utilizing insoluble extracellular electron acceptors. In anaerobic sediments, it has been hypothesized that MFCs enrich for organisms with superior direct extracellular electron

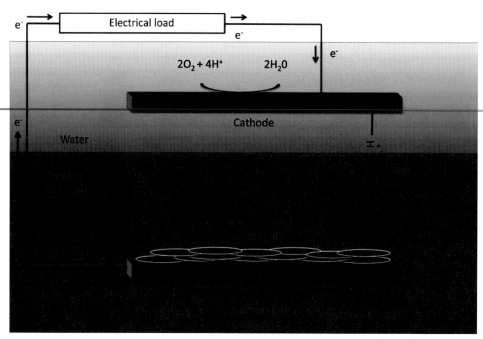

Fig. 2.1. Schematic of a benthic microbial fuel cell. The natural occurring microbial community oxidizes organic products and transfer electrons to the anode. Electrons are transferred to the cathode and react with oxygen creating an electron flow.

transfer characteristics (Bond *et al.*, 2002). This enrichment process causes a shift in the community on the electrode surface that is distinct from the original community found in the marine sediment (Tender *et al.*, 2002; Bond and Lovley, 2003; Holmes *et al.*, 2004). In marine sediments, this community shift repeatedly reveals that, over time, members of the Geobacteraceae dominate (Tender *et al.*, 2002; Bond and Lovley, 2003; Holmes *et al.*, 2004). Perhaps unsurprisingly, Geobacteraceae is a family commonly associated with the ability to reduce insoluble electron acceptors, such as iron and manganese oxides, in the subsurface environment (Lovley *et al.*, 2004).

Members of Geobacteraceae also dominate anode-associated communities when a variety of environmental inoculums and organic carbon sources are used. These include utilizing inocula from sewerage sludge, waste water, agriculture waste or sediment, as well as utilizing dairy waste, industrial waste water, complex sewerage water, agricultural run-off or simple organics as carbon sources (Bond *et al.*, 2002; Tender *et al.*, 2002; Bond and Lovley, 2003; Holmes *et al.*, 2004; Jung and Regan, 2007; Torres *et al.*, 2009; Freguia *et al.*, 2010; Kiely *et al.*, 2010a). Other species reported to dominate MFC systems include *Gammaproteobacteria, Betaproteobacteria, Rhizobiales* and *Clostridia* (Kim *et al.*, 2006, 2007; Rismani-Yazdi *et al.*, 2007; Ishii *et al.*, 2008; Chae *et al.*, 2009). Notably, while a large and diverse range of bacteria has been isolated from the anode surface, most of these studies used culture-independent techniques to characterize communities while relatively few of these microorganisms were isolated and examined in pure culture (Franks *et al.*, 2010).

Due to the dominance of the Geobacteraceae on the electrode surface and the extensive studies of the iron-reducing abilities of these organisms, initial isolation studies used media containing insoluble Fe(III) as an electron acceptor. Interestingly, while iron-reducing organisms are among the most effective current-producing organisms (Lovley, 2006c; Wei and Zhang, 2007; Xing *et al.*, 2008; Yi *et al.*, 2009), not all iron-reducing organisms are able to produce power in an MFC (Lovley, 2008b; Malki *et al.*, 2008; Richter *et al.*, 2008). In studies where isolates were cultivated from different media under aerobic conditions, representatives of Alphaproteobacteria, Gammaproteobacteria, the phylum Firmicutes, the family Flavobacteriaceae and the phylum Acintobacteria have been isolated from a marine enriched MFC (Vandecandelaere *et al.*, 2010). Testing is still required to determine whether these isolates are capable of power production individually or require syntrophic partnerships. Within mixed electrode associated communities, many microbial species may be able to use electron mediators produced by other organisms. The gram-positive organism *Brevibacillus* sp. PTHI was found to enrich on a MFC anode concurrent with the growth of several *Pseudomonas* sp. (Pham *et al.*, 2008a). In pure culture, *Brevibacillus* sp. PTHI was unable to produce power unless cocultured with a *Pseudomonas* sp. or through the addition of supernatant from a *Pseudomonas* sp. MFC (Pham *et al.*, 2008a, b). These results highlight the complexities of interspecies interactions that may occur within anode-associated communities.

The abundance of a given bacterial species within an anode-associated biofilm may not always be an indication of superior power-producing abilities. Due to complex interactions that occur within bacterial biofilms and a diverse range of complex organic sources used in MFCs, the selective pressure on an electrode is not as simple as just power production. For example, extensive degradation of complex organic compounds may be required before products are available for the bacterial species that can interact with an electrode are available. While several dominant species in MFCs have been found to produce equal or greater current density in pure culture than their original mixed communities, this is not always the case (Nevin *et al.*, 2008b; Kiely *et al.*, 2010b; Watson and Logan, 2010). Further understanding of how organisms on an electrode may be working as a consortium for the degradation of complex compounds in mixed communities and how those interactions impact of the complex processes involved in electron transfer is required.

Degradation of Waste and Contaminant Products using MFC Technology

An often overlooked aspect of MFC technology is the ability for a wide variety of complex organic compounds to be completely degraded to CO_2. This ability makes MFC technology ideally suited for the treatment of waste and bioremediation purposes. Industrial waste streams containing acetate, glucose, starch, cellulose, lignocellulosic biomass, wheat straw, pyridine, phenol, p-nitrophenol, domestic waste water, brewery waste, land-fill leachate, chocolate industry waste, mixed fatty acids and petroleum contaminates have been successfully treated at laboratory scales with MES technology while producing small quantities of current (Kim et al., 2000; Bond and Lovley, 2003; Liu and Logan, 2004; You et al., 2006b, 2009; Feng et al., 2008; Morris and Jin, 2008; Nevin et al., 2008b; Ren et al., 2008; Freguia et al., 2009b; Galvez et al., 2009; Lu et al., 2009; Pant et al., 2009; Patil et al., 2009; Zhang et al., 2009; Zhu and Ni, 2009). In these instances, the production of electricity is often insignificant in value to the treatment of the waste product.

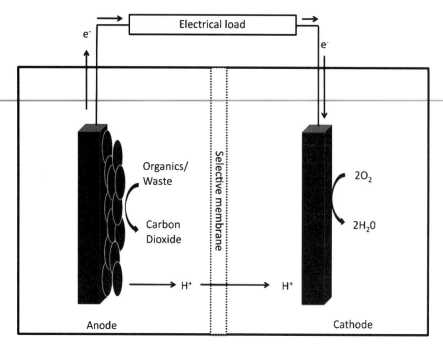

Fig. 2.2. Operation of a microbial fuel cell. Organic compounds are oxidized to CO_2 at the anode surface. Electrons are transferred to the cathode and combine with protons transferred across a proton-selective membrane.

MESs are able to remove the electron acceptor limitation that is often the impediment to degradation of aforementioned waste streams. MESs remove this limitation by the anode functioning as an electron sink during the oxidation of the organic compounds to CO_2. These systems reduce the cost of having to employ an energy-intensive process such as sparging, aeration or mixing in a reductant such as oxygen. In addition, these systems reduce the growth of unwanted organisms that can use the reductant for other processes. Such specificity increases the efficiency of the system. Furthermore, less biomass is generated within MFC systems than other waste-treatment systems. While still producing electricity, the ability to degrade the environmental pollutants may be of more value than the low level of power production itself.

Geobacter species have been found to be important for the anaerobic degradation of petroleum components and land-fill leachate contaminates in groundwater (Lovley *et al.*, 1989; Anderson *et al.*, 1998; Rooney-Varga *et al.*, 1999; Roling *et al.*, 2001; Lin *et al.*, 2005). The process of natural attenuation of petroleum spills is linked to Fe(III) oxidation by *Geobacter* species and may be accelerated through increasing access to the Fe(III), Fe(III) chelators or electron shuttles (Lovley *et al.*, 1994, 1996a, b). However, contaminated sites in the subsurface often lack suitable electron acceptors in large quantities (Lovley and Chapelle, 1995; Lovley, 1997; Reddy *et al.*, 2002; Frysinger *et al.*, 2003) and aeration, addition of chelators or electron shuttles is impractical. The contaminant is therefore often able to persist in the environment. An electrode is able to alleviate this limitation.

Benzoate and toluene can be oxidized by *Geobacter metallireducens* using an electrode as a final electron acceptor (Bond *et al.*, 2002; Zhang *et al.*, 2010). An electrode can also increase the rate of benzene, naphthalene and toluene degradation if used as an electron acceptor in contaminated soil (Zhang *et al.*, 2010). It has been highlighted that since electrodes are often made from graphite, which is a porous material, the electrode itself will absorb the contaminant in the subsurface, further optimizing this process. With the electrode acting as an unlimited reductant, the degradation process can be significantly increased and can selectively allow for the growth of a microbial community capable of oxidizing the contaminant. Furthermore, electrodes can be buried in marine sediments with long-term stability and promotion of microbial oxidation (Reimers *et al.*, 2001; Tender *et al.*, 2002; Ryckelynck *et al.*, 2005). This should hold true for other sediment and inocula types. Another major advantage to utilizing MFC to accelerate bioremediation is the ability to deploy these systems *in situ* with a minimal requirement for disruption or mechanical disturbance of the environment. In these situations, the low level of power production may be irrelevant, or used for low-power monitoring devices associated with the remediation site.

The ability to power small, low-current devices has seen some interesting proposals in the field of MFCs. These include the powering of medical devices, such as pacemakers, using glucose and oxygen from the blood (Calabrese Barton *et al.*, 2004; Minteer *et al.*, 2007). Currently, surgery is required to replace the batteries of these devices. In theory, an implanted device powered by an MFC would never need further surgery. *In vitro* and *in vivo* experiments have produced electricity with small medical devices from abiotic fuel cells based on noble catalysts, activated carbon and organic fuel (Kim *et al.*, 2003; Kerzenmacher *et al.*, 2008). Enzymatic fuel cells have also generated power but at significantly less density (Calabrese Barton *et al.*, 2004; Minteer *et al.*, 2007). Although a number of MFCs with both pure and mixed cultures are able to produce current from glucose, MFC devices that are small and yet powerful enough to produce the required electricity do not currently exist. In addition, the presence of bacteria within such a medical device creates a high potential for blood poisoning should the device break. To avoid the need to implant a bacterially powered device, MFCs have also been powered through the use of white blood cells as the anodic catalyst (Mingui *et al.*, 2006). These devices produced small current densities but it has yet to be determined if the blood cells can directly interact with an electrode or if they require the use of an intermediate (Justin *et al.*, 2005). Although such devices hold great promise, it may be a while before their application is realized.

On larger scales, a number of proposals have been made for industrial use of MFC technology. One such is the removal of fermentation inhibitors accumulated in process waters after pre-treatment of cellulosic biomass (Borole *et al.*, 2009). Also, pilot-scale reactors are being developed for power generation using brewery waste and waste water (Logan, 2010). Although these proposals appear promising, industrial scale implementation of MFC technology has yet to materialize. It is estimated that commercial scale power production is still several years away because of mass transfer limitations inherent with increased scaling of MFC technology (Logan, 2009).

Mechanisms of Microbial Extracellular Electron Transfer

The ability of microorganisms to interact with electrodes has stimulated great interest in the molecular mechanisms of electron transport between electrodes and bacteria. A number of models have been proposed to explain this phenomenon and have often been developed from research regarding the reduction of insoluble minerals, metals and metal oxides in the subsurface environment. These can be divided into direct and indirect electron transfer systems (Lovley, 2006b; Lovley and Nevin, 2008; Franks *et al.*, 2010). Microbes may utilize either mechanism independently or both together depending on their genetic capability and environmental conditions. Direct electron transfer is mediated by microbial nanowires and/or outer surface cytochromes. Indirect electron transfer involves soluble compounds that serve as electron shuttles (Lovley, 2006c, 2008a; Franks and Nevin, 2010). Redox active compounds can serve as mediators for electron transfer between the cell and extracellular electron acceptors such as Fe(III) oxide, Mn oxide and the anodes of MFCs (Lovley, 2008b). Bacteria may produce compounds for electron transfer or use naturally occurring compounds when available.

Mediator-producing bacteria include *Shewanella* spp., *Geothrix fermentans*, *Pseudomonas* spp. (Bouhenni *et al.*, 2010; Franks *et al.*, 2010) and *E. coli* (Lovley, 2006b; Franks *et al.*, 2010). Mediators may be capable of reducing a substrate, regardless of the ability of the bacteria that produced them to do so. Phenazine produced by a *Pseudomonas* species was shown to reduce Fe(II)oxide even though the organism does not respire Fe(III). An *E. coli* strain produces a hydroquinone derivative, which can transfer electrons to the anode of an MFC (Qiao *et al.*, 2008). *Shewanella oneidensis* produces riboflavins that mediate electron transfer between the cells and Fe(III) oxide or the electrode (Lovley, 2006d; Bouhenni *et al.*, 2010; Franks *et al.*, 2010). Exogenous compounds, such as humic substances, can stimulate iron reduction by serving as electron shuttles between microbes and minerals (Lovley, 2006c). Other microorganisms, such as *Geobacter sulfurreducens*, cannot produce electron mediators but can utilize them if available (Lovley, 2008a, b; Lovley and Nevin, 2008; Franks *et al.*, 2010). Lastly, abiotic mediators, such as thionine, benzylviologen, 2,6-dichlorophenolindophenol, 2-hydroxy-1,4-naphthoquinone and various phenazines, phenothiazines, phenoxoazines, iron chelates and neutral red, are added to microbial fuel cells of a number of bacterial species to allow transfer of electrons to the anode (Franks *et al.*, 2010). As previously mentioned, the addition of exogenous electron shuttles is expensive, at times toxic and unlikely to be a sustainable practice for the commercialization of the MES technology.

Microbial Nanowires

Several organisms have been reported to transfer electrons directly to an anode surface, including *Shewanella* spp., *Geobacter* spp., *Aeromonas hydrophilia*, *Clostridium* spp., *Rhodoferax ferrireducens* and *Desulfobulbus propionicus* (Franks and Nevin, 2010). Of these, the mechanisms of *Geobacter* and *Shewanella* have been examined in some depth and will be the focus of this discussion. Pure cultures of *G. sulfurreducens* are capable of current densities similar to biofilms of mixed communities. They form relatively thick (>50 mm) biofilm in which all cells, including cells not in contact with the anode, contribute to current production (Reguera *et al.*, 2006; Lovley, 2008b; Nevin *et al.*, 2008a). *Geobacter* does not produce compounds that can serve as electron shuttle, raising the question of how these cells can achieve long-range electron transfer (Lovley, 2008a). Microbial nanowires have been proposed as a possible agent for transfer of electrons between cells and insoluble

Fe(III) oxides (Reguera et al., 2005; Gorby et al., 2006). These nanowires are protein filaments, also called pili, that are 3–5 nm thick and extend from the cell upwards of 20 μm. They were found to be conductive in *G. sulfurreducens* using atomic force microscopy fitted with a conductive tip (Reguera et al., 2005). The nanowires were not conductive, rather insulated, if covered by nonpilin proteins, presumably outer surface cytochromes. Thus, it is speculated that outer surface cytochromes were not contributing to the conductivity measurements. Furthermore, *pilA*, the gene that codes for the protein subunit that assembles into type IV pili, was shown to be involved in nanowire production (Reguera et al., 2005). When this gene was deleted, no pili were formed but the ability to make pili was restored by reintroducing the gene into the cell.

Direct evidence showing electrons can flow from *Geobacter* to extracellular electron acceptors along nanowires is still lacking. However, *Shewanella oneidensis* MR-1 nanowires were found to be electrically conductive along its length (El-Naggar et al., 2008). Further investigation is also needed to elucidate the mechanism by which electrons can flow through the protein filament. Nevertheless, a number of reports regarding *Geobacter* implicate nanowires in extracellular electron transport to either Fe(III) oxides or the anode. These include: (i) a close association between pili and Fe(III) oxides (Lovley 2006a); (ii) induction of pili occurs at optimal growth temperatures when using Fe(III) oxides but not soluble electron acceptors; (iii) *pilA* deletion results in diminished ability to reduce insoluble Fe(III) oxide; and (iv) *pilA* mutant produces significantly less power than wild-type in the poised microbial fuel cells (Lovley 2006c; Franks et al., 2010). A *G. sulfurreducens* PilA mutant formed a monolayer of cells on the anode surface similar to biofilms of low-current density microbial fuel cells (Bond and Lovley, 2003). The low levels of current produced by the *pilA* knockout mutant were likely due to cells in direct contact with the anode and attributed to electron transfer via membrane-bound *c*-type cytochromes (Lovley, 2008b). Interestingly, adaptive selection for high current production within a microbial fuel cell yielded a variant of *G. sulfurreducens* capable of increased current densities, called KN400 (Yi et al., 2009). This strain produced a thinner biofilm, had decreased levels of outer surface cytochrome and expressed higher number of nanowires around the cell. In another adaptive evolution study, selective pressure for rapid Fe(III) oxide reduction not only led to increased (ten-fold) levels of Fe(III) oxide reduction but to increased expression of PilA and the periplasmic *c*-type cytochrome PgcA (Tremblay et al., 2010). Two types of mutations that occurred were identified. One mutation was a single base-pair change or a single nucleotide insertion in a GEMM riboswitch upstream of *PgcA*. The other mutation was in GSU1771 that codes for a SARP-like regulator that, when interrupted, was shown to increase levels of PilA. When both types of mutations were re-introduced into the wild-type strain, the mutated strain reduced Fe(III) comparable adapted strain and displayed increased levels of PgcA and as much as a 34-fold increase in PilA expression (Tremblay et al., 2010).

Recently, direct electron transfer was proposed in a syntrophic coculture consisting of two *Geobacter* species in a medium containing ethanol and fumarate (Summers et al., 2010). *Geobacter metallireducens* is an ethanol-oxidizing Fe(III) reducer that cannot utilize fumarate. It can ferment the ethanol to acetate and hydrogen gas in the presence of a hydrogen consumer. *Geobacter sulfurreducens* is unable to metabolize ethanol and was added as a hydrogen-consuming partner. It is capable of using fumarate as the electron acceptor. The coculture formed aggregates after repeated transfers. Aggregate formation was quicker when the syntrophic partner had a deficient uptake hydrogenase, ruling out interspecies hydrogen transfer. The aggregates were also conductive. Direct interspecies electron transfer was proposed to explain the ethanol consumption in the culture. The same study highlighted the importance of PilA and the outer surface cytochrome OmcS. When each of these genes were deleted in the syntrophic partner, the coculture was unable to grow (Summers *et al*., 2010).

Conductive microbial nanowires have also been reported in *S. oneidensis* using scanning tunnelling microscopy (Gorby *et al*., 2006). These filaments are 3–5 nm thick, and 10 μm long (Gorby *et al*., 2006). Thus, they tend to be thicker and shorter in length than *Geobacter*. Curiously, these nanowires form bundles that are from 50 nm to over 150 nm in diameter. In a recent study, *S. oneidensis* MR-1 nanowires were found to be electrically conductive along their length by nanofabricated electrodes patterned on top of individual nanowires, and conducting probe atomic force microscopy along a single nanowire bridging a metallic electrode and the conductive atomic force microscopy tip (El-Naggar *et al*., 2008). Both reports provide evidence that, unlike *Geobacter*, cytochromes are required for conductivity. Mutants lacking the genes coding for the cytochromes MtrC and OmcA produced filaments visually similar to nanowires but were not conductive (El-Naggar *et al*., 2008).

Outer Membrane *c*-Type Cytochromes

Geobacter nanowires were demonstrated to be type IV pili and their presence is associated with increased current densities and the ability to reduce insoluble Fe(III). In contrast, *S. oneidensis* that lack type IV pili generated more current in microbial fuel cells than the wild type (Bouhenni *et al*., 2010). This contradicts an earlier report showing a Shewanella *pilD* mutant had diminished metal reduction and current production in an MFC (Bretschger *et al*., 2007). *PilD* codes for a type IV prepilin peptidase that implicates type IV pili. However, in this organism *PilD* also processes Msh, and T2SS (type 2 secretion system) prepilins. Since outer surface cytochromes are localized by T2SS (Shi *et al*., 2008), the *PilD* mutant phenotype may be due to either the loss of pili or the loss of surface cytochromes (Bouhenni *et al*., 2010). Genetic analysis was used to examine the role of pili, flagella or cytochomes MtrC and OmcA in direct electron transfer to anodes and reduction of hydrous ferric oxide. A number of mutants, $\Delta pilM$-Q, $\Delta mshH$-Q, $\Delta pilM$-$Q/\Delta mshH$-Q, Δflg and $\Delta mtrC/\Delta omcA$, were constructed and compared. Only the absence of *c* cytochromes from the outer surface of MR-1 cells, and not the loss of pili, generated significantly less current than the wild type in a microbial fuel cell and lack of Fe(III) reduction. Thus, most extracellular electron transfer in *S. oneidensis* occurs through direct contact with outer membrane cytochromes MtrC and OmcA (El-Naggar *et al*., 2008; Gorby *et al*., 2006; Bouhenni *et al*., 2010).

In *Shewanella* sp., transfer of electrons to the outside of the cell requires the proteins MtrA, MtrB and MtrC, which form a complex. MtrB is a trans outer membrane b-barrel protein that serves as a sheath within which MtrA and MtrC exchange electrons (Hartshorne *et al*., 2009). MtrC is an extracellular element that mediates extracellular electron transfer. OmcA was localized between the cell and the mineral in *Shewanella* cells grown on insoluble Fe(III), while MtrC was displayed more uniformly around the cell surface (Lower *et al*., 2009).

In *G. sulfurreducens*, a number of *c*-type cytochromes have been shown to be involved in Fe(II) oxide reduction or current production. The outer membrane proteins OmcS and

OmcE can readily be sheared from the outer surfaces of cells. Deletion of either of their corresponding genes result in cells that cannot reduce Fe(III) oxides but can reduce soluble Fe(III) (Mehta *et al.*, 2005). OmcB when deleted had the same phenotype (Leang *et al.*, 2003). These cytochromes are not required for current production in high current-producing biofilms. Microarray analysis comparing a high current-producing biofilm with a biofilm grown within an identical system using fumarate as the electron acceptor showed the highest changes in expression in *pilA* and a putative gene directly downstream from *pilA*, followed by cytochrome, *omcB* and *omcZ* (Nevin *et al.*, 2009). Mutations in each of these genes as well as *omcS* and *omcE* demonstrated that only deletion of *pilA* and *omcZ* inhibited current production. Interestingly, *pilA* did not have a structural role within the biofilm since the control was a nonconductive biofilm.

These results contrast with a previous report comparing growth of *G. sulfurreducens* with an electrode as the sole electron acceptor in a microbial fuel cell to soluble Fe(III) citrate (Holmes *et al.*, 2006a). *OmcS* was found to be highly expressed and essential for current production. *OmcE* was also involved in electron transfer to the electrode but *OmcB* and *PilA* did not have higher transcript levels on electrodes. The two reports may not be directly comparable. This study used a low current density MFC and compared biofilm growth to planktonic growth on soluble iron. Nevertheless, these results highlight the role of gene plasticity and shifts in electron flow under different growth/current-producing conditions.

The essential role of *omcZ* and *pilA* in current production was observed in an MFC when a carbon cloth anode was used in place of graphite (Richter *et al.*, 2009). It was proposed that *OmcZ* was important for electron transfer through the bulk of the biofilm, *OmcB* was important for electron transfer between the electrode and the biofilm, *PilA* was important for both electrode and biofilm electron transfer and *OmcS* possessed a secondary role for electron transfer through the biofilm (Richter *et al.*, 2009). *OmcS* was recently suggested to facilitate electron transfer from pili to Fe(III) oxides. Immunogold localization revealed that OmcS localized along the pili with spacing between OmcS molecules too large for the molecules to interact with each other (Leang *et al.*, 2010).

Microbial Oxidation of an Electrode

Microorganisms have been found not only to use an electrode as an electron acceptor, such as in MFCs, but also to be capable of accepting electrons from an electrode. Organisms capable of receiving electrons necessary for growth solely from an electrode are termed electrophs. Overall, MEC systems are similar and opposite to that of an MFC. MECs reverse the process normally associated with an MFC by facilitating microorganisms to accept electrons from an electrode for the reduction of organic and inorganic compounds (Lovley, 2010). However, the physiological mechanism for electrotrophs utilizing an electrode as an electron donor may differ from that of electricigens utilizing an electrode as an electron acceptor (Strycharz *et al.*, 2010b).

The ability to accept electrons from an electrode, both directly and indirectly, has broadened the potential applications for an electrode-associated organism (Rabaey, K. *et al.*, 2005, 2007; Thrash *et al.*, 2007; Foley *et al.*, 2010). Using the electrode as a reducing equivalent, microorganisms can reduce a variety of compounds, including CO_2 (Nevin *et al.*, 2010). Indeed, electrotrophs are capable of converting CO_2 to multicarbon organic chemicals and fuels with electricity acting as a reductant (Lovley, 2010). Systems utilizing electrophs and solar cell technology as an electron source can be likened to photosynthesis in plants. This is due to the fact that energy from sunlight, water and CO_2 is converted to organic compounds and oxygen in both processes. However, MECs are more efficient than photosynthetic systems because no excess biomass is produced and a higher proportion of the energy from sunlight can be converted to organic compounds (Lovley, 2010). Other proposed applications for MEC technology include bioremediation, increasing power production of MFCs, streamlining wastewater treatment and biofuels production.

Indirect and direct interactions between electrodes and microorganism in MECs

Microbes may indirectly interact with an electrode in an MEC through electron shuttles or elemental cycling. Both natural and artificial electron shuttles are utilized for the electron transfer from an electrode to microorganisms. Methyl violygen, anthraquinone-2,6-disulfonate (AQDS) and neutral red are currently the most commonly used artificial electron shuttles in MECs (Rosenbaum et al., 2011). While a variety of electron shuttles have been shown to be capable of transferring electrons from an electrode to microorganisms, much like in MFCs, the cost, toxicity and need of constant electron shuttle addition due to their instability does not make them seem suitable on a commercial scale (Thrash and Coates, 2008). Overall, electron shuttles are unlikely to become authorized for use in natural environments (Steinbusch et al., 2010).

Instead, naturally occurring biogeochemical cycles, such as elemental cycling, have been exploited in combination with electrodes for the remediation and bioleaching of metals from ores (Ergas et al., 2006; Cheng et al., 2007). Termed biocathodes, they have been examined as an efficient means of supplying electrons to the microorganisms to drive this process. For example, *Acidothiobacillus ferrioxidans* may be capable of secondarily oxidizing the cathode via the oxidation of ferrous iron (Ter Heijne et al., 2007) and *Leptothrix discophora* may oxidize manganese compounds (Rhoads et al., 2005). These oxidized metals could, in turn, oxidize the electrode. The microorganism on the biocathode would then re-oxidize the reduced metal. However, no direct observance of iron or manganese cycling in these systems was mentioned. Moreover, such an indirect cycling of metals may not be applicable to all systems.

The ability to directly transfer electrons from an electrode is thought to be an ideal solution for remediation purposes due to the direct delivery of electrons to the organisms most capable of remediating a contaminated site without encouraging competing species (Skadberg et al., 1999; Lovley, 2010; Strycharz et al., 2010a). In addition, contaminated areas may be specifically targeted long-term, and contaminating compounds may be removed with the electrode (Gregory and Lovley, 2005; Tandukar et al., 2009; Lovley, 2010; Strycharz et al., 2010a). For example, uranium may be immobilized by microbial reducing soluble U(VI) to insoluble U(IV) (Finneran et al., 2002; Anderson et al., 2003). Utilizing an electrode as an electron donor, *G. sulfurreducens* is capable of catalyzing the adsorption of U(IV) to the electrode surface (Gregory and Lovley, 2005). Although large-scale deployment has not been examined, practical application of this system in a field study site is currently being examined (Lovley, 2010). Other compounds that may be remediated directly by cathodic biofilms include heavy metals such as Cr(VI) (Tandukar et al., 2009), recalcitrant hydrocarbon contaminants (Zhang et al., 2010), denitrification (Gregory et al., 2004; Clauwaert et al., 2007) and chlorinated compounds (Aulenta et al., 2007, 2010; Lovley, 2010; Strycharz et al., 2010a). In practice, heavily polluted sites may use an array of both cathodes and anodes to deliver and accept electrons to the microbial community as required for bioremediation.

Biocathodes for Improved Performance of MFCs

After numerous studies of anodic communities and improvements, the cathode appears to be the largest limiting factor to increasing current production in MFCs (Lovley and Nevin, 2008). Unfortunately, abiotic cathode catalysts are either expensive rare-earth metals like Pt or Ti, or are toxic, such as hexaferricyanide (Rabaey *et al*., 2003; You *et al*., 2006a). Therefore, the utilization of aerobic biocathodes has begun to be explored in more depth (Clauwaert *et al*., 2007; Logan, 2009; Freguia *et al*., 2010). Such aerobic biocathodes could be coupled with MFCs to decrease cathodic limitation in those systems. Other cross-overs between MFC and MEC technology may be seen with wastewater treatment. Utilization of MFC at wastewater treatment facilities has been proposed as an energy- and cost-efficient mechanism of waste treatment (Logan, 2009; Foley *et al*., 2010) and process monitoring (Clauwaert *et al*., 2007). Although waste water is often examined as a direct source of hydrogen from fermentation processes (Angenent *et al*., 2004), MECs may present an equally promising opportunity for hydrogen production from waste (Rozendal *et al*., 2008).

Initial studies examined cathodic hydrogen production via electron shuttles (Tatsumi *et al*., 1999; Lojou *et al*., 2002). However, hydrogen is energetically costly to produce from an electrode (Thrash and Coates, 2008). Therefore, recent research has shown that it is possible to produce hydrogen directly from a cathode through microbial catalysis (Rozendal *et al*., 2008). Hydrogen can also be produced through algal photosystems (Hallenbeck and Benemann, 2002; Hankamer *et al*., 2007; Chisti, 2008) or from acetate (Logan *et al*., 2008). Although hydrogen is an energy-dense fuel, there is currently no industrial infrastructure for mass hydrogen utilization and hydrogen is not easily transported (Antoni *et al*., 2007). Moreover, even with the addition of a microbial catalyst, hydrogen production from an electrode is fairly inefficient. A much more efficient process is generating organic compounds from electricity, termed microbial electrosynthesis (ME).

Microbial Electrosynthesis

The use of ME to produce organic compounds is documented in the case of methanogens (Cheng *et al*., 2009; Villano *et al*., 2010). *Methanobacterium palustre* and two mixed methanogenic communities are capable of sequestering CO_2 while oxidizing the cathode, thereby producing methane (Cheng *et al*., 2009; Villano *et al*., 2010). Their efficiencies were 80% and 85% while poising the system at -1 and -0.65 V (versus a Standard Hydrogen Electrode, SHE), respectively (Cheng *et al*., 2009; Villano *et al*., 2010). Although these results are encouraging, there are a few key problems with the utilization of methanogens via ME for methane production: (i) methane does not offer a high energy density relative to hydrogen or higher-carbon liquid biofuels and is difficult to transport (Lewis and Nocera 2006; Antoni *et al*., 2007; Cheng *et al*., 2009; Villano *et al*., 2010); and (ii) methanogens are difficult to maintain on an industrial scale due to their oxygen intolerance (Antoni *et al*., 2007), methane production is not readily reproducible in laboratory-scale experiments (Lovley and Nevin, 2008) and methanogens may be utilizing hydrogen, rather than the cathode directly, to produce methane (Lovley, 2010). If this latter point is true, then the same downfalls of utilizing electrodes for hydrogen production apply for methane production as well. However, acetogens can also utilize CO_2 as an electron acceptor (Durre, 2008), are more tolerant to the levels of oxygen contamination that are common at an industrial scale (Durre, 2008) and have been shown to be capable of directly utilizing an electrode as an electron donor (Nevin et al., 2010).

Acetogens utilize the Wood-Ljungdahl pathway for growth while producing acetate from a variety of compounds. Most acetogens are capable of heterotrophic growth, or chemolithoautotrophic growth on hydrogen or carbon monoxide as electron donors. Carbon dioxide is always the electron acceptor. It is believed that the plethora of compounds that may be used and produced by acetogens, including some aromatic compounds, allows for these anaerobes to survive in changing environmental conditions (Drake, 1994).

The specific compounds produced by acetogens depend upon the species, but include acids (e.g. acetate, butyrate, lactate and acetoin) (Rogers et al., 2006; Drake et al., 2008), solvents (e.g. ethanol, butanol, acetone and 2-propanol) (Durre, 2008; Lee et al., 2008) and fine chemicals (e.g. cysteine and corrinoids) (Koesnandar et al., 1991). Most acetogens are capable of not only producing an acid, but at least one solvent (Drake, 1994). This is due to the fact that acetogens switch their metabolism during acetogenic growth depending on environmental cues (Woods, 1993; Drake, 1994; Drake et al., 2008). These cues include a decrease in pH, excess butyrate and/or acetoacetate, excess CO or H_2 gas or addition of electron shuttles, such as methyl viologen (Kim et al., 1988; Woods, 1993). Acetogens produce primarily acids during their exponential growth phase (Andersch et al., 1983; Hartmanis and Gatenbeck, 1984).

The vast potential of acetogens to provide a source of fossil fuel-independent organic compounds led to a number of these phylogenetically diverse microorganisms to be examined for their ability of utilizing the electrode as an electron donor. The first was *Sporomusa ovata*, a gram-negative mesophile that primarily produces acetate, and some ethanol, when grown on hydrogen and CO_2 (Moller et al., 1984). When grown under ME conditions, a monolayer of cells of *S. ovata* produces acetate and 2-oxobutyrate with a coulombic efficiency of 86 ± 21% of electrons donated from an outside source (Nevin et al., 2010). Other pure cultures of acetogens have varying coulombic efficiencies and production rates (Table 2.1; Nevin et al., 2011). Acetogenic mixed communities may be less efficient at organic compound production than pure culture (Nevin et al., 2011).

However, not all acetogens are capable of utilizing electrons from the cathode. Thus far, *Acetobacterium woodii* is the only acetogen tested to be unable to accept electrons from an electrode. Understanding the mechanism behind the capability to accept electrons from the electrode could enable for greater rates of compound yields, more specific compounds and better selection of microorganisms for this process. However, it is not currently clear what enables microorganisms to accept electrons from the electrode.

Mechanisms for Electron Transfer from an Electrode

Although the physiological mechanism for the reduction of the electrode through acetate utilization has been studied in depth using *S. oneidensis* (Lies et al., 2005; Gorby et al., 2006) and *G. sulfurreducens* (Holmes et al., 2006; Nevin et al., 2009) as model organisms, very little has been done to date to understand the physiological mechanism of cathodic oxidation. This is in part due to the fact that utilization of the anode does not preclude the

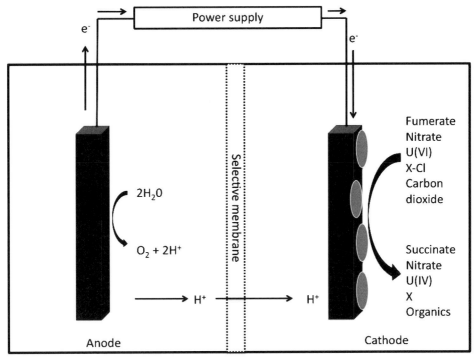

Fig.2.3. Schematic of using an electrode to provide microorganisms with electrons for the reduction of various organic and inorganic compounds. Strategies to provide electrons for the reaction include the oxidation of water or the reduction of organic compounds by microorganisms in a linked microbial fuel cell.

ability of an organism to oxidize the cathode (Strycharz *et al.*, 2010b). Of the acetogens examined, *A. woodii* is the only acetogen to utilize a Na^+-dependent ATPase and also the only acetogen tested to be incapable of electrode oxidation (Nevin *et al.*, 2011). However, the ability to accept electrons from the electrode by more Na^+-dependent acetogens should be investigated to clarify whether, as a group, Na^+-dependent acetogens are unable to utilize the electrode as an electron. In any event, the physiological mechanism of energy generation may not be the sole determining factor in whether a microbe can utilize an electrode as an electron donor.

The recently published genome of *Clostridium ljungdahlii* indicates no cytochromes are present in the membrane of this gram-positive acetogen (Köpke *et al.*, 2010), preventing it from using the model proton-dependent ATPase mechanism of acetogenic energy conservation that was established by *Moorella thermoaceticum* (Drake, 1994). Instead, it is thought to use the Rnf system, which is present in *Clostridium kluyveri* (Herrmann *et al.*, 2008; Seedorf *et al.*, 2008; Kopke *et al.*, 2010). In addition, the absence of cytochromes in this electrotroph suggests that there may be either more than one mechanism of accepting electrons directly from the electrode or that a conserved mechanism does not involve cytochromes, as previously supposed.

Indeed, when accepting electrodes from an electrode in order to reduce fumarate to succinate, *G. sulfurreducens* does not appear to utilize the same mechanism as for anode

Table 2.1. Acetogenic species identified as capable of accepting electrons from the cathode to produce compounds at −400 mV versus SHE.

Acetogenic species	Primary compound production	Trace compounds	Total coulombic efficiency
Sporomusa ovata[a]	Acetate	2-oxobutyrate	86 ± 21%
Sporomusa sphaeroides[b]	Acetate	2-oxobutyrate	84 ± 26%
Sporomusa silvacetica[b]	Acetate	2-oxobutyrate	48 ± 6%
Clostridia ljungdahlii[b]	Acetate	2-oxobutyrate, formate	82 ± 10%
Clostridia aceticum[b]	Acetate, 2-oxobutyrate	–	53 ± 4%
Moorella thermoaceticum[b]	Acetate	2-oxobutyrate	85 ± 7%

[a] Nevin *et al.*, 2010; [b] Nevin *et al.*, 2011.

reduction (Strycharz *et al.*, 2010b). Both *pilA* and *omcZ* did not appear to be upregulated in cathodic biofilm versus one grown in a system with no current generation. Moreover, the intracellular cytochrome GSU3274 appears to be required for current consumption, although its function in cathodic oxidation is not currently understood (Strycharz *et al.*, 2010b). However, the gene product of GSU3274 may be affecting the expression and activity of redox-active proteins already expressed within and on the surface of the cell, including outer membrane cytochromes (Lovley, personal communication). Research is currently being conducted in order to clarify the function of GSU3274, as well as the function of a number of hypothetical proteins differentially expressed in cathodic *G. sulfurreducens* biofilms.

Conclusion

MESs have the potential to become an important green technology for the degradation of waste, conversion of organic compounds to electricity, the production of carbon neutral compounds and conversion of electrical energy to organic compounds. While the field has rapidly advanced in recent years, large advances are still required before their widespread use becomes practical The diversity of microorganisms discovered to be capable of interacting with electrodes is in stark contrast to our understanding of the molecular mechanisms involved in these processes. While a number of mechanism have been identified, including nanowires, outer membrane *c*-type cytochromes and electron shuttles, these mechanisms still require further work to provide a continuous molecular pathway from microbial metabolism to an electrode. For this reason it can be speculated that further studies of the microbiology of microbial electric systems is required for continued advances within the field.

References

Andersch, W., Bahl, H. and Gottschalk, G. (1983) Level of enzymes involved in acetate, butyrate, acetone and butanol formation by *Clostridium acetobutylicum*. *Applied Microbiology and Biotechnology* 18, 327–332.

Anderson, R.T., Rooney-Varga, J., Gaw, C.V. and Lovley, D.R. (1998) Anaerobic benzene oxidation in the Fe(III)-reduction zone of petroleum-contaminated aquifers. *Environmental Science and Technology* 32, 1222–1229.

Anderson, R.T., Vrionis, H.A., Ortiz-Bernad, I., Resch, C.T., Long, P.E., Dayvault, R., Karp, K., Marutzky, S., Metzler, D.R., Peacock, A., White, D.C., Lowe, M. and Lovley, D.R. (2003) Stimulated *in situ* activity of *Geobacter* species to remove uranium from the groundwater of a uranium-contaminated aquifer. *Applied Environmental Microbiology* 69, 5884–5891.

Angenent, L.T., Karim, K., Al-Dahhan, M.H., Wrenn, B.A. and Domiguez-Espinosa, R. (2004) Production of bioenergy and biochemicals from industrial and agricultural wastewater. *Trends in Biotechnology* 22, 477–485.

Antoni, D., Zverlov, V.V. and Schwarz, W.H. (2007) Biofuels from microbes. *Applied Microbiology & Biotechnology* 77, 23–35.

Aulenta, F., Catervi, A., Majone, M., Panero, S., Reale, P. and Rossetti, S. (2007) Electron transfer from a solid-state electrode assisted by methyl viologen sustains efficient microbial reductive dechlorination of TCE. *Enviromental Science and Technology* 41, 2554–2559.

Aulenta, F., Reale, P., Canosa, A., Rossetti, S., Panero, S. and Majone, M. (2010) Characterization of an electro-active biocathode capable of dechlorinating trichloroethene and cis-dichloroethene to ethene. *Biosensors and Bioelectronics* 25, 1796–1802.

Bond, D. and Lovley, D.R. (2002) Reduction of Fe(III) oxide by methanogens in the presence and absence of extracellular quinones. *Environmental Microbiology* 4, 115–124.

Bond, D.R. and Lovley, D.R. (2003) Electricity production by *Geobacter sulfurreducens* attached to electrodes. *Applied Environmental Microbiology* 69, 1548–1555.

Bond, D.R., Holmes, D., Tender, L. and Lovley, D.R. (2002) Electrode-reducing microorganisms that harvest energy from marine sediments. *Science* 295, 483–483.

Borole, A.P., Mielenz, J.R., Vishnivetskaya, T.A. and Hamilton, C.Y. (2009) Controlling accumulation of fermentation inhibitors in biorefinery recycle water using microbial fuel cells. *Biotechnology and Biofuels* 2, 7.

Bouhenni, R.A., Vora, G.J., Biffinger, J.C., Shirodkar, S., Brockman, K., Ray, R., Wu, P., Johnson, B.J., Biddle, E.M., Marshall, M.J., Fitzgerald, L.A., Little, B.J., Fredrickson, J.K., Beliaev, A.S., Ringeisen, B.R. and Saffarini, D.A. (2010) The role of *Shewanella oneidensis* MR-1 outer surface structures in extracellular electron transfer. *Electroanalysis* 22, 856–864.

Bretschger, O., Obraztsova, A. and Sturm, C.A. (2007) An exploration of current production and metal oxide reduction by *Shewanella oneidensis* MR-1 wild type and mutants. *Applied and Envrionmental Microbiology* 73, 7003–70012

Calabrese Barton, S., Gallaway, J. and Atanassov, P. (2004) Enzymatic biofuel cells for implantable and microscale devices. *Chemical Reviews* 104, 4867–4886.

Chae, K.J., Choi, M.J., Lee, J.W., Kim, K.Y. and Kim, I.S. (2009) Effect of different substrates on the performance, bacterial diversity, and bacterial viability in microbial fuel cells. *Bioresource Technology* 100, 3518–3525.

Cheng, S., Dempsey, B.A. and Logan, B.E. (2007) Electricity generation from synthetic acid-mine drainage (AMD) water using fuel cell technologies. *Environmental Science and Technology* 41, 8149–8153.

Cheng, S., Xing, D., Call, D.F. and Logan, B.E. (2009) Direct biological conversion of electrical current into methane by electromethanogenesis. *Environmental Science and Technology* 43, 3953–3958.

Chisti, Y. (2008) Biodiesel from microalgae beats bioethanol. *Trends in Biotechnology* 26, 126–131.

Clauwaert, P., Rabaey, K., Aelterman, P., De Schamphelaire, L., Pham, T.H., Boeckx, P., Boon, N. and Verstraete, W. (2007) Biological denitrification in microbial fuel cells. *Environmental Science & Technology* 41, 3354–3360.

Clauwaert, P., Aelterman, P., Pham, T., Schamphelaire, L., Carballa, M., Rabaey, K. and Verstracte, W. (2008) Minimizing losses in bio-electrochemical systems: the road to applications. *Applied Microbiology and Biotechnology* 79, 901–913.

Drake, H.L. (1994) Acetogenesis, acetogenic bacteria, and the acetyl-CoA 'Wood/Ljungdahl' pathway: past and current perspectives. In: Drake, H.L. (ed.) *Acetogenesis*. Chapman & Hall, New York, pp. 3–60.

Drake, H.L., Gobner, A.S. and Daniel, S.L. (2008) Old acetogens, new light. *Annual New York Academy of Sciences* 1125, 100–128.

Durre, P. (2008) Fermentative Butanol Production: Bulk Chemical and Biofuel. *Annual New York Academy of Sciences* 1125, 353–362.

El-Naggar, M.Y., Gorby, Y.A., Xia, W. and Nealson, K.H. (2008) The molecular density of states in bacterial nanowires. *Biophysics Journal* 95, L10–2.

Ergas, S., Harrison, J., Bloom, J., Forloney, K., Ahlfeld, D., Nusslein, K. and Yuretich, R. (2006) Natural attenuation of acid mine drainage by acidophilic and acidotolerant Fe(III)- and sulfate-reducing bacteria. *American Chemical Society Symposium* 940, 105–127.

Feng, Y., Wang, X., Logan, B. and Lee, H. (2008) Brewery wastewater treatment using air-cathode microbial fuel cells. *Applied Microbiology and Biotechnology* 78, 873–880.

Finneran, K.T., Forbush, H.M., Van Praagh, C.V. and Lovley, D.R. (2002) *Desulfitobacterium metallireducens* sp. nov., an anaerobic bacterium that couples growth to the reduction of metals and humic acids as well as chlorinated compounds. *International Journal of Systematic and Evolutionary Microbiology* 52, 1929–35.

Foley, J.M., Rozendal, R.A., Hertle, C.K., Lant, P.A. and Rabaey, K. (2010) Life cycle assessment of high-rate anaerobic treatment, microbial fuel cells and microbial electrolysis cells. *Environmental Science & Technology* 44, 3629–3637.

Franks, A.E. and Nevin, K.P. (2010) Microbial fuel cells, a current review. *Energies* 3, 899–919.

Franks, A.E., Malvankar, N. and Nevin, K.P. (2010) Bacterial biofilms: the powerhouse of a microbial fuel cell. *Biofuels* 1, 589–604.

Freguia, S., Masuda, M., Tsujimura, S. and Kano, K. (2009a) Lactococcus lactis catalyses electricity generation at microbial fuel cell anodes via excretion of a soluble quinone. *Bioelectrochemistry* 76, 14–8.

Freguia, S., The, E.H., Boon, N., Leung, K.M., Keller, J. and Rabaey, K. (2009b) Microbial fuel cells operating on mixed fatty acids. *Bioresource Technology* 101, 1233–8.

Freguia, S., The, E.H., Boon, N., Leung, K.M., Keller, J. and Rabaey, K. (2010) Microbial fuel cells operating on mixed fatty acids. *Bioresource Technology* 101, 1233–1238.

Frysinger, G.S., Gaines, R.B., Xu, L. and Reddy, C.M. (2003) Resolving the unresolved complex mixture in petroleum-contaminated sediments. *Environmental Science & Technology* 37, 1653–1662.

Galvez, A., Greenman, J. and Ieropoulos, I. (2009) Landfill leachate treatment with microbial fuel cells; scale-up through plurality. *Bioresource Technology* 100, 5085–5091.

Gorby, Y.A., Yanina, S., McLean, J.S., Rosso, K.M., Moyles, D., Dohnalkova, A., Beveridge, T.J., Chang, I.S., Kim, B.H., Kim, K.S., Culley, D.E., Reed, S.B., Romine, M.F., Saffarini, D.A., Hill, E.A., Shi, L., Elias, D.A., Kennedy, D.W., Pinchuk, G., Watanabe, K., Ishii, S., Logan, B., Nealson, K.H. and Fredrickson, J.K. (2006) Electrically conductive bacterial nanowires produced by *Shewanella oneidensis* Strain MR-1 and other microorganisms. *Proceedings of the National Academy of Sciences of the United States of America* 103, 11358–11363.

Gregory, K. and Lovley, D. (2005) Remediation and recovery of uranium from contaminated subsurface environments with electrodes. *Environmental Science & Technology* 39, 8943–8947.

Gregory, K., Bond, D. and Lovley, D. (2004) Graphite electrodes as electron donors for anaerobic respiration. *Environmental Microbiology* 6, 596–604.

Hallenbeck, P.C. and Benemann, J.R. (2002) Biological hydrogen production; fundamentals and limiting processes. *International Journal of Hydrogen Energy* 27, 1185–1193.

Hankamer, B., Lehr, F., Rupprecht, J., Mussgnug, J.H., Posten, C. and Kruse, O. (2007) Photosynthetic biomass and H_2 production by green algae: from bioengineering to bioreactor scale-up. *Physiologia Plantarum* 131, 10–21.

Hartmanis, M.G.N. and Gatenbeck, S. (1984) Intermediary metabolism in *Clostridium acetobutylicum*: levels of enzymes involved in the formation of acetate and butyrate. *Applied Microbiology and Biotechnology* 47, 1277–1283.

Hartshorne, R.S., Reardon, C.L., Ross, D., Nuester, J., Clarke, T.A., Gates, A.J., Mills, P.C., Fredrickson, J.K., Zachara, J.M., Shi, L., Beliaev, A.S., Marshall, M.J., Tien, M., Brantley, S., Butt, J.N. and Richardson, D.J. (2009) Characterization of an electron conduit between bacteria and the extracellular environment. *Proceedings of the National Academy of Sciences of the United States of America* 106, 22169–22174.

Herrmann, G., Jayamani, E., Mai, G. and Buckel, W. (2008) Energy conservation via electron-transferring flavoprotein in anaerobic bacteria. *Journal of Bacteriology* 190, 784–791.

Holmes, D.E., Bond, D.R., O'Neil, R.A., Reimers, C.E., Tender, L.R. and Lovley, D.R. (2004) Microbial communities associated with electrodes harvesting electricity from a variety of aquatic sediments. *Microbial Ecology* 48, 178–190.

Holmes, D.E., Chaudhuri, S.K., Nevin, K.P., Mehta, T., Methé, B.A., Liu, A., Ward, J.E., Woodard, T.L., Webster, J. and Lovley, D.R. (2006) Microarray and genetic analysis of electron transfer to electrodes in *Geobacter sulfurreducens*. *Environmental Microbiology* 8, 1805–1815.

Ishii, S., Watanabe, K., Yabuki, S., Logan, B.E. and Sekiguchi, Y. (2008) Comparison of electrode reduction activities of *Geobacter sulfurreducens* and an enriched consortium in an air-cathode microbial fuel cell. *Applied and Environmental Microbiology* 74, 7348–7355.

Jung, S. and Regan, J.M. (2007) Comparison of anode bacterial communities and performance in microbial fuel cells with different electron donors. *Applied Microbiology and Biotechnology* 77, 393–402.

Justin, G.A., Zhang, Y., Sun, M. and Sclabassi, R. (2005) An investigation of the ability of white blood cells to generate electricity in biofuel cells. In: *Bioengineering Conference, 2005. Proceedings of the IEEE 31st Annual Northeast'*, pp. 277–278.

Kerzenmacher, S., Ducree, J., Zengerle, R. and von Stetten, F. (2008) Energy harvesting by implantable abiotically catalyzed glucose fuel cells. *Journal of Power Sources* 182, 1–17.

Kiely, P.D., Cusick, R., Call, D.F., Selembo, P.A., Regan, J.M. and Logan, B.E. (2010a) Anode microbial communities produced by changing from microbial fuel cell to microbial electrolysis cell operation using two different wastewaters. *Bioresource Technology* 102, 388–394.

Kiely, P.D., Rader, G., Regan, J.M. and Logan, B.E. (2010b) Long-term cathode performance and the microbial communities that develop in microbial fuel cells fed different fermentation endproducts. *Bioresource Technology* 102, 361–366.

Kim, B.H., Chang, I.S. and Gadd, G.M. (2007) Challenges in microbial fuel cell development and operation. *Applied Microbiology and Biotechnology* 76, 485–494.

Kim, G.T., Webster, G., Wimpenny, J.W., Kim, B.H., Kim, H.J. and Weightman, A.J. (2006) Bacterial community structure, compartmentalization and activity in a microbial fuel cell. *Journal of Applied Microbiology* 101, 698–710.

Kim, H.-H., Mano, N., Zhang, Y. and Heller, A. (2003) A miniature membrane-less biofuel cell operating under physiological conditions at 0.5 V. *Journal of The Electrochemical Society* 150, A209–A213.

Kim, J., Bajpai, R. and Iannotti, E.L. (1988) Redox potential in acetone-butanol fermentations. *Applied Biochemistry and Biotechnology* 18, 175–186.

Kim, N., Choi, Y., Jung, S. and Kim, S. (2000) Effect of initial carbon sources on the performance of microbial fuel cells containing *Proteus vulgaris*. *Biotechnology & Bioengineering* 70, 109–14.

Koesnandar, A., Nishio, N. and Nagai, S. (1991) Enzymatic reduction of cystine into cysteine by cell-free extract of *Clostridium thermoaceticum*. *Journal of Fermentative Bioengineering* 72, 11–14.

Köpke, M., Held, C., Hujer, S., Liesegang, H., Wiezer, A., Wollherr, A., Ehrenreich, A., Liebl, W., Gottschalk, G. and Dürre, P. (2010) *Clostridium ljungdahlii* represents a microbial production platform based on syngas. *Proceedings of the National Academy of Sciences of the United States of America* 107, 13087–13092.

Leang, C., Coppi, M.V. and Lovley, D.R. (2003) OmcB, a c-type polyheme cytochrome, involved in Fe(III) reduction in *Geobacter sulfurreducens*. *Journal of Bacteriology* 185, 2096–2103.

Leang, C., Qian, X., Mester, T. and Lovley, D. (2010) Alignment of the c-Type Cytochrome OmcS along pili of *Geobacter sulfurreducens*. *Applied and Environmental Microbiology* 76, 4080–4084.

Lee, H.S., Parameswaran, P., Kato-Marcus, A., Torres, C.I. and Rittmann, B.E. (2008) Evaluation of energy-conversion efficiencies in microbial fuel cells (MFCs) utilizing fermentable and non-fermentable substrates. *Water Resources* 42, 1501–1510.

Lewis, N.S. and Nocera, D.G. (2006) Powering the planet: chemical challenges in solar energy utilization. *Proceedings of the National Academy of Sciences of the United States of America* 103, 15729–15735.

Lies, D.P., Hernandez, M.E., Kappler, A., Mielke, R.E., Gralnick, J.A. and Newman, D.K. (2005) *Shewanella oneidensis* MR-1 uses overlapping pathways for iron reduction at a distance and by direct contact under conditions relevant for biofilms. *Applied Environmental Microbiology* 71, 4414–4426.

Lin, B., Braster, M., van Breukelen, B.M., van Verseveld, H.W., Westerhoff, H.V. and Roling, W.F.M. (2005) Geobacteraceae community composition is related to hydrochemistry and biodegradation in an iron-reducing aquifer polluted by a neighboring landfill. *Applied Environmental Microbiology* 71, 5983–5991.

Liu, H. and Logan, B.E. (2004) Electricity generation using an air-cathode single chamber microbial fuel cell in the presence and absence of a proton exchange membrane. *Environmental Science & Technology* 38, 4040–4046.

Logan, B.E. (2009) Exoelectrogenic bacteria that power microbial fuel cells. *Nature Review Microbiology* 7, 375–381.

Logan, B. (2010) Scaling up microbial fuel cells and other bioelectrochemical systems. *Applied Microbiology and Biotechnology* 85, 1665–1671.

Logan, B.E. and Regan, J.M. (2006) Microbial fuel cells – challenges and applications. *Environmetal Science & Technology* 40, 5172–80.

Logan, B.E., Call, D., Cheng, S., Hamelers, H.V., Sleutels, T.H., Jeremiasse, A.W. and Rozendal, R.A. (2008) Microbial electrolysis cells for high yield hydrogen gas production from organic matter. *Environmetal Science & Technology* 42, 8630–8640.

Lojou, E., Durand, M.C., Dolla, A. and Bianco, P. (2002) Hydrogenase activity control at *Desulfovibrio vulgaris* cell-coated carbon electrodes: biochemical and chemical factors – influencing the mediated bioelectrocatalysis. *Electroanalysis* 14, 913–922.

Lovley, D.R. (1997) Potential for anaerobic bioremediation of BTEX in petroleum-contaminated aquifers. *Journal of Industrial Microbiology* 18, 75–81.

Lovley, D. (2006a) Dissimilatory Fe(III)- and Mn(IV)-Reducing rokaryotes. *Prokaryotes* 2, 635–658.

Lovley, D. (2006b) Microbial fuel cells: novel microbial physiologies and engineering approaches. *Current Opinion in Biotechnology* 17, 327–332.

Lovley, D.R. (2006c) Bug juice: harvesting electricity with microorganisms. *Nature Reviews Microbiology* 4, 497–508.

Lovley, D.R. (2006d) Microbial energizers: fuel cells that keep on going. *Microbe* 7, 1.

Lovley, D.R. (2006e) Taming electricigens: how electricity-generating microbes can keep going, and going – faster. *The Scientist* 20, 46.

Lovley, D.R. (2008a) Extracellular electron transfer: wires, capacitors, iron lungs, and more. *Geobiology* 6, 225–231.

Lovley, D.R. (2008b) The microbe electric: conversion of organic matter to electricity. *Current Opinion in Biotechnology* 19, 564–571.

Lovley, D. (2010) Powering microbes with electricity: direct electron transfer from electrodes to microbes. *Environmental Microbiology Reports*, 1–9.

Lovley, D.R. and Chapelle, F.H. (1995) Deep subsurface microbial processes. *Reviews in Geophsyics* 33, 365–381.

Lovley, D.R. and Nevin, K.P. (2008) Electricity production with electricigens. In: Wall, J. *et al.* (eds) *Bioenergy*. ASM Press, Washington, DC, pp. 295–306.

Lovley, D.R. and Nevin, K.P. (2011) A shift in the current: new applications and concepts for microbe-electrode electron exchange. *Current Opinion in Biotechnology* 22, 441–448.

Lovley, D.R., Baedecker, M.J., Lonergan, D.J., Cozzarelli, I.M., Phillips, E.J.P. and Siegel, D.I. (1989) Oxidation of aromatic contaminants coupled to microbial iron reduction. *Nature* 339, 297–299.

Lovley, D.R., Woodward, J.C. and Chapelle, F.H. (1994) Stimulated anoxic biodegradation of aromatic hydrocarbons using Fe(III) ligands. *Nature* 370, 128–131.

Lovley, D.R., Woodward, J.C. and Chapelle, F.H. (1996a) Rapid anaerobic benzene oxidation with a variety of chelated Fe(III) forms. *Applied Environmental Microbiology* 62, 288–291.

Lovley, D.R., Coates, J.D., Blunt-Harris, E.L., Phillips, E.J.P. and Woodward, J.C. (1996b) Humic substances as electron acceptors for microbial respiration. *Nature* 382, 445–448.

Lovley, D.R., Holmes, D.E. and Nevin, K.P. (2004) Dissimilatory Fe(III) and Mn(IV) reduction. *Advances in Microbial Physiology* 49, 219–286.

Lower, B.H., Yongsunthon, R., Shi, L., Wildling, L., Gruber, H.J., Wigginton, N.S., Reardon, C.L., Pinchuk, G.E., Droubay, T.C., Boily, J.F. and Lower, S.K. (2009) Antibody recognition force microscopy shows that outer membrane cytochromes OmcA and MtrC are expressed on the exterior surface of *Shewanella oneidensis* MR-1. *Applied Environmental Microbiology* 75, 2931–2935.

Lu, N., Zhou, S.-G., Zhuang, L., Zhang, J.T. and Ni, J.R. (2009) Electricity generation from starch processing wastewater using microbial fuel cell technology. *Biochemical Engineering Journal* 43, 246–251.

Malki, M., De Lacey, A.L., Rodriguez, N., Amils, R. and Fernandez, V.M. (2008) Preferential use of an anode as an electron acceptor by an acidophilic bacterium in the presence of oxygen. *Applied Environmental Microbiology* 74, 4472–4476.

Mehta, T., Coppi, M.V., Childers, S.E. and Lovley, D. (2005) Outer membrane c-type cytochromes required for Fe(III) and Mn(IV) oxide reduction in *Geobacter sulfurreducens*. *Applied and Environmental Microbiology* 71, 8634–8641.

Mingui, S., Justin, G.A., Roche, P.A., Jun, Z., Wessel, B.L., Yinghe, Z. and Sclabassi, R.J. (2006) Passing data and supplying power to neural implants. *Engineering in Medicine and Biology Magazine, IEEE* 25, 39–46.

Minteer, S.D., Liaw, B.Y. and Cooney, M.J. (2007) Enzyme-based biofuel cells. *Current Opinion in Biotechnology* 18, 228–234.

Moller, B., Oflmer, R., Howard, B.H., Gottschalk, G. and Hippe, H. (1984) *Sporomusa*, a new genus of gram-negative anaerobic bacteria including *Sporomusa sphaeroides* spec. nov. and *Sporomusa ovata* spec. nov. *Archives of Microbiology* 139, 388–396.

Morris, J.M. and Jin, S. (2008) Feasibility of using microbial fuel cell technology for bioremediation of hydrocarbons in groundwater. *Journal of Environmental Science and Health, Part A: Toxic/Hazardous Substances and Environmental Engineering* 43, 18–23.

Nevin, K., Richter, H., Covalla, S., Johnson, J., Woodard, T., Orloff, A., Jia, H., Zhang, M. and Lovley, D. (2008a) Power output and columbic efficiencies from biofilms of *Geobacter sulfurreducens* comparable to mixed community microbial fuel cells. *Environmental Microbiology* 10, 2505–2514.

Nevin, K.P., Richter, H., Covalla, S.F., Johnson, J.P., Woodard, T.L., Orloff, A.L., Jia, H., Zhang, M. and Lovley, D.R. (2008b) Power output and columbic efficiencies from biofilms of *Geobacter sulfurreducens* comparable to mixed community microbial fuel cells. *Environmental Microbiology* 10, 2505–2514.

Nevin, K.P., Kim, B.C., Glaven, R.H., Johnson, J.P., Woodard, T.L., Methé, B.A., Didonato, R.J., Covalla, S.F., Franks, A.E., Liu, A. and Lovley, D.R. (2009) Anode biofilm transcriptomics reveals outer surface components essential for high density current production in *Geobacter sulfurreducens* fuel cells. *PLoS ONE* 4, e5628.

Nevin, K.P., Woodard, T.L., Franks, A.E., Summers, Z.M. and Lovley, D. (2010) Microbial electrosynthesis: feeding microbes electricity to convert carbon dioxide and water to multicarbon extracellular organic compounds. *mBio* 1, 1–5.

Nevin, K.P., Hensley, S.A., Franks, A.E., Summers, Z.M., Ou, J., Woodard, T.L., Snoeyenbos-West, O.L. and Lovley, D.R. (2011) Electrosynthesis of organic compounds from carbon dioxide is catalyzed by a diversity of acetogenic microorganisms. *Applied and Environmental Microbiology* 77, 2882–2886.

Pant, D., Van Bogaert, G., Diels, L. and Vanbroekhoven, K. (2009) A review of the substrates used in microbial fuel cells (MFCs) for sustainable energy production. *Bioresource Technology* 101, 1533–1543.

Patil, S.A., Surakasi, V.P., Koul, S., Ijmulwar, S., Vivek, A., Shouche, Y.S. and Kapadnis, B.P. (2009) Electricity generation using chocolate industry wastewater and its treatment in activated sludge based microbial fuel cell and analysis of developed microbial community in the anode chamber. *Bioresource Technology* 100, 5132–5139.

Pham, T.H., Boon, N., Aelterman, P., Clauwaert, P., De Schamphelaire, L., Vanhaecke, L., De Maeyer, K., Höfte, M., Verstraete, W. and Rabaey, K. (2008a) Metabolites produced by *Pseudomonas* sp. enable a Gram-positive bacterium to achieve extracellular electron transfer. *Applied Microbiology and Biotechnology* 77, 1119–1129.

Pham, T.H., Boon, N., De Maeyer, K., Hofte, M., Rabaey, K. and Verstraete, W. (2008b) Use of *Pseudomonas* species producing phenazine-based metabolites in the anodes of microbial fuel cells to improve electricity generation. *Applied Microbiology and Biotechnology* 80, 985–93.

Potter, M.C. (1911) Electrical effects accompanying the decomposition of organic compounds. *Proceedings of the Royal Society of London* 84, 260–276.

Qiao, Y., Li, C., Bao, S., Lu, Z. and Hong, Y. (2008) Direct electrochemistry and electrocatalytic mechanism of evolved *Escherichia coli* cells in microbial fuel cells. *Chemical Communications* 1290–1292.

Rabaey, I., Ossieur, W., Verhaege, M. and Verstraete, W. (2005) Continuous microbial fuel cells convert carbohydrates to electricity. *Water Science and Technology* 52, 515–523.

Rabaey, K. and Rozendal, R.A. (2010) Microbial electrosynthesis, revisiting the electrical route for microbial production. *Nature Review Microbiology* 8, 706–716.

Rabaey, K., Lissens, G., Siciliano, S.D. and Verstraete, W. (2003) A microbial fuel cell cabable lf converting glucose to electricity at high rate and efficiency. *Biotechnology Letters* 25, 1531–1535.

Rabaey, K., Lissens, G. and Verstraete, W. (2005) Microbial fuel cells: performances and perspectives. In: Lens, P.N., Westermann, P., Haberbauer, M. and Moreno, A. (eds) *Biofuels for Fuel Cells: biomass fermentation towards usage in fuel cells*. IWA Publishing, London, pp. 1–30).

Rabaey, K., Rodríguez, J., Blackall, L.L. and Keller, J. (2007) Microbial ecology meets electrochemistry: electricity-driven and driving communities. *The ISME Journal* 1, 9–18.

Reddy, C.M., Eglinton, T.I., Hounshell, A., White, H.K., Xu, L., Gaines, R.B. and Frysinger, G.S. (2002) The West Falmouth oil spill after thirty years: the persistence of petroleum hydrocarbons in marsh sediments. *Environmental Science & Technology* 36, 4754–4760.

Reguera, G., McCarthy, K.D., Mehta, T., Nicoll, J.S., Tuominen, M.T. and Lovley, D. (2005) Extracellular electron transfer via microbial nanowires. *Nature* 435, 1098–1111.

Reguera, G., Nevin, K., Nicoll, J., Covalla, S., Woodard, T. and Lovley, D. (2006) Biofilm and nanowire production leads to increased current in *Geobacter sulfurreducens* fuel cells. *Applied and Environmental Microbiology* 72, 7345–7348.

Reimers, C.E., Tender, L.M., Fertig, S. and Wang, W. (2001) Harvesting energy from the marine sediment-water interface. *Environmental Science & Technology* 35, 192–5.

Ren, Z., Steinberg, L.M. and Regan, J.M. (2008) Electricity production and microbial biofilm characterization in cellulose-fed microbial fuel cells. *Water Science and Technology* 58, 617–622.

Rhoads, A., Beyenal, H. and Lewandowski, Z. (2005) Microbial fuel cell using anaerobic respiration as an anodic reaction and biomineralized manganese as a cathodic reactant. *Environmental Science & Technology* 39, 4666–4671.

Richter, H., McCarthy, K., Nevin, K.P., Johnson, J.P., Rotello, V.M. and Lovley, D.R. (2008) Electricity generation by *Geobacter sulfurreducens* attached to gold electrodes. *Langmuir* 24, 4376–4379.

Richter, H., Nevin, K.P., Jia, H., Lowy, D., Lovley, D. and Tender, L. (2009) Cyclic voltammetry of biofilms of wild type and mutant *Geobacter sulfurreducens* on fuel cell anodes indicates possible roles of OmcB, OmcZ, type IV pili, and protons in extracellular electron transfer. *Energy & Environmental Science* 2, 506–516.

Rismani-Yazdi, H., Christy, A.D., Dehority, B.A., Morrison, M., Yu, Z. and Tuovinen, O.H. (2007) Electricity generation from cellulose by rumen microorganisms in microbial fuel cells. *Biotechnology and Bioengineering* 97, 1398–1407.

Rogers, P., Chen, J. and Zidwick, M.J. (2006) Organic acid and solvent production part I: Acetic, lactic, gluconic, succinic and polyhydroxyalkanoic acids. *Prokaryotes* 1, 511–755.

Roling, W.F.M., van Breukelen, B.M., Braster, B.L. and van Verseveld, H.W. (2001) Relationships between microbial community structure and hydrochemistry in a landfill leachate-polluted aquifer. *Applied Environmental Microbiology* 67, 4619–4629.

Rooney-Varga, J.N., Anderson, R.T., Fraga, J.L., Ringelberg, D. and Lovley, D.R. (1999) Microbial communities associated with anaerobic benzene degradation in a petroleum-contaminated aquifer. *Applied Environmental Microbiology* 65, 3056–3064.

Rosenbaum, M., Aulenta, F., Villano, M. and Angenent, L.T. (2011) Cathodes as electron donors for microbial metabolism: which extracellular electron transfer mechanisms are involved? *Bioresource Technology* 102, 324–333.

Rozendal, R., Sleutels, T., Hamelers, H. and Buisman, C. (2008) Effect of the type of ion exchange membrane on performance, ion transport, and pH in biocatalyzed electrolysis of wastewater. *Water Science & Technology* 57, 1757.

Ryckelynck, N., Stecher, H. and Reimers, C. (2005) Understanding the anodic mechanism of a seafloor fuel cell: interactions between geochemistry and microbial activity. *Biogeochemistry* 76, 113–139.

Seedorf, H., Fricke, W.F., Veith, B., Brüggemann, H., Liesegang, H., Strittmatter, A., Miethke, M., Buckel, W., Hinderberger, J., Li, F., Hagemeier, C., Thauer, R.K. and Gottschalk, G. (2008) The genome of *Clostridium kluyveri*, a strict anaerobe with unique metabolic features. *Proceedings of the National Academy of Sciences of the United States of America* 105, 2128–2133.

Shi, L., Deng, S., Marshall, M.J., Wang, Z., Kennedy, D.W., Dohnalkova, A.C., Mottaz, H.M., Hill, E.A., Gorby, Y.A., Beliaev, A.S., Richardson, D.J., Zachara, J.M. and Fredrickson, J.K. (2008) Direct involvement of type II secretion system in extracellular translocation of *Shewanella oneidensis* outer membrane cytochromes MtrC and OmcA. *Journal of Bacteriology* 190, 5512–5516.

Skadberg, B., Geoly-Horn, S.L., Sangamalli, V. and Flora, J.R.V. (1999) Influence of pH, current and copper on the biological dechlorination of 2,6-dichlorophenol in an electrochemical cell. *Water Research* 33, 1997–2010.

Steinbusch, K.J.J., Hamelers, H.V.M., Schaap, J.D., Kampman, C., and Buisman, C.J.N. (2010) Bioelectrochemical ethanol production through mediated acetate reduction by mixed cultures. *Environmental Science & Technology* 44, 513–517.

Strycharz, S., Gannon, S., Boles, A., Franks, A.E., Nevin, K.P. and Lovley, D. (2010a) Reductive dechlorination of 2-chlorophenol by *Anaeromyxobacter dehalogenans* with an electrode serving as the electron donor. *Environmental Microbiology Reports* 2, 289–294.

Strycharz, S., Glaves, R.H., Coppi, M.V., Gannon, S., Perpetua, L.A., Liu, A., Nevin, K.P. and Lovley, D. (2010b) Gene expression and deletion analysis of mechanisms for electron transfer from electrodes to *Geobacter sulfurreducens*. *Bioelectrochemistry* 1–31.

Summers, Z., Fogarty, H.E., Leang, C., Franks, A.E., Malvankar, M. and Lovley, D.R. (2010) Cooperative exchange of electrons within aggregates of an evolved syntrophic co-culture. *Science* 330, 1413–1415.

Tandukar, M., Huber, S.J., Onodera, T. and Pavlostathis, S.G. (2009) Biological chromium (VI) reduction in the cathode of a microbial fuel cell. *Environmental Science & Technology* 43, 8159–8165.

Tatsumi, H., Takagi, K., Fujita, M., Kano, K. and Ikeda, T. (1999) Electrochemical study of reversible hydrogenase reaction of *Desulfovibrio vulgaris* cells with methyl viologen as an electron carrier. *Analytical Chemistry* 71, 1753–1759.

Tender, L.M., Reimers, C.E., Stecher, H.A., Holmes, D.E., Bond, D.R., Lowy, D.A., Pilobello, K., Fertig, S.J. and Lovley, D.R. (2002) Harnessing microbially generated power on the seafloor. *Nature Biotechnology* 20, 821–825.

Tender, L.M., Gray, S.A., Groveman, E., Lowy, D.A., Kauffman, P., Melhado, J., Tyce, R.C., Flynn, D., Petrecca, R. and Dobarro, J. (2008) The first demonstration of a microbial fuel cell as a viable power supply: powering a meteorological buoy. *Journal of Power Sources* 179, 571–575.

Ter Heijne, A., Hamelers, H. and Buisman, C. (2007) Microbial fuel cell operation with continuous biological ferrous iron oxidation of the catholyte. *Environmental Science & Technology* 41, 4130–4134.

Thrash, J.C. and Coates, J.D. (2008) Review: direct and indirect electrical stimulation of microbial metabolism. *Environmental Science & Technology* 42, 3923–3931.

Thrash, J.C., Van Trump, J.I., Weber, K.A., Miller, E., Achenbach, L.A. and Coates, J.D. (2007) Electrochemical stimulation of microbial perchlorate reduction. *Environmental Science & Technology* 41, 1740–1746.

Torres, C.I., Krajmalnik-Brown, R., Parameswaran, P., Marcus, A.K., Wanger, G., Gorby, Y.A. and Rittmann, B.E. (2009) Selecting anode-respiring bacteria based on anode potential: phylogenetic, electrochemical, and microscopic characterization. *Environmental Science & Technology* 43, 9519–9524.

Tremblay, P., Summers, Z.M., Glaven, R.H., Nevin, K.P., Zengler, K., Barrett, C., Qiu, Y., Palsson, B.O. and Lovley, D. (2010) A c-type cytochrome and a transcriptional regulator responsible for enhanced extracellular electron transfer in *Geobacter sulfurreducens* revealed by adaptive evolution. *Environmental Microbiology* 13, 13–23.

Vandecandelaere, I., Nercessian, O., Faimali, M., Segaert, E., Mollica, A., Achouak, W., De Vos, P. and Vandamme, P. (2010) Bacterial diversity of the cultivable fraction of a marine electroactive biofilm. *Bioelectrochemistry* 78, 62–66.

Villano, M., Aulenta, F., Ciucci, C., Ferri, T., Giuliano, A. and Majone, M. (2010) Bioelectrochemical reduction of CO_2 to CH_4 via direct and indirect extracellular electron transfer by a hydrogenophilic methanogenic culture. *Bioresource Technology* 101, 3085–3090.

Watson, V.J. and Logan, B.E. (2010) Power production in MFCs inoculated with *Shewanella oneidensis* MR-1 or mixed cultures. *Biotechnology and Bioengineering* 105, 489–498.

Wei, D. and Zhang, X. (2007) Current production by a deep-sea strain *Shewanella* sp. DS1. *Current Microbiology* 55, 497–500.

Williams, K.H., N'Guessan, A.L., Druhan, J., Long, P.E., Hubbard, S.S., Lovley, D.R. and Banfield, J.F. (2010) Electrodic voltages accompanying stimulated bioremediation of a uranium-contaminated aquifer. *Journal of Geophysical Research* 115, G00–G05.

Woods, D.R. (1993) Biochemistry and regulation of acid and solvent production in clostridia. In: *The Clostridia and Biotechnology*. Stoneham, Massachusetts, pp. 25–50.

Xing, D., Zuo, Y., Cheng, S., Regan, J.M. and Logan, B.E. (2008) Electricity generation by *Rhodopseudomonas palustris* DX-1. *Environmental Science and Technology* 42, 4146–4151.

Yi, H., Nevin, K.P., Kim, B.-C., Franks, A.E., Klimes, A., Tender, L.M. and Lovley, D.R. (2009) Selection of a variant of *Geobacter sulfurreducens* with enhanced capacity for current production in microbial fuel cells. *Biosensors and Bioelectronics* 24, 3498–3503.

You, S., Zhao, Q., Zhang, J., Jiang, J. and Zhao, S. (2006a) A microbial fuel cell using permanganate as the cathodic electron acceptor. *Journal of Power Sources* 162, 1409–1415.

You, S.-J., Zhao, Q.L. and Jiang, J.Q. (2006b) Biological wastewater treatment and simultaneous generating electricity from organic wastewater by microbial fuel cell. *Huan Jing Ke Xue* 27, 1786–90.

You, S.-J., Ren, N.-Q., Zhao, Q.-L., Wang, J.-Y. and Yang, F.-L. (2009) Power generation and electrochemical analysis of biocathode microbial fuel cell using graphite fibre brush as cathode material. *Fuel Cells* 9, 588–596.

Zhang, T., Gannon, S.M., Nevin, K.P., Franks, A.E. and Lovley, D.R. (2010) Stimulating the anaerobic degradation of aromatic hydrocarbons in contaminated sediments by providing an electrode as the electron acceptor. *Environmental Microbiology Reports* 12, 1011–1020.

Zhang, Y., Min, B., Huang, L. and Angelidaki, I. (2009) Generation of electricity and analysis of microbial communities in wheat straw biomass-powered microbial fuel cells. *Applied Environmental Microbiology* 75, 3389–3395.

Zhu, X. and Ni, J. (2009) Simultaneous processes of electricity generation and p-nitrophenol degradation in a microbial fuel cell. *Electrochemistry Communications* 11, 274–277.

Chapter 3

A Comparative Assessment of Bioelectrochemical Systems and Enzymatic Fuel Cells

Deepak Pant, Gilbert Van Bogaert, Ludo Diels and Karolien Vanbroekhoven

Introduction

Globally, the demand for energy is rising continuously, however fossil fuel reserves are declining and their use is obivously linked to climate change. Governments are now setting challenging targets to increase the production of energy and transport fuel from sustainable sources. Several researchers have emphasized that energy is the biggest scientific and technological problem facing our planet in the next 50 years (Singh *et al.*, 2010). According to Lewis (2007), taking the number of joules of energy consumed by humans in a typical year, and dividing that by the number of seconds in a year, yields an average burn rate of about 13 trillion watts, or 13 TW. This is the amount of energy consumed worldwide to run our planet. Thus it is energy and not the dollar which is the currency of the world. It is the joule that drives every economy and gives people a way out of poverty. Energy generation from 'negative-value' waste streams can simultaneously help meet the world's energy needs, reduce pollution, and reduce costs associated with water and wastewater treatment (Liu, 2009).

Fuel cells have long been the objective of research as an alternative source of energy due to their ability to convert chemical energy to electrical energy. The first ever fuel cell was created in 1839 by William R. Groves, who called it a gaseous voltaic battery. The device consisted of two platinum electrodes immersed on one end in a solution of sulfuric acid and at the other end separately sealed in containers of oxygen and hydrogen (Andújar and Segura, 2009). Oxygen was reduced at the cathode and hydrogen was oxidized at the anode. The protons from the oxidation of hydrogen reacted with the oxygen ions to produce water and a potential. By combining pairs of electrodes connected in series, Groves produced a higher voltage drop, thus creating the first fuel cell. Even though a lot has changed since then in terms of reaction medium, choice of electrode material and the manner in which the fuel (gas) is transported to the electrode, the basic principle remains the same and is still being carried out in the form of modern fuel cells. The first fuel cell with practical applications was developed by William W. Jacques in 1896 (Appleby, 1990).

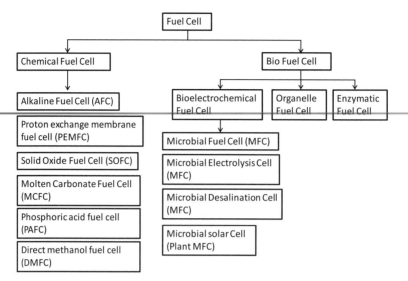

Fig. 3.1. Classification of chemical and biofuel cells.

Over the years, this field of chemical fuel cells saw tremendous developments, much of which has already been written and discussed (Andújar and Segura, 2009).

In this chapter, the other type of fuel cells, namely 'biofuel cells' and their major classes i.e. bioelectrochemical fuel cells and enzymatic fuel cells, are discussed. Biofuel cells are a subset of fuel cells that employ biocatalyst and normally operate under mild conditions (20–40°C). Figure 3.1 shows the major divisions in various types of fuel cells.

Biological Fuel Cells

Biological fuel cells (BioFCs) have been defined as devices capable of directly transforming chemical to electrical energy via electrochemical reactions involving biochemical pathways (Bullen *et al.*, 2006). In general, BioFCs are electrochemical devices in which organic material is biologically oxidized at an anode, producing carbon dioxide (CO_2), electrons and protons. Electrons are shuttled from one electrode to the other through an external electric circuit while protons diffuse through the electrolyte to the cathode. Although not always used, a membrane is typically used to separate the anode and cathode reactions as well as facilitate proton transfer. Finally, electrons and protons are electrochemically reduced with help of a chemical or biocatalyst at the cathode, to form water, H_2 or some specific chemical. Because current production is very small, the voltage drop (V_{drop}) across an external resistor (R_{ext}) is measured and the current is calculated according to Ohm's law. The BioFC can be further categorized into bioelectrochemical systems (BES) or enzymatic fuel cells (EFC), depending on the respective catalyst used in the system, i.e. a living cell (often bacteria) or enzyme. The main mechanisms involved in different kinds of fuel cells are detailed in Fig. 3.2 (Bullen *et al.*, 2006).

The earliest description of the ability of bacteria to transport electrons extracellularly was described by Potter in 1911. Since then there have been several attempts to bring this technology to a more commercial and viable scale. However, it is only in recent years that the research in this field has grown by leaps and bounds (Franks *et al.*, 2010; Pant *et al.*,

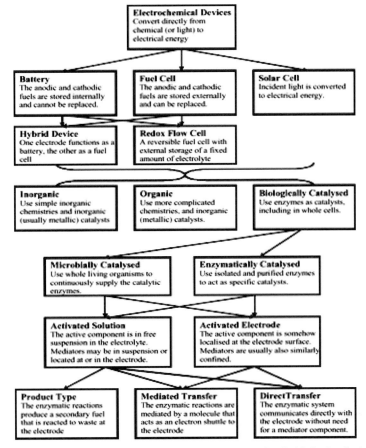

Fig. 3.2. Classification of electrochemical cells and more specifically biofuel cells based on their working principle (Bullen et al., 2006, reprinted with permission).

2010b). The development of BioFCs represents an interdisciplinary challenge, where biologists, chemists, material scientists and electrical engineers find an arena for innovative research (Willner, 2009).

Bioelectrochemical systems: from waste to watts

BESs can be further subdidved into microbial fuel cells (MFCs), microbial electrolysis cell (MEC), microbial desalination cells (MDCs) and microbial solar cells (MSC), depending on their mode of application. In fact, Harnisch and Schröder (2010) recently coined the term MXC for these systems, the X standing for the different types and applications. For the sake of simplicity, the discussion in this chapter is confined to MFCs only. A number of studies and reviews are already available for other type of BESs. These include MEC (Logan *et al.*, 2008), MDC (Cao *et al.*, 2009) and MSC (Strik *et al.*, 2011). Figure 3.3 shows the basic scheme of a two-chambered MFC with different electron transport mechanisms.

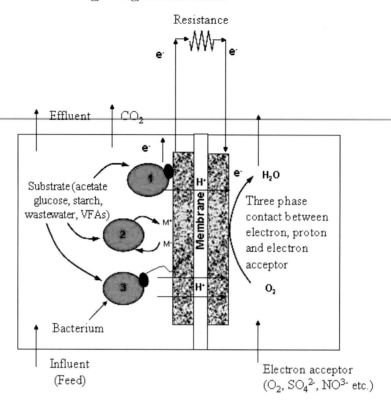

Fig. 3.3. Schematic of a two-chamber MFC with possible modes of electron transfer; (1) direct electron transfer (via outer membrane cytochromes); (2) electron transfer through mediators; and (3) electron transfer through nanowires. (Pant et al., 2010b, reprinted with permission.)

The popularity of the MFC technology has risen exponentially during the last few years because there is a hope that MFCs will allow harvesting of the energy stored in waste water directly in the form of electricity (Lefebvre et al., 2011). The main advantage of MFC is their typically longer lifetime (up to 5 years) (Moon et al., 2005). Typically, the chemical energy contained in an organic substrate is directly converted into electrical power, hydrogen gas, or methane in BESs (Clauwaert et al., 2008).

Their main limitation is the low power densities, the power generated per unit electrode surface area due to several major limitations such as slow transport across cellular membranes (Palmore and Whitesides, 1994). Other losses associated with these systems are ohmic voltage losses, activation overpotentials, concentration overpotentials and coulombic losses. All of them are described in detail by Clauwaert et al. (2008).

Another common variation of MFC is a microbial electrolysis cell (MEC) in which the main product formed is hydrogen. This is not a spontaneous reaction and hence is not an electricity (power) producing but power consuming system. The standard reduction potential, E_o', for H_2 production is lower than the reduction potential at the anode. In practice, this potential difference is even greater due to ohmic resistance through the system and overpotentials at the electrode–electrolyte interface. Consequently, energy from an external power source must be supplied to the cell in order to drive electrons from the anode to the cathode (Logan *et al*., 2008). Recently, Foley *et al*. (2010) using life cycle assessment showed that an MEC provides significant environmental benefits over MFC through the displacement of chemical production by conventional means.

Enzymatic fuel cells: energy of enzymatic reactions

Enzymes are biological catalysts with very good prospects for application in chemical industries due to their high activity under mild conditions, high selectivity and specificity (Hernandez and Fernandez-Lafuente, 2011). EFCs possess several positive attributes for energy conversion, including renewable catalysts, flexibility of fuels (including renewables) and the ability to operate at room temperature. However, they remain limited by short lifetimes, low power densities and inefficient oxidation of fuels (Minteer *et al*., 2007).

Compared to MFCs, the EFCs typically possess orders of magnitude higher power densities (although still lower than conventional fuel cells), but can only partially oxidize the fuel and have limited lifetimes (typically 7–10 days) owing to the fragile nature of the enzyme (Kim *et al*., 2006). Enzymes are also much more specific, thus eliminating the need for a membrane separator (Minteer *et al*., 2007). The main differences between MFC and EFC are summarized in Table 3.1. The use of single enzymes (or enzyme cascades) allows permits defined reaction pathways on the electrode surface and overcomes the limited output performance of microbial biofuel cells, which is considered to be due to mass transfer resistances across the cell membranes (Ivanov *et al*., 2010).

Basically, enzymes are proteins that typically have short lifetimes (8 h to 2 days) in buffer solution, although their active lifetimes can be extended to 7–20 days by immobilization on electrode surfaces via entrapment, chemical bonding and crosslinking (Kim *et al*., 2006). The earliest EFCs operating with purified enzymes harvesting hydrogen from a substrate through the paired action of a hydrogenase and a dehydrogenase enzyme have been demonstrated (Woodward *et al*., 1996; O'Neill and Woodward, 2000). A typical EFC is shown in Fig. 3.4. Enzymatic cells have been divided into mediated electron transfer (MET) and direct electron transfer (DET), with DET covering only systems where the electron tunnels directly from the active site fixed in the enzyme to the electrode, and MET encompassing all forms of regenerative mediation whether diffusive or non-diffusive (Barton *et al*., 2004). The difference between the two and various studies involving both the modes have been reviewed extensively by Bullen *et al*. (2006). In general, enzymatic biocatalysts are limited by the number of reactions they have evolved to catalyse, which in turn limits their applications for these systems. Enzyme electrodes restrict operation mostly within physiological conditions due to the usual instability of the enzymes in non-aqueous environments and at high temperatures. Another obstacle of enzymes in EFCs is their inhibition of activity by many different chemical compounds (Sarma *et al*., 2009).

Microbial biorefineries: factories of the future?

Initially, the only applications of BioFCs were for bioremediation (Gil *et al*., 2003) and as a biosensor for biological oxygen demand (BOD) (Kim *et al*., 2003). Though an unlimited supply of fuel (waste water from the starch plant) was available, keeping the sensor viable for 5 years, it took up to 10 h to measure the BOD of a sample. Later,

Ieropolous et al. (2004) demonstrated autonomous robots called 'gastrobots' based on the BioFC platform.

Table 3.1. The main differences between MFCs and EFCs.

Microbial fuel cells (MFCs)	Enzymatic fuel cells (EFCs)
Bacteria are the main catalyst	Enzymes are the main catalyst
Main application in treating low strength wastewater	Main application in powering biomedical devices
Low power and current density	High power and current density
Complete oxidation of fuel/substrate	Incomplete oxidation of fuel/substrate
Longer life time – up to several months/year	Shorter lifetime – up to a few hours to few days

Other suggested applications include generation of electrical power from body fluids (e.g. blood) by implantable devices that can operate pacemakers, hearing aids or prosthetic units, the generation of electrical power from biomass and waste organic materials, and the dvelopment of self-powered biosensors (Willner, 2009).

Unlike anaerobic digestion (AD), an MFC at present consumes more energy for its operation than what can be harvested; however, MFC has several advantages over conventional activated sludge (CAS), such as the possibility to use gaseous oxygen from the atmosphere using an air cathode (Park and Zeikus, 2003), which can potentially greatly reduce operation costs in an MFC-based wastewater treatment plant (Lefebvre et al., 2011).

Some of the achievable implementations of the EFC in the short term seem to be the implantable power supplies on a limited scale based on the 'two fibre' biofuel cell design (Mano et al., 2003a), biosensors that sustain themselves on nutrient content extracted from the sampled medium (Kim et al., 2003), and portable power cells based on an MEA configured biofuel cell (Akers and Minteer, 2003).

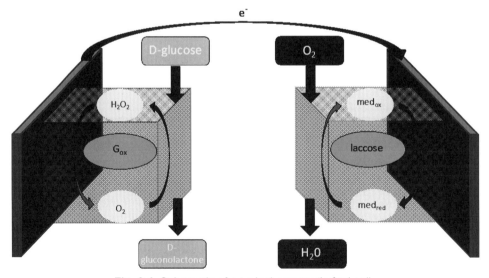

Fig. 3.4. Schematic of a typical enzymatic fuel cell.

Microbes and Enzymes Involved

Shewanella putrefaciens was the bacteria in which the ability of bacteria to transfer electrons directly to an electrode surface without the need for an external mediator was first reported (Kim *et al*., 1999). Recent advances have led to generation of new knowledge and better understanding of the intricate genetic and biochemical mechanisms involved in biosynthesis of secondary metabolites. Taxonomic profiles of electrode-reducing microbial communities from numerous MFC systems have been reported (Jung and Regan, 2007; Aelterman *et al*., 2008; Clauwaert *et al*., 2008). The bacterial communities that develop show great diversity and typically depend on the enrichment conditions used to colonize the electrode surface (Lovley, 2008). However, among this group of electrode-respiring microorganisms, bacteria from the phylum Proteobacteria dominate anode communities.

According to an eight-system comparative study (Aelterman *et al*., 2008), 64% of the anode population belonged to the class of α-, β-, γ-, or δ-Proteobacteria, the most studied of these belonging to the families of *Shewanella* and Geobacteraceae (Debabov, 2008). The complete genome of *Shewanella oneidensis* was sequenced in 2002 (Heidelberg *et al*., 2002) and subsequently that of *Geobacter sulfurreducens* in 2003 (Methe *et al*., 2003); both organisms serve as excellent models to elucidate the mechanisms of electron transfer between bacteria and electrode.

Redox enzymes (also known as oxidoreductases) are extensively used to construct amperometric enzyme electrodes. However, hydrolytic enzymes like lipases are also used along with the redox enzymes for added specificity or to sense a substrate where no oxidoreductase is known (Sarma *et al*., 2009). Redox enzymes usually lack direct electron transfer communication between their active redox centres and electrode support (Willner *et al*., 2009). On the basis of electron transfer establishment between enzymes and electrodes, the redox enzymes have been divided into three groups (Heller, 1992). The first group has enzymes with nicotinamide adenine dinucleotide (NADH/NAD$^+$) or nicotinamide adenine dinucleotide phosphate (NADPH/NADP$^+$) redox centres, which are often weakly bound to the protein of the enzyme. The weak binding allows the redox centres to diffuse away from the enzyme, acting as carriers of electrons – one of their natural functions is cellular electron transfer. In biofuel cell systems this allows the cofactor to also function as a natural redox mediator (Bullen *et al*., 2006). The specific conditions for this enzyme system are design of electrode for two-electron transfer, prevention of diffusion away of NADH/NAD$^+$ and prevention of hydrolysis of NADH/NAD$^+$ or NADPH/NADP$^+$ at a rapid rate. The second group of enzymes are those with part of the redox centre located at or near the periphery of the protein shell (e.g. peroxidases). These enzymes are 'designed' to transfer or accept electrons on contact and are thus able to communicate directly with electrodes. The challenge here is of orienting the enzyme on the electrode for maximum activity, both for rapid electron transfer and also for diffusional access of the substrate to the enzyme (Bullen *et al*., 2006). The third enzyme type contains those with a strongly bound redox centre deeply bound in a protein or glycoprotein shell. This type of enzyme neither communicates readily with electrodes, nor releases the active centre to act as an electron shuttle. Direct electron transfer from the active centre is either extremely slow or impossible, requiring the use of mediator molecules capable of penetrating into the enzyme to transport charge. Glucose oxidase (GOx) is an example of this type of enzyme (Bullen *et al*., 2006). Figure 3.5 shows the main electron transfer mechnisms in EFCs. Only a few enzymes have active centres close enough to the surface or intramolecular electron transfer mechanisms to directly interact with the electrode (direct electron transfer). Most enzymes require mediators to shuttle the electrons between the active centre and electrode (mediated electron transfer) (Rubenwolf *et al.*, 2011).

The oxidation of methanol to CO_2 and water was demonstrated by Palmore *et al.* (1998) in a multi-enzyme fuel cell. The methanol was successively oxidized by alcohol- (ADH), aldehyde-(AldDH) and formate-(FDH) dehydrogenases, dependent upon NAD^+ reduction. A fourth enzyme, diaphorase, regenerated the NAD^+, reducing benzylviologen (1,1'-dibenzyl-4,4'-bipyridinium dichloride, $BV^{+/2+}$). The benzylviologen was oxidized at a graphite plate anode (2 cm^2 area) to complete the electron transfer half-reaction. The multiple enzyme system used in this case offers a chance to completely oxidize a substrate but at the cost of greater complexity. However, enzyme-based systems can only fully oxidize glucose by using several enzymes, and the product of the first stage of the oxidation has the potential to polymerize and foul the electrode surface.

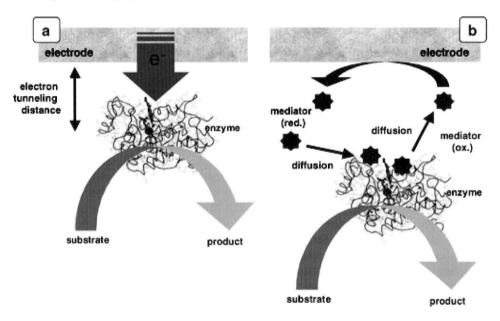

Fig. 3.5. Alternative electron-transfer mechanisms. (a) Direct electron transfer (tunneling mechanism) from electrode surface to the active site of an enzyme. (b) Electron transfer via redox mediator (Barton *et al.*, 2004, reprinted with permission).

Anodic reactions

The oxidation of a substrate, such as acetate, at the anode produces electrons, protons and ideally only CO_2 if all available electrons from the organic fuel are recovered. The development of processes that can use bacteria to produce electricity represents a plausible method for bioenergy production as the bacteria are self-replicating, and thus the catalysts for organic matter oxidation are self-sustaining. Sharma and Kundu (2010) recently reviewed the biocatalysts used in MFCs.

Current enzymatic biofuel cells have low efficiency, as only a single type of enzyme is employed and can only partially oxidize the fuel. This is in direct contrast to living cells that can completely oxidize biofuels (e.g. ethanol, lactate and glucose) to CO_2 and water (Minteer *et al.*, 2007). Table 3.2 shows the main anodic catalysts in both MFCs and EFCs.

Table 3.2. Major types of anodic catalysts used in MFCs and EFCs and their applications.

Mainly used bacteria as anodic catalyst in MFCs	Mainly used enzymes as anodic catalyst in EFCs
Geobacter sulfureducens for acetate oxidation (Reguera *et al.*, 2006)	Pyrolloquinoline quinone (PQQ) for oxidation of NADH and NADPH to $NAD(P)^+$ (Willner *et al.*, 1998a)
Shewanella oneidensis for acetate, lactate, and glucose oxidation (Biffinger *et al.*, 2009)	Alcohol and aldehyde dehydrogenase (AldDH) for oxidation of ethanol (Palmore *et al.*, 1998)
Saccharomyces cerevisiae for oxidation of glucose (Veer Raghavulu *et al.*, 2011)	Glucose oxidase for glucose oxidation (Willner *et al.*, 1998b)
Shewanella putrefaciens oxidizing d,l-lactate (Crittenden *et al.*, 2006)	Glucose dehydrogenase (sGDH) for oxidizing glucose to gluconate (Tsujimura *et al.*, 2002)
Pseudomonas aeruginosa for glucose oxidation (Rabaey *et al.*, 2005)	Cellobiose dehydrogenase (CDH) for oxidation of saccharides (glucose, lactose, cellobiose) (Tasca *et al.*, 2008)
Klebsiella pneumoniae for glucose oxidation (Menicucci *et al.*, 2006)	Cascade of PQQ-dependent alcohol dehydrogenase, PQQ-dependent aldehyde dehydrogenase and oxalate oxidase (Arechederra and Minteer, 2009)
Rhodopseudomonas palustris for acetate oxidation (Xing *et al.*, 2008)	
Escherichia coli for glucose oxidation (Qiao *et al.*, 2008)	

Cathodic reactions

In MFCs, the cathodic reduction can be classified into aerobic or anaerobic reactions depending on the source of the final electron acceptor available. In aerobic cathodes, oxygen is the terminal electron acceptor. The reduction of oxygen is the most dominant electrochemical reaction at the surface of cathode electrodes. Unlimited availability and high standard redox potential make oxygen an exceptional electron acceptor (Rismani-Yazdi *et al.*, 2008). Different catalysts have been used on the cathode surface to lower the cathodic activation overpotential and increase the current output of MFCs. These include metal-based such as platinum (Pt)-coated cathode and/or biocatalysts. Biocathodes have advantages over abiotic cathodes in terms of lower cost of construction and operation. Metal catalysts or artificial electron mediators could be made superfluous in MFCs with biocathodes, because microorganisms can function as catalysts to assist the electron transfer (He and Angenent, 2006). Biocathodes may also be utilized to produce useful products or remove unwanted compounds (Rabaey and Rozendal, 2010; Steinbusch *et al.*, 2010).

Applying a biocathode, Clauwaert *et al.* (2007a) utilized nitrate as electron acceptor to get an output around 10 W m^{-3}. When the anode of an acetate-oxidizing tubular microbial fuel cell was combined with an open air biocathode for electricity production, the maximum power production was around 83 and 65 W m^{-3} for batch fed and continuous systems, respectively (Clauwaert *et al.*, 2007b).

Enzymatic biocathodes have been discussed earlier (Palmore and Whitesides, 1994). Due to its ability to reduce oxygen, immobilized laccase has been used as a biocathode. This enzyme is classified as an oxidoreductase that uses diphenols and other substances as electron donors. Studies have shown that there are no significant differences in the electrochemical reduction of oxygen when using laccase itself when compared to the immobilized form (Tarasevich et al., 2003). On the other hand, peroxidase uses as electron donator the Fe^{3+} through a covalent link with the porphyrin haem (Katz et al., 1999). Because of the active redox centre (Fe^{3+}/Fe^{2+}), peroxidase can in principle be successfully immobilized, keeping its bioelectroactivity (Wang et al., 2005). Immobilization of horseradish peroxidase in a composite using an ionic liquid and carbon has been reported (Chen et al., 2007). The modified electrode presented electroactivity towards oxygen and hydrogen peroxide reduction in a phosphate buffer (pH = 7). Table 3.3 shows the main cathodic catalysts in both MFCs and EFCs.

Table 3.3. Major types of cathodic catalysts used in MFCs and EFCs and their applications.

Mainly used bacteria as cathodic catalyst in MFCs	Mainly used enzymes as cathodic catalyst in EFCs
Leptothrix discophora for oxygen reduction (Rhoads *et al.*, 2005)	Laccase for reduction of oxygen to water (Palmore and Kim, 1999)
Geobacter sulfurreducens to reduce uranium(VI) to uranium(IV) (Gregory and Lovley, 2005)	Microperoxidase for reduction of hydrogen peroxide to water (Willner *et al.*, 1998a)
Geobacter lovleyi for reduction of fumarate to succinate (Strycharz *et al.*, 2008)	Bilirubin oxidase (BOx) for reduction of oxygen to water (Tsujimura *et al.*, 2002)
Nitrosomonas sp. for reduction of nitrite to NO (Chen *et al.*, 2008)	Horseradish peroxidase (HRP) for reduction of hydrogen peroxide (Pizzariello *et al.*, 2002)
Actinobacillus succinogenes to reduce CO_2 to methane (Park *et al.*, 1999)	Multi-copper oxidase (MCO) catalysing a four-electron reduction of oxygen (Miura *et al.*, 2009)

The Role of Material Development

Materials for MFCs

In recent years, it has been proposed that the growth in terms of power densities in terms of biocatalyst has hit a plateau and the next big growth will come from improved materials used in these systems (Rabaey and Rozendal, 2010). This includes improved electrodes for anodes and cathodes (Zhang et al., 2009; Pant et al., 2010a), seperators (Li et al., 2011) and newer designs of the cells (Wang et al., 2011). The role of new materials in developing the next generation of bioelectrochemical systems was recently discussed by Logan (2010).

Criteria for materials selection and anode and cathode configurations are different. The anode needs to provide a suitable environment for microbes to attach and deposit electrons, while the cathode has to facilitate the occurrence of electron-accepting reactions; in other words, it should possess catalytic ability in most cases. The main requirements for anodes are high specific surface area, high porosity, limited propensity to fouling or corrosion, and good microorganism adaptivity (Logan, 2008; Lefebvre et al., 2011). The most versatile anode material is carbon-based due to its superior conductivity, chemical stability, structural strength, favourable surface properties for biofilm development, and versatility for creating a large surface area (Liu, 2009). Different forms of carbon have been used to increase the surface area and microorganism coupling ability (Bond *et al*., 2002). These include plain graphite (Venkata Mohan *et al*., 2008), woven graphite (Park and Zeikus, 2003), graphite rods (Liu *et al*., 2010), graphite foam (Chaudhuri and Lovley, 2003), graphite felt (Biffinger *et al*., 2007), graphite granules (Aelterman *et al*., 2006), graphite fibre brush (Logan *et al*., 2007), reticulated vitreous carbon (He *et al*., 2006), granular-activated carbon (Jiang and Li, 2009), carbon paper (Liu and Logan, 2004), carbon cloth (Liu *et al*., 2005; Pant *et al*., 2011b) and a carbon nanotube/polyaniline composite (Qiao *et al*., 2007).

Earlier, making an overview of power density versus electrode surface area in MFCs from published literature, Dewan *et al*. (2008) reported that the maximum power density generated by an MFC is not directly proportional to the surface area of the anode, but is instead proportional to the logarithm of the surface area of the anode. In other words, in MFCs power density decreases with increasing surface area of the current-limiting electrode and that when scaling up microbial fuel cells, one cannot assume that power density will remain constant with the increased electrode surface area. However, it was reported recently that enlarging the surface area of the electrode increases the total reaction rate, hence increases the amount of collected current (Di Lorenzo *et al*., 2010).

The cathode has been identified as the most challenging aspect of the MFC design due to the specific need to have a three-phase contact between air (oxygen), water (protons) and solid (electrons) (Logan, 2010). Also, it is the cathode that is more likely to limit power generation than the anode (Fan *et al*., 2008; Rismani-Yazdi *et al*., 2008). Efforts have been made to replace the platinum catalyst used for oxygen reduction with non-precious metals (Zhang *et al*., 2009; Pant *et al*., 2010a) and metal-organic compounds based on Co and Fe (Zhao *et al*., 2005; Cheng *et al*., 2006). Further, the biocathodes as discussed earlier are also considered as a promising alternative. Biocathodes are indeed the way forward in the quest to implement MFCs for practical applications, such as wastewater treatment and sediment MFCs, because of potential cost savings, waste removal and operational sustainability. At the same time, they are less robust then abiotic oxygen-reducing cathodes.

In the past decade, a variety of separators have been extensively explored for MFCs, including cation exchange membrane (CEM), anion exchange membrane (AEM), bipolar membrane (BPM), microfiltration membrane (MFM), ultrafiltration membrane (UFM), salt bridge, glass fibres, porous fabrics and other coarse-pore filter materials. However, the majority of these existing membranes were originally designed for chemical fuel cells and are not necessarily suitable for MFCs. Proton transfer has become a main limitation for most MFCs due to the competitive migration of other cations at a high concentration. Thus, development of proton-specific membranes according to the MFC characteristics may be a good solution (Li *et al*., 2011). It is thus very important to develop separators able to resist corrosion of various industrial and domestic waste waters, endure biofouling in long-term operation and prevent oxygen permeation that diminish system performance, though these may increase their cost.

Materials for EFCs

To be used as catalysts, enzymes must be conveniently immobilized in a conductive substrate. The objective to immobilize an enzyme is to obtain a biocatalyst with activity and stability that is kept unaffected during the operation, when compared to its free form. The procedure of immobilization of a protein can be achieved for instance via cross-linking, van der Waal's interactions and hydrogen bonding, among others. Ideally, an immobilized enzyme must show a superior catalytic activity, keeping at the same time its structural properties and the active sites' integrity (Colmati et al., 2007). Immobilization of enzymes and proteins using standard protocols is usually called 'random immobilization'. When properly designed, the immobilization of enzymes and proteins has also been a very powerful tool to improve enzyme or protein stability, and in certain cases even their activity or selectivity (Hernandez and Fernandez-Lafuente, 2011). Also, the target of enzyme immobilization strategies is to increase stability and enzyme lifetime, while addressing low efficiency by attempting to incorporate cascades of enzymes to ensure complete oxidation of fuels (Zulic and Minteer, 2010).

The selection of appropriate combinations of materials, such as enzyme, electron transport mediator, binding and encapsulation materials, conductive support matrix and solid support, for construction of enzyme-modified electrodes governs the efficiency of the electrodes in terms of electron transfer kinetics, mass transport, stability and reproducibility (Sarma et al., 2009). For enzyme-based systems the areas of electrodes are usually small – up to a few centimetres – and hence the power output is not large enough for many potential applications. To achieve higher currents the enzyme loading on the electrode therefore has to be increased and the simplest way to achieve this is to use a higher area electrode (Bullen et al., 2006). Minteer et al. (2007) discussed the requirements of an EFC electrode. This is partly based on the nature of the enzymes as discussed earlier. The anodes should be three-dimensional, should be able to optimize the need for surface area, which generally means smaller pores that increase the reactive surface area and thus the current generated, and the need for larger pores, which generally supports the mass transport of liquid-phase fuel. Also, the successful immobilization of multienzyme systems that can completely oxidize the fuel to CO_2 is needed. Finally, the anode must support efficient charge transfer mechanisms, whether it be direct or mediated, and balance electron transfer with proton transfer.

Porous chitosan scaffolds is considered one such material with desired control over the degree of dimensionality and directionality (Liu et al., 2005). A non-compartmentalized, miniature biofuel cell was developed, based on two carbon fibres coated with redox polymer, which entrap the enzyme and act as electron conductors, connecting glucose oxidase to the anode and laccase or bilirubin oxidase to the cathode (Mano et al., 2002, 2003b). Different studies reported on the assembly of electrically contacted enzyme electrodes by the direct absorption of the biocatalyst on the electrode materials. For instance, laccase was electrically wired by its direct immobilization on graphite electrodes (Blanford et al., 2007), and BioFCs were constructed by the use of electrically contacted anodes and cathodes that included directly immobilized biocatalysts on the electrode surface (Coman et al., 2008). The adsorption of enzymes on single-walled carbon nanotubes (SWCNTs) was attempted to directly electrically contact the enzymes with the electrodes, and to prepare bioelectrocatalytic cathodes and anodes. The adsorption of laccase or BOx on CNTs led to effective cathodes for the four-electron O_2 reduction to H_2O (Yan et al., 2006; Gao et al., 2007). Surface modification of pyrolytic graphite electrodes with molecular promoter units was reported to tightly bind laccase (from *Pycnoporus cinnabarinus*) to the graphitic electrode, and to electrically wire the enzyme towards the electrocatalysed reduction of O_2 (Blanford *et al.*, 2007). Recently, Sarma *et al.* (2009) reviewed the advances made in material science for enzyme-based electrodes. These materials include conducting electro-active, biocompatible polymers (CEPs) such as polypyrrole (PPy) and its derivatives, functionalized polymers such as modified chitosan and Nafion, composite materials, which are the combination of powdered electronic conductor (generally carbon powder) and a binding agent such as ferrocene-doped silica nanoparticles, sol–gel materials and nanomaterials such as carbon nanotubes (CNTs) and silicon nanoparticle and nanoporous poly(aniline) (PANIs).

Future Perspectives

Enhancing power generation while at the same time lowering material costs is one of the main challenges to making MFC technology feasible for practical wastewater treatment (Liu, 2009). A very important factor that needs to considered for the applications of these systems is costs associated with them. For MFCs, some figures have already been mentioned (Rozendal *et al.*, 2008; Foley *et al.*, 2010; Fornero *et al.*, 2010; Pant *et al.*, 2011a) and most of them suggest that with increasing research in this area the cost are reducing substantially even though much needs to be done before they become economically sustainable. The use of costly precious metal catalysts, such as platinum, should be prohibited for applications such as wastewater treatment, and cheaper alternative catalysts should be found before MFCs can be scaled-up to full plant application.

With enzymatic cells in particular, the lifetime of the biocatalysts, currently of the order of a month in favourable cases, needs to be increased to the order of at least a year to find commercial favour. Improved electron transfer mechanisms have been developed that can successfully compete with oxygen as a terminal electron acceptor, allowing oxidizing enzyme electrodes to function in oxygenated solutions (Chen *et al.*, 2001). The development of the different methods to electrically contact redox enzymes with electrodes led to tremendous efforts to use these enzyme-functionalized electrodes as amperometric biosensors (Willner *et al.*, 2009). Immobilization strategies should significantly reduce the cost of biofuel cells and improve their longer term performance, as loss of mediator has been a strongly limiting factor, particularly for NAD-dependent enzyme systems (although the tendency of NAD(H) to hydrolyse has not been addressed) (Bullen *et al.*, 2006). The challenge of continuous renewal of the inactivated enzyme catalyst in an electrochemical cell (Zulic and Minteer, 2010) is also being tackled by *in situ* microbial secretion of enzymes. For example, if wood-degrading fungi that secrete redox enzymes to the medium are incorporated into a biofuel cell, they can continuously produce fresh enzymes to replace inactivated ones (Rubenwolf *et al.*, 2011).

Conclusion

Both BioFC and EFC technologies are still in their infancy. Looking into their properties, mode of operation, materials used and potential applications, it can be said that instead of being competitors they are more likely to be complementary to each other. While MFCs are more efficient due to full oxidation of substrate and longer life, EFCs are also catching up with the development of multi-enzyme systems for complete oxidation of fuel. Better immobilization strategies are enhancing the lifetime of enzymatic electrodes. On the other hand, EFCs have higher power output and shorter life, but the new generation of MFCs are capable of producing almost equal power as EFCs with selective adaptation of strains. Existing technologies offer potential for practical application and MFCs have already proven their versatility in using a range of substrates to produce power. By utilizing pathways that nature uses to recycle the energy from renewable biomass, MFCs are operated in an environmentally benign manner, offering clean and sustainable energy. For MECs producing hydrogen to be commercially competitive, they must be able to synthesize H_2 at rates that are sufficient to power fuel cells of sufficient size to do practical work. Further research and development aimed at increasing rates of synthesis and final yields of current and H_2 in BESs are essential. Problems of level and stability of total power output have too often been neglected, with test systems being very small scale, and often set up with an initial load of fuel that is allowed to deplete, placing an intrinsic limit on output duration. The proper use of different compositions of binder and immobilization matrix, electron transport mediators, biomaterials and biocatalysts, and solid supports as electron collectors in the construction of enzyme electrodes is critical to generate optimum current from the enzymatic redox reactions and the eventual success of EFCs.

References

Aelterman, P., Rabaey, K., The Pham, H., Clauwaert, P., Boon, N. and Verstraete, W. (2006) Continuous electricity generation at high voltages and currents using stacked microbial fuel cells. *Environmental Science and Technology* 40, 3388–3394.

Aelterman, P., Rabaey, K., De Schamphelaire, L., Clauwaert, P., Boon, N. and Verstraete, W. (2008) Microbial fuel cells as an engineered ecosystem. In: Wall, J.D., Harwood, C.S., Demain, A.L. (eds) *Bioenergy*. ASM Press, Washington, DC, pp. 307–322.

Akers, N.L. and Minteer, S.D. (2003) Towards the development of a membrane electrode assembly (MEA) style biofuel cell. *ACS Division of Fuel Chemistry, Preprints* 48(2), 895–896.

Andújar, J.M. and Segura, F. (2009) Fuel cells: history and updating. A walk along two centuries. *Renewable and Sustainable Energy Reviews* 13, 2309–2322.

Appleby, J. (1990) From Sir William Grove to today: fuel cells and the future. *Journal of Power Sources* 29(1-2), 3–11.

Arechederra, R.L. and Minteer, S.D. (2009) Complete oxidation of glycerol in an enzymatic biofuel cell. *Fuel Cells* 9(1), 63–69.

Barton, S.C., Gallaway, J. and Atanassov, P. (2004) Enzymatic biofuel cells for implantable and microscale devices. *Chemical Reviews* 104(10), 4867–4886.

Biffinger, J.C., Ray, R., Little, B. and Ringeisen, B.R. (2007) Diversifying biological fuel cell designs by use of nanoporous filters. *Environmental Science and Technology* 41, 1444–1449.

Biffinger, J.C., Ribbens, M., Ringeisen, B., Pietron, J., Finkel, S. and Nealson, K. (2009) Characterization of electrochemically active bacteria utilizing a high-throughput voltage-based screening assay. *Biotechnology Bioengineering* 102, 436–444.

Blanford, C.F., Heath, R.S. and Armstrong, F.A. (2007) A stable electrode for high-potential, electrocatalytic O_2 reduction based on rational attachment of a blue copper oxidase to a graphite surface. *Chemical Communications* 1710–1712.

Bond, D.R., Holmes, D.E., Tender, L.M. and Lovley, D.R. (2002) Electrode-reducing microorganisms that harvest energy from marine sediments. *Science* 295(5554), 483–485.

Bullen, R.A., Arnot, T.C., Lakeman, J.B. and Walsh, F.C. (2006) Biofuel cells and their development. *Biosensors and Bioelectronics* 21, 2015–2045.

Cao, X., Huang, X., Liang, P., Xiao, K., Zhou, Y., Zhang, X. and Logan, B.E. (2009) A new method for water desalination using microbial desalination cells. *Environmental Science and Technology* 43(18), 7148–7152.

Chaudhuri, S.K. and Lovley, D.R. (2003) Electricity generation by direct oxidation of glucose in mediatorless microbial fuel cells. *Nature Biotechnology* 21, 1229–1232.

Chen, G.W., Choi, S.J., Lee, T.H., Lee, G.Y., Cha, J.H. and Kim, C.W. (2008) Application of biocathode in microbial fuel cells: cell performance and microbial community. *Applied Microbiology and Biotechnology* 79, 379–388.

Chen, H., Wang, Y., Liu, Y., Wang, Y., Qi, L. and Dong, S.J. (2007) Direct electrochemistry and electrocatalysis of horseradish peroxidase immobilized in Nafion-RTIL composite film. *Electrochemistry Communications* 9, 469.

Chen, T., Barton, S.C., Binyamin, G., Gao, Z.Q., Zhang, Y.C., Kim, H.H. and Heller, A. (2001) A miniature biofuel cell. *Journal of American Chemical Society* 123(35), 8630–8631.

Cheng, S., Liu, H. and Logan, B.E. (2006) Power densities using different cathode catalysts (Pt and CoTMPP) and polymer binders (Nafion and PTFE) in single chamber microbial fuel cells. *Environmental Science and Technology* 40, 364–369.

Clauwaert, P., Rabaey, K., Aelterman, P., De Schamphelaire, L., Pham, T.H., Boeckx, P., Boon, N. and Verstraete, W. (2007a) Biological denitrification in microbial fuel cells. *Environmental Science and Technology* 41, 3354–3360.

Clauwaert, P., Van der Ha, D., Boon, N., Verbeken, K., Verhaege, M., Rabaey, K. and Verstraete, W. (2007b) Open air biocathode enables effective electricity generation with microbial fuel cells. *Environmental Science and Technology* 41, 7564–7569.

Clauwaert, P., Aelterman, P., Pham, T.H., De Schamphelaire, L., Carballa, M., Rabaey, K. and Verstraete, W. (2008) Minimizing losses in bio-electrochemical systems: the road to applications. *Applied Microbiology and Biotechnology* 79(6), 901–913.

Colmati, F., Yoshikova, S.A., Silva, V.LV.B., Varela, H. and Gonzalez, E.R. (2007) Enzymatic based biocathode in a polymer electrolyte membrane fuel cell. *International Journal of Electrochemical Science* 2, 195–202.

Coman, V., Vaz-Dominguez, C., Ludwig, R., Harreither, W., Haltrich, D., De Laccey, A.L., Ruzgas, T., Gorton, L. and Shleev, S. (2008) A membrane-, mediator-, cofactor-less glucose/oxygen biofuel cell. *Physical Chemistry Chemical Physics* 10, 6093–6096.

Crittenden, S.R., Sund, C.J. and Sumner, J.J. (2006) Mediating electron transfer from bacteria to a gold electrode via a self-assembled monolayer. *Langmuir* 22(23), 9473–9476.

Debabov, V.G. (2008) Electricity from microorganisms. *Mikrobiologiia* 77(2), 149–57.

Dewan, A., Beyenal, H. and Lewandowski, Z. (2008) Scaling up microbial fuel cells. *Environmental Science and Technology* 42, 7643–7648.

Di Lorenzo, M., Curtis, T.P., Head, I.M. and Scott, K. (2010) Effect of increasing anode surface area on the performance of a single chamber microbial fuel cell. *Chemical Engineering Journal* 156, 40–48

Fan, Y., Sharbrough, E. and Liu, H. (2008) Quantification of the internal resistance distribution of microbial fuel cells. *Environmental Science and Technology* 42(21), 8101–8107.

Foley, J.M., Rozendal, R.A., Hertle, C.K., Lant, P.A. and Rabaey, R. (2010) Life cycle assessment of high-rate anaerobic treatment, microbial fuel cells, and microbial electrolysis cells. *Environmental Science and Technology* 44, 3629–3637.

Fornero, J.J., Rosenbaum, M. and Angenent, L.T. (2010) Electric power generation from municipal, food, and animal wastewaters using microbial fuel cells. *Electroanalysis* 22(7–8), 832–843.

Franks, A.E., Malvankar, N. and Nevin, K.P. (2010) Bacterial biofilms: the powerhouse of a microbial fuel cell. *Biofuels* 1(4), 589–604.

Gao, F., Yan, Y., Su, L., Wang, L. and Mao, L. (2007) An enzymatic glucose/O_2 biofuel cell: preparation, characterization and performance in serum. *Electrochemistry Communications* 9, 989–996.

Gil, G.C., Chang, I.S., Kim, B.H., Kim, M., Jang, J.K., Park, H.S. and Kim, H.J. (2003) Operational parameters affecting the performance of a mediator-less microbial fuel cell. *Biosensors and Bioelectronics* 18(4), 327–334.

Gregory, K.B. and Lovley, D.R. (2005) Remediation and recovery of uranium from contaminated subsurface environments with electrodes. *Environmental Science and Technology* 39, 8943–8947.

Harnisch, F. and Schröder, U. (2010) From MFC to MXC: chemical and biological cathodes and their potential for microbial bioelectrochemical systems. *Chemical Society Reviews* 39, 4433–4448.

He, Z. and Angenent, L.T. (2006) Application of bacterial biocathodes in microbial fuel cells. *Electroanalysis* 18(19–20), 2009–2015.

He, Z., Wagner, N., Minteer, S.D. and Angenent, L.T. (2006) An upflow microbial fuel cell with an interior cathode: assessment of the internal resistance by impedance spectroscopy. *Environmental Science and Technology* 40, 5212–5217.

Heidelberg, J.F., Paulsen, I.T., Nelson, K.E., Gaidos, E.J., Nelson, W.C., Read, T.D., Eisen, J.A., Seshadri, R., Ward, N., Methe, B., Clayton, R.A., Meyer, T., Tsapin, A., Scott, J., Beanan, M., Brinkac, L., Daugherty, S., DeBoy, R.T., Dodson, R.J., Durkin, A.S., Haft, D.H., Kolonay, J.F., Madupu, R., Peterson, J.D., Umayam, L.A., White, O., Wolf, A.M., Vamathevan, J., Weidman, J., Impraim, M., Lee, K., Berry, K., Lee, C., Mueller, J., Khouri, H., Gill, J., Utterback, T.R., McDonald, L.A., Feldblyum, T.V., Smith, H.O., Venter, J.C., Nealson, K.H. and Fraser, C.M. (2002) Genome sequence of the dissimilatory metal ion-reducing bacterium *Shewanella oneidensis*. *Nature Biotechnology* 20(11), 1118–123.

Heller, A. (1992) Electrical connection of enzyme redox centers to electrodes. *Journal of Physical Chemistry* 96(9), 3579–3587.

Hernandez, K. and Fernandez-Lafuente, R. (2011) Control of protein immobilization: coupling immobilization and site-directed mutagenesis to improve biocatalyst or biosensor performance. *Enzyme and Microbial Technology* 48(2), 107–122.

Ieropolous, I., Melhuish, C. and Greenman, J. (2004) Energetically autonomous robots. *Proceedings of the Eighth Intelligent Autonomous Systems Conference (IAS-8)*, Amsterdam, pp. 128–135.

Ivanov, I., Vidaković-Koch, T. and Sundmacher, K. (2010) Recent advances in enzymatic fuel cells: experiments and modeling. *Energies* 3(4), 803–846.

Jiang, D. and Li, B. (2009) Granular activated carbon single-chamber microbial fuel cells (GAC-SCMFCs): a design suitable for large-scale wastewater treatment processes. *Biochemical Engineering Journal* 47(1–3), 31–37.

Jung, S. and Regan, J.M. (2007) Comparison of anode bacterial communities and performance in microbial fuel cells with different electron donors. *Applied Microbiology and Biotechnology* 77(2), 393–402.

Katz, E., Filanovsky, B. and Willner, I. (1999) A biofuel cell based on two immiscible solvents and glucose oxidase and microperoxidase-11 monolayer-functionalized electrodes. *New Journal of Chemistry* 23, 481–487.

Kim, B.H., Kim, H.J., Hyun, M.S. and Park, D.S. (1999) Direct electrode reaction of Fe(III) reducing bacterium, *Shewanella putrefaciens*. *Journal of Microbiology* 9, 127–131.

Kim, B.H., Chang, I.S., Gil, G.C., Park, H.S. and Kim, H.J. (2003) Novel BOD (biological oxygen demand) sensor using mediator-less microbial fuel cell. *Biotechnology Letters* 25(7), 541–545.

Kim, J., Jia, H. and Wang, P. (2006) Challenges in biocatalysis for enzyme based biofuel cells. *Biotechnology Advances* 24, 296–308.

Lefebvre, O., Uzabiaga, A., Chang, I.S., Kim, B.H. and Ng, H.Y. (2011) Microbial fuel cells for energy self-sufficient domestic wastewater treatment – a review and discussion from energetic consideration. *Applied Microbiology and Biotechnology* 89(2), 259–270.

Lewis, N.S. (2007) Powering the planet. *MRS Bulletin* 32, 808–820.

Li, W.-W., Sheng, G.-P., Liu, X.-W. and Yu, H.Q. (2011) Recent advances in the separators for microbial fuel cells. *Bioresource Technology* 102, 244–252.

Liu, H. (2009) Microbial fuel cell: novel anaerobic biotechnology for energy generation from wastewater. In: Khanal, S.K. (ed.) *Anaerobic Biotechnology for Bioenergy Production: Principles and Applications*. Wiley-Blackwell, Oxford, UK, pp. 221–246.

Liu, H. and Logan, B.E. (2004) Electricity generation using an air-cathode single chamber microbial fuel cell in the presence and absence of a proton exchange membrane. *Environmental Science and Technology* 38, 4040–4046.

Liu, H., Cheng, S. and Logan, B.E. (2005) Power generation in fed-batch microbial fuel cells as a function of ionic strength, temperature, and reactor configuration. *Environmental Science and Technology* 39, 5488–5493.

Liu, Y., Wang, M., Zhao, F., Xu, Z. and Dong, S. (2005) The direct electron transfer of glucose oxidase and glucose biosensor based on carbon nanotubes/chitosan matrix. *Biosensors and Bioelectronics* 21, 984–988.

Liu, Y., Harnisch, F., Fricke, K., Schröder, U., Climent, V. and Feliu, J.M. (2010) The study of electrochemically active microbial biofilms on different carbon-based anode materials in microbial fuel cells. *Biosensors and Bioelectronics* 25(9), 2167–2171.

Logan, B.E. (2008) *Microbial Fuel Cells*. John Wiley & Sons, New York.

Logan, B.E. (2010) Scaling up microbial fuel cells and other bioelectrochemical systems. *Applied Microbiology and Biotechnology* 85, 16665–21671.

Logan, B.E., Cheng, S., Watson, V. and Estadt, G. (2007) Graphite fiber brush anodes for increased power production in air-cathode microbial fuel cells. *Environmental Science and Technology* 41(9), 3341–3346.

Logan, B.E., Call, D., Cheng, S., Hamelers, H.V.M., Sleutels, T.H.J.A., Jeremiasse, A.W. and Rozendal, R.A. (2008) Microbial electrolysis cells for high yield hydrogen gas production from organic matter. *Environmental Science and Technology* 42, 8630–8640.

Lovley, D.R. (2008) The microbe electric: conversion of organic matter to electricity. *Current Opinion in Biotechnology* 19(6), 564–571.

Mano, N., Mao, F. and Heller, A. (2002) A miniature biofuel cell operating in a physiological buffer. *Journal of American Chemical Society* 124(44), 12962–12963.

Mano, N., Mao, F. and Heller, A. (2003a) Characteristics of a miniature compartment-less glucose–O_2 biofuel cell and its operation in a living plant. *Journal of American Chemical Society* 125(21), 6588–6594.

Mano, N., Mao, F., Kim, Y., Shin, W., Bard, A.J. and Heller, A. (2003b) Oxygen is electroreduced to water on a 'wired' enzyme electrode at a lesser overpotential than on platinum. *Journal of American Chemical Society* 125(50), 15290–15291.

Menicucci, J., Beyenal, H., Marsili, E., Veluchamy, R.A., Demir, G. and Lewandowski, Z. (2006) Procedure for determining maximum sustainable power generated by microbial fuel cells. *Environmental Science and Technology* 40, 1062–1068.

Methe, B.A., Nelson, K.E., Eisen, J.A., Paulsen, I.T., Nelson, W., Heidelberg, J.F., Wu, D., Wu, M., Ward, N., Beanan, M.J., Dodson, R.J., Madupu, R., Brinkac, L.M., Daugherty, S.C., DeBoy, R.T., Durkin, A.S., Gwinn, M., Kolonay, J.F., Sullivan, S.A., Haft, D.H., Selengut, J., Davidsen, T.M., Zafar, N., White, O., Tran, B., Romero, C., Forberger, H.A., Weidman, J., Khouri, H., Feldblyum, T.V., Utterback, T.R., Van Aken, S.E., Lovley, D.R. and Fraser, C.M. (2003) Genome of *Geobacter sulfurreducens*: metal reduction in subsurface environments. *Science* 302(5652), 1967–1979.

Minteer, S.D., Liaw, B.Y. and Cooney, M.J. (2007) Enzyme-based biofuel cells. *Current Opinion in Biotechnology* 18, 228–234.

Miura, Y., Tsujimura, S., Kurose, S., Kamitaka, Y., Kataoka, K., Sakurai, T. and Kano, K. (2009) Direct electrochemistry of CueO and its mutant at residues to and near type I Cu for oxygen-reducing biocathode. *Fuel Cells* 9(1), 70–78.

Moon, H., Chang, I.S. and Kim, B.H. (2005) Continuous electricity production from artificial wastewater using a mediator-less microbial fuel cell. *Bioresource Technology* 97, 621–627.

O'Neill, H. and Woodward, J. (2000) Construction of a bio-hydrogen fuel cell: utilisation of environmental sources of carbohydrates. In: *DARPA Advanced Energy Technologies Energy Harvesting Program*. Arlington, Virginia.

Palmore, G.T.R. and Kim, H.-H. (1999) Electro-enzymatic reduction of dioxygen to water in the cathode compartment of a biofuel cell. *Journal of Electroanalytical Chemistry* 464, 110–117.

Palmore, G.T.R. and Whitesides, G.M. (1994) Microbial and enzymatic biofuel cells. In: Himmel, E., Baker, J.O. and Overend, R.P. (eds) *Enzymatic Conversion of Biomass for Fuels Production*. American Chemical Society, Washington, DC, pp. 271–290.

Palmore, G.T.R., Bertschy, H., Bergens, S.H. and Whitesides, G.M. (1998) A methanol/dioxygen biofuel cell that uses NAD^+-dependent dehydrogenases as catalysts: application of an electro-enzymatic method to regenerate nicotinamide adenine dinucleotide at low overpotentials. *Journal of Electroanalytical Chemistry* 443(1), 155–161.

Pant, D., Van Bogaert, G., De Smet, M., Diels, L. and Vanbroekhoven, K. (2010a) Use of novel permeable membrane and air cathodes in acetate microbial fuel cell. *Electrochimica Acta* 55, 7709–7715.

Pant, D., Van Bogaert, G., Diels, L. and Vanbroekhoven, K. (2010b) A review of the substrates used in microbial fuel cells (MFCs) for sustainable energy production. *Bioresource Technology* 101, 1533–1543.

Pant, D., Singh, A., Van Bogaert, G., Alvarez Gallego, Y., Diels, L. and Vanbroekhoven, K. (2011a) An introduction to the life cycle assessment (LCA) of bioelectrochemical systems (BES) for sustainable energy and product generation: relevance and key aspects. *Renewable and Sustainable Energy Reviews* 15, 1305–1313.

Pant, D., Van Bogaert, G., Porto-Carrero, C., Diels, L. and Vanbroekhoven, K. (2011b) Anode and cathode materials characterization for a microbial fuel cell in half cell configuration. *Water Science and Technology* 63(10), 2457–2461

Park, D.H. and Zeikus, J.G. (2003) Improved fuel cell and electrode designs for producing electricity from microbial degradation. *Biotechnology Bioengineering* 81(3), 348–355.

Park, D.H., Laivenieks, M., Guettler, M.V., Jain, M.K. and Zeikus, J.G. (1999) Microbial utilization of electrically reduced neutral red as the sole electron donor for growth and metabolite production. *Applied and Environmental Microbiology* 65, 2912–2917.

Pizzariello, A., Stred'ansky, M. and Miertus, S. (2002) A glucose/hydrogen peroxide biofuel cell that uses oxidase and peroxidase as catalysts by composite bulk-modified bioelectrodes based on a solid binding matrix. *Bioelectrochemistry* 56(1–2), 99–105.

Potter, M.C. (1911) Electrical effects accompanying the decomposition of organic compounds. *Royal Society (Formerly Proceedings of the Royal Society) B* 84, 260–276.

Qiao, Y., Li, C.M., Bao, S.J. and Bao, Q.L. (2007) Carbon nanotube/polyaniline composite as anode material for microbial fuel cells. *Journal of Power Sources* 170(1), 79–84.

Qiao, Y., Li, C.M., Bao, S.J., Lu, Z. and Hong, Y. (2008) Direct electrochemistry and electrocatalytic mechanism of evolved *Escherichia coli* cells in microbial fuel cells. *Chemical Communications* 11, 1290–1292.

Rabaey, K. and Rozendal, R.A. (2010) Microbial electrosynthesis – revisiting the electrical route for microbial production. *Nature Biotechnology* 8, 706–716.

Rabaey, K., Boon, N., Hofte, M. and Verstraete, W. (2005) Microbial phenazine production enhances electron transfer in biofuel cells. *Environmental Science and Technology* 39, 3401–3408.

Reguera, G., Nevin, K.P., Nicoll, J.S., Covalla, S.F., Woodard, T.L. and Lovley, D.R. (2006) Biofilm and nanowire production leads to increased current in *Geobacter sulfurreducens* fuel cells. *Applied and Environmental Microbiology* 72, 7345–7348.

Rhoads, A., Beyenal, H. and Lewandowski, Z. (2005) Microbial fuel cell using anaerobic respiration as an anodic reaction and biomineralized manganese as a cathodic reactant. *Environmental Science and Technology* 39, 4666–4671.

Rismani-Yazdi, H., Carver, S.M., Christy, A.D. and Tuovinen, O.H. (2008) Cathodic limitations in microbial fuel cells: an overview. *Journal of Power Sources* 180, 683–694.

Rozendal, R.A., Hamelers, H.V.M., Rabaey, K., Keller, J. and Buisman, C.J.N. (2008) Towards practical implementation of bioelectrochemical wastewater treatment. *Trends in Biotechnology* 26, 450–459.

Rubenwolf, S., Kerzenmacher, S., Zengerle, R. and von Stetten, F. (2011) Strategies to extend the lifetime of bioelectrochemical enzyme electrodes for biosensing and biofuel cell applications. *Applied Microbiology and Biotechnology* 89(5), 1315–1322.

Sarma, A.K., Vatsyayan, P., Goswami, P. and Minteer, S.D. (2009) Recent developments in material science for developing enzyme electrodes. *Biosensors and Bioelectronics* 24, 2313–2322.

Sharma, V. and Kundu, P.P. (2010) Biocatalysts in microbial fuel cells. *Enzyme and Microbial Technology* 47, 179–188.

Singh, A., Pant, D., Korres, N.E., Nizami, A.S., Prasad, S. and Murphy, J.D. (2010) Key issues in life cycle assessment (LCA) of ethanol production from lignocellulosic biomass: challenges and perspectives. *Bioresource Technology* 101, 5003–5012.

Steinbusch, K.J.J., Hamelers, H.V.M., Schaap, J.D., Kampman, C. and Buisman, C.J.N. (2010) Bioelectrochemical ethanol production through mediated acetate reduction by mixed cultures. *Environmental Science and Technology* 44, 513–517.

Strik, D.P.B.T.B., Timmers, R.A., Helder, M., Steinbusch, K.J.J., Hamelers, H.V.M. and Buisman, C.J.N. (2011) Microbial solar cells: applying photosynthetic and electrochemically active organisms. *Trends in Biotechnology* 29(1), 41–49.

Strycharz, S.M., Woodard, T.L., Johnson, J.P., Nevin, K.P., Sanford, R.A., Loffler, F.E. and Lovley, D.R. (2008) Graphite electrode as a sole electron donor for reductive dechlorination of tetrachlorethene by *Geobacter lovleyi*. *Applied and Environmental Microbiology* 74, 5943–5947.

Tarasevich, M.R., Bogdanovskaya, V.A. and Kapustin, A.V. (2003) Nanocomposite material laccase/dispersed carbon carrier for oxygen electrode. *Electrochemistry Communications* 5, 491–496.

Tasca, F., Gorton, L., Harreither, W., Haltrich, D., Ludwig, R. and Noll, G. (2008) Highly efficient and versatile anodes for biofuel cells based on cellobiose dehydrogenase from *Myriococcum thermophilum*. *Journal of Physical Chemistry C* 112, 13668–13673.

Tsujimura, S., Kano, K. and Ikeda, T. (2002) Glucose/O_2 biofuel cell operating at physiological conditions. *Electrochemistry* 70(12), 940–942.

Veer Raghavulu, S., Kannaiah Goud, R., Sarma, P.N. and Venkata Mohan, S. (2011) *Saccharomyces cerevisiae* as anodic biocatalyst for power generation in biofuel cell: Influence of redox condition and substrate load. *Bioresource Technology* 102, 2751–2757.

Venkata Mohan, S., Mohanakrishna, G., Purushotham Reddy, B., Sarvanan, R. and Sarma, P.N. (2008) Bioelectricity generation from chemical wastewater treatment in mediatorless (anode) microbial fuel cell (MFC) using selectively enriched hydrogen producing mixed culture under acidophilic microenvironment. *Biochemical Engineering Journal* 39(1), 121–130.

Wang, H.-Y., Bernarda, A., Huang, C.-Y., Lee, D.-J. and Chang, J.-S. (2011) Micro-sized microbial fuel cell: a mini-review. *Bioresource Technology* 102, 235–243.

Wang, M.K., Shen, Y., Liu, Y., Wang, T., Zhao, F., Liu, B.F. and Dong, S.J. (2005) Direct electrochemistry of microperoxidase 11 using carbon nanotube modified electrodes. *Journal of Electroanalytical Chemistry* 578, 121.

Willner, I. (2009) Biofuel cells: harnessing biomass or body fluids for the generation of electrical power. *Fuel Cells* 9(1), 5.

Willner, I., Arad, G. and Katz, E. (1998a) A biofuel cell based on pyrroloquinoline quinone and microperoxidase-11 monolayer-functionalized electrodes. *Bioelectrochemistry and Bioenergetics* 44(2), 209–214.

Willner, I., Katz, E., Patolsky, F. and Buckmann, A.F. (1998b) Biofuel cell based on glucose oxidase and microperoxidase-11 monolayer-functionalized electrodes. *Journal of Chemical Society Perkin Transactions* 2(8), 1817–1822.

Willner, I., Yan, Y.-M., Willner, B. and Tel-Vered, R. (2009) Integrated enzyme-based biofuel cells – a review. *Fuel Cells* 9(1), 7–24.

Woodward, J., Mattingly, S.M., Danson, M., Hough, D., Ward, N. and Adams, M. (1996) *In vitro* hydrogen production by glucose dehydrogenase and hydrogenase. *Nature Biotechnology* 14, 872–874.

Xing, D.F., Zuo, Y., Cheng, S., Regan, J.M. and Logan, B.E. (2008). Electricity generation by *Rhodopseudomonas palustris* DX-1. *Environmental Science and Technology* 42, 4146–4151.

Yan, Y., Zheng, W., Su, L. and Mao, L. (2006) Carbon-nanotube-based glucose/O_2 biofuel cells. *Advanced Materials* 18, 2639–2643.

Zhang, F., Cheng, S., Pant, D., Van Bogaert, G. and Logan, B.E. (2009) Power generation using an activated carbon and metal mesh cathode in a microbial fuel cell. *Electrochemistry Communications* 11, 2177–2179.

Zhao, F., Harnisch, F., Schröder, U., Scholz, F., Bogdanoff, P. and Herrmann, I. (2005) Application of pyrolysed iron (II) phthalocyanine and CoTMPP based oxygen reduction catalysts as cathode materials in microbial fuel cells. *Electrochemistry Communications* 7, 1405–1410.

Zulic, Z. and Minteer, S.D. (2010) Enzymatic fuel cells and their complementarities relative to BES/MFC. In: Rabaey, K., Angenet, L., Schröder, U. and Keller, J. (eds) *Bioelectrochemical Systems: from extracellular electron transfer to biotechnological application.* Integrated Environmental Technology Series, IWA Publishing, London, pp. 39–57.

Chapter 4

Electrical Energy from Microorganisms

Sheela Berchmans

Introduction

One of the earliest and most significant applications of electrochemistry was the storage and conversion of energy. A galvanic cell converts chemical energy to work and an electrolytic cell converts electrical work into *chemical free energy*. Devices that carry out the conversion of chemical energy to electrical energy are called *batteries*. In batteries, the chemical components are contained within the device itself. If the reactants are supplied from an external source for this conversion, the device is called a *fuel cell*. In batteries and fuel cells, electrical energy is stored in the form of chemical energy. It is possible to convert the chemical energy associated with the biological systems to electrical energy. The origin of membrane potentials and transmission of signal across nerve cells and electricity-producing electric eel fish have inspired us to tap the chemical energy available in biological systems in the form of electrical energy (Fig. 4.1). The evolution of microbial fuel cells (MFCs) is one such example in this direction. Biological batteries are present in a number of electric fish. The electric organs of these fish are modified muscle cells known as *electrocytes*, which are arranged in long stacks. A neural signal from the brain causes all the electrocytes in a stack to become polarized at the same time, in effect creating a battery made of series-connected cells. Most electric fish produce only a small voltage, which they use for navigation, which is similar to the way that bats use sound for echolocation of prey. A large adult electric eel, however, is able to produce a 600 V jolt that it employs to stun nearby prey.

The use of fossil fuels, especially oil and gas in recent years, has accelerated and this triggers a global energy crisis. Renewable bioenergy without a net carbon-dioxide (CO^2) emission is much desired. The consumption habits of modern consumer lifestyles are causing a huge worldwide waste problem.

Microbial fuel cells are devices that use live catalysts to generate electrical energy from organic matter present naturally in the environment or in waste (Fig. 4.2; Logan, 2008). The biocatalysts are microorganisms living at the surface of electrodes or in the electrolyte. The oxidation of organic matter by the microbes that are living in an anaerobic medium leads to the liberation of electrons, which constitutes the current in the circuit. In the absence of dioxygen, or other natural electron acceptors, some microbial species have developed the ability to transfer electrons to the anode of a fuel cell. The electric current is then discharged through a load in the external circuit and, at the cathode, an electron acceptor

BEWARE OF ELECTRIC EEL!

Fig. 4.1. Electric eel production of high voltage electric field.

such as dioxygen is reduced ultimately. A membrane separates the anode and cathode compartment, which avoids mixing of the electrolytes in the two compartments. Further, the anolyte is kept oxygen free in the presence of the membrane. The first demonstration of current generation from a microbial fuel cell came in 1912 with the publication of Potter (Potter, 1912). Since then, microbial fuel cell research has gained attention and popularity. Initially the investigations were scarce. A renewal of interest happened in the last decade because of our quest towards alternative and clean energy sources. Another drive for MFC research is the coupling of bioenergy production to the cleaning of waste waters. When realized in practice, this technology could reduce the cost of wastewater treatment plants by recovering and possibly supplying energy along with bioremediation of polluted water (Rabaey and Keller, 2008; Schaetzle *et al.*, 2008).

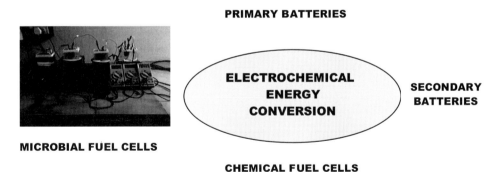

Fig. 4.2. Electrochemical devices for the generation of electrical energy.

Principle of Current Generation

Understanding electricity generation from an MFC requires a prior knowledge about how bacteria capture and generate energy. Bacteria grow by using the metabolic energy derived by catalysing chemical reactions and store energy in the form of adenosine triphosphate (ATP). In some bacteria, reduced substrates are oxidized and NADH, the reduced form of nicotinamide adenine dinucleotide (NAD), transfers electrons to respiratory enzymes. These electrons flow down a respiratory chain made up of a series of enzymes that move protons across an internal membrane thus creating a proton gradient. The protons flow back into the cell through the enzyme ATPase, creating 1 ATP molecule from 1 adenosine diphosphate for every 3–4 protons. The electrons are finally accepted by a soluble terminal electron acceptor, such as nitrate, sulfate, or oxygen. The maximum electron-accepting potential of the process is ~1.2 V based on the potential difference between the electron carrier (NADH) and oxygen under standard conditions. Bacteria can produce as much as 38 molecules of ATP per molecule of glucose by aerobic respiration. Some bacteria can use insoluble metaoxides, such as Fe(III) and Mn(IV), as electron acceptors. The redox potential of the internal site of the bacterial cell at which electrons make an exit from the cell and the redox potential of the terminal electron acceptor determines the potential difference/free energy available for current generation. Bacteria are able to transfer the electrons to the electrode in the anode compartment instead of the terminal electron acceptor like oxygen in the anodic compartment of MFCs. The link between the last redox intermediary site within the membrane from which the electrons are released and the terminal electron acceptor differs from one organism to another. Electron transfer to an electrode depends on the location of the intermediary site in the membrane structures of the cell and its ability to shuttle electrons out of the cell (Fig. 4.3).

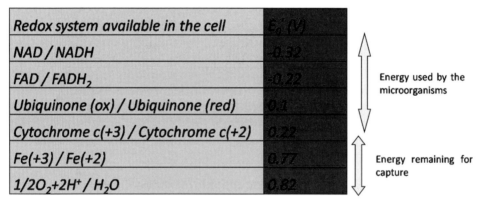

Fig. 4.3. Standard redox potentials versus NHE (pH = 7.0) for the different components of the electron transport chain. The energy available as redox potentials inside the cells is used up by the cell. The energy available at the exit of the electron and the terminal electron acceptor is available for current generation.

The essential physical components of the MFC are the anode, cathode and electrolyte (Fig. 4.4). In an MFC, bacteria catalyse the oxidation of the organic substrate used as the fuel, releasing some of the electrons produced from cell respiration to the anode, where they flow through an external circuit to the counter electrode (cathode) and generate current. For each electron that is produced, a proton must be conducted to the cathode through the membrane separator to maintain the current. Typically, the electrons and protons react with oxygen at the cathode, aided by a catalyst such as platinum, to form water (Liu and Logan, 2004; Logan, 2004; Logan and Regan, 2006). Chemicals other than oxygen that can act as electron transfer mediators such as ferricyanide can be used, resulting in greater overall potentials (Rabaey et al., 2003; Logan et al., 2006). A cation-exchange membrane separates the catholyte and the anolyte acting as a barricade that keeps chemicals and materials other than protons from reaching the cathode.

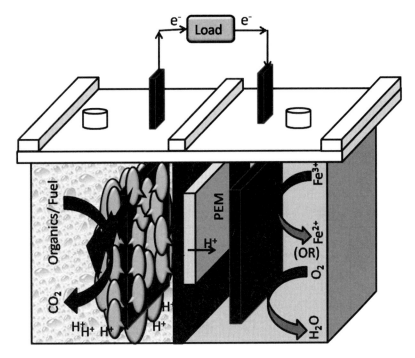

Fig. 4.4. Schematic diagram of a two compartment microbial fuel cell.

Electron Transfer Mechanisms

Direct electron transfer

One of the possible electron transfer paths between a microbe and an electrode is direct electron transfer, upon which the electro-active centre of the membrane enzyme is directly connected to the electrode or through biological nanowires of 2–3 mm called pili, made of fibrous protein structures (Fig. 4.5).

In the case of direct electron transfer from proteins, the electron transfer rate can be very low due to the camouflaging of the active site of the enzyme in the protein environment and the isolation of the enzyme from the electrode surface by its relative concealment into the bacterial membrane. For some exoelectrogens species, however, the redox enzymes involved in electron transfer to electrodes may be located at the outer surface of the microorganism membrane in a favourable orientation for electron transfer (Busalmen *et al.*, 2008). In these circumstances (e.g. with *Geobacter sulfurreducens*), cyclic voltammetry studies show that this allows electron transfer at a rather high rate (Fricke *et al.*, 2008). In addition, it is observed that the thin protruding 'nano wires' called pili facilitate direct electron transfer between the microbe and the electrode. A recent study of electricity production by *G. sulfurreducens* growing on graphite electrodes demonstrated that the 'geopilins' play a role in Fe(III) oxide reduction by this organism and they also contribute to maximum power output when *G. sulfurreducens* is growing as a multilayered biofilm (Reguera *et al.*, 2006). It is interesting to find that, if *G. sulfurreducens* grows on electrodes under conditions where cells do not pile on top of each other, the geopili do not appear to affect current production (Holmes et al., 2006).

Mediated electron transfer

The second mechanism is based on a mediated electron transfer, where the electron transport happens by means of soluble redox mediators: either exogenous ones (natural or synthetic) or endogenous electron shuttles. A shuttle is a compound that transports electrons from the bacteria by diffusion to the surface of the electrode and is itself oxidized. Then, this compound, in its oxidized state, diffuses back to the cells, and continues shuttling electrons between bacteria and the surface of the electrode. Electron transfer aided by mediators usually proceeds at much faster rates. Mediators such as dye molecules and humic substances have some effects on the mediator-less MFCs even though the anodophiles can transfer the electrons to the anode directly in the early stage of biofilm formation. Electron transfer mediators such as Mn^{4+} or neutral red (NR) integrated into the anode markedly enhance the performance of MFCs using anodophile *S. putrefaciens* (Park and Zeikus, 2002). Mediators play an important role in the electron transport for those microbes that are unable to transfer the electrons to the anode (Lovley *et al.*, 1996; Ieropoulos *et al.*, 2005a). *Actinobacillus succinogenes*, *Desulfovibrio desulfuricans*, *E. coli*, *Proteus mirabilis*, *Proteus vulgaris* and *Pseudomonas fluorescens* need exogenous mediators. However, external addition of mediators is not palatable to the process of water purification. Exogenous mediators are not necessary when the mediation process is carried out by shuttles produced by the bacteria as secondary metabolites (i.e. endogenous redox mediators) (Hernandez *et al.*, 2001). In the case of endogenous shuttles, the bacteria can better control the bioelectrochemical activity by being able to regulate both the release of electrons and the production of its own transporters. In the case of exogenous mediators, the adaptivity of the bacteria can rely just on keeping up with the rate of electrons production and on the availability of exogenous transporters added in the solution. Some of the endogenous mediators produced by bacteria are melanin, phenazines, flavins and quinones (Newman and Kolter, 2000;Turick *et al.*, 2002; Hernandez *et al.*, 2004; Von Canstein *et al.*, 2008). Bacteria known to produce electron shuttles in MFCs include members of *Shewanella*, *Pseudomonas* and *Escherichia* (Rabaey *et al.*, 2005; Marsili *et al.*, 2008a, b; Zhang *et al.*, 2008).

The importance of adjuvant microorganisms for a correct ecology of the consortium and, hence, for greater energy production has been discussed at length in recent literature (Aelterman *et al.*, 2006; Hanno *et al.*, 2007; Pham *et al.*, 2008).

Direct oxidation of catabolites

Another promising mechanism involves the direct oxidation at the anode of exported catabolites by the microbes, such as dihydrogen or formate (Rosenbaum *et al.*, 2006).
In the case of mixed cultures acting as biocatalysts all these mechanisms may be found to be operative. Mixed-culture MFCs usually provide better electric performance compared to the pure-culture counterparts (Jung and Regan, 2007). This situation may be regarded as a tight alliance among the mutually dependent and challenging bacterial species contributing to the

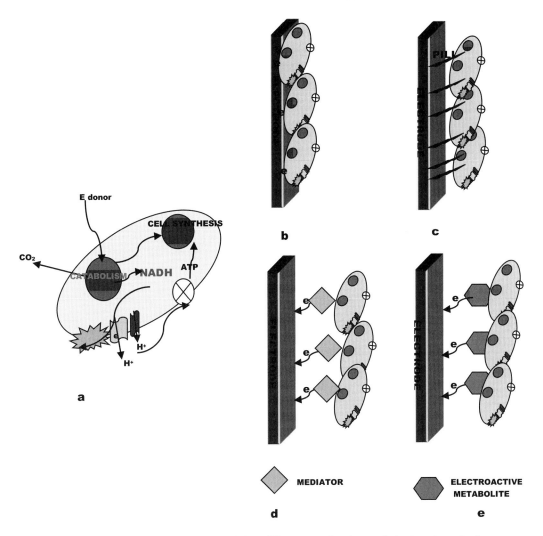

Fig. 4.5. Schematic diagram explaining different mechanisms of electron transfer from microorganisms: (a) different cellular processes involved in electron generation; (b) direct electron transfer; (c) electron transfer through 'pili'; (d) electron transfer through mediators; and (e) electron transfer through electro-active metabolites produced.

consortium aimed at the full fuel substrate degradation. The first group of fermentation bacteria break complex molecules into energy-rich reduced metabolites suitable for the anaerobic respiration of a second bacterial group. Finally, some bacteria in the latter group are able to carry out an extracellular respiration when provided with a proper anode material, while the remaining ones take advantage of co-existing bacterial strains to enhance the metabolic breakdown of the complex molecules. Such bacterial strains can be considered as adjuvant in nature.

Microorganisms Explored for Current Generation

The microbial diversity observed in MFCs showed the existence of a variety of microbial communities. Indeed, the nature and the diversity of the microorganisms present in MFCs is a function of several factors such as the origin of the sample, the nature of the fuel, the presence of redox mediator and the oxic conditions (Kim et al., 2000; Lee et al., 2003). The nature and diversity of the microorganisms will be different if, for example, the MFC contains activated sludge or marine sediment. In most situations, proteobacteria (Gram-negative) dominate the communities' composition and the ratios between α-, β-, γ- and δ-proteobacteria are very different according to the nature of the inoculum. It was shown for example that α-proteobacteria comprised 64.5% of the communities present in an MFC fed with artificial waste water and only 10.8% when fed by river water. This reflects on the different types of the communities' composition to the operational parameters of the fuel cell. In addition to proteobacteria, a number of other bacteria were also shown to be present. A number of recent publications discussed the screening and identification of microbes and the creation of a chromosome library for microorganisms that are able to generate electricity from degrading organic matters (Back et al., 2004; Holmes et al., 2004a, b; Logan et al., 2005).

The commonly used microorganisms in MFC research for the construction of mediatorless microbial fuel cells include members of *Shewanella*, *Rhodoferax* and *Geobacter*. *Geobacter* belongs to dissimilatory metal-reducing microorganisms, which produce biologically useful energy in the form of ATP during the dissimilatory reduction of metal oxides under anaerobic conditions in soils and sediments. The electrons are transferred to the final electron acceptor such as Fe_2O_3 mainly by a direct contact of mineral oxides and the metal-reducing microorganisms (Vargas et al., 1998; Lovley et al., 2004). The anodic reaction in mediatorless MFCs constructed with metal-reducing bacteria belonging mainly to the families of *Shewanella*, *Rhodoferax* and *Geobacter* is similar to that in this process because the anode acts as the final electron acceptor just like the solid mineral oxides. Figure 4.6 illustrates the chemical compounds proposed to be involved in the electron transportation from electron transporters in the intracellular matrix to the solid-state final electron acceptor (anode) in dissimilatory metal-reducing microorganisms. *Shewanella putrefaciens*, *G. sulfurreducens*, *G. metallireducens* and *R. ferrireducens* transfer electrons to the solid electrode (anode) mostly using this system. Though most of the mediatorless MFCs are operated with dissimilatory metal-reducing microorganisms, few exceptions were reported with *Clostridium butyricum* (Park et al., 2001; Oh and Logan, 2006), *Hansenula anomala* (Prasad et al., 2007), *Clostridium* sp. (Prasad et al., 2006), *Gluconobacter roseus* and *Acetobacter aceti* (Karthikeyan et al., 2009) and *Candida melibiosica* 2491 (Hubenova and Mitov, 2010).

The microorganisms employed in MFC research which need mediators are Actinobacillus succinogenes, Desulfovibrio desulfuricans, E. coli, Proteus mirabilis, Proteus vulgaris and Pseudomonas fluorescens. Microorganisms such as Pseudomonas aeruginosa can produce their own mediators. For example, P. aeruginosa produces pyocyanin molecules as endogenous mediator. When an MFC is inoculated with marine sediments or anaerobic sludge, mixed culture microbes are added in the anode chamber. Usually, mixed culture MFCs exhibit good performance. Complex mixed cultures (anodic microcosm) allow much wider substrate consumption. It means that the MFCs have much

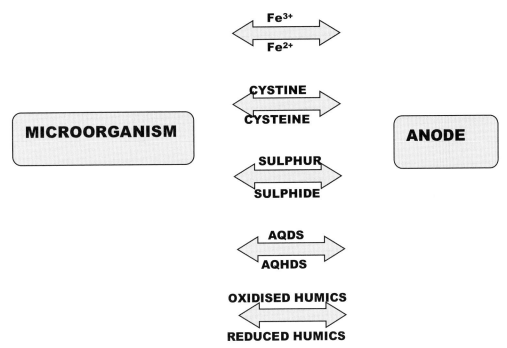

Fig. 4.6. List of various compounds serving as electron shuttles between a bio-electrochemically active microorganism and the anode.

broader substrate specificity when mixed than do pure cultures. In mixed culture MFCs (with anaerobic sludge) both electrophiles/anodophiles and small secondary metabolites use natural mediators together in the same chamber (Ieropoulos et al., 2005b). Yeast should be an ideal biocatalyst for microbial fuel cells. Most of them are non-pathogens and have high growth rates and some exhibit very wide substrate ranges and they are strong and easily handled. However, reports on the use of yeast in microbial fuel cells are limited. Potter was the first scientist to demonstrate that current can be generated from yeasts. More recently, Bennetto (1990) and Wilkinson (2000) employed *S. cerevisiae* cells as the anodic biocatalyst, methylene blue as the anodic mediator and potassium hexacyanoferrate as the cathodic electron acceptor. The open circuit voltage (OCV) observed was around 0.3 V. Walker and Walker (2006) demonstrated that *S. cerevisiae* can produce a maximum power density per projected electrode surface area of 32 mW m^{-2} at 0.3 V and at 55°C. In addition to the other electron transfer mechanisms mentioned previously (see Electron Transfer Systems), yeast have trans-plasma membrane electron transport systems (tPMET), also referred to as plasma membrane oxido-reductase systems (PMOR). These systems stretch out across the membrane and supply electrons from reduced cytoplasmic molecules such as NADH and NADPH to an external electron acceptor. The supplied electrons are, for example, used to prepare external nutrients for uptake as in the reduction of Fe^{3+} to Fe^{2+}. Some electrons are thus available at the cell membrane surface for either direct or mediated transfer to the electrode. However, the number of electrons getting out of the cell by this route are very small when compared to total number of electrons available from the catabolism of aerobically grown cells. However, this mechanism is not a suitable explanation for the direct electron transfer reported by Prasad et al. (2007). The yeast cell wall is very dense (Klis et al., 2006) and the distance from the exterior of the cell membrane to the outside of the cell wall is large (cell wall 100–200 nm, periplasmic space 35–45 A°) and thus direct contact between the cell membrane and an electrode does not seem possible. However, Wartmann et al. (2002) and Prasad et al. (2007) did provide evidence for the existence of a ferric reductase and ferricyanide reductase and cytochrome b2 on the exterior surface of the cell wall of *A. adeninivorans* and *Hansenula anomala*, respectively, and this may be the explanation for the observed results.

To conclude this section, mention should be made about plant microbial fuel cells (De Schamphelaine et al., 2008; Kaku et al., 2008; Strik et al., 2008) and light-dependent electrogenic activity of cyanobacteria. The principle is shown in Fig. 4.7. The first step involves carbon fixation by plants through photosynthesis. Some of the fixed carbon is used for plant growth but a fraction of it is exudated through the roots as rhizodeposits. These organic molecules would then in turn be used as fuel and oxidized back to CO_2 at the microbial anode of an MFC. Using a plant–MFC prototype, with the help of Reed mannagrass, Strik et al. were able to establish this principle and reported a maximum power output of 67 mW m^{-2} of anode. Similarly, a sediment-type MFC was recently shown to increase its power output by one order of magnitude in the presence of living rice plants. A solar-powered MFC, wherein living algae and biocatalysing bacteria work together to produce renewable electricity in a photosynthetic algal microbial fuel cell (PAMFC 1) has been reported. The algal species *Chlorella* were photosynthesizing and transforming light energy into chemical energy as biomass from which electricity was generated with electrochemically active bacteria at the graphite bioanode of the MFC. The bacteria present in the biofilm were electrochemically active bacteria and the maximal power output of PAMFC 1 was 79 times higher than the non-inoculated PAMFC 2. The overall efficiency of PAMFC 1 on transforming photosynthetic active radiation (PAR) light energy into electricity was 0.1%. The PAMFC 1 gross average electricity generation was 14 mW m^{-2}. The maximum power production per square metre surface area of the photobioreactor was 110 mW m^{-2} (David et al., 2008). A salt marsh species *Spartina anglica* generated current for up to 119 days in a plant MFC. This research shows the application of MFC technology in salt marshes for bioenergy production through plant MFC (Timmers et al., 2010).

Rhodobacter sphaeroides (Cho et al., 2008) has been used in a solar-powered electricity generator in a single-chamber microbial fuel cell. The MFC used flooded platinum-coated carbon paper anodes and cathodes of the same material, in contact with atmospheric oxygen. Power was measured by monitoring voltage drop across an external resistance. Biohydrogen production and *in situ* hydrogen oxidation were identified as the main mechanisms for electron transfer to the MFC circuit. The nitrogen source affected MFC performance, with glutamate and nitrate-enhancing power production over ammonium. Power generation depended on the nature of the nitrogen source and on the availability of light. With light, the maximum power density was 790 mW m^{-2} (2.9 W m^{-3}). In the dark, power output was less than 0.5 mW m^{-2} (0.008 W m^{-3}). Also, sustainable electrochemical activity was possible in cultures that did not receive a nitrogen source.

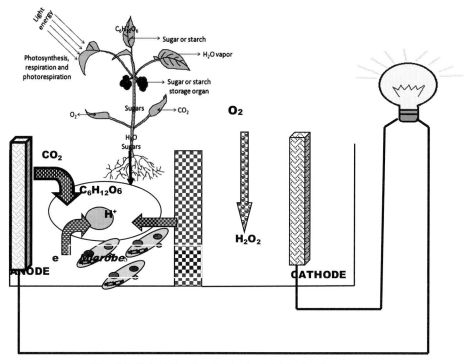

Fig. 4.7. Principle of a plant microbial fuel cell.

Anode materials

The anode material and its structure can directly affect bacteria attachment, electron transfer and substrate oxidation. The choice of anodic materials depends on their conductivity, biocompatibility and chemical stability in the reactor solution. Various materials, including non-corrosive stainless steel (Dumas *et al.*, 2007), plain graphite (Rabaey *et al.*, 2003), carbon paper (Liu *et al.*, 2005a), carbon cloth, felt, or foam (Chaudhuri and Lovley, 2003), reticulated vitreous carbon (He *et al.*, 2005), graphite granules (Aelterman *et al.*, 2006; He and Angenent, 2006; He *et al.*, 2006) and graphite fibre brushes (Logan *et al.*, 2007) have been used as anodes, due to their stability in a microbial inoculum mixture, high conductivity and high specific surface area. Different methodologies have been pursued to enhance the electron transport at the anode–microbe electrical interface. For example, Park and Zeikus showed that the maximum power output of *E. coli*-based MFCs increased up to 152 mW m^{-2}, a value three orders of magnitude larger than the power achieved with an unmodified graphite felt electrode by embedding Mn(IV) and Fe(III) covalently linked to neutral red (NR) to mediate electron transfer from microbes to the anode material (Park and Zeikus, 2003). Alternatively, a large power density was achieved by ammonia gas treatment of carbon cloth at elevated temperatures. Polyaniline (PANI)–Pt composites have been used to improve current generation, owing to their electrical conductivity and the environmental stability of polyaniline (Schröder *et al.*, 2003; Niessen *et al.*, 2004b; Lowy *et al.*, 2006). Even though such approaches have contributed to some extent to the improvement of anode performance, it is well known that the use of Pt on the electrodes is the most effective approach to enhance electrocatalysis. In fact, Pt is the most widely used electrocatalyst in the fuel cell field, being a very efficient catalyst towards the electrochemical oxidation of organic molecules. However, the use of costly Pt as an electrocatalyst has been already criticized as a major drawback for MFC mass production. Major efforts have been focused on overcoming the problem by either minimizing the Pt loading amount or finding Pt substitute materials. Several routes have been explored in these directions. To minimize the Pt loading, some authors deposited a Pt nanolayer on a carbon paper anode and cathode via an e-beam evaporator, obtaining effective and low-cost Pt electrodes (Park *et al.*, 2007). Tungsten carbide anodes have reached performance levels comparable to Pt in MFC applications (Rosenbaum *et al.*, 2006, 2007). Anode performance can be boosted also by increasing the surface area and the biocompatibility of the substrates. Scott *et al.* (2007) prepared graphite felt anodes modified with: (i) C/PANI composites; (ii) carbon nanofibres; or (iii) nitric acid carbon activation. They demonstrated that the modified anode materials have superior performance in terms of power density when compared to the unmodified graphite felt. This improvement was probably due to an increase in surface area, which proliferates the number of sites for microbial colonization, as well as the anode biocompatibility, the latter being related to surface functionalization by the quinoid groups of PANI or by the formation of quinone groups on the carbon surface by the carbon activation treatment. A biocompatibility test related to the carbon structures was performed in an MFC incorporating *E. coli* (as bacteria) and methylene blue (as electron mediator). The test revealed that there is a greater biocompatibility with nanotubes than with regular carbon fibres or other carbon forms (Morozan *et al.*, 2007). Graphite–polytetrafluoroethylene composite films as anodes for *E. coli*-based MFCs were evaluated and it was observed that the polytetrafluoroethylene (PTFE) content in the anode film affected the catalytic activity of the electrochemically activated *E. coli* (Niesen *et al.*, 2004a, b; Park *et al.*, 2004; Zhang *et al.*, 2007). They obtained a power density of 760 mW m^{-2} with a composite anode containing 30% PTFE in the absence of exogenous mediators.

Cathode materials

The cathode is responsible for transmitting the electrons to the terminal electron acceptor, i.e. oxygen in most cases, and is currently the major stumbling block preventing the application of MFCs for electricity generation. Like conventional fuel cells, the non-catalysed cathodes are rather inefficient due to the high overpotential of the oxygen reduction reaction. Hydrogen peroxide, an intermediate product obtained during oxygen reduction, has also been used as a cathode oxidant (Park *et al.*, 2004; Tartakovsky and Giout, 2006). Due to slow oxygen reduction rates on the surface of carbon/graphite electrode catalysts, artificial electron mediators are generally required (Shukla *et al.*, 2004). Platinum is the most popular catalyst for oxygen reduction in MFCs due to its exceptional catalytic ability (Liu and Logan, 2004; Oh *et al.*, 2004; Pham *et al.*, 2004). Platinum application, however, is limited because of the excessive cost and possible poisoning of the electrode surface (Bard and Faulkner, 2001). In addition, increase of pH, which is found to be common in the catholyte of the MFCs, reduces the catalytic activity of platinum (Zhao *et al.*, 2006). Transition metals, such as iron and cobalt, are suitable electron mediators between the cathode and the oxygen, because of their variable redox states and fast electron transfer rate. Park and Zeikus (Park and Zeikus, 2002, 2003) impregnated Fe(III) compounds in the cathode electrode, where Fe(III) was reduced to Fe(II) by electrons generated from the anode reaction, and Fe(II) was, subsequently, reoxidized to Fe(III) by oxygen. In this way, electrons were transferred from the cathode electrode to the terminal electron acceptor (oxygen) with iron compounds as the electron mediators. The Fe(III)-cathode significantly improved the power output of MFCs compared to woven graphite cathode, and showed a greater potential for commercial application. A cobalt-based material, cobalt tetra methoxy phenyl porphyrin (CoTMPP), has been evaluated by two research groups. Both groups concluded that the use of CoTMPP resulted in a similar performance as a platinum cathode (Zhao *et al.*, 2005; Cheng *et al.*, 2006). In addition, pyrolyzed iron(II) phthalocyanine and CoTMPP were compared and found that CoTMPP was slightly better than iron compounds, probably due to a stronger back binding between oxygen and cobalt (Zhao *et al.*, 2005). Other compounds, such as cobalt oxide and molybdenum/ vanadium, have also been used to improve the reaction rates on active carbon or titanium cathodes of an MFC (Habermann and Pommer, 1991). Lead dioxide was also considered as an alternative cathode catalyst in a double-chamber MFC utilizing glucose. Even though this kind of catalyst leads to higher power generation and is associated with lower production costs, its stabilization on the cathode should be improved to prevent dissolution (Morris *et al.*, 2007). Nanostructured materials can also be employed, to facilitate the oxygen reduction reaction at the cathode surface. Freguia *et al.* (2007) proposed an approach to limit the oxygen reduction over potential, which focused on surface area rather than on various catalysts. Thus, instead of lowering the activation energy through a catalyst, they used a non-catalysed material with a high specific surface area (i.e. highly porous granular graphite). Power outputs as high as 21 W m^{-3} or 50 W m^{-3} (over cathode total volume and cathode liquid volume, respectively) were achieved in an MFC fed continuously with acetate. Catholytes such as ferricyanide (Oh *et al.*, 2004; Venkata Mohan *et al.*, 2008) or permanganate (You *et al.*, 2006) were also investigated for optimizing cathodic reactions. Such redox mediators serve as a terminal electron acceptor instead of oxygen, allowing for a power output as large as 258 W m^{-3} (Aelterman *et al.*, 2006).

Biocathodes

The use of biocathodes is highly advantageous over abiotic cathodes for many reasons. First, the cost of construction and operation of MFCs may be lowered. Metal catalysts or artificial electron mediators could be made redundant in MFCs with biocathodes, because microorganisms can function as catalysts to assist the electron transfer. In addition, microorganisms, such as algae, can produce oxygen through photosynthetic reactions, excluding the cost for an external oxygen supply. Second, biocathodes may improve MFC sustainability, because problems with poisoning of platinum or consumption and replacement of electron mediator will be eliminated. Third, the microbial metabolism in biocathodes may be utilized to produce useful products or remove unwanted compounds. For example, the microbial reduction of Fe(III) and Mn(IV), which can function as terminal electron acceptors in the cathode, is an alternative method to extract those metals from minerals (Lovley, 1991; He *et al.*, 2006). Biocathodes can also be involved in the nitrogen removal process during wastewater treatment by reducing nitrate compounds (i.e. denitrification).

The growth of microorganisms in the cathode chamber with the subsequent formation of biofilms on the cathode is difficult to avoid in MFCs, especially when the electrolyte membrane is absent (Hasvold *et al.*, 1997; Holmes *et al.*, 2004a; Liu *et al.*, 2005b). Rather than preventing microbes from depositing on the cathode, bacteria could be used as biocatalysts to accept electrons from the cathode substrate (Rismani-Yazdi *et al.*, 2008). Among recent studies on biological cathodes, Gregory *et al.* (2004) demonstrated that bacteria can take up electrons from a graphite electrode for nitrate and fumarate reduction. Recently, it has been shown that a bioanode oxidizing acetate could be combined with a biocathode, reducing nitrate to nitrogen gas (Clauwaert *et al.*, 2007a). Seawater biofilms growing on a stainless-steel cathode were found to be able to catalyse the oxygen reduction, using the electrons supplied by the cathode (Bergel *et al.*, 2005). Dumas *et al.* (2007) checked the effectiveness of a stainless steel cathode covered with a seawater biofilm formed during MFC operations, for its utilization in sediment MFCs. *Leptothrix discophora* SP-6, i.e. a type of manganese (Mn)-oxidizing bacteria, which has been known to build up Mn oxides from the aqueous environment and biomineralized Mn oxides. This indicates its potential use as a cathodic reactant in a new generation of MFCs featuring a biocathode. Aerobic biocathodes have been evaluated in fresh water with manganese as an electron shuttle between a graphite electrode and *L. discophora* (Rhoads *et al.*, 2005). Also, an acetate oxidizing tubular MFC was built by combining the anode with an open air biocathode in freshwater conditions (Clauwaert *et al.*, 2007b). The results from the latter group indicated that cathode-driven microbial growth is possible at high cathodic potentials (above 0 mV), which excludes the occurrence of cathodic hydrogen evolution and successive bacterial hydrogen consumption. As a follow up on this theme, Rabaey *et al.* (2008) investigated open-air carbon cathodes colonized by bacterial strains that were able to reduce oxygen. Although best results were achieved with a mixed population mainly of Proteobacteria and Bacteroidetes, the use of selected pure microbial cultures yielded a three-fold increase in power density compared to the same non-inoculated MFC used as a control. The marked decrease in activation losses proved that bacteria acted as true catalysts for the oxygen reduction reaction. More work is needed to optimize this result and overcome the technical difficulties, such as pH control and surface modification of cathode substrates. Further studies will result in interesting possibilities for sustainable and low cost MFCs.

Applications

Wastewater treatment

Simultaneous wastewater treatment and current generation is the most probable application of MFCs. An MFC would be used in a treatment system as a replacement for the existing energy-demanding bioreactor (such as an activated sludge system), resulting in a net energy-producing system. However, low coulombic efficiencies, low kinetic rates and scale-up and materials issues are the greatest hurdles in the development of MFCs for wastewater treatment. The technology can be made cost effective, if, besides BOD removal, the system is used for the production of valuable chemical products instead of the low-value electricity. Thus, bioelectrochemical systems (BES) are gaining great importance as innovative technological devices for the renewable generation of chemical products (Rozendal *et al.*, 2006, 2009; Cheng *et al.*, 2009) and for the sequestration of CO_2 (Cao *et al.*, 2009a). They also have shown promise as biosensors (Tront *et al.*, 2008; Di Lorenzi *et al.*, 2009), desalination devices (Cao *et al.*, 2009b) and for recalcitrant chemical conversion (Angenent *et al.*, 2002; Pham *et al.*, 2009).

Environmental sensors

MFCs can possibly be used to power environmental sensors used for collecting data on the natural environment, which can be helpful in understanding and modelling ecosystem responses. Such devices are placed in river and deep-water environments where it is difficult to routinely access the system to replace batteries. Sediment fuel cells are being developed to monitor environmental systems such as creeks, rivers and the ocean (Reimers *et al.*, 2001; Tender *et al.*, 2002). Power for these devices can be provided by organic matter in the sediments. Power densities are low in sediment fuel cells because of both the low organic matter concentrations and their high intrinsic internal resistance. Systems developed to date are limited to producing <30 mW m^{-2}. However, the low power density can be offset by energy storage systems that release data in bursts to central sensors (Shantaram *et al.*, 2005).

Researchers used biological Mn(II) oxidation in the cathode of a sediment MFC to power wireless sensors (Shantaram *et al.*, 2005). This MFC is different from a conventional MFC because it used magnesium alloy as a sacrificial anode, instead of the microbial oxidation of organic fuel. Thus, the anode metal has to be replaced after consumption, limiting the lifetime of the MFC. This MFC produced a maximum voltage of 2.1 V, because of the high redox potential of magnesium oxidation. The voltage was further amplified to 3.3 V, which was sufficient to power a wireless sensor in the environment. The low-temperature environment (Roskie Creek in Bozeman, Montana, USA) where the MFC was installed, affected the microbial activity of the cathode, limiting the power output of this MFC. The study demonstrated, for the first time, the application of MFCs to power small electronic sensors in the environment.

Hydrogen production

MFCs can be readily modified to produce hydrogen instead of electricity. Under normal operating conditions, protons released by the anodic reaction migrate to the cathode to combine with oxygen to form water. Hydrogen generation from the protons and the electrons produced by the metabolism of microbes in an MFC is thermodynamically unfavourable. Liu *et al.* (2005b) applied an external potential to increase the cathode potential in an MFC circuit and thus surmounted the thermodynamic barrier. In this mode, protons and electrons produced by the anodic reaction are combined at the cathode to form hydrogen. The required external potential for an MFC is theoretically 110 mV, much lower than the 1210 mV required for direct electrolysis of water at neutral pH, because some energy comes from the biomass oxidation process in the anodic chamber. MFCs can potentially produce about 8–9 mol H_2 mol^{-1} glucose compared to the typical 4 mol H_2 mol^{-1} glucose achieved in conventional fermentation (Liu *et al.*, 2005b). Therefore, MFCs provide a renewable hydrogen source that can contribute to the overall hydrogen demand in a hydrogen economy (Holzman, 2005).

Renewable electricity production from biomass

In the near future, MFCs will have to race against the more mature renewable-energy technologies, such as wind and solar power. The operating costs needed for electricity production with MFCs will probably be too great if the substrate for the MFC is grown as a crop in a manner similar to that for ethanol production from maize. Renewable energy production from waste biomass is likely to be a more viable route for near-term energy recovery. Great interest exists in using wood-based materials for renewable energy production. Steam explosion is currently the most cost-effective treatment process for the production of soluble sugars from solid lignocellulosic materials, such as agricultural residues and hardwoods (Sun and Cheng, 2002). The use of a neutral hydrolysate, produced by steam explosion of maize stover in an MFC, was recently shown to produce an electrical energy of 933 mW m^{-2} in MFC tests (Zuo *et al.*, 2006). Thus, MFC technologies appear to be technically feasible for energy recovery from similar waste biomass materials.

Biosensors

MFCs have been proposed as BOD sensors, as the current production can be related to the BOD concentration in a solution (Kim *et al.*, 2003). In addition, for biocatalysed cathodic processes like nitrate or oxygen reduction, the current production might be a useful online concentration measurement (Clauwaert *et al.*, 2007a, b). The inhibition of electrical current, on the other hand, can be a good indicator for toxic substances. Introduction of an electrode in liquid streams containing biodegradable organic compounds can be a means to detect microbial contamination if current production is monitored.

Conclusions

The analysis of the literature reports reveal that MFCs represent a promising technology for renewable energy production and their most probable upcoming application is simultaneous wastewater treatment and electricity production. The other specialized applications that are demonstrated are as power sources for environmental sensors and environmental bioremediation. MFC technologies could also find applications in the production of H$_2$. The ability of a diverse range of bacteria to function cooperatively in an MFC is a phenomenon to be understood in depth to enhance our knowledge of the microbial ecology of biofilms and bacteria. The MFC technology has to compete with the mature methanogenic anaerobic digestion technology that has seen wide commercial applications. The possibilities of coupling MFC technology with the anaerobic digestion technology has to be explored seriously. The bottlenecks for the development of MFC technology, which include low coulombic efficiencies, slow kinetic rates and nonlinear power density increase during scale up efforts, are to be addressed in detail in the coming years. The technology can be made more cost effective, if besides BOD removal, the system is used for the production of valuable chemical products instead of low-value electricity. Another interesting question to be addressed is whether we can convert a species such as *E. coli* into an efficient electricigen (bacteria that can efficiently couple to electrodes are called electricigens) for improved power output and bacterial communication with the electrode. One of the possibilities of finding such an electricigen lies in the reodox active protein content of the electricigen outer membrane. Redox-active proteins over-expressed in the outer membrane at the exit point of an electron transfer reduction pathway can lead to efficient electron transfer between electrode and the microorganism. Another possibility is in the ability of electricigens to reduce a substrate like an inorganic metal ion Fe(III) and its oxides. We can direct redox enzymes/proteins to the outer membrane of any bacteria and powerfully wire these proteins to an accepting metal electrode to create a clear-cut electron transfer pathway. These methods will aid us to engineer a super electricigen and hence to the development of the next generation of microbial fuel cells.

References

Aelterman, P., Rabaey, K., Pham, T.H., Boon, N. and Verstraete, W. (2006) Continuous electricity generation at high voltages and currents using stacked microbial fuel cells. *Environmental Science and Technology* 40(10), 3388–3394.

Angenent, L.T., Zheng, D., Sung, S. and Raskin, L. (2002) Microbial community structure and activity in a compartmentalized anaerobic bioreactor. *Water Environment Research* 74(5), 450–461.

Back, J.H., Kim, M.S., Cho, H., Chang, I.S., Lee, J., Kim, K.S., Kim, B.H., Park, Y.I. and Han, Y.S. (2004) Construction of bacterial artificial chromosome library from electrochemical microorganisms. *FEMS Microbiology Letters* 238(1), 65–70.

Bard, A.J. and Faulkner, L.R. (eds) (2001) *Electrochemical Methods: Fundamentals and Applications*. John Wiley and Sons, New York.

Bennetto, H.P. (1990) Electricity generation by micro-organisms. *Biotechnology Education* 1(4), 163–168.

Bergel, A., Feron, D. and Mollica, A. (2005) Catalysis of oxygen reduction in PEM fuel cell by seawater biofilm. *Electrochemistry Communications* 7 (9), 900–904.

Busalmen, J.P., Esteve-Nunez, A. and Feliu, J.M. (2008) C-Type cytochromes wire electricity-producing bacteria to electrodes. *Angewandte Chemie International Edition* 47, 4874–4877.

Cao, X.X., Huang, X., Liang, P., Boon, N., Fan, M.Z., Zhang, L. and Zhang, X.Y. (2009a) A completely anoxic microbial fuel cell using a photo-biocathode for cathodic carbon dioxide reduction. *Energy and Environmental Science* 2, 498–501.

Cao, X., Huang, X., Liang, P., Xiao, K., Zhou, Y., Zhang, X. and Logan, B.E. (2009b) A new method for water desalination using microbial desalination cells. *Environmental Science and Technology* 43(18), 7148–7152.

Chaudhuri, S.K. and Lovley, D.R. (2003) Electricity generation by direct oxidation of glucose in mediatorless microbial fuel cells. Nature *Biotechnology* 21, 1229–1232.

Cheng, S., Liu, H. and Logan, B.E. (2006) Power densities using different cathode catalysts (Pt and CoTMPP) and polymer binders (Nafion and PTFE) in single chamber microbial fuel cells. *Environmental Science and Technology* 40, 364–369.

Cheng, S., Xing, D., Call, D.F. and Logan, B.E. (2009) Direct biological conversion of electrons into methane by electromethanogenesis. *Environmental Science and Technology* 43(10), 3953–3958.

Cho, Y.K., Donohue, T.J., Tejedor, I., Anderson, M.A., McMahon, K.D. and Noguera, D.R. (2008) Development of a solar-powered microbial fuel cell. *Journal of Applied Microbiology* 104(3), 640–650.

Clauwaert, P., Rabaey, K., Aelterman, P., de Schamphelaire, L., Pham, T.H., Boeckx, P., Boon, N. and Verstraete, W. (2007a) Biological denitrification in microbial fuel cells. *Environmental Science and Technology* 41(9), 3354–3360.

Clauwaert, P., Van der Ha, D., Boon, N., Verbeken, K., Verhaege, M., Rabaey, K. and Verstraete, W. (2007b) Open air biocathode enables effective electricity generation with microbial fuel cells. *Environmental Science and Technology* 41(21), 7564–7569.

David, P.B., Strik, T.B., Terlouw, H., Hamelers, H.V.M. and Buisman, C.J.N. (2008) Renewable sustainable biocatalyzed electricity production in a photosynthetic algal microbial fuel cell (PAMFC). *Applied Microbiology and Biotechnology* 81, 659–668.

De Schamphelaire, L., van den Bossche, L., Dang, H.S., Hofte, M., Boon, N., Rabaey, K. and Verstraete, W. (2008) Microbial fuel cells generating electricity from rhizodeposits of rice plants. *Environmental Science and Technology* 42, 3053–3058.

Di Lorenzo, M., Curtis, T.P., Head, I.M. and Scott, K. (2009) A single-chamber microbial fuel cell as a biosensor for wastewaters. *Water Research* 43(13), 3145–3154.

Dumas, C., Mollica, A., Féron, D., Basséguy, R., Etcheverry, L. and Bergel, A. (2007) Marine microbial fuel cell: use of stainless steel electrodes as anode and cathode materials. *Electrochimica Acta* 53, 468–473.

Freguia, S., Rabaey, K., Yuan, Z. and Keller, J. (2007) Non-catalyzed cathodic oxygen reduction at graphite granules in microbial fuel cells. *Electrochimica Acta* 53(2), 598–603.

Fricke, K., Harnisch, F. and Schröder, U. (2008) On the use of cyclic voltammetry for the study of anodic electron transfer in microbial fuel cells. *Energy and Environmental Sciences* 1, 144–147.

Gregory, K.B., Bond, D.R. and Lovley, D.R. (2004) Graphite electrodes as electron donors for anaerobic respiration. *Environmental Microbiology* 6(6), 596–604.

Habermann, W. and Pommer, E.H. (1991) Biological fuel cells with sulphide storage capacity. *Applied Microbiology and Biotechnology* 35, 128–133.

Hanno, R., Martin, L., Nevin, K.P. and Lovley, D.R. (2007) Lack of electricity production by *Pelobacter carbinolicus* indicates that the capacity for Fe(III) oxide reduction does not necessarily confer electron transfer ability to fuel cell anodes. *Applied and Environmental Microbiology* 73(16), 5347–5353.

Hasvold, O., Henriksen, H., Melvaer, E., Citi, G., Johansen, B.O., Kjonigsen, T. and Galetti, R. (1997) Sea-water battery for subsea control systems. *Journal of Power Sources* 65, 253–261.

He, Z. and Angenent, L.T. (2006) Application of bacterial biocathodes in microbial fuel cells. *Electroanalysis* 18, 2009–2015.

He, Z., Minteer, S.D. and Angenent, L.T. (2005) Electricity generation from artificial wastewater using an upflow microbial fuel cell. *Environmental Science and Technology* 39(14), 5262–5267.

He, Z., Wagner, N., Minteer, S.D. and Angenent, L.T. (2006) An upflow microbial fuel cell with an interior cathode: assessment of the internal resistance by impedance spectroscopy. *Environmental Science and Technology* 40(17), 5212–5217.

Hernandez, M.E. and Newman, D.K. (2001) Extracellular electron transfer. *Cellular and Molecular Life Sciences* 58, 1562–1571.

Hernandez, M.E., Kappler, A. and Newman, D.K. (2004) Phenazines and other redox-active antibiotics promote microbial mineral reduction. *Applied Environmental Microbiology* 70, 921–928.

Holmes, D.E., Bond, D.R. and Lovley, D.R. (2004a) Electron transfer by *Desulfobulbus propionicus* to Fe(III) and graphite electrodes. *Applied Environmental Microbiology* 70, 1234–1237.

Holmes, D.E., Bond, D.R., O'Neil, R.A., Reimers, C.E., Tender, L.R. and Lovley, D.R. (2004b) Microbial communities associated with electrodes harvesting electricity from a variety of aquatic sediments. *Microbial Ecology* 48, 178–190.

Holmes, D.E., Chaudhuri, S.K., Nevin, K.P., Mehta, T., Methé, B.A., Liu, A., Ward, J.E., Woodard, T.L., Webster, J. and Lovley, D.R. (2006) Microarray and genetic analysis of electron transfer electrodes in *Geobacter sulfurreducens*. *Environmental Microbiology* 8(10), 1805–1815.

Holzman, D.C. (2005) Microbe power! *Environmental Health Perspectives* 113, A754–A757.

Hubenova, Y. and Mitov, M. (2010) Potential application of *Candida melibiosica* in biofuel cells. *Bioelectrochemistry* 78, 57–61.

Ieropoulos, I., Greenman, J., Melhuish, C. and Hart, J. (2005a) Energy accumulation and improved performance in microbial fuel cells. *Journal of Power Sources* 145, 253–256.

Ieropoulos, I.A., Greenman, J., Melhuish, C. and Hart, J. (2005b) Comparative study of three types of microbial fuel cell. *Enzyme and Microbial Technology* 37, 238–245.

Jung, S. and Regan. J.M. (2007) Comparison of anode bacterial communities and performance in microbial fuel cells with different electron donors. *Applied Microbiology and Biotechnology* 77, 393–402.

Kaku, N., Yonezawa, N., Kodama, Y. and Watanabe, K. (2008) Plant/microbe cooperation for electricity generation in a rice paddy field. *Applied Microbiology and Biotechnology* 79, 43–49.

Karthikeyan, R., Sathishkumar, K., Murugesan, M., Berchmans, S. and Yegnaraman, V. (2009) Bioelectrocatalysis of *Acetobacter aceti* and *Gluconobacter roseus* for current generation. *Environmental Science and Technology* 43(22), 8684–8689.

Kim, B.H., Chang, I.S., Gil, G.C., Park, H.S. and Kim, H.J. (2003) Novel BOD (biological oxygen demand) sensor using mediator-less microbial fuel cell. *Biotechnology Letters* 25, 541–545.

Kim, N., Choi, Y., Jung, S. and Kim. S. (2000) Effect of initial carbon sources on the performance of microbial fuel cells containing *Proteus vulgaris*. *Biotechnology Bioengineering* 70, 109–114.

Klis, F.M., Boorsma, A. and De Groot, P.W.J. (2006) Cell wall construction in *Saccharomyces cerevisiae*. *Yeast* 23, 185–202.

Lee, J., Phung, N.T., Chang, I.S., Kim, B.H. and Sung, H.C. (2003) Use of acetate for enrichment of electrochemically active microorganisms and their 16S rDNA analyses. *FEMS Microbiology Letters* 223, 185–191.

Liu, H. and Logan, B.E. (2004) Electricity generation using an air-cathode single chamber microbial fuel cell in the presence and absence of a proton exchange membrane. *Environmental Science and Technology* 38, 4040–4046.

Liu, H., Cheng, S.A. and Logan, B.E. (2005a) Production of electricity from acetate or butyrate using a single-chamber microbial fuel cell. *Environmental Science and Technology* 39, 658–662.

Liu, H., Grot, S. and Logan, B.E. (2005b) Electrochemically assisted microbial production of hydrogen from acetate. *Environmental Science and Technology* 39(11), 4317–4320.

Logan, B.E. (2004) Extracting hydrogen and electricity from renewable resources. *Environmental Science and Technology* 38, 160A–167A.

Logan, B.E. (2008) *Microbial Fuel Cells*. John Wiley and Sons, New York.

Logan, B.E. and Regan, J.M. (2006) Microbial challenges and fuel cells – applications. *Environmental Science and Technology* 40, 5172–5180.

Logan, B.E., Murano, C., Scott, K., Gray, N.D. and Head, I.M. (2005) Electricity generation from cysteine in a microbial fuel cell. *Water Research* 39, 942–952.

Logan, B.E., Hamelers, B., Rozendal, R., Schröder, U., Keller, J., Freguia, S., Aelterman, P., Verstraete, W. and Rabaey, K. (2006) Microbial fuel cells: methodology and technology. *Environmental Science and Technology* 40(17), 5181–5192.

Logan, B., Cheng, S., Watson, V. and Estadt, G. (2007) Graphite fiber brush anodes for increased power production in air-cathode microbial fuel cells. *Environmental Science and Technology* 41, 3341–3346.

Lovley, D.R. (1991) Dissimilatory Fe(III) and Mn(IV) reduction. *Microbiology Review* 55, 259–287.

Lovley, D.R., Coates, J.D., Blunt-Harris, E.L., Phillips, E.J.P. and Woodward, J.C. (1996) Humic substances as electron acceptors for microbial respiration. *Nature* 382, 445–448.

Lovley, D.R., Holmes, D.E. and Nevin, K.P. (2004) Dissimilatory Fe(III) and Mn(IV) reduction. *Advances in Microbial Physiology* 49, 219–286.

Lowy, D.A., Tender, L.M., Zeikus, J.G., Park, D.H. and Lovley, D.R. (2006) Harvesting energy from the marine sediment-water interface II kinetic activity of anode materials. *Biosensors and Bioelectronics* 21, 2058–2063.

Marsili, E., Baron, D.B., Shikhare, I.D., Coursolle, D., Gralnick, J.A. and Bond, D.R. (2008a) *Shewanella* secretes flavins that mediate extracellular electron transfer. *Proceedings of the National Academy of Sciences USA* 105, 3968–3973.

Marsili, E., Rollefson, J.B., Baron, D.B., Hozalski, R.M. and Bond, D.R. (2008b) Microbial biofilm voltammetry: direct electrochemical characterization of catalytic electrode-attached biofilms. *Applied and Environmental Microbiology* 74, 7329–7337.

Morozan, A., Stamatin, I., Stamatin, L., Dumitru, A. and Scott, K. (2007) Carbon electrodes for microbial fuel cells. *Journal of Optoelectronics and Advanced Materials* 9(1), 221–224.

Morris, J.M., Jin, S., Wang, J., Zhu, C. and Urynowicz, M.A. (2007) Lead dioxide as an alternative catalyst to platinum in microbial fuel cells. *Electrochemistry Communications* 9, 1730–1734.

Newman, D.K. and Kolter, R. (2000) A role for excreted quinones in extracellular electron transfer. *Nature* 405, 94–97.

Niessen, J., Schroder, U. and Scholz, F. (2004a) Exploiting complex carbohydrates for microbial electricity generation – a bacterial fuel cell operating on starch. *Electrochemistry Communications* 6, 955–958.

Niessen, J., Schröder, U., Rosenbaum, M. and Scholz, F. (2004b) Fluorinated polyanilines as superior materials for electrocatalytic anodes in bacterial fuel cells. *Electrochemistry Communications* 6, 571–575.

Oh, S. and Logan, B.E. (2006) Proton exchange membrane and electrode surface areas as factors that affect power generation in microbial fuel cells. *Applied Microbiology and Biotechnology* 70, 162–169.

Oh, S., Min, B. and Logan, B.E. (2004) Cathode performance as a factor in electricity generation in microbial fuel cells. *Environmental Science and Technology* 38, 4900–4904.

Park, D.H. and Zeikus, J.G. (2002) Impact of electrode composition on electricity generation in a single-compartment fuel cell using *Shewanella putrefaciens*. *Applied Microbiology and Biotechnology* 59, 58–61.

Park, D.H. and Zeikus, J.G. (2003) Improved fuel cell and electrode designs for producing electricity from microbial degradation. *Biotechnology and Bioengineering* 81, 348–355.

Park, D.H., Park, Y.K. and Choi, E.S. (2004) Application of single-compartment bacterial fuel cell (SCBFC) using modified electrodes with Metal ions to wastewater treatment reactor. *Journal of Microbiology and Biotechnology* 14, 1120–1128.

Park, H.I., Mushtaq, U., Perello, D., Lee, I., Cho, S.K., Star, A. and Yun, M. (2007) Effective and low-cost platinum electrodes for microbial fuel cells deposited by electron beam evaporation. *Energy Fuels* 21, 2984–2990.

Park, H.S., Kim, B.H., Kim, H.S., Kim, H.J., Kim, G.T., Kim, M., Chang, I.S., Park, Y.K. and Chang, H.I. (2001) A novel electrochemically active and Fe(III)-reducing bacterium phylogenetically related to *Clostridium butyricum* isolated from a microbial fuel cell. *Anaerobe* 7(6) 297–306.

Pham, H., Boon, N., Marzorati, M. and Verstraete. W. (2009) Enhanced removal of 1,2-dichloroethane by anodophilic microbial consortia. *Water Research* 43, 2936–2946.

Pham, T.H., Jang, J.K., Chang, I.S. and Kim, B.H. (2004) Improvement of cathode reaction of a mediatorless microbial fuel cell. *Journal of Microbiology and Biotechnology* 14(2), 324–329.

Pham, T.H., Boon, N., Aelterman, P., Clauwaert, P., De Schamphelaire, L., Vanhaecke, L., De Maeyer, K., Höfte, M., Verstraete, W. and Rabaey, K. (2008) Metabolites produced by *Pseudomonas* sp. enable a Grampositive bacterium to achieve extracellular electron transfer. *Applied Microbiology and Biotechnology* 77, 1119–1129.

Potter, M.C. (1912) Electrical effects accompanying the decomposition of organic compounds. *Proceedings of the Royal Society (London) series B* 84, 260–276.

Prasad, D., Sivaram, T.K., Berchmans, S. and Yegnaraman, V. (2006) Microbial fuel cell constructed with a micro-organism isolated from sugar industry effluent. *Journal of Power Sources* 160, 991–996.

Prasad, D., Arun, S., Murugesan, M., Padmanaban, S., Satyanarayanan, R.S., Berchmans S. and Yegnaraman, V. (2007) Direct electron transfer with yeast cells and construction of a mediatorless fuel cell. *Biosensors and Bioelectronics* 22, 2604–2610.

Rabaey, K. and Keller, J. (2008) Microbial fuel cell cathodes: from bottleneck to prime opportunity? *Water Science and Technology* 57(5), 655–659.

Rabaey, K., Lissens, G., Siciliano, S.D. and Verstraete. W. (2003) A microbial fuel cell capable of converting glucose to electricity at high rate and efficiency. *Biotechnology Letters* 25, 1531–1535.

Rabaey, K., Boon, N., Höfte, M. and Verstraete, W. (2005) Microbial phenazine production enhances electron transfer in biofuel cells. *Environmental Science and Technology* 39, 3401–3408.

Rabaey, K., Read, S.T., Clauwaert, P., Freguia, S., Bond, P.L., Blackall, L.L. and Keller, J. (2008) Cathodic oxygen reduction catalyzed by bacteria in microbial fuel cells. *International Society for Microbial Ecology Journal* 2(5), 519–527.

Reguera, G., Nevin, K.P., Nicoll, J.S., Covalla, S.F., Woodard, T.L. and Lovley, D.R. (2006) Biofilm and nanowire production leads to increased current in *Geobacter sulfurreducens* fuel cells. *Applied and Environmental Microbiology* 72, 7345–7348.

Reimers, C.E., Tender, L.M., Fertig, S. and Wang, W. (2001) Harvesting energy from the marine sediment–water interface. *Environmental Science and Technology* 35(1), 192–195.

Rhoads, A., Beyenal, H. and Lewandowski, Z. (2005) Microbial fuel cell using anaerobic respiration as an anodic reaction and biomineralized manganese as a cathodic reactant. *Environmental Science and Technology* 39, 4666–4671.

Rismani-Yazdi, H., Carver, S.M., Christy, A.D. and Tuovinen, O.H. (2008) Cathodic limitations in microbial fuel cells: an overview. *Journal of Power Sources* 180, 683–694.

Rosenbaum, M., Zhao, F., Schröder, U. and Scholz, F. (2006) Interfacing electrocatalysis and biocatalysis with tungsten carbide: a high-performance, noble-metal-free microbial fuel cell. *Angewandte Chemie International Edition* 45, 6658–6661.

Rosenbaum, M., Zhao, F., Quaas, M., Wulff, H., Schröder, U. and Scholz, F. (2007) Evaluation of catalytic properties of tungsten carbide for the anode of microbial fuel cells. *Applied Catalysis B – Environmental* 74, 262–270.

Rozendal, R.A., Hamelers, H.V.M., Euverink, G.J.W., Metz, S.J. and Buisman, C.J.N. (2006) Principle and perspectives of hydrogen production through biocatalyzed electrolysis. *International Journal of Hydrogen Energy* 31, 1632–1640.

Rozendal, R.A., Leone, E., Keller, J. and Rabaey, K. (2009) Efficient hydrogen peroxide generation from organic matter in a bioelectrochemical system. *Electrochemical Communications* 11, 1752–1755.

Schaetzle, O., Barrière, F. and Baronian, K. (2008) Bacteria and yeasts as catalysts in microbial fuel cells: electron transfer from micro-organisms to electrodes for green electricity. *Energy and Environmental Science* 1, 607–620.

Schröder, U., Niessen, J. and Scholz, F. (2003) A generation of microbial fuel cells with current outputs boosted by more than one order of magnitude. *Angewandte Chemie International Edition* 42, 2880–2883.

Scott, K., Rimbu, G.A., Katuri, K.P., Prasad, K.K. and Head, I.M. (2007) Application of modified carbon anodes in microbial fuel cells. *Process Safety and Environmental Protection* 85(5), 481–488.

Shantaram, A., Beyenal, H., Raajan, R., Veluchamy, A. and Lewandowski, Z. (2005) Wireless sensors powered by microbial fuel cells. *Environmental Science and Technology* 39(13), 5037–5042.

Shukla, A.K., Suresh, P., Berchmans, S. and Rajendran, A. (2004) Biological fuel cells and their applications. *Current Science India* 87, 455–468.

Strik, D.P.B.T.B., Hamelers, H.V.M., Snel, J.F.H. and Buisman, C.J.N. (2008) Green electricity production with living plants and bacteria in a fuel cell. *International Journal of Energy Research* 32(9), 870–876.

Sun, Y. and Cheng, J. (2002) Hydrolysis of lignocellulosic materials for ethanol production: a review. *Bioresource Technology* 83(1), 1–11.

Tartakovsky, B. and Guiot, S.R. (2006) A comparison of air and hydrogen peroxide oxygenated microbial fuel cell reactors. *Biotechnology Progress* 22, 241–246.

Tender, L.M., Reimers, C.E., Stecher, H.A., Holmes, D.E., Bond, D.R., Lowy, D.A., Pilobello, K., Fertig, S.J. and Lovley, D.R. (2002) Harnessing microbially generated power on the seafloor. *Nature Biotechnology* 20, 821–825.

Timmers, R.A., David, P.B., Strik, T.B., Hubertus, V., Hamelers, M. and Buisman, C.J.N. (2010) Long term performance of a plant microbial fuel cell with *Spartina anglica*. *Environmental Biotechnology* 86, 973–981

Tront, J.M., Fortner, J.D., Plötze, M., Hughes, J.B. and Puzrin, A.M. (2008) Microbial fuel cell biosensor for *in situ* assessment of microbial activity. *Biosensors and Bioelectronics* 24, 586–590.

Turick, C.E., Tisa, L.S. and Caccavo Jr, F. (2002) Melanin production and use as a soluble electron shuttle for Fe(III) oxide reduction and as a terminal electron acceptor by *Shewanella* algae BrY. *Applied and Environmental Microbiology* 68, 2436–2444.

Vargas, M., Kashefi, K., Blunt–Harris, E.L. and Lovley, D.R. (1998) Microbiological evidence for Fe(III) reduction on early Earth. *Nature* 395, 65–67.

Venkata Mohan, S., Saravanan, R., Raghuvulu, S.V., Mohanakrishna, G. and Sarma, P.N. (2008) Bioelectricity production from wastewater treatment in dual chambered microbial fuel cell (MFC) using selectively enriched mixed microflora: effect of catholyte. *Bioresource Technology* 99, 596–603.

Von Canstein, H., Ogawa, J., Shimizu, S. and Lloyd, J.R. (2008) Secretion of flavins by *Shewanella* species and their role in extracellular electron transfer. *Applied and Environmental Microbiology* 74, 615–623.

Walker, A.L. and Walker, C.W. (2006) Biological fuel cell and an application as a reserve power source. *Journal of Power Sources* 160(1), 123–129.

Wartmann, T., Stephan, U.W., Bube, I., Böer, E., Melzer, M., Manteuffel, R., Stoltenburg, R., Guengerich, L., Gellissen, G. and Kunze, G. (2002) Post–translational modifications of the AFET3 gene product – a component of the iron transport system in budding cells and mycelia of the yeast *Arxula adeninivorans*. *Yeast* 19, 849–862.

Wilkinson, S. (2000) 'Gastrobots' – benefits and challenges of microbial fuel cells in food powered robot applications. *Autonomous Robots* 9, 99–111.

You, S.J., Zhao, Q.L., Zhang, J.N., Jiang, J.Q. and Zhao, S.Q. (2006) A microbial fuel cell using permanganate as the cathodic electron acceptor. *Journal of Power Sources* 162(2), 1409–1415.

Zhang, T., Zeng, Y., Chen, S., Ai, X. and Yang, H. (2007) Improved performances of *E. coli*-catalyzed microbial fuel cells with composite graphite/PTFE anodes. *Electrochemistry Communications* 9, 349–353.

Zhang, T., Cui, C., Chen, S., Yang, H. and Shen, P. (2008) The direct electrocatalysis of *Escherichia coli* through electroactivated excretion in microbial fuel cell. *Electrochemistry Communications* 10, 293–297.

Zhao, F., Harnisch, F., Schröder, U., Scholz, F., Bogdanoff, P. and Herrmann, I. (2005) Application of pyrolysed iron(II) phthalocyanine and CoTMPP based oxygen reduction catalysts as cathode materials in microbial fuel cells. *Electrochemistry Communications* 7, 1405–1410.

Zhao, F., Harnisch, F., Schroder, U., Scholz, F., Bogdanoff, P. and Herrmann, I. (2006) Challenges and constraints of using oxygen cathodes in microbial fuel cells. *Environmental Science and Technology* 40, 5193–5199.

Zuo, Y., Maness, P.-C. and Logan, B.E. (2006) Electricity production from steam exploded corn stover biomass. *Energy Fuels* 20, 1716–1721.

Chapter 5

Rumen Microbial Fuel Cells

Chin-Tsan Wang, Che-Ming J. Yang and Yung-Chin Yang

Introduction

Converting renewable biomass into electricity by microbial fuel cells (MFCs) can produce clean and transportable energy, with minimal impact on the environment. The capacity and efficacy of MFC have been extensively evaluated on laboratory scales (Rabaey and Verstraete, 2005). Thus far, electron suppliers from biomass for MFCs have been primarily limited to those soluble and rapidly metabolized organic compounds such as simple carbohydrates (Rabaey et al., 2003; Chae et al., 2009), small organic acids (Liu et al., 2005a; Chae et al., 2009), starch (Niessen et al., 2004) and amino acids (Logan et al., 2005). Although good performance has been obtained, these materials are valuable and have other economical uses. Indeed, when used to produce other biomass energy such as ethanol, they give better energy unit yields. Another potential limitation for these substrates for MFCs is that they are hydrolysed at a rapid rate by microbes, and fermentation or metabolic end-products can build up quickly. The accumulation of these products has been shown to affect microbial ecology (Mohan et al., 2007) and the electrical property of the MFC system (Logan et al., 2006).

Transformation of Plant Fibre into Electricity by MFCs

Plant fibre carbohydrates, including wastes from agricultural and industrial activities, are the most abundant and renewable biomass on Earth (Niessen et al., 2005). There are three types of existing renewable energy technologies related to plant fibre biomass: bio-ethanol (Chen, W.H. et al., 2008), bio-hydrogen (Huang, 2008) and power generation. At relatively lower cost than other substrates described above, direct use of fibre to produce electricity would enable MFC to utilize a full spectrum of organic compounds and potentially become a more sustainable source of energy supply.

In a plant fibre-fed MFC system, the biocatalyst is fibre-hydrolysing microbes, such as rumen microbes (Rismani-Yazdi et al., 2007; Chen, 2010; Wang et al., 2011, 2012) and *Clostridium* spp. (Ren et al., 2007). The microbes produce enzymes to hydrolyse fibre with the release of hydrogen ions and electrons required for the MFC system.

Compared to other organic substrates, plant fibre is relatively insoluble and is a large polymer with diverse and complex structure, which varies greatly with species and plant living environment (Malherbe and Cloet, 2002). The biodegradation of fibre coupled with electrical output by MFCs requires cooperated actions of various microorganisms. Consequently, as summarized in Table 5.1, contemporary attempts for power generation utilize mostly purified plant fibre sources as cellulose (Niessen *et al.*, 2005; Ren *et al.*, 2007; Rismani-Yazdi *et al.*, 2007; Rezaei *et al.*, 2008). Although high efficiency of power output has been acknowledged, such fibre substrates require considerable pretreatment, which would add to the cost of power output.

Pretreatment of plant fibre removes hemicellulose and lignin, which hinder cellulose hydrolysis by microorganisms (His *et al.*, 2002). Cellulose after treatment can become soluble, which would further increase susceptibility to microbial degradation. The rate of substrate hydrolysis has been demonstrated as a constraint for power output by MFCs (Rezaei *et al.*, 2007; Huang and Logan, 2008; Mathis *et al.*, 2008; Jadhav and Ghangrekar, 2009; Rezaei *et al.*, 2009). Table 5.1 illustrates variations in electrical performance by MFCs under various experimental conditions, which could reflect solubility of cellulose employed. However, some studies argued that the loss of the activation polarization, Ohm polarization and concentration polarization of the cell system may have greater impact on power output (Zhang and Halme, 1995; Logan *et al.*, 2005; Rismani-Yazdi *et al.*, 2007; Huang *et al.*, 2009; Wang, X. *et al.*, 2009).

In contrast, little work has been conducted to evaluate native plant fibre for generating electricity by MFCs (Rezaei *et al.*, 2009; Chen, 2010; Wang *et al.*, 2011, 2012). This is likely due to the fact that native fibre is virtually insoluble and is much less degradable than purified fibre by microorganisms.

Power output via native plant fibre by rumen MFCs

The rumen of ruminant animals harbours a variety of symbiotic microorganisms, consisting of bacteria, protozoa and fungi. These microorganisms produce assorted enzymes and, working in concert, can effectively degrade structurally complex plant fibre and starch under anaerobic conditions (Hobson and Stewart, 1997; Krause *et al.*, 2003). Reduced metabolites such as organic acids (acetate, propionate and butyrate) are produced and used in term by ruminants as an energy source. The organic acids can also be further catabolized to carbon dioxide (CO^2) by rumen microorganisms.

During these processes of anaerobic metabolism of organic matter, reducing equivalent is produced, which is accompanied by the release and translocation of protons and electrons (Offner and Sauvant, 2006). These products could theoretically be connected to power generation. Such potential is further manifested by the fact that in normal conditions, the ruminal milieu is anaerobic with an oxidation reduction potential (ORP) that is markedly negative (Marden *et al.*, 2005), indicating a strong power of reduction.

Resent research work has demonstrated that purified cellulose could be converted into electricity by rumen microorganisms in MFCs, in the absence of exogenous electron transfer mediators (Rismani-Yazdi *et al.*, 2007). These results indicate that rumen microorganisms could degrade cellulose in MFC conditions and process electrical-chemical properties to reduce the anode, with simultaneous generation of electricity.

In nature, cellulose is always inevitable associated with other components present in the fibre. To date, direct use of native fibre for power generation in rumen MFCs (RMFC) has not been adequately investigated. Our preliminary work shows that electricity could be produced from forage plant fibre (Bermuda grass straw) via transformation by rumen

Table 5.1. Plant fibre substrate applied in MFCs.

C/N [a]	Anode electrode (m^2)	Culture	Substrate (g l^{-1})	P (mW m^{-2})	Reference
2*	Platinum sheet and net (—)	C. cellulolyticum	Cellulose powder (3)	—	Niessen et al. (2005)
	Graphite brush (0.22)	C. thermocellum Anaerobic sludge	Cellulose (1)	35	Rezaei et al. (2007)
	Graphite plates (0.0084)	Rumen microbes	Microcrystalline cellulose (7.5)	55	Rismani-Yazdi et al. (2007)
	Graphite plates (0.00152)	C. cellulolyticum	Carboxymethyl cellulose [b] (CMC ; 1 / 2)	1.16 / —	Ren et al. (2007)
		Co-culture	CMC (1 / 2)	143 ± 7.2 / 151	
			MN301 (1) [c]	59.2 ± 3.5	
		Mixed culture (sludge)	CMC (2)	42.2 ± 6.1	
			MN301 (2)	33.7 ± 4.9	
	Graphite rods (0.00039)	C. cellulolyticum	Cellulose (—)	—	Sund et al. (2007)
	Bundled graphite fibres (—)	G. sulfurreducens PCA	Cellulose (6)	—	Ishii et al. (2008b)
	Carbon electrodes (0.00004)	Palm oil sludge (strain Bb)	CMC (5)	1840	Aslizah et al. (2007)
			Ethyl cellulose (Eth cel; 5)	1300	
			Native cellulose (Nat cel; 5)	3300	
			Empty fruit bunch (EFB; 5)	1365	
		Palm oil mill effluent (strain P9)	CMC (5)	787.5	
			Eth cel (5)	892.5	
			Nat cel (5)	1400	
			EFB (5)	1470	
	Graphite brush electrode (0.22)	Anaerobic sludge	Microcrystalline insoluble cellulose with cellulase (1.3) [e]	98 ± 0.05	Rezaei et al. (2008)
	Graphite plates (0.0016)	C. cellulolyticum G. sulfurreducens [d]	CMC (1)	153	Ren et al. (2008)
			MN301 (1)	83	
	Carbon paper (0.0042)	Waste water	Wheat straw hydrolysate (1) [e]	123	Zhang et al. (2009)

	Graphite plates (0.02024)	Rumen microbes	Bermuda grass (3.3)	66.2 / 273.5 [f]	Chen (2010)
**	Ammonia-treated carbon cloth (1.13)	Enterobacter cloacae ATCC 13047[T]	Pure cellulose of plant origin (4)	5.4 ± 0.3	Rezaei et al. (2009)
		Enterobacter cloacae ER		4.9 ± 0.01	
		mixed culture		18 ± 2.2	
1 ***	Carbon cloth (0.0045)	Mixture of sediment and sand	Cellulose (1)	83 ± 3	Rezaei et al. (2007)
	Plain carbon paper (0.001125)	H-C culture with domestic waste water (20%, v/v)	Corn stover powder (CSP; 1)	333	Wang, X. et al. (2009)
			Corn stover residual solids (CSRS; 1)	390	
		H-C culture	CSP (1)	—	
****	Carbon paper (0.00071)	Domestic waste water [g]	Neutral hydrolysates (1) [e]	371 ± 13	Zuo et al. (2006)
			Acid hydrolysates (1) [e]	367 ± 13	
	Graphite-fibre brush (—) [h]	Paper recycling water unamended waste water	Cellulose (1.464) [i]	672 ± 27	Huang and Logan (2008)
*****	Stainless steel (0.004049/ 0.005101) [j]	Rumen microbes	Bermuda grass (3.3)	0.021102 / 0.014492	Chen (2010)

[a] chamber number; [b] soluble; [c] insoluble; [d] co-culture; [e] g-COD l^{-1}; [f] $K_3Fe(CN)_6$/$KMnO_4$ catholyte; [g] 5 ml, 0.3 g-COD l^{-1}; [h] 5418 $m^2 m^{-3}$; [i] $g\ l^{-1}$ initial TCOD; [j] with/without obstacle at Re = 496.18
* H-type; ** U-type; *** bottle-type air cathode; **** tube-type air cathode; ***** plate-type air cathode

microorganisms in MFCs. Although comparatively high efficiency could be obtained, the capacity of power output was low (average 305 mV; Chen, 2010).

Biotic Factors Affecting Power Output by RMFCs

Degradation rate of plant fibre

Plant fibre varies greatly in the inherent rate of degradation by ruminal microorganisms (Yang, 2002; Chang, 2005). Fibre hydrolysis can be the rate-limiting step for power generation in fibre-fed MFCs. It is anticipated that fibre sources with a faster rate of breakdown in MFCs by rumen microorganisms would also lead to greater production of reducing equivalents available for electricity output. Research work using a defined binary culture by Ren et al. (2007) illustrated that amorphous and microcrystalline cellulose, comparable to native cellulose, produced less electricity by MFCs in comparison to carboxyl methyl cellulose, which is soluble and presumably degraded faster.

Several studies with other microorganisms have also pointed out that the rate of purified cellulose breakdown could affect power output (His *et al.*, 2002; Zuo *et al.*, 2006; Ren *et al.*, 2007; Huang and Logan, 2008; Rezaei *et al.*, 2008; Wang, X. *et al.*, 2009). However, the correlation between cellulose degradation and power generation remains inclusive.

End-products of substrate fermentation

The major end-products from substrate fermentation in the rumen are short chain fatty acids (SCFA), mainly acetate followed by propionate and butyrate. These reduced products are potential fuels for power output in MFCs. Rumen bacteria which could oxidize SCFA and reduce the electrode have been detected in cellulose-fed MFCs (Rismani-Yazdi *et al.*, 2007).

Production of electricity from acetate or butyrate in MFCs with other microorganisms has also been demonstrated (Bond and Lovley, 2003; Min and Logan, 2004; Liu *et al.*, 2005a). Acetate was found to produce more electricity than butyrate (Liu *et al.*, 2005a) and converted to more oxidized product in the presence of the anode (Bond and Lovley, 2003).

Although SCFA can serve as a source of fuel for electrical generation, accumulation of them has been found to affect microorganisms in MFCs (Mohan *et al.*, 2007; Jeong *et al.*, 2008) and functional properties of MFCs (Logan *et al.*, 2006; Jeong *et al.*, 2008; Wang *et al.*, 2012). Therefore, optimal SCFA and other fermentation end-products for power output by RMFC require further investigation. In particular, quantitative analyses on fermented organic acids in relation to reduction potential and cell electrical property are warranted.

Profiles of substrate fermentation

Rumen microbial communities and their fermentation pathways alter when switching ruminants from forage diets to increasing proportions of grain (Tajima *et al.*, 2000). Therefore, changes in diet or substrate for ruminal microorganisms are the major impacts on rumen SCFA profiles and fermentation efficiency (Guan *et al.*, 2008; Yu *et al.*, 2010). When substrates contain less forage (mostly fibre) and more grain (mostly starch), the concentration of total SCFA and the proportion of propionate increase, but acetate proportion decreases (García-Martínez *et al.*, 2005; Wang *et al.*, 2012).

The ratio of forage to concentrate for ruminant diet also affects ORP in the rumen (Mishra *et al.*, 1970; Wang *et al.*, 2012). Chen (2010) observed that the ratio of acetate to propionate in the rumen is highly correlated to ORP in a negative fashion ($r = -0.712$). This is consistent with the fact that acetate-dominated fermentation pattern produces more reducing equivalents in comparison to propionate-type of fermentation (Russell and Wallace, 1997). A more negative ORP would be apt to reduce the anode. Regardless, how relative alterations in SCFA profile affect RMFC performance is currently inclusive.

Composition and function of rumen microorganisms

Within the rumen, microorganisms exist as attached either to rumen epithelium or feed particles, and as free floating cells in fluid fraction (Chen et al., 1998). In addition, distinct distribution of microbial communities forms among ruminal locations (Yang and Varga, 1989). The establishment of these microorganisms is an integral part of normal rumen function and the degradation of resistant plant fibre.

It is expected that to utilize plant fibre by rumen microorganisms for electricity production via biotransformation with efficacy would depend on: (i) the hydrolytic ability by fibre-associated microbial consortia; (ii) the catabolic rate on fibre fermentation products by liquid-associated microbial consortia; and (iii) electron-transmitting microbial consortia attached on the anode. All these individual metabolic sectors cannot be accomplished by any single microbe.

Work by Rismani-Yazdi et al. (2007) illustrates that the inoculum source and substrate type could affect composition of attached and suspended microbial communities enriched in rumen MFC. Similarly in other MFCs, substrate composition and electrochemical conditions have been shown to alter composition of MFC-adapted microbes (Jung and Regan, 2007; Reimers et al., 2007; Aelterman et al., 2008c; Chae et al., 2009). Therefore, it appears that microbial composition and function can vary due to differences in fibre substrate composition and changes in MFC environment.

How changes in microbial composition in MFCs affect power output is not yet clear (Back et al., 2004; Phung et al., 2004; Logan and Regan, 2006; Freguia et al., 2007). In the rumen, the existence of protozoa may affect the variations in ruminal pH, ORP and SCFA (Abe and Kumeno, 1973; Mathieu et al., 1996; Wang et al., 2012). Chen (2010) observed that, in the presence of protozoa, ruminal ORP was more negative, and higher maximal voltage output (595 versus 480 mV) from rumen MFCs fed Bermuda grass straw was recorded. Identification of changes in microbial consortia under various substrate and MFC conditions should help to understand microbial physiology and ecology in MFCs and increase efficiency of electrical output from fibre by MFCs.

Abiotic Factors Affecting Power Output of RMFCs

An MFC is a renewable energy device that converts energy available in organic compounds to electricity via the catalysation of microorganisms. Its power output has been greatly improved in recent years, but the maximum power is still several orders of magnitudes lower than that of chemical fuel cells (CFCs). At present, the development and application of MFCs are decided by factors such as bacteria, electrode material, transportation at proton exchange membrane, etc. These result in a necessity to elevate the power generation of the cell.

Electrode materials that can increase system efficiency and the electron transfer from bacteria to electrode should be helpful for elevating the power generation efficiency of MFCs. The anodes associated with microbial metabolism and the cathodes related to oxygen reduction reaction are often the limiting factors that affect power performance. A conductive film-modified anode material was demonstrated to have a beneficial impact on enhancing the power density of an MFC by increasing the surface area and the biocompatibility of the electrode substrate.

An RMFC is one type of microbial fuel cell. In principle, rumen bacteria are cultivated in an anode trough into which a cellulose source is introduced and there oxidization takes place. The electrons and protons generated during the oxidation process reach the cathode through wire and proton exchange membrane, respectively. Then, oxygen reduction on the cathode completes the power generation process and water synthesis.

In RMFC, the bacteria in the anode transfer electrons by respiration or fermentation. There are two types of electron transfer model between microorganism and anode. One is that an electron transfers from cell membrane to anode directly without any electron mediator as shown in Fig. 5.1. In this case, the *Rhodoferax ferrireducens* and *Geobacter sulfurreducens* are the typical two. They can reduce the dissimilatory metal under an anaerobic environment. The electron transfers from cell membrane to anode by direct attaching of microorganism on the metal oxide surface (Lovley *et al.*, 2004).

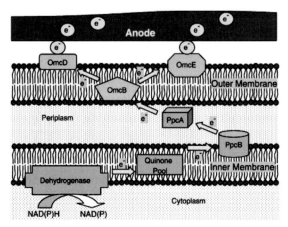

Fig. 5.1. Illustration of electron transfer from microorganism (*Geobacter* species) directly to anode without any electron mediator (Lovley *et al.*, 2004).

Another type of electron transfer is shown as Fig. 5.2 (Lovley *et al.*, 1996), with the electron transfer occurring at a biofilm formed by the bacteria on the anode surface. The attachment of bacteria on the anode is achieved by producing a matrix in the biofilm. The matrix is composed of extracellular proteinase, sugar and cells, and is enriched with matter that can potentially transport electrons.

It has been proved that the matrix contains conductive nanowires, which can accelerate electron transport. In this case, *Actinobacillus succinogenes*, *E. coli*, *Proteus mirabilis*, etc. are the typical species, which need the electron mediator for achieving electron transfer. Power production in MFCs depends mainly upon factors such as bacteria, water-power retention time, organic charge, feeding matrix, electron transfer, inner resistance, electrode material, proton exchange membrane, etc.

Requirements for the electrode in MFCs

Lower power generation ability limits practical application of MFCs. Therefore, to elevate power generation is one goal of MFC development. The anode is the attachment substrate for bacteria in MFCs. It affects not only the amount of bacterial adsorption but

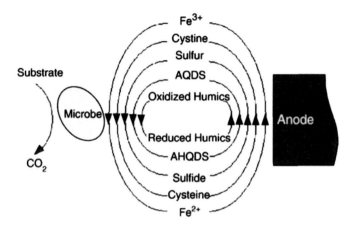

Fig. 5.2. Illustration of electron transfers from microorganism to anode through the electron mediator (biofilm) (Lovley *et al.*, 1996).

also the direction of electron transfer from microorganism to anode (Rabaey and Verstraete, 2005; Du *et al.*, 2007). Therefore, the anode plays an important role in the power generation of MFCs.

Previous studies and literature (Lee *et al.*, 2003; Kim *et al.*, 2004; Rabaey *et al.*, 2004; Logan *et al.*, 2006; Davis and Higson, 2007; Mohan *et al.*, 2008; Chung and Okabe, 2009; Prathap *et al.*, 2009) pointed out the three key factors of MFC performance: (i) total organic substrate (fuel); (ii) biomass accumulation on anode; and (iii) potential of biofilm-anode. For the third factor, if the biofilm on the anode plate is too thick, it does not favour transfer of electrons to the anode. If the biofilm is too thin, the bacteria of the film cannot quickly consume fuel to generate power (Lovley *et al.*, 1996; Ieropoulos *et al.*, 2005).

The reaction on the electrode is controlled by the electron transformation. Therefore, the material comprising the electrode is important to the reaction rate. For cathode material, the reduction rate of hydrogen ion to hydrogen varies with different materials (Kong *et al.*, 2010). This is due to the different catalytic effects of the metals on hydrogen generation. For example, platinum and palladium are very effective catalysts for hydrogen generation. In addition, both anode and cathode with high electronic conductivity could enhance the reaction rate of MFCs and also the performance.

The anode of an MFC should be conductive, biocompatible and chemically stable. It may be metal or non-metal. A majority of non-metal anodes are made of close carbons or fibrous carbons. The former include graphite plate, graphite stick and graphite powder. The latter include carbon carpet, carbon cloth, carbon paper, carbon fibre and carbon cotton. Graphite has typically been the material of choice for the construction of anodes of MFCs. Other conductive materials may be preferable, either because they enhance electron transfer between the microorganisms and the anode material or because they are better adapted to specific applications. The high electron conductivity metal could be employed as the anode. Metal anodes include stainless steel, copper, alloy, etc. Although metal is a good conductor it is not bio-compatible and chemically stable; it should be under control when applied. How to prevent the oxidation of metal and the toxicity of metal is a key point.

Except for the nature of material, the electrode surface area is also another important factor of MFC performance. This is because the electrode reaction takes place at the interface between the electrode and liquid electrolyte, and the reaction rate is proportional to electrode surface area (Logan et al., 2007). Therefore, developing an electrode with high specific surface area is very important. Furthermore, producing anodes with nano and porous surface leads to the increasing of specific surface area and is useful for producing a thinner and large area biofilm on the anodes. Such a biofilm could accelerate the electron transfer. Meanwhile, higher power would be generated by the large amount of bacteria processing the fuel consumption.

Proton exchange membrane (PEM) has a great influence on the proton transfer efficiency of MFCs (Rabaey and Verstraete, 2005; Rozendal et al., 2006), and will affect the internal resistance and concentration polarization loss of cells (Logan et al., 2006). Nafion proton exchange membrane manufactured by Du Pont is more often used because of its high proton selective permeability (Du et al., 2007). In the research by Rozendal et al. (2006), the balance of positive ions by using Nafion proton exchange membrane was tested. The ratio of proton exchange membrane surface area to battery size has a great influence on cell performance. The larger contact area between proton exchange membrane and the electrode, the higher performance of the MFCs, so as to reduce the internal resistance of the MFC (Oh and Logan, 2007). In addition to the Nafion, Park et al. (2000) employed kaolin to produce porcelain septum to replace the Nafion proton exchange membrane. The porcelain septum installed in the MFCs with wastewater sediments can yield power density with 788 mW m^{-2}. The porcelain septum is cheaper than commercial film (Nafion), but it is not easy to make.

Electrode materials in MFCs

There are two sorts of common MFC anode materials. One is the original plate type, such as carbon paper, graphite, soft graphite, metal plate, etc. The other is the compound plate type, material powder mixed with binder and then compressed to form a plate, such as a carbon anode made from carbon powder by press-mould process with polytetrafluoroethene (PTFE) as the binder (Zhang et al., 2007). A high performance MFC anode should be easily attached by bacteria, transfer electrons easily from bacteria to anode, have a low inner resistance, have a high electronic conductivity and be voltage stable.

Among the commonly used materials at present, graphite plate and graphite stick are cheap and easily available, but their effects are not better than carbon carpet and any other fibrous materials. This can be attributed to the lower surface area of graphite plate and graphite stick than carbon carpet, which has a surface area of 0.47 m^2 g^{-1}. The order in surface area is carbon carpet > carbon cotton > graphite powder. The generated electric power is increased with the surface area of the anode. In general, an electrode coated with platinum or platinum black powder is better than one made of graphite granule, graphite carpet, carbon black, carbon cloth, etc.

In an MFC system with *E. coli* as the microorganism, a carbon cloth electrode coated with platinum black powder is better in electricity density than the one without coating. The former has an electricity density of 0.84 mA cm^{-2} and the latter has 0.02 mA cm^{-2}. However, platinum or platinum black powder are very expensive, and platinum is possibly toxic to bacteria (Trinh et al., 2009). It is necessary to seek methods that avoid toxic substances produced by platinum. In recent years, it has been discovered that a layer of polyaniline (conduction layer) coated on platinum is able to avoid the production of toxic substances on platinum and retain the activity of platinum, and cause no interference in the growth of bacteria. Compare with the platinum-coated carbon cloth, coating of polyaniline on a platinum carbon cloth electrode can elevate the electricity density to 1.45 mA cm^{-2}. Due to the better catalysis ability of platinum for oxygen ionization, it can accelerate the reduction of oxygen on the cathode (Jang et al., 2004; Oh et al., 2004; Moon et al., 2006).

Gold is a potentially attractive anode material for some MFC applications because it is highly conductive and because gold provides a high degree of versatility for electrode manufacture. However, a previous study (Richter et al., 2008) suggested that bare gold is a poor electrode material for the anode of MFCs. Current production with gold electrodes was low and increased 100-fold when the gold surface was coated with a surface-associated monolayer (SAM) of 11-mercapto-undecanoic acid (Crittenden et al., 2006), even though the SAM would be expected to have insulating properties. These results indicate that the gold surface was either toxic to the cells or otherwise poorly suited to interact with electron-transfer cell components. Redox-active proteins, such as cytochromes, may adsorb strongly to gold, resulting in denaturation and loss of their electron-transfer capabilities.

There are also other research findings related to MFC anodes. Rosenbaum et al. (2007) studied WC-Nafion/graphite anode. They showed that the tungsten carbide displayed electric catalysis. WC can efficiently oxidize the products from fermentation, such as chlorine, formates, lactates, etc. This can enhance the electric generation ability of MFC. Electrode resistance is also one factor for power generation. To increase the power generation, experimental catalytic anodes were made of low inner resistance metals or metal oxides, which were dispersed in carbon or conductive polymer substrate (Maksimov et al., 1998; Gloguen et al., 1999). Lowy et al. (2006) made graphite anode via Fe_3O_4, Fe_3O_4 and Ni^{2+}, and graphite-ceramic anode via Mn^{2+} and Ni^{2+} in a similar way. The power density of MFCs were 1.7~2.2 times that without modification.

Carbon nanotube (CNT) has specific pore structure, high mechanic strength, toughness, large specific surface area, high thermostability, chemical inertness and high electric conductivity. The electrons in the surface are highly reactive. Electrons can migrate easily between the CNT and the surrounding matter. CNT is becoming an ideal material for electrodes. Qiao et al. (2007) prepared an anode by the mixture of CNT and polyaniline, and the maximal output of MFC was 42 mW m^{-2}. This indicates that addition of CNT can elevate power generation of MFCs. Morozan et al. (2007) pointed out that an anode with CNT displayed good biocompatibility.

The charge in the surface of the electrode is increased by pretreating the electrode with high-temperature ammonia. This speeds attachment of bacteria and increases the attached bacteria number. Electron transfer between bacteria and electrode is therefore elevated. Cheng and Logan (2007) prepared an anode by using amine-treated carbon cloth. The anode surface charge density increased and power generation elevated to near 2000 mW m^{-2}. The power generation was 300 mW m^{-2}, more than MFC using normal carbon cloth as the anode. Logan et al. (2007) achieved 2400 mW m^{-2} of power density by treating a graphite anode with ammonia. With regard to the effects of electrode area on power generation rate, Oh et al. (2004) proved that power density would be elevated 22% by increasing cathode surface area three times more. In addition, an increase of one-third of cathode surface area increased the voltage by 11%. This indicates that increasing the cathode surface area will increase the power density of the MFC.

In summary, for the electric conductivity of electrode materials, an anode or cathode that has a high electric conductivity can elevate reaction rate in MFCs. Therefore, a material with a high electric conductivity could be selected as anode and the cathode material should have good catalytic capacity in hydrogen generation. It is certainly important to develop metals with high current collecting property and that are anticorrosive for the long-term durability of MFCs. For the surface area of the electrode, because electrode reaction takes place at the interface between electrode and solution, reaction rate is proportional to the electrode surface area. Therefore, developing electrodes that have high specific surface areas are very important. On the other hand, too thick biofilms are not good for electron transfer, and they are short of bacteria for consuming power to generate power when the biofilms are too thin. This problem may be solved by employing anodes with high specific surface area that are achieved through the nano and porous surface modification.

Cell configuration and operational condition

The design and operation have been shown to affect redox potential (Bretschger et al., 2010), metabolic activities required for microbial growth (Cheng et al., 2008) and the performance of MFCs (Liu et al., 2008). Several considerations are discussed pertaining to rumen MFCs.

Mixing mechanism: flow-field plate design

The contents in the rumen are churned continuously to be pushed forward and against by ruminal contractions. Such action results in three-dimensional mixing to facilitate contact between microorganisms and feed substrate and rumen wall, and minimize fermentation products from building up elsewhere in the rumen. For example, as stated above, the accumulation of SCFA can be toxic to rumen microorganisms (Mohan et al., 2007; Jeong et al., 2008). Adopting designs to mimic ruminal mixing is expected to play a vital role for maintaining a functional MFC. This has particularly been the case for dual-chamber MFCs, in which precipitation of microbes and floating of insoluble substrates could affect the performance of the MFC.

Because mixing *in situ* may not be applicable, a flow-field plate has been shown to distribute reaction fluid more evenly (Wang, C.T. et al., 2009, 2011). Chen (2010) coupled flow-field plates to single-chamber MFCs with mixed ruminal microorganisms. The flow-field plates were designed bionically to create various liquid flow patterns at various Reynolds numbers before entering the MFC. Such operation resulted in different cell performance, with improvement by 16% over previous flow-field plates (Dicks, 2006). Improving the efficiency of fluid mixing presumably could reduce concentration polarization in MFCs. Application of flow-field on electrical property was conducted on other occasions (Min and Logan, 2004; Ter Heijne et al., 2006), but the efficiency was not clear.

Use of oxidizers in cathode solution

Oxidizers, such as $K_3(Fe(CN)_6)$, applied to the cathode chamber can increase the cell power output (Logan et al., 2005; Rezaei et al., 2008) and decrease the cell internal resistance (Rabaey et al., 2003, 2007; Logan et al., 2005, 2007). If oxidizer concentration is sufficiently high, the replacement interval can be extended (Logan et al., 2005; Jia et al., 2008), though it can be toxic (Jia et al., 2008).

As mentioned earlier, the ORP of ruminal fluid is already very negative. Selection of proper oxidizers for the cathode to run the cell system smoothly can influence power output. Chen (2010) compared $KMnO_4$ with $K_3(Fe(CN)_6)$ as oxidizers in cathode solution for MFCs with rumen microbes. The performance was better with $KMnO_4$ (746.8 mW m^{-2}) than with $K_3(Fe(CN)_6)$ (352.3 mW m^{-2}). In this work, when $KMnO_4$ was used as the catholyte, two cells of RMFCs were stacked and able to light an LED (Chen, 2010). Such difference could be explained by differentials in redox potential gradient present in MFCs (Rabaey and Verstraete, 2005; Fig. 5.3).

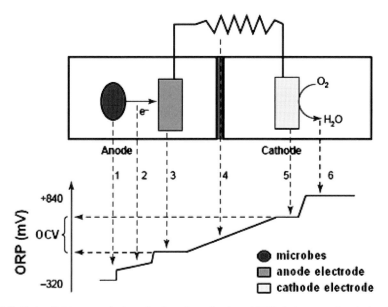

Fig. 5.3. Potential losses during electron transfer in a MFC. 1: Loss owing to bacterial electron transfer. 2: Losses owing to electrolyte resistance. 3: Losses at the anode. 4: Losses at the MFC resistance (useful potential difference) and membrane resistance losses. 5: Losses at the cathode. 6: Losses owing to electron acceptor reduction. (Rabaey and Verstraete, 2005).

Circuit and current

Ishii et al. (2008a) indicated that, compared with an open circuit MFC, the high current density operation could reduce methane generation, while increasing the concentration of short-chain fatty acids, such as acetic acid or propionic acid. Similarly, methanogenesis occurs during rumen microbial fermentation, which consumes hydrogen at the expense of

SVFA formation (Van Soest, 1982). Decreasing methane, a highly reduced compound, would conserve electron and hydrogen ion availability for power generation in MFCs.

Research work has revealed that electron flow may correlate to microbial diversity in MFCs (Holmes *et al.*, 2004; Reimers *et al.*, 2007; Aelterman *et al.*, 2008a, b, c; Cheng *et al.*, 2008; Ishii *et al.*, 2008a; Erable *et al.*, 2009; Virdis *et al.*, 2009; White *et al.*, 2009). However, it remains unclear how current in the cell affects rumen microbes (Bretschger *et al.*, 2010).

Current status with RMFCs

As shown in Table 5.1, both Rismani-Yazdi *et al.* (2007) and Chen (2010) utilized dual-chamber microbial fuel cells to study power output with rumen microorganisms. Insoluble but purified plant fibre was used as substrate by them, as opposed to native plant fibre substrate (Chen, 2010). Carbon plates were used as electrodes for both studies, but differed in reaction surface areas (0.0084 versus 0.02024 m^2). The cathode solution of $K_3(Fe(CN)_6)$ adopted was at different concentrations (0.05 versus 0.5 M). Two kinds of proton exchange membrane with different surface area were mounted, the Ultrex proton-exchange membrane (CMI-7000; 28.3 cm^2) and Nafion 117 (136.3 cm^2), respectively.

A better cell performance (66.2 versus 55 mW m^{-2}) was observed by Chen (2010) than by Rismani-Yazdi *et al.* (2007). Such difference could be due to the fact that Chen (2010) used an electrode with a larger reaction area, which would have allowed microbes to generate biofilm more easily. Biofilm plays an important role in assisting electron transfer from substrate on to the electrode plate (Chae *et al.*, 2009). Also, a sufficiently large reaction surface area decreases the activation polarization in the cell (Logan *et al.*, 2005).

Second, the proton exchange membrane used by Chen (2010) was also larger in reaction area. Although such an increase in area is not immune to microbial-derived biological blocking (Li *et al.*, 2009), which may suppress cell performance, it can reduce the resistance of ions passing through the proton exchange membrane (Rismani-Yazdi *et al.*, 2008) or lessen the Ohm polarization in the cell (Logan *et al.*, 2005).

Third, a higher concentration of oxidizer in cathode solution as $K_3(Fe(CN)_6)$ was applied by Chen (2010). An adequately high concentration favours cell power output (Logan *et al.*, 2005; Rezaei *et al.*, 2008) and decreases the cell internal resistance (Rabaey *et al.*, 2003, 2004; Logan *et al.*, 2005).

Conclusions

Comparing general MFCs with RMFCs

Ruminal microorganisms have evolved to adapt in the rumen environment and actively break down plant fibre therein for growth and survival. Plants have also evolved, but toward resistance by microbial degradation. Ruminal degradation of plant fibre is a rather complex process. Therefore, the magnitude of degradation relies heavily on the amounts and the make-up of rumen microorganisms. Technologies have been developed to culture rumen microorganisms in simulated conditions outside the rumen. Methods to enhance fibre degradation by ruminal microorganisms *in vitro* with acetate-dominating type of fermentation pattern are well documented. Recent technologies have been applied to manipulate rumen microbial fermentation to decrease methane (a highly reduced compound). Such changes in fermentation profiles should spare hydrogen with concurrent generation of reducing equivalents. If such conditions could be realized in MFCs, the generated reducing power should translate into an elevated electricity output.

Therefore the prerequisite for feasible transformation of plant fibre into power output by rumen microorganisms in MFCs requires sustained survival and growth of rumen microorganisms. Only after biotic and abiotic considerations relevant to rumen conditions have been taken into account for designing rumen MFCs should the potential merit of converting plant fibre into electricity be practical.

Future perspectives

At present, power output from fibre of plant origin by RMFCs is relatively low. Linking RMFCs to treatment of fibrous wastes would seem applicable at the initial stage of RMFC development. By simultaneously treating wastes and generating electricity, the cost of waste treatment could be spread out and the added value of RMFCs should be appreciated. When the capacity of RMFCs is fully amplified, an independent power production system from plant fibre can be established.

References

Abe, M. and Kumeno, F. (1973) *In vitro* simulation of rumen fermentation: apparatus and effects of dilution rate and continuous and protozoal population. *Journal of Animal Science* 36, 941–948.

Aelterman, P., Freguia, S., Keller, J., Verstraete, W. and Rabaey, K. (2008a) The anode potential regulates bacterial activity in microbial fuel cells. *Applied Microbiology and Biotechnology* 78, 409–418.

Aelterman, P., Rabaey, K., De Schamphelaire, L., Clauwaert, P., Boon, N. and Verstraete, W. (2008b) Microbial fuel cells as an engineered ecosystem. In: Wall, J.D., Harwood, C.S. and Demain, A.L. (eds) *Bioenergy*. ASM, Washington, DC.

Aelterman, P., Versichele, M., Marzorati, M., Boon, N. and Verstraete, W. (2008c) Loading rate and external resistance control the electricity generation of microbial fuel cells with different three dimensional anodes. *Bioresource Technology* 99, 8895–8902.

Aslizah, M.A., Adibah, Y. and Zaharah, I. (2007) Development of microbial fuel cell using cellulose-degrading bacteria. *ICENV*.

Back, J.H., Kim, M.S., Cho, H., Chang, I.S., Lee, J., Kim, K.S., Kim, B.H., Park, Y.I. and Han, Y.S. (2004) Construction of bacterial artificial chromosome library from electrochemical microorganisms. *FEMS Microbiology Letters* 238, 65–70.

Bond, D.R. and Lovley, D.R. (2003) Electricity production by *Geobacter sulfurreducens* attached to electrodes. *Applied Environmental Microbiology* 69, 1548–1555.

Bretschger, O., Osterstock, J.B., Pinchak, W.E., Ishii, S. and Nelson, K.E. (2010) Microbial fuel cells and microbial ecology applications in ruminant health and production research. *Microbial Ecology* 59, 415–427.

Chae, K.J., Choi, M.J., Lee, J.W., Kim, K.Y. and Kim, I.S. (2009) Effect of different substrates on the performance, bacterial diversity, and bacterial viability in microbial fuel cells. *Bioresource Technology* 100, 3518–3525.

Chang, C.T. (2005) Effect of adding a commercial fibrolytic enzyme and soluble protein on ruminal fermentation, enzyme activity, and degradation of bermudagrass hay. Masters thesis, Department of Animal Science and Biotechnology, Tunghai University, Taichung, Taiwan. (In Chinese)

Chen, K.-J., McAllister, T.A., Popp, J.D., Hristov, A.N., Mir, Z. and Shin, H.T. (1998) A review of bloat in feedlot cattle. *Journal of Animal Science* 76, 299–308.

Chen, W.H., Guo, J.L., Huang, W.S. and Wang, G.B. (2008) The development of cellulosic ethanol technology. *Agricultural Biotechnology* 9, 62–69. (In Chinese)

Chen, Z.S. (2010) Evaluation and analysis of applying native fiber substrates in rumen microbial fuel cell. Masters thesis, Department of Mechanical and Electro-Mechanical Engineering, National Ilan University, Ilan, Taiwan. (In Chinese)

Cheng, K.Y., Ho, G. and Cord-Ruwisch, R. (2008) Affinity of microbial fuel cell biofilm for the anodic potential. *Environmental Science and Technology* 42, 3828–3834.

Cheng, S. and Logan, B.E. (2007) Ammonia treatment of carbon cloth anodes to enhance power generation of microbial fuel cells. *Electrochemistry Communications* 9, 492–496.

Chung, K. and Okabe, S. (2009) Continuous power generation and microbial community structure of the anode biofilms in a three-stage microbial fuel cell system. *Applied Microbiology and Biotechnology* 83, 965–977.

Crittenden, S.R., Sund, C.J. and Sumner, J.J. (2006) Mediating electron transfer from bacteria to a gold electrode via a self-assembled monolayer. *Langmuir* 22, 9473–9476.

Dicks, A.L. (2006) The role of carbon in fuel cells. *Journal of Power Sources* 156, 128–141.

Davis, F. and Higson, S.P.J. (2007) Biofuel cells – recent advances and applications. *Biosensors and Bioelectronics* 22, 1224–1235.

Du, Z., Li, H. and Gu, T. (2007) A state of the art review on microbial fuel cells: a promising technology for wastewater treatment and bioenergy. *Biotechnology Advances* 25, 464–482.

Erable, B., Roncato, M.A., Achouak, W. and Bergel, A. (2009) Sampling natural biofilms: a new route to build efficient microbial anodes. *Environmental Science and Technology* 43, 3194–3199.

Freguia, S., Rabaey, K., Yuan, Z. and Keller, J. (2007) Electron and carbon balances in microbial fuel cells reveal temporary bacterial storage behavior during electricity generation. *Environmental Science and Technology* 41, 2915–2921.

García-Martínez, R., Ranilla, M.J., Tejido, M.L. and Carro, M.D. (2005) Effects of disodium fumarate on *in vitro* rumen microbial growth, methane production and fermentation of diets differing in their forage:concentrate ratio. *British Journal of Nutrition* 94, 71–77.

Gloguen, F., Leger, J.M., Lamy, C., Marmann, A., Stimming, U. and Vogel, R. (1999) Platinum electrodeposition on graphite: electrochemical study and STM imaging. *Electrochimica Acta* 44, 1805–1816.

Guan, I.L., Nkrumah, J.D., Basarab, J.A. and Moore, S.S. (2008) Linkage of microbial ecology to phenotype: correlation of rumen microbial ecology to cattle's feed efficiency. *FEMS Microbiology Letters* 288, 85–91.

His, P.T., Pai, C.C., Li, Y.C., Liu, H.L., Huang, K.H. and Tseng, K.M. (2002) Study on current status of lignin and cellulose biodegradation in composting process. *Techniques and Equipment for Environmental Pollution Control* 3. (In Chinese)

Hobson, P.N. and Stewart, C.S. (1997) *The Rumen Microbial Ecosystem*. Blackie Academic & Professional, New York.

Holmes, D.E., Bond, D.R., O'Neil, R.A., Reimers, C.E., Tender, L.R. and Lovley, D.R. (2004) Microbial communities associated with electrodes harvesting electricity from a variety of aquatic sediments. *Microbial Ecology* 48, 178–190.

Huang, L. and Logan, B.E. (2008) Electricity generation and treatment of paper recycling wastewater using a microbial fuel cell. *Applied Microbiology and Biotechnology* 80, 349–355.

Huang, L., Cheng, S., Rezaei, F. and Logan, B.E. (2009) Reducing organic loads in wastewater effluents from paper recycling plants using microbial fuel cells. *Environmental Technology* 30, 499–504.

Huang, Q.U. (2008) Hydrogen production from cellulose in biomass energy development. *Agricultural Biotechnology* 13, 54–60. (In Chinese)

Ieropoulos, I.A., Greenman, J., Melhuish, C. and Hart, J. (2005) Comparative study of three types of microbial fuel cell. *Enzyme and Microbial Technology* 37, 238–245.

Ishii, S., Hotta, Y. and Watanabe, K. (2008a) Methanogenesis versus electrogenesis: morphological and phylogenetic comparisons of microbial communities. *Bioscience, Biotechnology, and Biochemistry* 72, 286–294.

Ishii, S., Shimoyama, T., Watanabe, K. and Hotta, Y. (2008b) Characterization of a filamentous biofilm community established in a cellulose-fed microbial fuel cell. *BMC Microbiology* 8, 6.

Jadhav, G.S. and Ghangrekar, M.M. (2009) Performance of microbial fuel cell subjected to variation in pH, temperature, external load and substrate concentration. *Bioresource Technology* 100, 717–723.

Jang, J.K., Pham, T.H., Chang, I.S., Kang, K.H., Moon, H., Cho, K.S. and Kim, B.H. (2004) Construction and operation of a novel mediator- and membrane-less microbial fuel cell. *Process Biochemistry* 39, 1007–1012.

Jeong, C.M., Choi, J.D.R., Ahn, Y. and Chang, H.N. (2008) Removal of volatile fatty acids (VFA) by microbial fuel cell with aluminium electrode and microbial community identification with 16S rRNA sequence. *Korean Journal of Chemical Engineering* 25, 535–541.

Jia, Y.H., Tran, H.T., Kim, D.H., Oh, S.J., Park, D.H., Zhang, R.H. and Ahn, D.H. (2008) Simultaneous organics removal and bio-electrochemical denitrification in microbial fuel cells. *Bioprocess and Biosystems Engineering* 31, 315–321.

Jung, S. and Regan, J. (2007) Comparison of anode bacterial communities and performance in microbial fuel cells with different electron donors. *Applied Microbiology and Biotechnology* 77, 393–402.

Kim, B.H., Park, H.S., Kim, H.J., Kim, G.T., Chang, I.S., Lee, J. and Phung, N.T. (2004) Enrichment of microbial community generating electricity using a fuel-cell-type electrochemical cell. *Applied Microbiology and Biotechnology* 63, 672–681.

Kong, X., Sun, Y., Yuan, Z., Li, D., Li, L. and Li, Y. (2010) Effect of cathode electron-receiver on the performance of microbial fuel cell. *International Journal of Hydrogen Energy* 55, 7224–7227.

Krause, D.O., Denman, S.E., Mackie, R.I. and Morrison, M. (2003) Opportunities to improve fiber degradation in the rumen: microbiology, ecology, and genomics. *FEMS Microbiological Review* 27, 663–693.

Lee, J., Phung, N.T., Chang, I.S., Kim, B.H. and Sung, H.C. (2003) Use of acetate for enrichment of electrochemically active microorganisms and their 16S rDNA analyses. *FEMS Microbiology Letters* 223, 185–191.

Li, Z., Zhang, X., Zeng, Y. and Lei, L. (2009) Electricity production by an overflow-type wetted-wall microbial fuel cell. *Bioresource Technology* 100, 2551–2555.

Liu, H., Cheng, S. and Logan, B.E. (2005a) Production of electricity from acetate or butyrate using a single-chamber microbial fuel cell. *Environmental Science and Technology* 39, 658–662.

Liu, Z., Li, H., Liu, J. and Su, Z. (2008) Effects of inoculation strategy and cultivation approach on the performance of microbial fuel cell using marine sediment as bio-matrix. *Journal of Applied Microbiology* 104, 1163–1170.

Logan, B.E. and Regan, J.M. (2006) Electricity-producing bacterial communities in microbial fuel cells. *Trends in Microbiology* 14, 512–518.

Logan, B.E., Murano, C., Scott, K., Gray, N.D. and Head, I.M. (2005) Electricity generation from cysteine in a microbial fuel cell. *Water Resource* 39, 942–952.

Logan, B.E., Hamelers, B., Rozendal, R., Schroder, U., Keller, J., Freguia, S., Aelterman, P., Verstraete, W. and Rabaey, K. (2006) Microbial fuel cells: methodology and technology. *Environmental Science and Technology* 40, 5181–5192.

Logan, B.E., Cheng, S., Watson, V. and Estadt, G. (2007) Graphite fiber brush anodes for increased power production in air-cathode microbial fuel cells. *Environmental Science Technology* 41, 3341–3346.

Lovley, D.R., Coates, J.D., Blunt-Harris, E.L., Philips, E.J.P. and Woodward, J.C. (1996) Humic substances as electro acceptors for microbial respiration. *Nature* 382, 445–448.

Lovley, D.R., Holmes, D.E. and Nevin, K.P. (2004) Dissimilatory Fe(III) and Mn(IV) reduction. *Advances in Microbial Physiology* 49, 219–286.

Lowy, D.A., Tender, L.M., Zeikus, J.G., Park, D.H. and Lovley, D.R. (2006) Harvesting energy from the marine sediment-water interface II Kinetic activity of anode materials. *Biosensors & Bioelectronics* 21, 2058–2063.

Marden, J.P., Bayourthe, C., Enjalbert, F. and Moncoulon, R. (2005) A new device for measuring kinetics of ruminal pH and redox potential in dairy cattle. *Journal of Dairy Science* 88, 277–281.

Maksimov, Y.M., Podlovchenko, B.I. and Azarchenko, T.L. (1998) Preparation and electrocatalytic properties of platinum microparticles incorporated into polyvinylpyridine and Nafion films. *Electrochimica Acta* 43, 1053–1059.

Malherbe, S. and Cloete, T.E. (2002) Lignocellulose biodegradation: fundamentals and applications. *Environmental Science and Bio/Technology* 1, 105–114.

Mathieu, F., Jouany, J.P., Senaud, J., Bohatier, J., Bertin, G. and Mercier, M. (1996) The effect of *Saccharomyces cerevisiae* and *Aspergillus oryzae* on fermentations in the rumen of faunated and defaunated sheep; protozoal and probiotic interactions. *Reproduction and Nutrition Development* 36, 271–287.

Mathis, B.J., Marshall, C.W., Milliken, C.E., Makkar, R.S., Creager, S.E. and May, H.D. (2008) Electricity generation by thermophilic microorganisms from marine sediment. *Applied Microbiology and Biotechnology* 78, 147–155.

Min, B. and Logan, B.E. (2004) Continuous electricity generation from domestic wastewater and organic substrates in a flat plate microbial fuel cell. *Environmental Science and Technology* 38, 5809–5814.

Mishra, M., Martz, F.A., Stanley, R.W., Johnson, H.D., Campbell, J.R. and Hilderbrand, E. (1970) Effect of diet and ambient temperature-humidity on ruminal pH, oxidation reduction potential, ammonia and lactic acid in lactating cows. *Journal of Animal Science* 30, 1023–1028.

Mohan, S.V., Mohanakrishna, G. and Sarma, P.N. (2008) Effect of anodic metabolic function on bioelectricity generation and substrate degradation in single chambered microbial fuel cell. *Environmental Science and Technology* 42, 8088–8094.

Mohan, S.V., Raghavulu, S.V., Srikanth, S. and Sarma, P.N. (2007) Bioelectricity production by mediatorless microbial fuel cell under acidophilic condition using wastewater as substrate: influence of substrate loading rate. *Current Science* 92, 1720–1726.

Moon, H., Chang, I.S. and Kim, B.H. (2006) Continuous electricity production from artificial wastewater using a mediator-less microbial fuel cell. *Bioresource Technology* 97, 621–627.

Morozan, A., Stamatin, L., Nastase, F., Dumitru, A., Vulpe, S., Nastase, C., Stamatin, I. and Scott, K. (2007) The biocompatibility microorganisms-carbon nanostructures for applications in microbial fuel cells. *Physical Status Solidi (a)* 204, 1797–1803.

Niessen, J, Schröder, U. and Scholz, F. (2004) Exploiting complex carbohydrates for microbial electricity generation – a bacterial fuel cell operating on starch. *Electrochemical Communication* 6, 955–958.

Niessen, J., Schröder, U., Harnisch, F. and Scholz, F. (2005) Gaining electricity from *in situ* oxidation of hydrogen produced by fermentative cellulose degradation. *Letters in Applied Microbiology* 41, 286–290.

Offner, A. and Sauvant, D. (2006) Thermodynamic modeling of ruminal fermentations. *Animal Research* 55, 343–365.

Oh, S., Min, B. and Logan, B.E. (2004) Cathode performance as a factor in electricity generation in microbial fuel cells. *Environmental Science and Technology* 38, 4900–4904.

Oh, S.E. and Logan, B. (2006) Proton exchange membrane and electrode surface areas as factors that affect power generation in microbial fuel cells. *Applied Microbiology and Biotechnology* 70, 162–169.

Oh, S.E and Logan, B.E. (2007) Voltage reversal during microbial fuel cell stack operation. *Journal of Power Sources* 167, 11–17.

Park, D.H. and Zeikus, J.G. (2003) Improved fuel cell and electrode designs for producing electricity from microbial degradation. *Biotechnology and Bioengineering* 81, 348–355.

Park, D.H., Kim, S.K., Shin, I.H. and Jeong, Y.J. (2000) Electricity production in biofuel cell using modified graphite electrode with neutral red. *Biotechnology Letters* 22, 1301–1304.

Phung, N.T., Lee, J., Kang, K.H., Chang, I.S., Gadd, G.M. and Kim, B.H. (2004) Analysis of microbial diversity in oligotrophic microbial fuel cells using 16s rDNA sequences. *FEMS Microbiology Letters* 233, 77–82.

Prathap, P., César, I.T., Hyung-Sool, L., Rosa, K.B. and Bruce, E.R. (2009) Syntrophic interactions among anode respiring bacteria (ARB) and non-ARB in a biofilm anode: electron balances. *Biotechnology and Bioengineering* 103, 513–523.

Qiao, Y., Li, C.M., Bao, S.J. and Bao, Q.L. (2007) Carbon nanotube/polyaniline composite as anode material for microbial fuel cells. *Journal of Power Sources* 170, 79–84.

Rabaey, K. and Verstraete, W. (2005) Microbial fuel cell: novel biotechnology for energy generation. *Trends in Biotechnology* 23, 291–298.

Rabaey, K., Lissens, G., Siciliano, S.D. and Verstraete, W. (2003) A microbial fuel cell capable of converting glucose to electricity at high rate and efficiency. *Biotechnology Letters* 25, 1531–1535.

Rabaey, K., Boon, N., Siciliano, S.D., Verhaege, M. and Verstraete, W. (2004) Biofuel cells select for microbial consortia that self-mediate electron transfer. *Applied Environmental Microbiology* 70, 5373–5382.

Rabaey, K., Rodriguez, J., Blackall, L.L., Keller, J., Gross, P., Batstone, D., Verstraete, W. and Nealson, K.H. (2007) Microbial ecology meets electrochemistry: electricity-driven and driving communities. *ISME Journal* 1, 9–18.

Reimers, C.E., Stecher, H.A. III, Westall, J.C., Alleau, Y., Howell, K.A., Soule, L., White, H.K. and Girguis, P.R. (2007) Substrate degradation kinetics, microbial diversity, and current efficiency of microbial fuel cells supplied with marine plankton. *Applied Environmental Microbiology* 73, 7029–7040.

Ren, Z.Y., Ward, T.E. and Regan, J.M. (2007) Electricity production from cellulose in a microbial fuel cell using a defined binary culture. *Environmental Science and Technology* 41, 4781–4786.

Ren, Z., Steinberg, L.M. and Regan, J.M. (2008) Electricity production and microbial biofilm characterization in cellulose-fed microbial fuel cells. *Water Science and Technology* 58, 617–622.

Rezaei, F., Richard, T.L., Brennan, R.A. and Logan, B.E. (2007) Substrate-enhanced microbial fuel cells for improved remote power generation from sediment-based systems. *Environmental Science and Technology* 41, 4053–4058.

Rezaei, F., Tom, L.R. and Logan, B.E. (2008) Enzymatic hydrolysis of cellulose coupled with electricity generation in a microbial fuel cell. *Biotechnology and Bioengineering* 101, 1163–1169.

Rezaei, F., Xing, D., Wagner, R., Regan, J.M., Richard, T.L. and Logan, B.E. (2009) Simultaneous cellulose degradation and electricity production by *Enterobacter cloacae* in an microbial fuel cell. *Applied Environmental Microbiology* 75, 3673–3678.

Richter, H., McCarthy, K., Nevin, K.P., Johnson, J.P., Rotello, V.M. and Lovley, D.R. (2008) Electricity generation by *Geobacter sulfurreducens* attached to gold electrodes. *Langmuir* 24, 4376–4379.

Rismani-Yazdi, H., Christy, A.D., Dehority, B.A., Morrison, M., Yu, Z. and Tuovinen, O.H. (2007) Electricity generation from cellulose by rumen microorganisms in microbial fuel cells. *Biotechnology and Bioengineering* 97, 1398–1407.

Rismani-Yazdi, H., Carver, S.M., Christy, A.D. and Tuovinen, O.H. (2008) Cathodic limitations in microbial fuel cells: an overview. *Journal of Power Sources* 180, 683–694.

Rosenbaum, M., Zhao, F., Quaas, M. and Wulff, H. (2007) Evaluation of catalytic properties of tungsten carbide for the anode of microbial fuel cells. *Applied Catalysis B: Environmental* 74, 261–269.

Rozendal, R.A., Hamelers, H.V.M. and Buisman, C.J.N. (2006) Effects of membrane cation transport on pH and microbial fuel cell performance. *Environmental Science & Technology* 40, 5206–5211.

Russell, J.B. and Wallace, R.J. (1997) Energy yielding and consuming reactions. In: Stewart, C.S. and Hobson, P.N. (eds) *The Rumen Microbial Ecosystem*. Blackie Academic & Professional, London, pp. 246–282.

Sund, C.J., McMasters, S., Crittenden, S.R., Harrell, L.E. and Sumner, J. (2007) Effect of electron mediators on current generation and fermentation in a microbial fuel cell. *Applied Microbiology and Biotechnology* 76, 561–568.

Tajima, K., Arai, S., Ogata, K., Nagamine, T., Matsui, H., Nakamura, M., Aminov, R.I. and Benno, Y. (2000) Rumen bacterial community transition during adaptation to high-grain diet. *Anaerobe* 6, 273–284.

Ter Heijne, A., Hamelers, H.V.M., De Wilde, V., Rozendal, R.A. and Buisman, C.J.N. (2006) A bipolar membrane combined with ferric iron reduction as an efficient cathode system in microbial fuel cells. *Environmental Science and Technology* 40, 5200–5205.

Trinh, N., Park, J. and Kim, B.W. (2009) Increased generation of electricity in a microbial fuel cell using *Geobacter sulfurreducens*. *Korean Journal of Chemical Engineering* 26, 748–753.

Van Soest, P.J. (1982) Rumen microbes. In: *Nutritional Ecology of the Ruminant*. O & B Books, Oregon, pp. 152–177.

Virdis, B., Rabaey, K., Yuan, Z.G., Rozendal, R.A. and Keller, J. (2009) Electron fluxes in a microbial fuel cell performing carbon and nitrogen removal. *Environmental Science and Technology* 43, 5144–5149.

Wang, C.T., Chang, C.P., Shaw, C.K. and Cheng, J.Y. (2009) Fuel cell bionic flow slab design. *Journal of Fuel Cell Science and Technology* 6, 011009 1–5.

Wang, C.T., Yang, C.M.J., Chen, Z.S. and Tseng, S. (2011) Effect of biometric flow channel on the power generation at different Reynolds numbers in the single chamber of rumen microbial fuel cells (RMFCs). *International Journal of Hydrogen Energy* 36, 9242–9251.

Wang, C.T., Yang, C.M.J. and Chen, Z.S. (2012) Rumen microbial volatile fatty acids in relation to oxidation reduction potential and electricity generation from straw in microbial fuel cells. *Biomass and Bioenergy* 37, 318–329.

Wang, X., Feng, Y., Wang, H., Qu, Y., Yu, Y., Ren, N., Li, N., Wang, E., Lee, H. and Logan, B.E. (2009) Bioaugmentation for electricity generation from corn stover biomass using microbial fuel cells. *Environmental Science and Technology* 43, 6088–6093.

White, H.K., Reimers, C.E., Cordes, E.E., Dilly, G.F. and Girguis, P.R. (2009) Quantitative population dynamics of microbial communities in plankton-fed microbial fuel cells. *ISME Journal* 3, 635–646.

Yang, C.-M.J. (2002) Response of forage fiber degradation by ruminal microorganisms to branched-chain volatile fatty acids, amino acids, and dipeptides. *Journal of Dairy Science* 85, 1183–1190.

Yang, C.-M.J. and Varga, G.A. (1989) Effect of sampling site on protozoa, and fermentation end products in the rumen of dairy cows. *Journal of Dairy Science* 72, 1492–1498.

Yu, C.W., Chen, Y.S., Cheng, Y.-S., Cheng, Y.-II., Yang, C.M.J. and Chang, C.-T. (2010) Effects of fumarate on ruminal ammonia accumulation and fiber digestion *in vitro*, and nutrient utilization in dairy does. *Journal of Dairy Science* 93, 701–710.

Zhang, T., Zeng, Y., Chen, S., Ai, X. and Yang, H. (2007) Improved performances of *E. coli*-catalyzed microbial fuel cells with composite graphite/PTFE anodes. *Electrochemistry Communications* 9, 349–353.

Zhang, X.C. and Halme, A. (1995) Modelling of a microbial fuel cell process. *Biotechnology Letters* 17, 809–814.

Zhang, Y., Min, B., Huang, L. and Angelidaki, I. (2009) Electricity generation and microbial community analysis of wheat straw biomass powered microbial fuel cells. *Applied Environmental Microbiology* 75, 3389–3395.

Zuo, Y., Maness, P.C. and Logan, B.E. (2006) Electricity production from steam-exploded corn stover biomass. *Energy Fuels* 20, 1716–1721.

Chapter 6

Systems Microbiology Approach to Bioenergy

Qasim K. Beg and Ritu Sarin

Introduction

The renewable energy derived from biomass or living matter is called 'bioenergy' and can be generated using cellular machinery in several ways. The term bioenergy is often used as a synonym to biofuels. Renewable energy is one of the most efficient ways to achieve sustainable development. Current socio-economic and scientific affairs including high fuel prices and fossil fuel depletion are prompting world leaders and scientists to look for alternative energy resources in order to meet the demands of future generations. The unprecedented need for fuels together with increasing socio-economic issues presents a challenge in advancing bioenergy production. In addition, the ever increasing consumption of gasoline coupled with tremendous greenhouse gas emissions is forcing severe climate changes, further diminishing the quality of our environment and the air we breathe. With the recent disasters such as in the 2010 Deep Horizon oil spill in the Gulf of Mexico, it has become very clear that global oil reserves and new gasoline discoveries will not be enough to meet the worldwide demand of natural gas in years to come. Today, more than ever before, the environmental issues related to both social and economic impact of energy are dominating international energy summits and energy policy makers.

Scientists have spent numerous years using traditional approaches to understand the potential of microorganisms for bioenergy generation. The following chapter focuses on the use of an upcoming and advanced field of systems microbiology in two major avenues of bioenergy development: microbial fuel cells (MFCs) and biohydrogen production. First, we discuss the basics and background of these two promising technologies followed by how systems microbiology approach is being used towards better understanding of these two state-of-the-art technologies.

Systems Approach to Bioenergy

Recent advances in the study of microorganisms using microbial genome sequencing, functional genomics, proteomics and metabolic engineering have greatly improved our abilities to conduct systems-level studies of microbial metabolism, furthering the use of the obtained knowledge to identify and answer the fundamental questions of a system. In the last decade we have seen a slow evolution of traditional microbiology and molecular biology techniques into an advanced area of systems biology, and during this time a lot has changed in the approach of how researchers have thought and adapted themselves with regards to developing new methods to analyse and interpret the humongous amount of data continuously generated over these years as a result of high-throughput technologies. Systems biology offers a study of interactions between different components of any biological system, and how these interactions give rise to the function and behaviour of that system. The dynamic field of systems biology comprises broad subject areas. Due to exploitation of microorganisms using a systems approach, it is sometimes also referred to as systems microbiology (Buchrieser and Cole, 2009; Kay and Wren, 2009; Janssens, 2010; Zhao et al., 2010). Systems microbiology can be applied for the production of bioelectricity, biofuels, solvents, chemicals, bioplastics, pharmaceuticals and many more (Mukhopadhyay et al., 2008). The entire approach of systems microbiology is based upon getting a holistic picture of a microbial system. The holistic views of the microbial systems relevant to bioenergy applications employing the interdisciplinary science of systems microbiology offer a solution to the bioenergy problems. Since systems microbiology helps us in understanding the behavioural patterns of microbial systems, it thus advances our ability to manipulate systems capable of producing bioenergy by maximizing the production levels in order to meet our ever-increasing demands.

Although regarded as a relatively new discipline, systems microbiology probably had laid its foundation partly in the 19th century with Pasteur's discovery of 'microbe attenuation'. The aim of systems microbiology is to understand the structure, dynamics and interactions of whole cells rather than the function of individual modules (Barabasi and Oltvai, 2004; Price et al., 2004; Hatzimanikatis et al., 2005). By way of systems approach, one can identify cell's organizing principles and fundamental constraints that characterize the function of molecular interaction networks, and the limits of an organism's phenotypic diversity. One of the biggest challenges faced by systems microbiology is uncovering the facts related to transcriptional regulation of enzymes and metabolic changes in the cell. These challenges become more difficult due to the fact that metabolic pathways are not only controlled by the genes but that their regulation also involves a myriad of regulatory mechanisms, such as allosteric interactions, feed-back regulations, covalent modifications and post-translational mechanisms that affect enzyme activity and thus the overall metabolite synthesis in a living cell. These days, the understanding of interaction between transcriptional, translational, post-translational mechanisms and metabolism in cellular process of any organism from a systems approach has become the forefront of all systems microbiology research (Heinemann and Sauer, 2010; Reaves and Rabinowitz, 2011). Use of high throughput 'omics' techniques in systems microbiology delivers loads of data, which demands tremendous bioinformatics and computational analysis to understand cell's function (Loewe and Hillston, 2008; Gehlenborg et al., 2010; Sun et al., 2011). Use of constraints-based, systems-level modelling approaches help us to comprehensively understand the importance of individual pathways within the context of entire cell metabolism (Price et al., 2004; Reed et al., 2006; Becker et al., 2007; Beg et al., 2007). These computational and mathematical modelling approaches hold enormous promise for the rational design and manipulation of microbial systems to make them more efficient and economically attractive in real-world applications of bioenergy (Sun et al., 2010; Mahadevan et al., 2011; Westfall and Gardner, 2011).

Systems microbiology approaches, driven by genome sequencing and high-throughput functional genomics data, are revolutionizing single-cell-organism biology (Wang and

Bodovitz, 2010). The biggest challenge now is to use the data in conjunction with modelling techniques to extend our knowledge of microbial systems and to be able to experimentally verify *in silico* predictions. The Office of Science at the United States Department of Energy (DOE) is engaged in developing and improving environmental technologies for biofuel production and harnessing biomass power from renewable biobased materials (http://genomicscience.energy.gov). One of the DOE's programmes (called Genomic Science Program, formerly Genomics: GTL) is using systems approach for improving the quality of research in several sectors including bioenergy. This programme uses microbial and plant genomic data, high-throughput analysis, modelling and simulation to develop a predictive understanding of biological systems behaviour relevant to solving energy and environmental challenges including bioenergy, bioremediation and carbon cycling. The programme is taking the route of systems approach to integrate and utilize a wide variety of available datasets and computational methods to systematically address modelling of microorganisms. A vast amount of experimental data is available and the DOE is working towards developing a systems biology knowledgebase (Kbase) data information management system. The discussion of entire work of DOE and its present and future projects is beyond the scope of this chapter (for more information, see http://genomicscience.energy.gov).

Microbial Fuel Cells

Microorganisms have considerable potential, which can be explored to produce bioenergy in a wide variety of contexts. Decades of research in microbiology has shown that low amounts of bioelectricity (term only used when electricity is produced using biological process) can be generated directly from the degradation of organic matter and lignocellulosic biomass. Nowadays, researchers all over the world are trying to take advantage of this knowledge to generate electricity in a device called a microbial fuel cell (MFC). The recent energy crisis has re-ignited interests in MFCs among academic and industry researchers as a way to produce electricity and hydrogen. MFCs are fast becoming one of the most efficient methods of energy production. One of the main advantages of MFCs is their zero carbon emission into the ecosystem. Not only are MFCs being used in wastewater treatment facilities to break down organic matter, they are also being studied for applications as biosensors for biological oxygen demand monitoring (Angenent *et al.*, 2004; Moon *et al.*, 2004; Rabaey *et al.*, 2005; Logan *et al.*, 2006; Lovley, 2006; Kim *et al.*, 2009). MFCs are the devices that take advantage of specialized metabolic capacities of microbes to produce electric current from a wide range of organic substrates mainly found in waste material. Microorganisms consume the waste material, preferably carbohydrates, and produce electric current. The carbohydrates can be found in diverse sources, such as food industry or agricultural waste or domestic waste water. At present the real-world applications of MFCs are limited due to low power density level. Several laboratories are engaged in different aspects of MFC research worldwide and efforts are being made to improve their power output performance, operating costs and size they occupy in a research laboratory.

The idea of using a microbial cell for generating electricity was hatched in the early 20th century using *Escherichia coli* and *Saccharomyces* on platinum electrodes (Potter, 1911). However, the MFC research did not gain too much attention for several years until the 1980s, when it was discovered that current density and the power output could be greatly enhanced by the addition of electron mediators. Although the exact mechanism of microbial electricity generation still remains a scientific mystery, several hypotheses have been proposed. The most acceptable explanation is that an electron mediator is needed that helps microbes to transfer electrons to its surface, hence increasing the power generation at the anode. Mediators shuttle between the anode and the bacteria transferring the electrons, during this process taking up the electrons from microbes and discharging them at the surface of the anode. The most commonly used externally supplied synthetic mediators include dyes and metalloorganic compounds such as methylene blue, thionine, 2-hydroxy-1,4-naphthoquinone, phenazine-1-carboxamide, ferric EDTA and neutral red (Du *et al.*, 2007). However, from an environmental friendly perspective, the toxic nature and instability of these synthetic compounds limit their applications in MFCs. Alternatively, the best option is by using microbes or microbial consortia (for example from sludge or marine sediments) that can themselves produce intermediate metabolites (pyocyanin and phenazine) to act as mediators for electron transfer (Rabaey *et al.*, 2004, 2005; Pham *et al.*, 2008). In mediator-less MFCs, some bio-electrochemically active microbes can also transfer electrons directly to the anode by forming a biofilm across the surface of the anode (Bond and Lovley, 2003; Pham *et al.*, 2003; Min *et al.*, 2005). Most notable microbes used in MFC research are: *Geobacter sulfurreducens* (Bond and Lovley, 2003), *Geobacter metallireducens* (Nevin and Lovley, 2000; Min *et al.*, 2005), *Shewanella putrefaciens* (Hou *et al.*, 2009; Kiely *et al.*, 2010), *Shewanella oneidensis* (Bretschger *et al.*, 2007; Fredrickson *et al.*, 2008; Marsili *et al.*, 2008; Baron *et al.*, 2009; Rosenbaum *et al.*, 2010, 2011), *Rhodoferax ferrireducens* (Chaudhuri and Lovley, 2003) and *Aeromonas hydrophila* (Pham *et al.*, 2003). Large number of substrates such as starch, cellulose, glucose, lactate, acetate, sucrose, wheat straw, phenol, domestic and industrial waste, mixed fatty acids and molasses to name a few can be used as feed in MFCs (Du *et al.*, 2007; Franks and Nevin, 2010; Pant *et al.*, 2010).

Systems Microbiology and Microbial Fuel Cells

Microorganisms are excellent model systems for studying complex transcriptional, regulatory and metabolic networks. We are in the phase of MFC research where we are beginning to understand how individual microbial species or microbial consortia interact with an electrode as electron donor using global gene expression coupled with metabolite analysis. Scientists are now addressing the systems approach-related questions in advancing the MFC research (Holmes *et al.*, 2006; Butler *et al.*, 2007; Nevin *et al.*, 2009; Franks, 2010; Franks *et al.*, 2010; Rosenbaum *et al.*, 2010). The model-based analysis of bioenergy-related microorganisms has begun to reveal functional modules in metabolic and transcriptional networks and predict cellular behaviour from genome-scale physicochemical constraints. The genome-scale models of *Geobacter* sp. (Sun *et al.*, 2009; Mahadevan *et al.*, 2011), *Pelobacter* sp. (Sun *et al.*, 2010) and *S. oneidensis* (Pinchuk *et al.*, 2010) are providing important insights into the strategies for improving biodegradation and MFC potential of these microorganisms. The examples of two micro-array-based gene expression studies described next are important steps for the design of MFCs and using systems approach to improve the function of MFCs. In a recent report, Rosenbaum *et al.* (2011, 2012) discuss how a systems approach of using two microbes (homolactic acid fermenter *Lactococcus lactis* and a metal-reducing *S. oneidensis*) can be used to gain insights into their interaction inside an MFC. Using microarray expression studies, they have compared the bioelectrochemical system (BES) performance of *S. oneidensis* in a pure-culture and in a co-culture with *L. lactis* at conditions that are pertinent to conventional BES operation. The presented results represent an important step in the understanding of metabolic and transcriptional network relationships in mixed microbial culture and are an important step in the advanced understanding of microbial interactions in BES communities. The next goal should be to design complex microbial communities, which specifically convert target substrates into electricity.

Another systems approach study to gain insights into the physiology of *G. sulfurreducens* involves the genetic analysis of electron transfer to electrodes using microarrays (Holmes *et al.*, 2006). The authors showed that the transcript expression of 474 genes was significantly different when two growth conditions (an electrode as the sole electron acceptor versus growth on ferric citrate) were compared. The transcript abundance of one key gene (*omcS*), which encodes an outer-membrane cytochrome, was 19-fold higher during growth on electrodes. These transcriptional profiles across two growth conditions demonstrated that cells growing on electrodes were subjected to less oxidative stress than cells growing on ferric citrate and that a number of genes annotated as encoding metal efflux proteins or proteins of unknown function may be important for growth on electrodes.

Future Appears Bright in MFC Research

Low power output in MFCs limits its commercial applications. Scale up of MFCs from technology developed in a laboratory to large-scale treatment of waste water has its own problems. Barely 3 years ago, for the first time, a practical device powered by MFC technology was reported (Tender *et al.*, 2008). A pilot-scale MFC has been constructed by Advanced Water Management Centre on the site of the Foster's brewery in Queensland, Australia, where brewery waste water is used as a feed (http://www.microbialfuelcell.org). More research on further commercial applications is currently underway in several laboratories worldwide and much still needs to be done before the technology can be commercialized and implemented in more industries or reach our households (see the webpage of the Geobacter project: http://www.geobacter.org). To improve the power output, microbes that can tremendously improve the electron transport rate on a biofilm are needed. It is believed that a synergistic microbial consortium that provides substrates to other microbes or mediators to transport electrons more efficiently by other organisms can be used (Logan, 2009; Kiely *et al.*, 2010; Rosenbaum *et al.*, 2010). Use of adaptive evolution in laboratory also holds great promise (Summers *et al.*, 2010). A simple microbial community analysis of anodes (De Schamphelaire *et al.*, 2010) from sediment MFC is a good start. Additionally, high throughput 'omics' approaches, such as proteogenomics (VerBerkmoes *et al.*, 2009) or metagenome analysis of microbial consortia from electrode similar to that used in other systems (Warnecke *et al.*, 2007; Biers *et al.*, 2009; Lazarevic *et al.*, 2009; Gonzalez-Pastor and Mirete, 2010; Tasse *et al.*, 2010; Vacharaksa and Finlay, 2010; Xie *et al.*, 2010; Gosalbes *et al.*, 2011) would be valuable in understanding the microbial interactions inside an MFC. Metagenomics is an interesting tool to study uncultured microbial communities, and it is difficult to accept the fact that more than 99% of Earth's microbial diversity still needs to be explored via culturing methods in scientific laboratories. Metagenomic approaches needs to be combined with systems microbiology to better understand the vast and diverse microbial world (Ward, 2006; Raes *et al.*, 2007; Vieites *et al.*, 2010). In a first of its kind study, a recent metagenomic article reports the existence of multiple syntrophic interactions between terephthalate-degrading methanogenic microbial communities grown in an anaerobic methane producing bioreactor (Lykidis *et al.*, 2011).

Biohydrogen

As the growing shortage of gasoline poses problem to both developed and developing nations, reduced biohydrogen presents itself as an attractive fuel. The existing method of hydrogen production is expensive and is dependent on either electrolysis of water, gasification of coal or steam reformation of natural gas (Nath and Das, 2004). Similar to the production of biogas (methane), biohydrogen is the molecular hydrogen produced by microbes. Biohydrogen offers a lucrative renewable energy alternative to fossil fuels. It offers the advantages of being cleaner in terms of producing water as an end product of combustion and has a higher conversion efficiency in fuel cells compared to current combustion engines (~60–70%) (Dunn, 2002). In addition to being cleaner it also does not contribute to global warming by being CO_2 neutral. Biohydrogen in nature can be produced either by photosynthesis or fermentation by microorganisms. Photosynthetic methods prove to be more challenging, being limited by the requirement for sunlight, thus favouring fermentative methods as a better option. The future of biohydrogen is however dependent on the development of storage techniques to enable proper storage, distribution and combustion of hydrogen. To meet the fuel supply of the various nations, industrial production of biohydrogen will depend on the photo-fermentation of organic acids (Tao *et al.*, 2007), since fermentation of organic acids produces high hydrogen rates. However, continuous fermentative process would be required as the fermentative process itself takes few days to achieve high production rates. Sewage waste water or agricultural wastes could provide the source of necessary organic acids for this purpose (Kapdan and Fikret, 2006). At present, no clear strategy for the production of biohydrogen by industry exists and future hope for industrial process development is contingent on streamlining or modifying existing laboratory production process. Traditionally in the laboratory, the process of biohydrogen synthesis is carried out by photosynthesis, dark fermentation, and combined photo- and dark fermentation. The production of biohydrogen in a laboratory is dependent upon the growth and physiological patterns of the microbes. There exist many limiting factors, owing to the metabolic, genetic and regulatory make-up of the microbe.

There are many advantages of using microbes in the production of hydrogen. Microbes generally require inexpensive growth media. In addition, growth or metabolic processes generating hydrogen could be carried out in short steps from biomass, hence leading to increased energy efficiency. Also metabolic pathways that generate H_2 are known or can be identified and regulated to exert better control over production. As stated above, hydrogen can be produced in nature by many microbes either in a photosynthetic or a fermentative process. Members of algal genera including cyanobacteria and green algae generate hydrogen as part of photosynthetic process by utilizing solar radiation to convert H_2O, or by reducing sulfur and organic compounds into molecular hydrogen, whereas members of family Enterobactereaceae including *Enterobacter aerogenes*, *Escherichia coli* and Clostridiaceae including *Clostridium* produce H_2 fermentatively (Madamwar *et al.*, 2000). Production of hydrogen involves the following enzymes, which are also considered to be potential targets of increasing hydrogen production by use of bioengineering:

1. Hydrogenases: these include uptake and reversible hydrogenases and are involved in the catalysis of hydrogen oxidation to H^+ or reduction of H^+ to hydrogen. Hydrogenases can have either Fe–S, Ni/Fe–S (selenocysteine) or non-metal centres (Madamwar *et al.*, 2000). Blue-green algae have Ni/Fe–S-hydrogenases (Appel and Schulz, 1998). While the uptake hydrogenases catalyse breakdown of molecular H_2 from nitrogen fixation reaction, reversible bidirectional hydrogenases carry out dual functions of both hydrogen uptake or breakdown and hydrogen evolution (Lambert and Smith, 1981). Reversible bidirectional hydrogenase is found in a wide majority of N_2 fixing and non-N_2 fixing cyanobacteria (Eisbrenner *et al.*, 1978; Lambert and Smith, 1981). Further, the non-requirement of ATP by this enzyme makes it more efficient in producing H_2 than nitrogenase (Angermayr *et al.*, 2009).

2. Nitrogenases: heterocysts of the filamentous cyanobacteria contain an O_2 sensitive enzyme called nitrogenase, capable of fixing N_2 to ammonia (Fay, 1992). During the process of reduction of N_2 to NH_3, hydrogen is liberated. Simultaneous production of hydrogen and oxygen in an argon atmosphere has been reported in a nitrogen-fixing cyanobacterium, *Anabaena cylindrica* (Benemann and Weare, 1974). However, two molecules of ATP are required per electron pair to produce hydrogen, thus reducing the efficiency of the entire process. Nitrogenase in nature is found as a metalloenzyme based either upon molybdenum, vanadium, iron or tungsten (Kentemich *et al.*, 1988; Kajii *et al.*, 1994; Kim and Rees, 1994).

Systems Microbiology and Biohydrogen

In this section, the emphasis is on the significance of systems microbiology in improvising upon the current biohydrogen production strategies in a laboratory and scope for future scale-up to industrial level and promise as a competitive fuel of the future. Cutting-edge research efforts are required on adapting and improving photosynthetic organisms for biohydrogen production to produce sufficient levels of biohydrogen. Photosynthetic microbes could be manipulated to carry out fermentative processes in order to achieve high production rates. Thus, it is necessary to understand the basic metabolism and physiological processes of these organisms that govern the growth and synthesis capabilities and efficiencies by applying a holistic approach. As discussed in earlier sections, systems microbiology helps to gain insights into the regulatory mechanisms and networking in an organism on the whole. It is not only dependent upon the processes within the cell but also in context of dynamic environment. It could thus be used to tailor these organisms according to production and economical needs. First step in the application of this area of biology is collecting data sets from 'omics' approaches together with imaging techniques (Kherlopian *et al.*, 2008) on various model organisms. Next is the development of labelling techniques (e.g. ^{13}C, ^{15}N) to study metabolic flux analysis of the cell (Rupprecht, 2009). Finally, with the help of bioinformatics, the data so obtained could be translated into a mathematical model that can be verified experimentally on different nutritional conditions, environmental perturbations and mutant strains. The model can be refined by feeding the obtained experimental data back into the model, which can be used as a guide to tailor the microbes for enhanced biohydrogen production.

A microbial cell in its natural environment comprises networks of complex metabolic pathways that are finely regulated in concert with its existing environment. As conditions around it change, so does the functioning of these pathways to maintain balance within and outside. Introducing changes in the microbe that lead to improved hydrogen production could also perturb the natural balance of that microbe. The microbial cells' behaviour under new conditions could be difficult to predict, thus there exists a need to understand organism's genetic, regulatory and metabolic make-up to enable the scientists to interfere with cellular activities that aim towards optimal biohydrogen production. Systems microbiology tools provide abundant data on the physiology and metabolic fluxes of hydrogen-producing microbes (Amador-Noguez et al., 2010; Miskovic and Hatzimanikatis, 2010). The ultimate goal of systems microbiology is to computationally model the complete metabolism of known hydrogen-producing microorganisms utilizing genome-scale models of cyanobacteria and green algae, design the tools for the discovery of various parameters, and their optimization at organism level to advance the knowledge of hydrogen-producing photosynthetic organisms. A typical systems microbiology approach in cyanobacterial or green algal microbes towards biohydrogen production is carried out in the following steps:

1. Generation of mutants by insertional mutagenesis or targeted gene disruption for the next round 'omics' study. The mutant selection is aimed at increasing the efficiency of photosynthetic energy conversion, regulating the flow of reducing equivalents towards energy storage (e.g. lipids and carbohydrates) and the expression/activity of hydrogenase. Focus could also be laid upon producing cyanobacterial mutant strains by engineering them to produce light-dependent nitrogenase-based hydrogen-producing mutants incapable of recycling H_2 such that light energy may drive H_2 production rather than N_2 fixation under intense light (Angermayr et al., 2009).
2. Characterization of these mutants using the combination of physiology and 'omics' technologies.
3. Biological data-driven constraint-based computational modelling to generate information on metabolic fluxes of an organism.
4. Identification of key metabolic nodes and limiting factors to further optimize the pathways or networks towards increased biohydrogen production using bioengineering approaches of RNAi, gene knockout or post-translational modifications.

The results of systems microbiology screening should serve as platform for constructing future photobioreactors or electrode-driven biomolecular devices that combine a light-driven water-splitting process as it occurs naturally in photosynthesis to produce biohydrogen. The cyanobacterial (e.g. *Synechocystis* sp.) and green alga (*Chlamydomas reinhardtii*) with entire genome sequence information are available as model organisms with detailed information on genetic manipulation and are serving as tools for metabolic engineering exploitation (Kaneko et al., 1996; Merchant et al., 2007; Vallon and Dutcher, 2008; Angermayr et al., 2009; Rupprecht, 2009).

Since systems microbiology approaches greatly rely on computational methods, hybrid bottom-up or top-down methodologies utilizing computational algorithms such as DENSE and PRP-Finder, respectively, have been used to identify phenotype-related proteins (e.g. Fe–Fe hydrogenase, Ni–Fe hydrogenase and glutamate synthase) and metabolic pathways related to biohydrogen production (Rocha *et al.*, 2010). DENSE capitalizes on the availability of partial 'prior knowledge' about the proteins involved in this process and adds on the available information with newly identified sets of functionally associated proteins present in individual phenotype-expressing microorganisms. However, PRP-Finder or the 'top-down' approach depends on high-throughput comparative analysis of multiple genome-scale metabolic networks to identify phenotype-related metabolic genes and pathways. Even if systems microbiology approaches identify the proteins or metabolic pathways, pertaining to biohydrogen production, few impediments in the pathway of bioengineering of algal biohydrogen systems can still exist. Heterologous expression of enzymes that are optimized to function under the environmental conditions likely to exist in scaled-up operations is difficult and hydrogenase sensitivity to O_2 requires the identification of oxygen-tolerant hydrogenase.

In the light of recent discoveries that have identified Fe–Fe hydrogenase (Posewitz *et al.*, 2004; Girbal *et al.*, 2005; McGlynn *et al.*, 2008), NiFe hydrogenase (Schubert *et al.*, 2007; Ludwig *et al.*, 2009a, b) and the development of heterologous expression systems for the expression of these enzymes (Lenz *et al.*, 2005; Sybirna *et al.*, 2008), the above mentioned problems could be solved. In a separate fermentative approach, *Saccharomyces cerevisae* was engineered to overproduce formate by overexpressing pyruvate formate lyase enzyme and *adhE* gene that was used as the metabolic precursor for hydrogen synthesis by formate hydrogen lyase complex of *E. coli* (Waks and Silver, 2009). To identify or generate oxygen-tolerant hydrogenase, gene-shuffling to artificially generate a library of hydrogenase to screen for O_2 tolerance (Nagy *et al.*, 2007) or PCR using degenerate primers to identify O_2 tolerant hydrogenase naturally are being employed (Boyd *et al.*, 2009). Native O_2-tolerant hydrogenases, such as *hydS* and *hydL* from *Thiocapsa roseopersicina*, could be engineered into sensitive organisms (Xu and Smith, 2008). ChlamCyc, a system biology database for *Chlamydomas reinhardtii* (Rupprecht, 2009) or Cyanobase for *Synechocystis* sp. PCC 6803 (Angermayr *et al.*, 2009) provide useful information for exploitation of H_2-producing pathways. Systems microbiology approaches along with successful metabolic engineering methodologies could thus not only improve the microorganisms' production flow, but also open up avenues for the production of other cheap and next-generation biofuels (Keasling and Chou, 2008; Dellomonaco *et al.*, 2010; Sims *et al.*, 2010; Costa and De Morais, 2011; Rismani-Yazdi *et al.*, 2011; Wackett, 2011).

Conclusions

The advent of systems microbiology has enabled traditional life science research to come a long way towards understanding appreciable number of microorganisms and their utilization in day to day commercial and environmental phenomena. Whereas the newly available renewable biofuel sources such as microbial fuel cells and biohydrogen are still under commercial development in various parts of the world, support from ongoing development of methods and techniques of interdisciplinary science of systems microbiology is already pitching in to tailor these biofuel-generating microbes. More than 30 years ago, the ethanol produced from sugarcane was released for commercial use in Brazil for the first time; however, still a major percentage of countries do not have technology or facilities to produce the same. At the current consumption rate, the existing Earth reserves for oil and gasoline will be exhausted within the next 40–60 years. This raises some significant questions addressing the energy crisis worldwide and the role of environmental microbiology or biotechnology research in leveraging it. As we stand in the middle of transition between the old and new era of systems microbiology research, we are in desperate need of solving growing human dependence on the world's existing fossil fuels. The situation seems to be worsened by consumers' agony of government-imposed regulations on production, distribution or use of natural resources. What nations need today is an economically priced environmental friendly abundant supply of fuel that reaches the consumers without harsh impositions. Therefore, before we run out of time, scientists across the world need to come up with the use of new high throughput methods, making use of a comprehensive systems microbiology approach to solve our dependence on energy needs. Certainly being an evolving science, most of the success for the use of systems microbiology lies in the hand of research and development laboratories across the world. Microbiology researchers across the world need to integrate their work with highly trained 'Systems Scientists' and vice versa. Together they can bridge the massive gap between them by becoming familiar with each other's demands, understanding and approaches to problem solving. Integrating experts from different expertise and disciplines will help in solving the grand challenge of bioenergy problems we are facing today. Generations after us should see a bright future.

References

Amador-Noguez, D., Feng, X.J., Fan, J., Roquet, N., Rabitz, H. and Rabinowitz, J.D. (2010) Systems-level metabolic flux profiling elucidates a complete, bifurcated tricarboxylic acid cycle in *Clostridium acetobutylicum*. *Journal of Bacteriology* 192, 4452–4461.

Angenent, L.T., Karim, K., Al-Dahhan, M.H., Wrenn, B.A. and Domiguez-Espinosa, R. (2004) Production of bioenergy and biochemicals from industrial and agricultural wastewater. *Trends in Biotechnology* 22, 477–485.

Angermayr, S.A., Hellingwerf, K.J., Lindblad, P. and De Mattos, M.J. (2009) Energy biotechnology with cyanobacteria. *Current Opinion in Biotechnology* 20, 257–263.

Appel, J. and Schulz, R. (1998) Hydrogen metabolism in organisms with oxygenic photosynthesis – hydrogenases as important regulatory devices for a proper redox poising? *Journal of Photochemistry and Photobiology B: Biology* 47, 1–11.

Barabasi, A.L. and Oltvai, Z.N. (2004) Network biology: understanding the cell's functional organization. *Nature Reviews Genetics* 5, 101–113.

Baron, D., Labelle, E., Coursolle, D., Gralnick, J.A. and Bond, D.R. (2009) Electrochemical measurement of electron transfer kinetics by *Shewanella oneidensis* MR-1. *Journal of Biological Chemistry* 284, 28865–28873.

Becker, S.A., Feist, A.M., Mo, M.L., Hannum, G., Palsson, B.O. and Herrgard, M.J. (2007) Quantitative prediction of cellular metabolism with constraint-based models: the COBRA Toolbox. *Nature Protocols* 2, 727–738.

Beg, Q.K., Vazquez, A., Ernst, J., De Menezes, M.A., Bar-Joseph, Z., Barabasi, A.L. and Oltvai, Z.N. (2007) Intracellular crowding defines the mode and sequence of substrate uptake by *Escherichia coli* and constrains its metabolic activity. *Proceedings of National Academy of Sciences, USA* 104, 12663–12668.

Benemann, J.R. and Weare, N.M. (1974) Hydrogen evolution by nitrogen-fixing *Anabaena cylindrica* cultures. *Science* 184, 174–175.

Biers, E.J., Sun, S. and Howard, E.C. (2009) Prokaryotic genomes and diversity in surface ocean waters: interrogating the global ocean sampling metagenome. *Applied and Environmental Microbiology* 75, 2221–2229.

Bond, D.R. and Lovley, D.R. (2003) Electricity production by *Geobacter sulfurreducens* attached to electrodes. *Applied and Environmental Microbiology* 69, 1548–1555.

Boyd, E.S., Spear, J.R. and Peters, J.W. (2009) [Fe] hydrogenase genetic diversity provides insight into molecular adaptation in a saline microbial mat community. *Applied and Environmental Microbiology* 75, 4620–4623.

Bretschger, O., Obraztsova, A., Sturm, C.A., Chang, I.S., Gorby, Y.A., Reed, S.B., Culley, D.E., Reardon, C.L., Barua, S., Romine, M.F., Zhou, J., Beliaev, A.S., Bouhenni, R., Saffarini, D., Mansfeld, F., Kim, B.H., Fredrickson, J.K. and Nealson, K.H. (2007) Current production and metal oxide reduction by *Shewanella oneidensis* MR-1 wild type and mutants. *Applied and Environmental Microbiology* 73, 7003–7012.

Buchrieser, C. and Cole, S.T. (2009) From functional genomics to systems (micro)biology. *Current Opinions in Microbiology* 12, 528–30.

Butler, J.E., He, Q., Nevin, K.P., He, Z., Zhou, J. and Lovley, D.R. (2007) Genomic and microarray analysis of aromatics degradation in *Geobacter metallireducens* and comparison to a *Geobacter* isolate from a contaminated field site. *BMC Genomics* 8, 180.

Chaudhuri, S.K. and Lovley, D.R. (2003) Electricity generation by direct oxidation of glucose in mediatorless microbial fuel cells. *Nature Biotechnology* 21, 1229–1232.

Costa, J.A. and De Morais, M.G. (2011) The role of biochemical engineering in the production of biofuels from microalgae. *Bioresource Technology* 102, 2–9.

De Schamphelaire, L., Cabezas, A., Marzorati, M., Friedrich, M.W., Boon, N. and Verstraete, W. (2010) Microbial community analysis of anodes from sediment microbial fuel cells powered by rhizodeposits of living rice plants. *Applied and Environmental Microbiology* 76, 2002–2008.

Dellomonaco, C., Fava, F. and Gonzalez, R. (2010) The path to next generation biofuels: successes and challenges in the era of synthetic biology. *Microbial Cell Factories* 9, 3.

Du, Z., Li, H. and Gu, T. (2007) A state of the art review on microbial fuel cells: a promising technology for wastewater treatment and bioenergy. *Biotechnology Advances* 25, 464–482.

Dunn, S. (2002) Hydrogen futures: toward a sustainable energy system. *International Journal of Hydrogen Energy* 27, 235–264.

Eisbrenner, G., Distler, E., Floener, L. and Bothe, H. (1978) The occurrence of hydrogenase in some blue-green algae. *Archives of Microbiology* 118, 177–184.

Fay, P. (1992) Oxygen relations of nitrogen fixation in cyanobacteria. *Microbiology Reviews* 56, 340–373.

Franks, A.E. (2010) Transcriptional analysis in microbial fuel cells: common pitfalls in global gene expression studies of microbial biofilms. *FEMS Microbiology Letters* 307, 111–112.

Franks, A.E. and Nevin, K.P. (2010) Microbial fuel cells, a current review. *Energies* 3, 899–919.

Franks, A.E., Nevin, K.P., Glaven, R.H. and Lovley, D.R. (2010) Microtoming coupled to microarray analysis to evaluate the spatial metabolic status of *Geobacter sulfurreducens* biofilms. *ISME Journal* 4, 509–519.

Fredrickson, J.K., Romine, M.F., Beliaev, A.S., Auchtung, J.M., Driscoll, M.E., Gardner, T.S., Nealson, K.H., Osterman, A.L., Pinchuk, G., Reed, J.L., Rodionov, D.A., Rodrigues, J.L., Saffarini, D.A., Serres, M.H., Spormann, A.M., Zhulin, I.B. and Tiedje, J.M. (2008) Towards environmental systems biology of *Shewanella*. *Nature Reviews Microbiology* 6, 592–603.

Gehlenborg, N., O'Donoghue, S.I., Baliga, N.S., Goesmann, A., Hibbs, M.A., Kitano, H., Kohlbacher, O., Neuweger, H., Schneider, R., Tenenbaum, D. and Gavin, A.C. (2010) Visualization of omics data for systems biology. *Nature Methods* 7, S56–S68.

Girbal, L., Von Abendroth, G., Winkler, M., Benton, P.M., Meynial-Salles, I., Croux, C., Peters, J.W., Happe, T. and Soucaille, P. (2005) Homologous and heterologous overexpression in *Clostridium acetobutylicum* and characterization of purified clostridial and algal Fe-only hydrogenases with high specific activities. *Applied and Environmental Microbiology* 71, 2777–2781.

Gonzalez-Pastor, J.E. and Mirete, S. (2010) Novel metal resistance genes from microorganisms: a functional metagenomic approach. *Methods in Molecular Biology* 668, 273–285.

Gosalbes, M.J., Durban, A., Pignatelli, M., Abellan, J.J., Jimenez-Hernandez, N., Perez-Cobas, A.E., Latorre, A. and Moya, A. (2011) Metatranscriptomic approach to analyze the functional human gut microbiota. *PLoS One* 6, e17447.

Hatzimanikatis, V., Li, C., Ionita, J.A., Henry, C.S., Jankowski, M.D. and Broadbelt, L.J. (2005) Exploring the diversity of complex metabolic networks. *Bioinformatics* 21, 1603–1609.

Heinemann, M. and Sauer, U. (2010) Systems biology of microbial metabolism. *Current Opinions in Microbiology* 13, 337–43.

Holmes, D.E., Chaudhuri, S.K., Nevin, K.P., Mehta, T., Methe, B.A., Liu, A., Ward, J.E., Woodard, T.L., Webster, J. and Lovley, D.R. (2006) Microarray and genetic analysis of electron transfer to electrodes in *Geobacter sulfurreducens*. *Environmental Microbiology* 8, 1805–1815.

Hou, H., Li, L., Cho, Y., De Figueiredo, P. and Han, A. (2009) Microfabricated microbial fuel cell arrays reveal electrochemically active microbes. *PLoS One* 4, e6570.

Janssens, B. (2010) Meeting report: systems biology of microorganisms. *Biotechnology Journal* 5, 641–645.

Kajii, Y., Kobayashi, M., Takahashi, T. and Onodera, K. (1994) A novel type of mutant of *Azotobacter vinelandii* that fixes nitrogen in the presence of tungsten. *Bioscience Biotechnology and Biochemistry* 58, 1179–1180.

Kaneko, T., Sato, S., Kotani, H., Tanaka, A., Asamizu, E., Nakamura, Y., Miyajima, N., Hirosawa, M., Sugiura, M., Sasamoto, S., Kimura, T., Hosouchi, T., Matsuno, A., Muraki, A., Nakazaki, N., Naruo, K., Okumura, S., Shimpo, S., Takeuchi, C., Wada, T., Watanabe, A., Yamada, M., Yasuda, M. and Tabata, S. (1996) Sequence analysis of the genome of the unicellular cyanobacterium *Synechocystis* sp. strain PCC6803. II. Sequence determination of the entire genome and assignment of potential protein-coding regions (supplement). *DNA Research* 3, 185–209.

Kapdan, I.K. and Fikret, K. (2006) Biohydrogen production from waste materials. *Enzyme and Microbial Technology* 38, 569–582.

Kay, E. and Wren, B.W. (2009) Recent advances in systems microbiology. *Current Opinions in Microbiology* 12, 577–581.

Keasling, J.D. and Chou, H. (2008) Metabolic engineering delivers next-generation biofuels. *Nature Biotechnology* 26, 298–299.

Kentemich, T., Dannenberg, G., Hundeshagen, B. and Bothe, H. (1988) Evidence for the occurring of the alternative vanadium-containing nitrogenase in the cyanobaterium *Anabena variabilis*. *FEMS Microbiology Letters* 51, 19–24.

Kherlopian, A.R., Song, T., Duan, Q., Neimark, M.A., Po, M.J., Gohagan, J.K. and Laine, A.F. (2008) A review of imaging techniques for systems biology. *BMC Systems Biology* 2, 74.

Kiely, P.D., Call, D.F., Yates, M.D., Regan, J.M. and Logan, B.E. (2010) Anodic biofilms in microbial fuel cells harbor low numbers of higher-power-producing bacteria than abundant genera. *Applied and Environmental Microbiology* 88, 371–380.

Kim, J. and Rees, D.C. (1994) Nitrogenase and biological nitrogen fixation. *Biochemistry* 33, 389–397.

Kim, M., Hyun, M.S., Gadd, G.M., Kim, G.T., Lee, S.J. and Kim, H.J. (2009) Membrane-electrode assembly enhances performance of a microbial fuel cell type biological oxygen demand sensor. *Environmental Technology* 30, 329–336.

Lambert, G.R. and Smith, G.D. (1981) The hydrogen metabolism of cyanobacteria (blue-green algae). *Biological Reviews* 56, 589–660.

Lazarevic, V., Whiteson, K., Huse, S., Hernandez, D., Farinelli, L., Osteras, M., Schrenzel, J. and Francois, P. (2009) Metagenomic study of the oral microbiota by illumina high-throughput sequencing. *Journal of Microbiological Methods* 79, 266–271.

Lenz, O., Gleiche, A., Strack, A. and Friedrich, B. (2005) Requirements for heterologous production of a complex metalloenzyme: the membrane-bound [NiFe] hydrogenase. *Journal of Bacteriology* 187, 6590–6595.

Loewe, L. and Hillston, J. (2008) Computational models in systems biology. *Genome Biology* 9, 328.

Logan, B.E. (2009) Exoelectrogenic bacteria that power microbial fuel cells. *Nature Reviews Microbiology* 7, 375–381.

Logan, B.E., Hamelers, B., Rozendal, R., Schroder, U., Keller, J., Freguia, S., Aelterman, P., Verstraete, W. and Rabaey, K. (2006) Microbial fuel cells: methodology and technology. *Environmental Science and Technology* 40, 5181–5192.

Lovley, D.R. (2006) Bug juice: harvesting electricity with microorganisms. *Nature Reviews Microbiology* 4, 497–508.

Ludwig, M., Cracknell, J.A., Vincent, K.A., Armstrong, F.A. and Lenz, O. (2009a) Oxygen-tolerant H_2 oxidation by membrane-bound [NiFe] hydrogenases of *Ralstonia* species. Coping with low level H_2 in air. *Journal of Biological Chemistry* 284, 465–477.

Ludwig, M., Schubert, T., Zebger, I., Wisitruangsakul, N., Saggu, M., Strack, A., Lenz, O., Hildebrandt, P. and Friedrich, B. (2009b) Concerted action of two novel auxiliary proteins in assembly of the active site in a membrane-bound [NiFe] hydrogenase. *Journal of Biological Chemistry* 284, 2159–2168.

Lykidis, A., Chen, C.L., Tringe, S.G., Mchardy, A.C., Copeland, A., Kyrpides, N.C., Hugenholtz, P., Macarie, H., Olmos, A., Monroy, O. and Liu, W.T. (2011) Multiple syntrophic interactions in a terephthalate-degrading methanogenic consortium. *ISME Journal* 5, 122–130.

Madamwar, D., Garg, N. and Shah, V. (2000) Cyanobacterial hydrogen production. *World Journal of Microbiology and Biotechnology* 16, 757–767.

Mahadevan, R., Palsson, B.O. and Lovley, D.R. (2011) *In situ* to *in silico* and back: elucidating the physiology and ecology of *Geobacter* spp. using genome-scale modeling. *Nature Reviews Microbiology* 9, 39–50.

Marsili, E., Baron, D.B., Shikhare, I.D., Coursolle, D., Gralnick, J.A. and Bond, D.R. (2008) *Shewanella* secretes flavins that mediate extracellular electron transfer. *Proceedings of National Academy of Sciences USA* 105, 3968–3973.

McGlynn, S.E., Shepard, E.M., Winslow, M.A., Naumov, A.V., Duschene, K.S., Posewitz, M.C., Broderick, W.E., Broderick, J.B. and Peters, J.W. (2008) HydF as a scaffold protein in [FeFe] hydrogenase H-cluster biosynthesis. *FEBS Letters* 582, 2183–2187.

Merchant, S.S. *et al.* (2007) The *Chlamydomonas* genome reveals the evolution of key animal and plant functions. *Science* 318, 245–250.

Min, B., Cheng, S. and Logan, B.E. (2005) Electricity generation using membrane and salt bridge microbial fuel cells. *Water Research* 39, 1675–1686.

Miskovic, L. and Hatzimanikatis, V. (2010) Production of biofuels and biochemicals: in need of an ORACLE. *Trends in Biotechnology* 28, 391–397.

Moon, H., Chang, I.S., Kang, K.H., Jang, J.K. and Kim, B.H. (2004) Improving the dynamic response of a mediator-less microbial fuel cell as a biochemical oxygen demand (BOD) sensor. *Biotechnology Letters* 26, 1717–1721.

Mukhopadhyay, A., Redding, A.M., Rutherford, B.J. and Keasling, J.D. (2008) Importance of systems biology in engineering microbes for biofuel production. *Current Opinions in Biotechnology* 19, 228–234.

Nagy, L.E., Meuser, J.E., Plummer, S., Seibert, M., Ghirardi, M.L., King, P.W., Ahmann, D. and Posewitz, M.C. (2007) Application of gene-shuffling for the rapid generation of novel [FeFe]-hydrogenase libraries. *Biotechnology Letters* 29, 421–430.

Nath, K. and Das, D. (2004) Improvement of fermentative hydrogen production: various approaches. *Applied Microbiology and Biotechnology* 65, 520–529.

Nevin, K.P. and Lovley, D.R. (2000) Lack of production of electron-shuttling compounds or solubilization of Fe(III) during reduction of insoluble Fe(III) oxide by *Geobacter metallireducens*. *Applied and Environmental Microbiology* 66, 2248–2251.

Nevin, K.P., Kim, B.C., Glaven, R.H., Johnson, J.P., Woodard, T.L., Methe, B.A., Didonato, R.J., Covalla, S.F., Franks, A.E., Liu, A. and Lovley, D.R. (2009) Anode biofilm transcriptomics reveals outer surface components essential for high density current production in *Geobacter sulfurreducens* fuel cells. *PLoS One* 4, e5628.

Pant, D., Van Bogaert, G., Diels, L. and Vanbroekhoven, K. (2010) A review of the substrates used in microbial fuel cells (MFCs) for sustainable energy production. *Bioresource Technology* 101, 1533–1543.

Pham, C.A., Jung, S.J., Phung, N.T., Lee, J., Chang, I.S., Kim, B.H., Yi, H. and Chun, J. (2003) A novel electrochemically active and Fe(III)-reducing bacterium phylogenetically related to *Aeromonas hydrophila*, isolated from a microbial fuel cell. *FEMS Microbiology Letters* 223, 129–134.

Pham, T.H., Boon, N., De Maeyer, K., Hofte, M., Rabaey, K. and Verstraete, W. (2008) Use of *Pseudomonas* species producing phenazine-based metabolites in the anodes of microbial fuel cells to improve electricity generation. *Applied Microbiology and Biotechnology* 80, 985–993.

Pinchuk, G.E., Hill, E.A., Geydebrekht, O.V., De Ingeniis, J., Zhang, X., Osterman, A., Scott, J.H., Reed, S.B., Romine, M.F., Konopka, A.E., Beliaev, A.S., Fredrickson, J.K. and Reed, J.L. (2010) Constraint-based model of *Shewanella oneidensis* MR-1 metabolism: a tool for data analysis and hypothesis generation. *PLoS Computational Biology* 6, e1000822.

Posewitz, M.C., King, P.W., Smolinski, S.L., Zhang, L., Seibert, M. and Ghirardi, M.L. (2004) Discovery of two novel radical S-adenosylmethionine proteins required for the assembly of an active [Fe] hydrogenase. *Journal of Biological Chemistry* 279, 25711–25720.

Potter, M.C. (1911) Electrical effects accompanying the decomposition of organic compounds. *Royal Society (Formerly Proceedings of the Royal Society) B* 84, 260–276.

Price, N.D., Reed, J.L. and Palsson, B.O. (2004) Genome-scale models of microbial cells: evaluating the consequences of constraints. *Nature Reviews Microbiology* 2, 886–897.

Rabaey, K., Boon, N., Hofte, M. and Verstraete, W. (2005) Microbial phenazine production enhances electron transfer in biofuel cells. *Environmental Science and Technology* 39, 3401–3408.

Rabaey, K., Boon, N., Siciliano, S.D., Verhaege, M. and Verstraete, W. (2004) Biofuel cells select for microbial consortia that self-mediate electron transfer. *Applied and Environmental Microbiology* 70, 5373–5382.

Raes, J., Foerstner, K.U. and Bork, P. (2007) Get the most out of your metagenome: computational analysis of environmental sequence data. *Current Opinions in Microbiology* 10, 490–498.

Reaves, M.L. and Rabinowitz, J.D. (2011) Metabolomics in systems microbiology. *Current Opinions in Biotechnology* 22, 17–25.

Reed, J.L., Patel, T.R., Chen, K.H., Joyce, A.R., Applebee, M.K., Herring, C.D., Bui, O.T., Knight, E.M., Fong, S.S. and Palsson, B.O. (2006) Systems approach to refining genome annotation. *Proceedings of National Academy of Sciences USA* 103, 17480–17484.

Rismani-Yazdi, H., Haznedaroglu, B.Z., Bibby, K. and Peccia, J. (2011) Transcriptome sequencing and annotation of the microalgae *Dunaliella tertiolecta*: pathway description and gene discovery for production of next-generation biofuels. *BMC Genomics* 12, 148.

Rocha, A.M., Hendrix, W., Schmidt, M.C., Mihelcic, J.R. and Samatova, N.F. (2010) Systems biology approach to discovery of phenotype-specific metabolic processes for engineering bacterial bio-hydrogen. *Scientific discovery through advanced computing program (SciDAC), July 11–15, 2010,* Chattanooga, Tennessee.

Rosenbaum, M., Cotta, M.A. and Angenent, L.T. (2010) Aerated *Shewanella oneidensis* in continuously fed bioelectrochemical systems for power and hydrogen production. *Biotechnology and Bioengineering* 105, 880–888.

Rosenbaum, M.A., Bar, H.Y., Beg, Q.K., Segre, D., Booth, J., Cotta, M.A. and Angenent, L.T. (2011) *Shewanella oneidensis* in a lactate-fed pure-culture and a glucose-fed co-culture with *Lactococcus lactis* with an electrode as electron acceptor. *Bioresource Technology* 102, 2623–2628.

Rosenbaum, M.A., Bar, H.Y., Beg, Q.K., Segre, D., Booth, J., Cotta, M.A. and Angenent, L.T. (2012) Transcriptional analysis of *Shewanella oneidensis* MR-1 with an electrode compared to Fe(III)citrate or oxygen as terminal electron acceptor. *PloS One* 7, e30827.

Rupprecht, J. (2009) From systems biology to fuel – *Chlamydomonas reinhardtii* as a model for a systems biology approach to improve biohydrogen production. *Journal of Biotechnology* 142, 10–20.

Schubert, T., Lenz, O., Krause, E., Volkmer, R. and Friedrich, B. (2007) Chaperones specific for the membrane-bound [NiFe]-hydrogenase interact with the Tat signal peptide of the small subunit precursor in *Ralstonia eutropha* H16. *Molecular Microbiology* 66, 453–467.

Sims, R.E., Mabee, W., Saddler, J.N. and Taylor, M. (2010) An overview of second generation biofuel technologies. *Bioresource Technology* 101, 1570–80.

Summers, Z.M., Fogarty, H.E., Leang, C., Franks, A.E., Malvankar, N.S. and Lovley, D.R. (2010) Direct exchange of electrons within aggregates of an evolved syntrophic coculture of anaerobic bacteria. *Science* 330, 1413–1415.

Sun, J., Sayyar, B., Butler, J.E., Pharkya, P., Fahland, T.R., Famili, I., Schilling, C.H., Lovley, D.R. and Mahadevan, R. (2009) Genome-scale constraint-based modeling of *Geobacter metallireducens*. *BMC Systems Biology* 3, 15.

Sun, J., Haveman, S.A., Bui, O., Fahland, T.R. and Lovley, D.R. (2010) Constraint-based modeling analysis of the metabolism of two *Pelobacter* species. *BMC Systems Biology* 4, 174.

Sun, S., Chen, J., Li, W., Altintas, I., Lin, A., Peltier, S., Stocks, K., Allen, E.E., Ellisman, M., Grethe, J. and Wooley, J. (2011) Community cyberinfrastructure for advanced microbial ecology research and analysis: the CAMERA resource. *Nucleic Acids Research* 39, D546–D551.

Sybirna, K., Antoine, T., Lindberg, P., Fourmond, V., Rousset, M., Mejean, V. and Bottin, H. (2008) *Shewanella oneidensis*: a new and efficient system for expression and maturation of heterologous [Fe–Fe] hydrogenase from *Chlamydomonas reinhardtii*. *BMC Biotechnology* 8, 73.

Tao, Y.C., Chen, Y., Wu, Y., He, Y. and Zhou, Z. (2007) High hydrogen yield from a two-step process of dark- and photo-fermentation of sucrose. *International Journal of Hydrogen Energy* 32, 200–206.

Tasse, L., Bercovici, J., Pizzut-Serin, S., Robe, P., Tap, J., Klopp, C., Cantarel, B.L., Coutinho, P.M., Henrissat, B., Leclerc, M., Dore, J., Monsan, P., Remaud-Simeon, M. and Potocki-Veronese, G. (2010) Functional metagenomics to mine the human gut microbiome for dietary fiber catabolic enzymes. *Genome Research* 20, 1605–1612.

Tender, L.M., Gray, S.M., Groveman, E., Lowy, D.A., Kauffman, P., Melhado, J., Tyce, R.C., Flynn, D., Petrecca, R. and Dobarro, J. (2008) The first demonstration of a microbial fuel cell as a viable power supply: powering a meteorological buoy. *Journal of Power Sources* 179, 571–575.

Vacharaksa, A. and Finlay, B.B. (2010) Gut microbiota: metagenomics to study complex ecology. *Current Biology* 20, R569–R571.

Vallon, O. and Dutcher, S. (2008) Treasure hunting in the *Chlamydomonas* genome. *Genetics* 179, 3–6.

VerBerkmoes, N.C., Denef, V.J., Hettich, R.L. and Banfield, J.F. (2009) Systems biology: functional analysis of natural microbial consortia using community proteomics. *Nature Reviews Microbiology* 7, 196–205.

Vieites, J.M., Guazzaroni, M.E., Beloqui, A., Golyshin, P.N. and Ferrer, M. (2010) Molecular methods to study complex microbial communities. *Methods in Molecular Biology* 668, 1–37.

Wackett, L.P. (2011) Engineering microbes to produce biofuels. *Current Opinions in Biotechnology* 22, 388–393.

Waks, Z. and Silver, P.A. (2009) Engineering a synthetic dual-organism system for hydrogen production. *Applied and Environmental Microbiology* 75, 1867–1875.

Wang, D. and Bodovitz, S. (2010) Single cell analysis: the new frontier in 'omics'. *Trends Biotechnology* 28, 281–290.

Ward, N. (2006) New directions and interactions in metagenomics research. *FEMS Microbiology Ecology* 55, 331–338.

Warnecke, F., Luginbühl. P., Ivanova, N., Ghassemian, M., Richardson, T.H., Stege, J.T., Cayouette, M., McHardy, A.C., Djordjevic, G., Aboushadi, N., Sorek, R., Tringe, S.G., Podar, M., Martin, H.G., Kunin, V., Dalevi, D., Madejska, J., Kirton, E., Platt, D., Szeto, E., Salamov, A., Barry, K., Mikhailova, N., Kyrpides, N.C., Matson, E.G., Ottesen, E.A., Zhang, X., Hernández, M., Murillo, C., Acosta, L.G., Rigoutsos, I., Tamayo, G., Green, B.D., Chang, C., Rubin, E.M., Mathur, E.J., Robertson, D.E., Hugenholtz, P. and Leadbetter, J.R. (2007) Metagenomic and functional analysis of hindgut microbiota of a wood-feeding higher termite. *Nature* 450, 560–565.

Westfall, P.J. and Gardner T.S. (2011) Industrial fermentation of renewable diesel fuels. *Current Opinion in Biotechnology* 22, 344–350.

Xie, G., Chain, P.S., Lo, C.C., Liu, K.L., Gans, J., Merritt, J. and Qi, F. (2010) Community and gene composition of a human dental plaque microbiota obtained by metagenomic sequencing. *Molecular Oral Microbiology* 25, 391–405.

Xu, Q. and Smith, H. (2008) Development of a novel recombinant cyanobacterial system for hydrogen production from water. *United States Department of Energy*. Annual Progress Report.

Zhao, Z., Kier, L.B. and Buck, G.A. (2010) Systems biology in the microbial world and beyond. *Chemistry and Biodiversity* 7, 1019–1025.

Chapter 7

Nanotechnology and Bioenergy: Innovations and Applications

Mrunalini V. Pattarkine

Introduction

In the last two decades, the international community has witnessed the most difficult energy market. Energy resources are vital for a globally sustainable economic growth. With natural resource depletion, the need for renewable and sustainable resources has become critical. Unless new, affordable and renewable energy supplies are made available, the energy demand-supply equation will collapse in near future. The need to improve current methods and bring in new and revolutionary breakthroughs in the energy field was never more apparent in the past. In addition to depleting natural resources, the undesirable contribution of fossil fuels towards global warming and economic impacts of using non-renewable energy sources have been concerns. Scientists are convinced of the need for new and revolutionary technologies to address energy issues.

Of the available renewable energy resources, bioenergy appears very promising. Bioenergy can be generated in liquid form as biofuels, gaseous form as natural gas, or as electricity produced by microbial fuel cells. Figure 7.1 represents a generic landscape for bioenergy generation, starting from biomass collection all the way to distribution to the end users.

Fig. 7.1. Biofuels supply chain (from http://www1.eere.energy.gov/biomass/pdfs/nbap.pdf).

In the middle of the first decade of this century, approximately 10% of global energy demand was met by biofuels (EIA, 'International Energy: Outlook', Energy Information Administration, Office of Integrated Analysis and Forecasting, US Department of Energy,

Washington 2006). By 2022, the US government has set an annual production goal of 36 billion gallons of renewable fuel (EIA, 'Energy Independence and Security Act of 2007', Energy Information Administration, Office of Integrated Analysis and Forecasting, US Department of Energy, Washington 2007). Table 7.1 outlines the major steps taken by the Department of Energy (DOE) as a plan for achieving energy independence.

Table 7.1. Timeline for the Action Plan for Biofuels by the US government.

Year	Plan	Goal
2006	Formation of Alternative Energy Initiative	Energy independence for the USA
2007	Twenty In Ten Initiative	Plan to reduce gasoline consumption by 20% in 10 years
2007	Renewable Fuel Standard (RFS) approved by congress	36 billion gallons per year of biofuel production goal by 2022
2007	Energy Independence and Security Act	
2008	2008 Farm Bill approved over $1 billion for biofuel projects funding	

To meet this ambitious goal, novel and 'non-traditional' sources of energy must be found to meet the global energy demand. These sources must not rely on fossil fuels and must be cost-effective to sustain basic economic growth (Baker Institute Study #30, 2005).

The need at this time is for the availability of sustainable technologies that can be scaled up cost-effectively. Bioenergy technologies can meet these criteria and therefore present an attractive solution. While considering renewable resources and bioenergy production, three factors can impact the final outcome and overall efficiency: (i) the type of biomass and the processes for pretreatment; (ii) conversion of the biomass feedstock into high energyproducts; and (iii) biofuel extraction and separation technologies. For each of these steps, the common requirement is novel, low-cost and low-energy technologies that would enable efficient and highly sustainable biofuel production.

Currently, efforts are concentrated on biofuel production using non-food feedstock such as domestic waste, biomass and non-edible oilseed crops, using thermochemical and biochemical processes. High production costs and lack of efficient technologies have impacted the outcomes for the second- and third-generation biofuels. The application of novel nanomaterials and nanotechnology is believed to offer solutions for existing economic barriers and improve biofuel production.

What is Nanotechnology?

Nanotechnology is the science of manufacturing structures at the nanometre scale. It allows one to synthesize materials with unique properties and customise their structures for specific applications. Nanotechnology exists in nature in form of all the nano-machinery of cellular systems and viruses. Figure 7.2 illustrates natural and man-made nanostructures and gives a perspective about their relative sizes.

These natural nanosystems demonstrate natural precision in molecular fabrication. The same knowledge and architecture can be exploited to create a new 'non-traditional' source of energy or energy-related technology. Current research efforts in nanotechnology have been dedicated to investigate its use for manufacturing (cost and energy efficient)

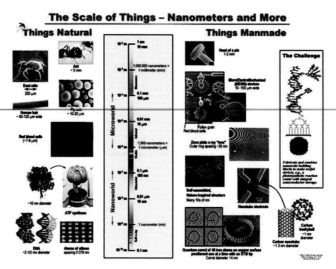

Fig. 7.2. Nanostructures found in nature and nanostructures that are man-made (adapted from http://www.science.doe.gov/bes/scale_of_things.html).

processes, minimizing negative impacts on the environment due to industry and transportation, as well as for offering improved technologies in healthcare, agricultural, food and pharmaceutical industries. With its nascent nature, many of the protocols in nanotechnology are relatively unrefined. But with the advancement in characterization and development tools, highly efficient and eco-friendly nanotechnology protocols will emerge in the next few years (Wegner and Jones, 2007). Nanotechnology can play a significant role by providing new materials for all aspects of bioenergy generations that need improvement over current technologies. It can provide high-performance materials for biomass conversion, fractionation and extraction; lightweight materials for use in making vehicles and safer and more efficient materials for storage of hydrogen fuels. Additionally, through the use of nanotechnology, green and clean and improved energy-generation protocols could be made available for geothermal, unconventional natural gas technologies, coal and carbon sequestration, nuclear, solar, wind, and hydraulics as well. Nevertheless, this chapter covers nanotechnology innovations specifically applied only to the bioenergy field. After careful consideration, these applications have been grouped into two categories: one category deals with applications for actual process of biomass transformation and treatment whereas the other considers applications for processes that are not actively involved in bioenergy production but impact the efficiency of the process, hence critical for bioenergy generation.

Nanotechnology is a new field and the terms use unique language or vocabulary. Since the focus of the chapter is to elaborate on applications of nanotechnology for bioenergy production, it is necessary to know certain basic terms used in nanotechnology (Table 7.2).

Nanotechnology has the potential for improving current technologies applied for biochemical as well as thermochemical processes for treatment and conversion of biomass to generate bioenergy in a variety of forms such as liquid biofuels, biohydrogen, biogas and electricity. These include improved materials for enzyme immobilization, materials with improved enzyme loading capacity, nanocatalysts, materials for storage of bioenergy

Table 7.2. Glossary of common nanotechnology terms (modified from Wegner and Jones, 2007).

Term	Definition
Nanoparticles	Particles with one or more dimensions at the nanoscale
Nanoscale	At this level, structures have at least one or more dimensions under 100 nm
Nanoscience	Field of science to study properties and manipulation of materials at atomic and molecular levels. Typically bulk properties and nanolevel properties differ significantly
Nanotechnology	The technology to fabricate and characterize and apply structures, devices and systems by controlling shape and size at the nanoscale
Nanostructured	Possessing structures that are at nanoscale
Engineered nanoparticles	Customised/designed nanostructures with specific properties or composition for a specific application
Nan fibre	Nan particles with two dimensions at the nanoscale and an aspect ratio greater than 3:1
Quantum dot	Fluorescent nanocrystals of semiconductor materials that exhibit size-dependent electronic and optical properties
Nanocomposites	Composites in which at least one of the phases has at least one dimension on the nanoscale
Nanophase	Discrete phase, within a material, which is at the nanoscale
Nanowire	A wire with diameter at nanometre scales. Nanowires can also be defined as structures that have lateral size constrained to tens of nanometres or less and an unconstrained longitudinal size. Nanowires can be coiled and stretched to reach full length
Nanoribbon	A nanoribbon has a flat structure with an unconstrained longitudinal size. The thickness is about tens of nanometres and the width can be of the order of 10 to 100 nanometres.

products, materials for separation and purification of liquid biofuels, materials for improved performance of microbial fuel cells and so on.

The chapter covers each of these areas in a detailed manner in the following sections. Many of these nano-enabled technologies have been adapted commercially, as shown in Table 7.3.

The list presented is far from complete as a number of other nano-enabled technologies are either available in the market or are likely to be launched soon; however, the idea is to present some of the technologies that have reached the market. Besides, there is much scope for others, which would shape the future.

Nano-applications for Bioenergy

Nanotechnology can contribute to the bioenergy production in multiple ways. In general, these applications can be grouped into the following classes:

1. Technologies directly involved in processing and production of bioenergy: processes that belong to this category are biomass pretreatment, biomass processing for extraction and separation of biofuels, technologies for storage of products such as biogas, biohydrogen and other liquid biofuels.

2. Technologies that indirectly support and improve the overall yield of bioenergy

Table 7.3. Commercial application of nanotechnologies in biofuel production.

Company	Product description
Headwaters	H-CAT®Conversion technology involving nanocatalysts for converting residual oil to energy and for coal to liquid fuels
Refinery Science	Nanocatalyst technology for upgrading crude oil and recombination of carbon and hydrogen into liquid fuels
Oxonica	Envirox™, a fuel-borne combustion nanocatalyst for diesel fuel
H2OIL	H2OIL's green nanotechnology F2-21® eeFuel® and eeLube® additives provide: increased fuel efficiency; dramatic fuel savings; reduction of harmful emissions; increased engine power; and prolonged engine life
Catlin	Heterogeneous biodiesel catalysis using nanocatalysts.
Agrivida	Bioengineered plants that produce enzymes to simplify the conversion of cellulose to ethanol

production: this class consists of parameters that do not directly constitute a part of the actual biomass treatment and conversion process, but comprise activities such as biomass transportation, biomass handling and waste heat recovery.

3. Technologies for waste to energy projects: these technologies are the latest in the innovation. These allow biological and organic waste to be treated for production of energy. Some of these are tied to simultaneous production of industrially important chemicals, making this field commercially attractive. In an effort to align the content mostly towards bioenergy, these waste to energy technologies are not considered in this chapter.

4. Additives for biofuels: these nanomaterials enhance the burning efficiency while reducing emissions from biofuels and are highly desirable in reducing greenhouse gas emissions and overall carbon footprints.

This chapter covers these categories, since the nano-enabled improvements in these fields have a direct impact on bioenergy production.

Technologies Directly Involved in Processing and Production of Bioenergy

There are several areas that are major steps in production of bioenergy from biomass. As stated earlier, the production of bioenergy from biomass can be achieved either by thermochemical or biochemical conversion methods. Current methodologies for both routes suffer certain bottlenecks and nanotechnology has offered solutions either in the form of improved nonmaterials for the processes or as cost-effective and more sustainable technologies. Figure 7.3 offers a snap-shot of various aspects of bioenergy production impacted by nanotechnology contributions.

The four major aspects of biomass to bioenergy conversion are: (i) biomass pretreatment and transformation; (ii) biofuel processing: extraction and separation; (iii) bioenergy product storage; and (iv) biohydrogen: production and harvesting. For each of these aspects, novel nano-integrated technologies are on the forefront of research and commercialization.

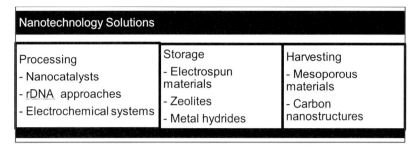

Fig.7.3. Contributions of nanotechnology in bioenergy production.

Nanotechnology for Biomass Transformation

Cellulose and lignocelluloses are among the most abundant renewable biomass available for biofuel production. Cellulose represents about 1.5×10^{12} tons of total annual biomass production(Wegner and Jones, 2007). The complex cellulosic biomass needs to be broken down into simple sugars for fermentation to liquid biofuels, a process called pre-treatment of feedstock. The pre-treated feedstock is then fermented to produce ethanol or other liquid biofuels. Using the lignocellulosic biomass for biofuel production thus requires efficient and sustainable industrial technologies for pre-treating this complex carbohydrate biomass. The traditional method has been that of using enzyme biocatalysts such as cellulases and several other hydrolases (Zhang *et al.*, 2010). Nevertheless, these traditional enzyme-based feedstock transformation technologies suffer from limitations of low catalytic efficiencies of enzymes (Li *et al.*, 2007), high cost, poor recovery of catalysts and a need for extreme conditions such as high temperatures and strong acids (Moxley *et al.*, 2008).

Enzyme immobilization has been used as a solution to attain increased specific activity and increased thermal and pH stability(Yuan *et al.*, 1999; Saville *et al.*, 2004). But these traditional enzyme immobilization methods have several drawbacks, such as: (i) reduced enzyme specific activity (relative to the free, soluble form) as a consequence of immobilization conditions; (ii) limited enzyme loading capacity onto the immobilization matrix; (iii) high cost of enzyme purification, low recovery and poor recycling of enzymes; (iv) applicable only for one-step transformations; and (v) dependence of immobilized enzymes on availability of co-factors for enzyme activity.

The problem of low specific activity of the immobilized enzyme has been resolved by loading increased amounts of the enzyme (Qhobosheane *et al.*, 2001). But even then, certain limitations due to availability of loading surface area still remain. Since the current immobilization technologies depend on surface-association mechanisms for immobilization, there are limitations to the enzyme-loading capacity of the matrix due to the monolayer adsorption processes applied for enzyme immobilization (Pugh *et al.*, 2010).

Nanotechnology solutions for improved enzyme immobilization

Nanotechnology offers solutions to all of the major aspects of enzyme immobilization through:(i) increased enzyme loading ability; (ii) innovative nanomaterials as immobilization matrices; and(iii) characterization of environments for nano-immobilized enzymes.

Nanotechnology for higher enzyme loading capacity

The traditional enzyme immobilization applies surface-associated methods. Immobilization conditions often result in reduction in specific enzyme activity, which can be compensated by loading high amount of enzymes in the immobilization beds. Under this situation, an immobilization technique that allows high loading capacity coupled to a high volume to surface area ratio would be ideal. This suffers limitations due to the surface-associated nature of the immobilization procedures. For any immobilization procedure, it is desirable that the matrix has a high volume to surface area ratio. When one considers the size for a particle, the smaller sized particle has a higher volume to surface area ratio when compared to the bigger particles.

Therefore, by utilizing nanoscaled structures, one can achieve significant increase in the surface area as the size approaches nanoscale. In addition to thus providing increased surface area for enzyme interaction, the nano-immobilization has several advantages, in pre-treatment stage (mostly for cellulosic biomass) as well as the production processes. Along with increased surface areas for enzyme loading (Cruz *et al*., 2010), nanomaterials help also to increase diffusion of substrates to enzyme molecules, leading to increased production rates (Table 7.4; Kim *et al*., 2004). They offer easier catalyst recovery and recycling (Ashok *et al*., 2009) along with the ability for a continuous operation (Hongfei *et al*., 2002). Last, the product processing and purification steps are also rendered easier with the use of nanomaterials. Significantly improved biocatalyst lifetime and stability are additional advantages when nanomaterials are used (Kim and Grate, 2003).

Table 7.4. Types of nanostructures and methodologies used for enzyme immobilization.

Nanomaterial	Immobilisation method	Reference
Nanoparticles	Surface attachment	Matsunaga and Kamyia, 1987; Shinkai *et al*., 1991; Crumbliss *et al*., 1992; Kondo *et al*., 1992
Nanofibres	Carrier-binding	Jia *et al*., 2002
Nanoporous matrix	Entrapment	Wang *et al*.,2001; Sotiropoulou *et al*.,2005.
Nanotubes	Adsorption and entrapment	Mitchell *et al*., 2002; Besteman *et al*., 2003; Rege *et al*., 2003; Yim *et al*., 2005
Nanofibrous membrane	Encapsulation	Wang and Hsieh, 2008
Magnetic nanoparticles	Surface loading	Johnson *et al*., 2009; Hohn *et al*., 2009

Novel nanomaterials for immobilization

Immobilization methods using surface attachment processes in the past have suffered limitations due to the monolayer adsorption mode of attachment. Innovations in the field, especially using nano-structured supports, have enabled enzyme aggregate coatings on the surface of carrier molecules as shown in Fig. 7.4 below (Kim *et al*., 2005). This approach enhanced enzyme activity nine-fold due to increased enzyme loading. In addition, it led to increased enzyme stability with essentially no measurable loss of activity over 1 month of observation under rigorous shaking conditions. These immobilized enzyme aggregates had high loading capacities, increased enzyme activity and stability. In another experiment, the same application was demonstrated in the case of β-glucosidase loaded on nanofibres to create multilayer enzyme aggregate coatings with a36-fold increase in the activity (Lee *et al.*, 2010). The only downside of this immobilized enzyme aggregate approach is that chances exist for an overall reduced activity due to multilayer structure blocking access of substrate molecules to the enzymes (Pugh *et al.*, 2010).

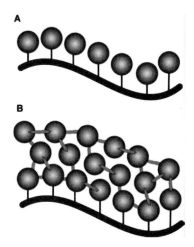

Fig. 7.4. Enzyme immobilization approaches. **A**: Conventional monolayer immobilization of enzyme molecules. **B**: Covalently cross-linked enzyme aggregate coatings. (Reprinted with permission from *The Canadian Journal of Chemical Engineering*, vol. 89, February 2011 published by the Canadian Society for Chemical Engineering.)

In 2007, Li and co-workers (Li *et al.*, 2007) reported manufacturing cellulase-coupled liposomes using aldehyde groups for surface attachment of cellulase (Fig. 7.5). This system had higher efficiency than the liposomes system with encapsulated cellulase. This system demonstrated higher retention of enzyme activity and reusability compared to the conventionally immobilized cellulase.

Nanomaterials have thus offered unique advantages for enzyme immobilization to overcome some key problems associated with the conventional technologies.

Nanotechnology for biomass pretreatment

Beyond enzyme immobilization, nanotechnology has also been applied for improving the efficiency and reducing the cost of feedstock transportation. In production of ethanol from cornstalk, the feedstock was broken down to nano-sized particles (Jennifer and Peter, 2006) for efficient and cost-effective transportation to the biofuel plants. In yet another nano-enabled approach for pretreatment of genetically engineered cellulosic biomass, Sticklen and fellow workers (Sticklen, 2009) proposed application of high shear-force nanomixers and nanodispersers in combination with turbines (Fig. 7.6). This led to a process, which eventually allowed elimination of some steps and by-passed the requirement of costly chemicals (current procedure), to allow for a simplified, on-site (nana-enabled) process for production of ethanol/butane from genetically engineered biomass. By eliminating the need

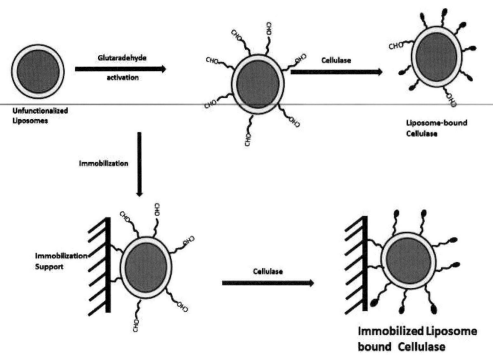

Fig. 7.5. Liposome-immobilized cellulases for application in bioconversion of cellulosic biomass (based on Li et al., 2007).

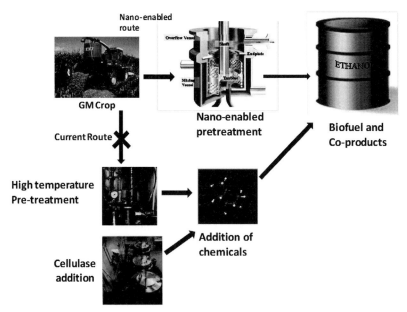

Fig. 7.6. Use of high shear force nanomixers and nanodispersers in combination with turbines for biomass pretreatment (based on Sticklen, symposium presentation, 2009).

for addition of costly cellulase enzyme for conversion of pre-treated biomass into ethanol, this process was made significantly cost-effective.

Use of lignocellulosic biomass offers apparent advantages in biofuel production due to its low cost and high availability, but the lignin component poses a major problem for dissolution (Zhang *et al.*, 2010). To overcome this, novel magnetic ferrite nanoparticles have been investigated. Recyclable enzyme constructs were constructed by immobilizing cellulase on magnetic nanoparticles using glucose oxidase as a model enzyme. Using co-precipitation and oxidation of $Fe(OH)_2$, magnetic nanoparticles of varying sizes (5 nm, 25 nm and 50 nm) were synthesized and the effect of particle size on diffusion efficiency was studied. With the help of amine groups, the particles were functionalized by 3-(amino propyl)-triethoxysilane followed by cross-linking with glutaraldehyde. The immobilized glucose oxidase retained activity after more than 40 days at various temperatures. The particles also demonstrated high recycling stability, since they retained almost 80% of activity during ten consecutive cycles for large (50 nm) and medium (25 nm) size enzyme magnetic nanoparticles (Ansari and Husain, 2011).

In an innovative application, Hohn and co-workers (Hohn *et al.*, 2009) provided a nanotechnology-based solution for preventing acid corrosion of equipment during acid catalysis of lignocellulose for bioethanol fermentation. They developed sulfonic acid cobalt spinel ferrite magnetic nanoparticles as catalysts for lignocellulose conversion, which had increased conversion efficiency. Additionally, they eliminated the problem of equipment corrosion by using these nanoparticles.

Other applications of nanomaterials for biomass pretreatment are as nanoparticles loaded with metal catalysts. Carbon nanotubes (CNTs) are ideal candidates for the formation of such nanostructures due to their well-defined hollow interiors, unusual mechanical and thermal stability, and high electron conductivity (Fig. 7.7). These cavities can accept entities to create nanocomposites with unique catalytic properties that are different from the bulk properties. When Rh particles were introduced inside CNTs, a striking enhancement of the catalytic activity was seen for the conversion of CO and H_2 to ethanol. The overall ethanol formation rate (30.0 mol $mol^{-1} Rh\ h^{-1}$) inside the nanotubes exceeds that on the outside of the nanotubes by more than an order of magnitude, although the latter is much more accessible (Pan *et al.*, 2007).

Similar applications of carbon nanotubes-based (CNT-based) catalytic nanoparticles have been developed for fuel cells and are discussed later in the chapter in the section on electrochemical systems.

Sun and co-workers (Sun *et al.*, 2010) fabricated Pyridylthio-modified multiwalled carbon nanotubes (pythio-MWNTs) by a reaction of the oxidized MWNTs with S-(2-aminoethylthio)-2-thiopyridine hydrochloride (Fig. 7.8). The composites of pyridylpythio-MWNTs were transferred onto substrate surfaces by the Langmuir-Blodgett (LB) method and were used as an immobilization matrix for hydrogenase (H2ase) to form a bionanocomposite of pyridylthio-MWNTs-H2ase. When tested for activity of the immobilized hydrogenase, the nanocomposites revealed high activity retention in the above matrix.

The nanocomposites support matrix was able to retain the hydrogenase with high catalytic activity. The nanocomposites immobilized hydrogenase demonstrated high stability as the LB films were stable and catalytically active for more than 2 months when stored at room temperatures.

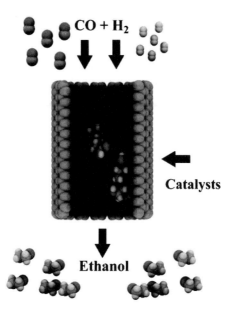

Fig. 7.7. Schematics explaining production of ethanol from syngas inside catalyst-loaded carbon nanotubes.

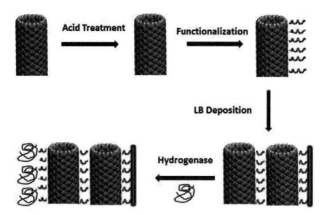

Fig. 7.8. Pyridylthio-modified multiwalled carbon nanotubes immobilized hydrogenase for biofuel applications (Modelled after Sun *et al.*, 2010).

Characterization of enzymatic nanoparticle catalysts for biofuel production

The application of enzyme-nanoparticles immobilization for pretreatment or transformation of feedstock offers very promising and innovative technology. These concepts could be made more applicable with innovative approaches that would allow predictions about the enzyme activities and efficiencies under a set of reaction conditions. At nanoscale, properties of matter are different from those of bulk matter. On similar lines, upon attachment with the nanomaterial surfaces, the immobilized enzymes may have altered characteristics as a consequence of the immobilization protocols, alteration in the microenvironment, and intra-molecular interaction within the nanomatrix (Fischer *et al.*, 2002; Asuri *et al.*, 2006; Karajanagi *et al.*, 2006; Cruz *et al.*, 2010). Ashok and co-workers concluded the limitations of available knowledge regarding the structure–function relationships for proteins and enzymes in immobilized states (Ashok *et al.*, 2009). Studies for establishing these structure–function parameters conventionally apply bio-analytical spectroscopic techniques such as fluorescence, circular dichroism, nuclear magnetic resonance etc. But when applied to nanoparticles, these are ineffective since the nanomatrix contributes to significant background signal (Ashok *et al.*, 2009; Cruz *et al.*, 2010). Since the immediate environments impact the activity of immobilized enzymes, approaches need to be developed to characterize and predict nano-immobilized enzymes with respect to their structure and function.

Non-enzymatic nanoparticle catalysts for biofuel production

Nanomaterial catalysts have been successfully applied for transesterification of fats and oils. The biodiesel derived from vegetable oils has physico-chemical properties similar to the petroleum-based biodiesel. Additionally this biodiesel has reduced emission of greenhouse gases compared to the fossil fuels. The conversion mainly involves alcoholysis and pyrolysis of vegetable oils into methyl- or ethyl-fatty acid esters and hydrocarbons, respectively. Homogeneous as well as heterogeneous catalysis has been applied for the conversions for a long time (Encinar *et al.*, 2002; Dorado *et al.*, 2004; Shah *et al.*, 2004; Haas, 2005; Kinney and Clemente, 2005).

These protocols, though successful, suffer several disadvantages such as lack of catalyst reusability, difficulties in handling, work only at elevated temperature, require multiple steps in the synthesis, and pose problems with scalability for large scale production (Reddy *et al.*, 2006).

For these conversions, nanoparticles have gained increasing attention due to their unique chemical (Sun and Klabunde, 1999; Jeevanandam and Klabunde, 2002) and physical properties (Hadjipanayis and Siegel, 1994; Fecht, 1996). The increased surface area, higher concentrations of reactive surface and corner defect sites, unusual lattice planes (Stoimenov *et al.*, 2002) make these an ideal choice for transesterification of fats and oils. Their ability to function at room temperatures allows for simple, energy-efficient and safe process designs. Additionally, the room temperature reaction conditions eliminate/reduce the possibility of side reactions that require high reaction temperatures (Stoimenov *et al.*, 2002).

Rubio-Arroyo *et al.*(2009) demonstrated a heterogeneous catalysis process involving tin oxide in a mesoporous matrix and sodium hydroxide for almost 99.05% conversion of sunflower oil into biofuel. In other studies by Boz *et al.* (2009) and Reddy *et al.* (2009), almost 99% conversion efficiencies were shown with nano-gamma-Al_2O_3 and nanocrystalline calcium oxide).

Transesterification of soybean oil into fatty acid methyl esters was shown to be much more efficient with use of calcium oxide nanoparticles.

Nanohybrid Catalystsas Emulsion Stabilizers

The bio-oil produced by pyrolysis of biomass is a liquid with complex properties in that it is only partially soluble in water and partially soluble in organic solvents (Huber and Dumesic, 2006). In such cases, application of phase-transfer catalysis can provide a solution. In such catalytic processes, catalysis is carried out using a bi-phasic solvent system comprised of two immiscible solvents (generally water and a water-immiscible organic solvent) stabilized by some surfactants such as quaternary ammonium salts as emulsifier (Stark, 1971). In these bi-phasic catalytic processes, a product may be unstable under reaction conditions in one solvent phase, but upon formation, can partition into the other phase stably. Such solubility-based partitioning of the products in the phase-transfer catalytic reactions can greatly simplify isolation and purification processes as they bypass the need for heat-intensive distillation processes to separate the hydrophilic components from the fuels. In case of the complex bio-oils, water content is almost 30%. The major drawback for the bi-phasic catalysis system is difficulty in separation of the surfactants from the final products. Applications of solid-particle emulsifiers have been documented in literature (Dai et al., 1996; Dinsmore et al., 2002), but these are not catalytic particles. Therefore, the option of solid-phase particles that have the combination of emulsion stabilizing as well as catalytic properties is highly desirable for biofuel production. Metal oxides are known to stabilize oil-in-water emulsions (Binks et al.,2002; Amalvy, 2004) whereas carbon nanotubes stabilize water-in-oil emulsions owing to their hydrophobicity (Wang and Hsieh, 2008). Crossley et al. (2010) synthesized hybrid nanoparticles by fusion of carbon nanotubes with silica. They were able to tune the hydrophilic–hydrophobic proportion of the particles by modifying the composition. Beyond biodiesel production from oils and fats, nanoparticles have also helped improve the technologies for production of hydrogen with chemical reactions. Moreover, they were able to achieve various degrees of hydrogenation activity in the organic phase by varying the formulation of the nanohybrids between SiO_2 and MgO as supporting materials for the nanoparticles. With such solid-stabilized nanohybrid systems, it is possible to design a continuous process with a layered oil-emulsion-water structure where one can achieve full conversion on the both sides of the emulsion phase, followed by removal of oil-soluble products from the top layer and water-soluble products from the bottom layer and the reaction happening in the emulsion.

Nanostructured Photo-catalysis for Lignocellulosic Biomass

Heterogeneous photo-catalysis covers a wide variety of reactions such as photo-reduction, hydrogen transfer, water splitting, organic synthesis, metal deposition etc. (Herrmann, 1999; Carp et al., 2004). The photo-catalysis reaction is initiated when photo-excited electrons are promoted from the filled valence band of a semiconductor metal catalyst to the empty conduction band gap of the semiconductor, which leaves behind a hole in the valence band. The photo-holes thus generated have a great potential to oxidize organic matter (Fujishima et al., 2000; Zhao and Yang, 2003;Zeltner and Tompkin, 2005). Major attention in this area is given to advanced oxidation processes that catalyse breakdown of organic matter otherwise resistant to breakdown using traditional methods and with no selectivity. These reactions utilise in situ generated free radicals, mainly $HO^{\bullet}$ generated by solar, chemical or other types of energy (Kudo et al., 2003; Bahnemann, 2004). Titania (TiO_2) is the most used photo-catalyst due to its high catalytic activity, low toxicity, good chemical stability and very low cost. The photo-catalyst in the form of nanotubes, nanowires and nanofibres have attracted significant attention lately (Yuan and Su, 2004; Xiong and Balkus, 2005;Zhong *et al*., 2005). The major limitation to the traditional forms of Titania as a photo-catalyst is its lack of photo-catalytic activity in visible light (Kisch and Macyk, 2002; Yamashita *et al*., 2004;Kitano *et al*., 2007). Use of photo-sensitizer and doping the Titania with metal/non-metal compounds has been used as a solution to solve this problem. Two major applications of Titania for biomass conversion have been for photo-catalytic production of hydrogen and for non-photo-catalytic conversion of biomass for solar gasification. Hydrogen can be produced either by using fast pyrolysis (Iwasaki, 2003; Li *et al*., 2004), steam gasification (Rapagna *et al*., 1998) or by supercritical conversion (Watanabe *et al*., 2002; Hao *et al*., 2003), but these processes use harsh conditions and are not cost-effective. Alternatively, hydrogen can also be produced by water splitting, but the efficiency of biomass conversion for hydrogen production has superior efficiency using Titania-photo-catalysts. Another application of photo-catalysis is in the solar gasification process. This reduces the amount of biomass that needs to be burnt in the gasification process, leading to thermal efficiency of the process.

Nanotechnology for Engineering Microbes for Biofuel Production

Enzyme immobilization using novel nanotechnologies offers unique advantages and solutions to several problems faced in traditional enzyme immobilization for biofuel production. Nanotechnology continues to provide novel matrices for enzyme immobilization, but still faces limitations due to the high cost of enzyme purification and subsequent scalability problems. The alternative to that comes in the form of genetic engineering of the microbial species, applied either as metabolic or protein engineering. These efforts have been dedicated in two main areas as: (i) genetic and protein engineering organisms for improved biocatalytic function in biofuel production; and (ii) engineering the biomass structure for easier and more efficient biofuel production (recalcitrance issues).

Using an antisense DNA approach, efforts have been made for altered genomic and proteomic make-up of microorganisms to generate efficient strains of bacteria or for production of engineered enzymes (Atsumi and Liao, 2008; Atsumi *et al*., 2008a; Fortman *et al*., 2008; Nielsen *et al*., 2009). This approach allows for selective up or down regulation of metabolic processes by targeting key biocatalyst concentrations either at transcriptional or translational stages. This approach was successfully demonstrated in the case of *n*-butanol production in *Clostridium acetobutylicum* (Tummala *et al*., 2003a,b,c). Efficient modulation of the genetic activity in these processes entirely depends on strategies for efficient delivery, uptake and cytoplasmic stability of the antisense molecules. Traditionally, in recombinant DNA technology, the successful transformation strategies apply careful selection from a wide range of established and well characterized vector–host systems. Easy commercial availability for a range of materials and tools adds to the efficiency of these processes. The cytoplasmic stability is achieved by chemical modification of the antisense nucleotides. Nevertheless, the microbial species used as platforms for genetic manipulations for biofuel production are not as well characterized (Bramucci *et al*., 2008; Ruhl *et al*., 2009) and much work needs to be done in this field.

The solution to the above problem comes in the form of nanoparticle–nucleotide conjugates. Such particles have been successful in transfection studies using mammalian systems (Brannon-Peppas and Blanchette, 2004; Davis and Cooper, 2007) and can be tried as transformation strategies in prokaryotic systems. Gold nanoparticle–nucleotide conjugates used by Rosi et al. (2006) showed good transfection efficiency and cytoplasmic stability. The need at present is for development of nanoparticle–nucleotide conjugates and optimization of conjugation and transformation protocols for use of nucleotide–nanoparticle conjugates for application in bacterial strains used for biofuel production.

Some researchers have tried applying metal-based nanomaterial–nucleotide conjugates for bacterial transformation. In this case, one needs to address issues due to potential metal toxicity and biocompatibity for the bacterial species used. Certain bacterial strains have an inherent ability to internalize and store metals as cytoplasmic nanoparticles (Mandal et al., 2006) and can offer an efficient platform. Critical investigations of the mechanisms applied for internalization of the metals by these bacteria would help in formulating strategies for developing optimized protocols for cytoplasmic delivery of nucleotide–nanoparticle conjugates to prokaryotic systems using metal-based nanoparticle–nucleotide conjugates.

The main problems in utilization of nanoparticle–nucleotide conjugates are the non-replicable nature of the conjugates and the lack of selectable property/marker on the conjugates (antibiotic selection, blue-white selection or complementation). Recombinant DNA technology has dealt with similar problems by using co-conjugation of target nucleotide sequences with a marker gene and similar strategies could be developed for nanoparticle–nucleotide conjugates in future.

Genetic engineering has also been applied for alteration of biomass composition for easier and more efficient transformation for biofuel production. The nanoscale wall structures within trees could be manipulated for easier and efficient utilization in biofuel production (Lucian and Rojas, 2009). The same authors proposed another approach to breakdown recalcitrant material. Typically, about 15–20% content from a lignocellulosic biomass is recalcitrant. For efficient fermentation and ethanol production, the lignin removal and breakdown of the recalcitrant biomass is critical. Nanocatalysts can be an innovative solution to the problem. Generally, in a catalytic process, the reactants are brought to the catalysts, but in this case, due to the nanosize, it is possible to deliver the catalytic nanomaterials to the site. Additionally, by careful design of the process, it would be possible to develop strategies for synthesis of water-soluble products and subsequent recovery of nanocatalysts. Last, biological treatment of the recalcitrant cellulose by applying designer nanocatalysts enzyme systems based on lignin-degrading enzymes and other enzymes such as expansins and hydrolases is another possibility (Lucia and Rojas, 2009).

Nanotechnology applications for fuel cells and hydrogen production

In considering bioenergy production, one has to consider liquid biofuels as well as energy generated as electrical current. Fuel cells or electrochemical systems are heavily utilizing nanotechnology for improved designs and higher efficiencies for electricity generation. The two types of bioelectrochemical (BEC) systems are microbial fuel cells (MFCs) and microbial electrolysis cells (MECs). The bioenergy produced by MFCs is in the form of electricity whereas MECs produce biohydrogen (Clauwaret et al., 2008). The electricity as well as the hydrogen generated is cleaner fuels compared to the fossil fuels. The more controlled nature of electrochemical reactions and higher combustion efficiency make these energy sources very promising and attractive technologies (Sherif *et al.*, 2005). Both systems rely upon the ability of bacterial species that are electrochemically able to use their oxidative metabolic pathways to catalyse and produce hydrogen from organic substrates (Pant *et al.*, 2010). The main barriers for commercial application of these systems for energy generation are the high costs of manufacturing the main components such as the electrodes, electrolytes, membranes and catalysts. Due to its ability of customising nanomaterial compositions for specific applications and power to conserve resources, nanotechnology is a very real and cost-effective solution to many of the problems stated above.

Microbial Fuel Cell (MFC)

Nanotechnology has several other applications that could directly impact the energy generated through electrochemical systems, for example systems for storage of electricity such as rechargeable batteries and super-capacitors.

Nanotechnology-enabled electrode systems

MFCs typically need a high content of platinum (Pt), roughly about 20–60 wt % Pt, as an anode catalyst in a thin layer form (Lee *et al.*, 2004). Recent progress has been made in minimizing this need for Pt to 0.1 mg Pt cm^{-2} to make the system cost effective. When the catalyst material is stable and uniformly dispersed on the electrode, the efficiency of the system improves. Carbon nanotubes have been the topic of heavy research due to their unique mechanical and electronic properties (Iijima, 1991; Dai *et al.*, 1996; Ebbesen *et al.*, 1996). CNTs have been explored for several energy-related applications such as carriers for catalysts (Planeix *et al.*, 1994), as a storage material for gases (Gadd *et al.*, 1997) and hydrogen (Dillon *et al.*, 1997; Lin, 2000) as nanoreactors for biocatalysis in energy generation. The CNTs can provide support to catalysts either by surface-attachment on outer walls of CNTs (Li *et al.*, 1997; Chen *et al.*, 2000; Zhang *et al.*, 2000) or they can be internalized in the hollow cores (Tsang *et al.*, 1994; Pradhan *et al.*, 1998; Rajesh *et al.*, 2000). While developing CNTs as supports for catalytic particles, anchoring them on the external walls of CNTs offers easier access for the reactants rather than the CNT-encapsulated catalysts. The major problem in developing CNT-supported catalysts is the inert nature of CNT surfaces, which makes it difficult to attach metal particles on the external walls (Fig. 7.9). The external surfaces of CNTs have been modified to introduce some anchoring/docking sites for the catalytic metal particles such as Pt (Liu *et al.*, 2002). Another issue for the process is that certain liquids require surface tensions of 100–200 mNm^{-1} to be able to deposit metal layers on the CNT external walls (Dujardin *et al.*, 1994). Researchers found a solution to this by functionalizing external walls of CNTs before depositing the metal layers. The outer walls of nanotubes are modified by oxidation using HNO$_3$ and then metals such as gold (Au) or Pt are layered on the oxidized CNT. Typically the CNT sample is ultra-sonicated in the presence of Pt and the oxidizing agent at high temperatures (120°C) (Figs 7.10, 7.11) for a couple of days (Liu *et al.*, 2002).

Unlike the single-wall carbon nanotubes (SWNT) used by Liu *et al.* (2002), Wang *et al.*(2004) synthesized multi-walled carbon nanotubes (MWNT) as support for catalysts for polymer electrolyte membrane fuel cells (PEMFC) and the study demonstrated almost 60%

reduction in the amount of catalyst compared to a traditional PEMFC. Another study by Rajalakshmi and Dhatartreyan (2008) compared Pt/nanotitania electrodes with a traditional Pt/C system to find out the catalytic and stability parameters. The Pt/nanotitania electrodes were superior with respect to several features such as catalytic activity, durability and thermal stability.

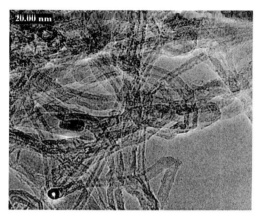

Fig. 7.9. TEM image of purified carbon nanotubes (CNTs) synthesized by chemical vapour deposition (CVD).

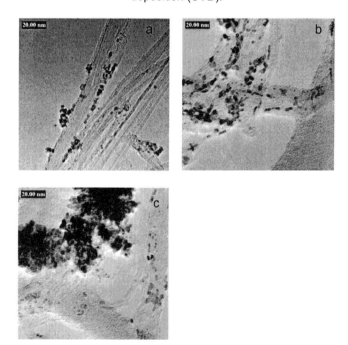

Fig. 7.10. TEM images of Pt coated on CNTs using Pt-electroless plating solutions of (a) pH8, (b) pH10.5 and (c) pH11. CNTs were treated by HNO_3 and subjected to the two-step sensitization procedure before Pt deposition.

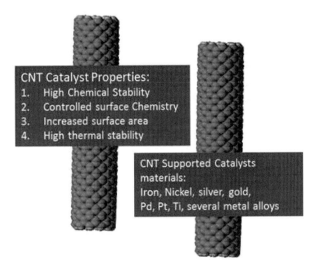

Fig.7.11. Functionalized carbon nanotubes as support for catalysts.

Traditionally, the MEC systems have utilized anode-respiring microorganisms to produce hydrogen at the anoxic cathode surfaces in the presence of applied external voltage. The hydrogen production traditionally used expensive Pt electrodes; nanotechnology provides a cost-effective solution in the form of Pt-free electrodes coated with carbon nanopowders. Figure 7.12 depicts cathode configurations in three different bioelectrochemical systems.

Nano-enabled membranes for fuel cells

In the case of MFCs, membranes play a significant role by improving hydrogen ion conductivity. At high temperatures, nanoscaled, hydrophilic inorganic materials such as lithium salts have been shown to improve the hydrogen ion conductivity. Nanoporous films were prepared by doping polybenzimidazoles with phosphoric acid to improve nanoporosity in order to increase membrane performance (Mecerreyes *et al.*, 2004). Conventional Naofin membranes were modified with Titania and tin oxide nanoparticles in order to achieve optimal efficiency at higher temperatures. Some of the proton and ethanol membrane cells have used silica membranes that were fabricated by depositing silica nanoparticles on poly(arylethersulfonate) for protein and ethanol fuel cells (Mango de Caravalho *et al.*, 2008).

Electrospun nanofibres for bioenergy

Electrospinning is a procedure that allows fabrication of nanostructures with controlled diameters. This is the only method currently available that can spin continuous fibres with nanometre diameters (<500–20 nm). Additionally, a variety of morphologies can be fabricated using this method. This is a versatile method to convert a wide range of polymer and ceramic materials into nanostructures of various morphologies. A wide range of the electrospun materials can be used for a variety of processes, but their applications as/in

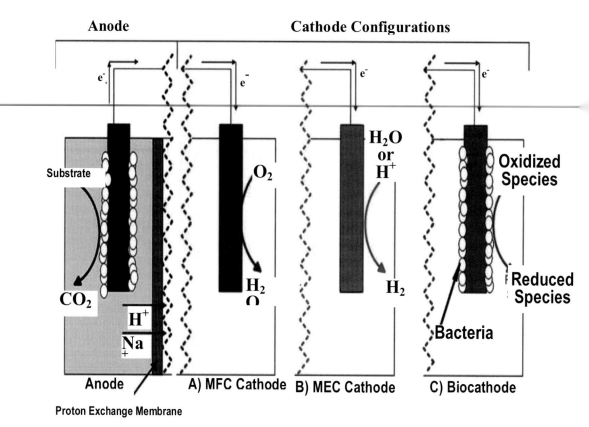

Fig. 7.12. Cathode configurations in A: microbial fuel cell, B: microbial electrochemical cell and C:biocathode with microbe-coatings (reprinted with permission from *The Canadian Journal of Chemical Engineering*, vol. 89, February 2011).

energy materials are significant. Due to the unique nanoscale structures and varied morphologies of the electrospun materials, bulk materials that are currently not applicable for energy applications have been suddenly transformed into something very desirable. These materials provide increased surface area and possess quantum-level effects (Laudenslager *et al.*, 2010).

Electrospinning is a fibre fabrication method practised since 1934. Recent advances in fluid dynamics allow electrospinning to predict diameters of the final fibres (Fridrikh *et al.*, 2003). Several excellent reviews about electrospun nanofibres are available (Hohman *et al.*, 2001; Huang *et al.*, 2003; Subbiah *et al.*, 2005; Ramakrishna *et al.*, 2006; Sigmund *et al.*, 2006).

In the electrospinning process, viscous fluids are forced through an electrically charged jet to form nanoparticles of controlled diameters. In this fabrication process, a pump moves a polymer or ceramic material solution through a syringe. The metallic needle is attached to a power supply, which charges the liquid droplet that forms at the end of needle as the solution flows through it. The electrically charged fluid turns into a thin jet, which gets pulled into a thin fibre towards a ground target and then collected. Various modifications of electrospinning procedures are practised currently for uniformity and orientation of the fibres (Deitzel *et al.*, 2001; Theron *et al.*, 2001; Li *et al.*, 2004; Sundarey *et al.*, 2004; Pan *et al.*, 2006; Yang *et al.*, 2007).

These electrospun materials have been developed for applications in electrodes, electrolytes and for hydrogen storage. The electrospun materials developed for electrodes have been mostly applied for lithium batteries and will not be discussed here. Detailed information of these applications could be found elsewhere in literature. However, since applications of electrospun materials for hydrogen storage align well with the focus of this chapter, they are discussed in the following section.

Electrospun materials for hydrogen storage

Hydrogen is the most clean-burning fuel due to its combustion process, but to develop a hydrogen-based economy, the biggest obstacle is hydrogen storage and transportation abilities. Transportation of hydrogen using compressed and liquid-based methods is neither practical nor a safe solution. Due to the low volumetric density of hydrogen, there is immediate need for economic methodologies that offer favourable energetics of hydrogen release and for novel materials with high hydrogen storage density (Orimo *et al.*, 2003).

Numerous procedures have been researched to include electrospun fibres to create porous carbon networks for hydrogen storage. Electrospun fibres of polymers such as polyacrylonitrile (PAN) are cross-linked and then activated by treatment with KOH to increase surface area. Studies of such activated and carbonized fibres demonstrated that the higher adsorption results due to formation of ultramicropores in these fibres (Im *et al.*, 2009).

Another process applied for increasing sorption-desorption rates is the catalysed hydrogen spill-over (Yagi *et al.*, 2008). In this process, hydrogen is first dissociated into elemental hydrogen and then transferred to a hydrogen supporting matrix. Im and co-workers (Im *et al.*, 2008) studied copper, vanadium, iron and magnesium as catalysts embedded into carbon fibres. As stated at the beginning of this section, electrospun nanofibres have applications in multiple fields. Table 7.5 below lists electrospun fibres that have applications as energy materials.

Nanotechnology for Harvesting, Separation and Storage of Biofuel Products

In the previous sections, applications of nanotechnology for conversion of biomass to bioenergy were discussed. This section presents applications of nanomaterials in harvesting, separation and storage of various bioenergy products such as liquid biofuels and gases such as biohydrogen and methane.

While creating sustainable biofuel production procedures, the two critical issues are effective conversion of biomass to biofuel and efficient separation and harvesting of the biofuel from the reaction mixtures. At present, the biofuels dominating the bioenergy sector are the short chain, aliphatic alcohols such as ethanol, butanol, iso-butanol and pentanol (Atsumi *et al.*, 2008b, 2010; Cann and Lio, 2008). Higher conversions of biomass to these alcohol-based biofuels somehow face a unique situation. Due to the cytotoxicity of the

Table 7.5. Various electrospun materials and their applications as energy materials (modified from Laudenslager et al., 2010).

Application	Structural component	Electrospun material	Reference
Fuel cells	Catalytic electrode	Nanowireselectrocatalysts of Pt and Pt/Rucarbonized fibres with Pt clusterspolycaprolactane membranes coated with Ag	Schechner et al., 2007; Li et al., 2008; Kim et al., 2009
	Membrane	Nafion-impregnated PVDF Sulfonated poly(etherketone)	Choi et al., 2008 Li et al., 2006
	Enzyme immobilization	Polystyrene nanofibers carrying α-chymotrypsin	Jia et al., 2002
Hydrogen storage	Highly porous layer for adsorption	Oxides of Fe, Mg, Cu in carbonized PAN fibers; Graphite nanofibres from PVDF; PAN/PMMA-derived carbon tubes	Zussman et al., 2006; Hong et al., 2007; Im et al., 2008
	Electrode	Carbonized solutions of PAN, polymers Ni-embedded carbon nanofibres	Im et al., 2008; Teng et al., 2010

short-chain alcohols, the processes are designed for dilute fermentation, which impacts the downstream processing yields, making the process less sustainable and ineffective with respect to operational costs (Straathof, 2003; Schugerl and Hubbuch, 2005). Energy costs for dewatering of the biomass and distillation purification of liquid biofuels typically add up to 7–10% of the total production costs (Galbe et al., 2007).

The need for selective, novel and efficient adsorbents for separation of biofuels is highly desirable. The customisable nature of the nanomaterials provides a very attractive and reasonable solution to synthesis of such adsorbents, though nanoporous carbon materials have been developed (Vinu et al., 2006).

Zeolites

Zeolites are microporous, aluminosilicate minerals commonly used as adsorbents in commercial processes. They have a 3-D silicate structure and depending on various framework types, zeolites can have very open porous structures. The crystallographic structure formed by tetrahedras of AlO_4 and SiO_4 are the basic building blocks for various zeolite structures, such as zeolites A and X, the most common commercial adsorbents. Typically, due to the alumina-content, these structures have cationic surfaces, which can be functionalized for a customized pore size of the particles which are on nanoscale (4×10^{-10} m)(http://www.grace.com/EngineeredMaterials/MaterialSciences/Zeolites/ZeoliteStructure.aspx). The cage structure, precise control over pore size and charge distribution make zeolites a very attractive option as an adsorbent for biofuel separation processes (Fig. 7.13). The ethanol produced in the bioenergy sector typically has residual moisture content (~ 4–6%). Removal of this water by distillation is an energy-intensive process (Cardona and Sanchez Toro, 2006), but the hydrophilic zeolites offer a low-energy intensive and cost-effective option with improved ability in selectively removing water from fuels (Wu et al., 2009).

Zeolites have a double advantage when applied to biodiesel production. In the case of biodiesel fuels, the presence of trace amounts of moisture is undesirable as it may result in production of glycerin as a by-product (Kusdiana and Saka, 2004). The removal of this trace amount of moisture improves the efficiency of the production process. Additionally, as biodiesel is a combination of methyl esters of fatty acids and water is a by-product of this esterification reaction, effective removal of water from the conversion reactions drives the reaction towards esterification leading to higher efficiency of biodiesel production (Lucena et al., 2008).

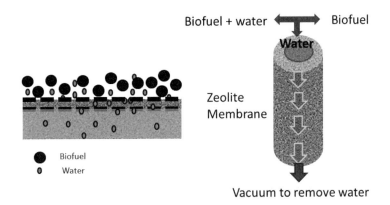

Fig. 7.13. Zeolite membranes for separation of biofuel products (based onhttp://news.mongabay.com/bioenergy/2007/07/mitsui-engineering-to-use-zeolite.html).

When zeolite particles are synthesized with higher Si/Al ratios, they develop hydrophobic features and have been used as selective adsorbents for separating biofuels from aqueous solutions (Carton et al., 1998; Oudshoorn et al., 2009; Saravanan et al., 2010). The high Si/Al composition allows these nanoparticles to have selective affinity for more hydrophobic fuels (Vankelecom et al., 1995).Numerous studies have indicated the use of membranes fabricated with zeolites with a higher content of Si. These membranes, known as MF-1 membranes, have desirable qualities not found in traditional polymer membranes. The selectivity for traditional polymer membranes for ethanol over water range from 4 to 11 whereas for the zeolite membranes, they range between 10 and 100 (Nagase et al., 1991a, b; Hickey et al., 1992). Additionally, these lend chemical stability as they do not swell in the presence of organic solvent (Bowen et al., 2004) and have high flux values for permeation compared to polymeric membranes.

Nanofarming for biofuel production using algal culture is a very innovative approach currently being explored (Lin et al., 2008). In this method developed at the Ames Laboratory (Gibson, 2009), mesoporous nanoparticles were used for continuous harvesting of biofuels from algal cultures without cell lysis (Fig. 7.14). These particles act as absorbent sponge-like material to selectively remove the lipids from the algal cell membranes, eliminating the need for cell lysis. This opens up possibilities for *in situ* transesterification of calcium/strontium oxide nanoparticles functionalized with catalytic properties (Liu et al., 2008).

Nanotechnology Applications for Hydrogen Storage

In 2005, the US Department of Energy (DOE) came out with a road map for developing hydrogen storage materials (DOE Roadmap, 2005). Current procedures to store hydrogen as pressurized gas mandate bulky and heavy pressure vessels in addition to the need for very low temperatures during storage (Serrano *et al.*, 2009). These procedures require a high amount of energy, rendering the entire process energy-ineffective. Improved systems for hydrogen storage and transportation are one of the most critical aspects for overcoming the current shortcomings.

Nanomaterials help meet the need for high surface area, optimized pore size and shape, high storage capacity, controlled desorption kinetics and the ability to perform at room temperatures. Carbon nanotubes (Dresselhaus *et al.*, 1999; Pinkerton *et al.*, 2000; Züttel *et al.*, 2002; Hao *et al.*, 2003; Schimmel *et al.*, 2003), graphite nanoparticles (RSC, 2005), carbon nanofibres and nanocatalyst-loaded mesoporous matrices (Jung *et al.*, 2006; Ramchandrana *et al.*, 2007; Sheppard and Buckley, 2008) are some of the materials being developed currently for hydrogen storage applications.

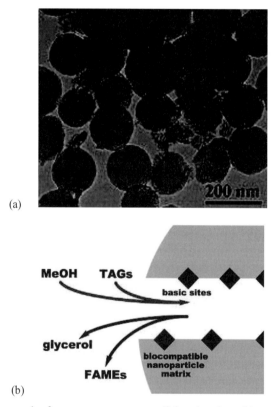

Fig 7.14.(a) A micrograph of mesoporous nanoparticles developed by researchers at Ames Laboratory to harvest biofuel oils from algae; (b) Generic scheme for the proposed function of mesoporous nanoparticles as adsorbents in nanofarming. (Reprinted with permission from *The Canadian Journal of Chemical Engineering*, vol. 89, February 2011 published by the Canadian Society for Chemical Engineering.)

When one considers the actual mechanism involved in hydrogen storage, chemisorption and physiorption are the two most common techniques developed for sustainable hydrogen storage systems. Nevertheless, the two mechanisms require completely opposite conditions for carrying out hydrogen storage. Chemisorption involves tight binding of hydrogen in addition to a need for above ambient temperatures (excess of 400 K) for operation. On the other hand, physiorption needs very low temperatures for operation (less than 100 K). Due to the inherent ability of nanotechnology to offer customized material architecture, several novel materials such as activated carbon, mesoporous matrices and metal hydrides have been developed as discussed in the following section.

Metal hydrides have been applied for hydrogen storage for a while and are the most cost-effective and safest mode amongst the traditional methods (pressurized liquid gas stored at cryogenic temperatures).While considering the efficiency of metal hydrides for hydrogen storage, the storage capacity, ability to offer multiple reversible storage cycles and kinetics of hydrogen adsorption/desorption are the parameters of critical significance. Hydrogen atoms are stored in the metal lattice of metal hydrides and can achieve efficiencies close to that of liquid hydrogen storage. The metal alloys currently explored include ZrV_2, Mg_2NI or $LaNi_5$ (Sherif *et al.*, 2005). The hydrogen storage ability of these hydrides is a direct function of the alloy structure. Nanostructured alloys such as Mg-Ni are considered as the potential optimal alloys for hydrogen storage in the near future (RSC, 2005). These are light-weight, low-cost and the interiors of these nanomaterials allow high adsorption of hydrogen without a pretreatment or activation. In an effort to improve the metal hydride performance, doping the metal hydrides with Ti nanoparticles improves the potential for metal hydrides to become a sustainable and low-cost system for storage and transport of hydrogen. Another novel technique combines porous materials with metal nanoparticles as demonstrated by Serrano *et al.* (2009). They activated the porous support and then loaded Pd nanoparticles on the surface of the activated porous material. Before loading the nanoparticles on the porous support, the Pt nanoparticles were functionalized and introduced onto the silica support using co-polymerization with tetra-alkoxysilanes in the presence of a surfactant. The surfactant treatment provided highly dispersed nanoparticles and was found to offer enhanced ability for hydrogen adsorption.

Physisorption is the other common mechanism explored for hydrogen storage but needs low temperatures for efficient adsorption to occur. Zeolites and mesoporous matrices have been researched for hydrogen storage at room temperatures as well as cryogenic temperatures. Silica derivatives have offered a good solution in this area. Organic groups incorporated in the silica walls or doping with metal oxides have led to increased hydrogen storage capacity (Jung *et al.*, 2006; Ramchandrana *et al.*, 2007; Sheppard and Buckley, 2008). Silica-derivatives have been synthesized based on incorporation of biomolecules such as lecithin to create novel silica architectures as seen with MCM-41. The lecithin addition triggered formation of a hexagonal arrangement of the MCM-41 particles with a completely novel architecture with concentric pores arranged hexagonally (García Martínez *et al.*, 2007).

Carbon nanostructures offer distinct advantages over current activated carbon materials. The conventional materials need cryogenic temperatures, and have wide pore size distributions. Most of the tested materials presented pore sizes higher than those required for hydrogen atoms or molecules (Schwartz and Amankwah,1993; Dillon and

Heben, 2001). Activated carbon nanostructures offer distinct advantages over traditional carbon materials for hydrogen storage applications. These offer controlled pore size distribution and the ability to customize appropriate surface chemistry in addition to tremendous increased surface area – all these features make these structures highly efficient for hydrogen storage. Single-wall (SWNT) as well as multi-wall (MWNT) carbon nanotubes (Züttel et al., 2002; Schimmel et al., 2003) and graphite nanofibres (Dresselhaus et al., 1999; Pinkerton et al., 2000; Hao et al., 2003) have been in use since 1998 (Ye et al., 1999). Hollow nanotubes were used as a hydrogenation agent by Nilsson group (Nikitin et al., 2008) where they report hydrogen storage capacities of 7 wt%. They further state that these capacities are a function of the carbon nanotube diameter and it is possible to achieve 100% hydrogenation using nanotubes with 2 nm diameter.

Ancillary Nanotechnology Applications

Biomass applications for bioenergy are established and research work is in progress to optimize the procedures for efficient, cost-effective, sustainable and environmentally friendly protocols. Nanotechnology has contributed towards many of the above areas as evident from the previous sections. Beyond actual contributions to the processes of conversion, extraction and storage, nanotechnology has the potential to impact some ancillary areas of bioenergy production such as transportation and handling of the biomass and bioenergy products. This section elaborates on those specific aspects of nanotechnology contributions to the field of bioenergy.

Current Problems

Currently there are several problems faced in transport, handling, conversion and recovery of heat from the waste areas in a biomass-based bioenergy production facility.

1. Transport: biomass used in energy generation has a high moisture content. This leads to problems of biomass freezing during transportation. Additionally, the container weight and size pose problems as the fuel density for biomass is less than that of competing fuels such as coal. The third problem arises as a result of variation in the rate of biomass delivery to the plant as a function of the size of the plant and the nature of the biomass.
2. Handling: the behaviour of the biomass in the boiler silo is a major factor as biomass may stick to the walls of the silo and subsequently there may be accumulation of biomass above the stuck layer. Hammer mills as well as the conveyer systems also experience major damage during biomass processing.
3. Biomass combustion: there are ash deposits as well as corrosion of the inside walls of the boiler, which mandate periodic cleaning and at times replacement of the boiler (Drbal et al., 1995). Additionally, at times, deactivation of the catalyst may result from biomass combustion (Strege et al., 2007).
4. Recovery of heat from waste: for power plants using steam turbines or flue gas condensation, the temperatures of the steam condensate and flue gas exhaust are higher than the ambient air temperature to which they are released. Ability to trap that energy stored in the temperature difference would add to overall energy efficiency of the plant (Drbal et al., 1995).

Nanotechnology Solutions to the Above Problems

As a solution to the major problems stated above, nanotechnology-based technologies are being explored, specifically as nanolayers, nanocomposites and nanostructured materials. All of these materials have at least one dimension in nanometres and could be one, two or three dimensional which would be referred to as nanostructured surfaces, tubes and particles, respectively (Ajayan *et al.*, 2003).

Nanotechnology-enabled Coatings

These are preferably monolayer coatings with a size from ~100 nm up to a few micrometres. The nanostructured coatings may be of varying thickness but gain their specific properties due to the nanostructures in the coating materials. The application methods for these coatings range from simple methods such as thermal spraying (plasma spraying) to more elaborate methods such as sputtering or molecular beam epitaxy. The types of the nanomaterials range from nanopowders (Wang *et al.*, 2000; Chen *et al.*, 2002; Chawla *et al.*, 2008) to nanocrystalline matrices (Branagan *et al.*, 2005; Mishra, 2009; Dosbaeva *et al.*, 2010). Based on their function, these are divided into mechanical wear-resistant coatings, anti-friction coatings, anti-icing coatings and anti-corrosion coatings.

Mechanical wear-resistant coatings have been commercially sold by several companies across the world. UDiamond is a coating material made up of diamond nanoparticles and used in electrolytic coatings. These coatings have been shown to improve hardness and wear resistance (Kalyansundaram and Molian, 2008). Another brand name is BALINIT, which offers a series of coatings for cutting and machining tools. One product from this company called FUTURA NANO is sold as an abrasion- and wear-resistant coating material (Kramb, 2011).

Chawla and co-workers (Chawla *et al.*, 2008) compared and listed a number of nanostructured aluminium and titanium compounds for coating manufacturing where they suggest a superior ability of plasma spraying over thermal spraying procedures for application of the coatings. Titanium compounds/alloys is yet another class of coating materials currently in use (Mishra, 2009; Dosbaeva *et al.*, 2010).

Regarding the ability to serve as anti-friction coatings, chromium nitride (Yamamoto *et al.*, 2006)is currently used. The layer is 20–100 nm thick with a friction coefficient of 0.2. In the case of tungsten carbide, the layer has a thickness of only a few nanometres (Liu *et al.*, 2007; Yamamoto *et al.*, 2007) with good anti-friction properties.

For boilers processing biomass, corrosion due to elements such as chlorine and sulfur are major concerns. This corrosion leads to expensive maintenance protocols and at times replacement of corroded parts. An excellent review by Gandy (2007) presents current developments in the field of corrosion-resistant nanomaterials. Due to their ability to form oxide scales, chromium and aluminium compounds have better corrosion-resistant properties than the conventional coatings. Nanocrystalline TiAlCrSiYn coatings (Dosbaeva *et al.*, 2010) and FeCrNiAl coatings (Liu *et al.*,1999) are some current commercial products available for anti-corrosion applications. Nanostructured coatings with steel alloys have been developed for boilers processing woody biomass and have demonstrated protection against chlorine corrosion (Branagan *et al.*, 2005).

Nanotechnology and Biofuel Additives

Solid nanoparticles have demonstrated an impact on the actual process of biofuel production, either as catalytic particles or as carriers for catalytic enzymes. Nevertheless, there are significant applications of liquid nanoparticles or nanodroplets (www. Economist,com/node/16271415). These liquid additives usually have surface active properties and help improve the fuel efficiency by monolayer coating on the mechanical parts in contact with the fuel. In the earlier part in this chapter, we have seen applications of nanoemulsions similar to these liquid nanomaterials in improving heterogeneous catalysis for biomass conversions. These so-called additive-based nanoemulsions result from the interaction of surfactants added to the fuels (fossil and biofuels as well) and trace amounts of moisture content from the fuels (Pascal et al., 2008). For conventional fuels, there is a trade-off between the formation of soot and levels of NO formed. By applying these nanoemulsions, that mandate is no longer valid and both the goals are achieved simultaneously resulting in complete fuel combustion, low fuel emissions and higher fuel efficiency. The authors Strey et al.(2004) hold a US patent for this application of additive-enhanced fuel performance. Their theory suggested for this emulsion-based fuel efficiency is that as the first step, the fuels (diesels and biodiesels) readily dissolve fatty acids (oleic acid) and nitrogen-containing compounds (amines). In the second step, these then interact with trace amounts of water from the emulsion without the need for stirring or sonication, typically required to solubilized water in a hydrophobic/non-polar medium. These nanodroplets of water in fact stabilize the interaction of water and hydrophobic factors creating a situation similar to a liquid sponge.

Another application of nanotechnology for biofuels is the use of solvent-resistant nanofiltration (SRNF) in processing petroleum for production of cleaner burning fuel. Traditionally, gasoline may contain components such as organometallics (copper and iron compounds), sulfur compounds (mercaptans and thiophenes) and certain aromatic impurities that may lead to fouling of engine parts during fuel combustion. The fuel in direct contact with the mechanical parts of the engine may leave a deposit in the form of thin films after evaporation. This residue can then have a significant build up in the parts and restrict air and fuel lines leading into the combustion cylinder. This eventually leads to incomplete combustion, which results in reduced power, poor fuel efficiencies and increased emissions. As a novel approach, Tarleton and Low (2008) explored the possibility of using nanomembranes formed from polydimethylsiloxane (PDMS) and polyacrylonitrile (PAN) to construct several membrane structures and use these as nanofilters for simulated fuel mixtures. These workers used fuel blends artificially enriched with copper and iron and followed the filtrates with elemental analysis. The filtrates were significantly low in the concentrations of these metals indicating effective filtration of the same due to the nanofilters. The nanofilters were also effective in removing oxygenates added to fuels.

Nanotechnology for Fossil Fuels

This may be the most important but the least glamorous application of nanotechnology as applied to fossil fuels. Conventional recovery processes typically leave around 60% of the oil in the ground (Montgomery and Carla, 1997). Nanotechnology allows one to track that fraction of fossil fuel that may otherwise remain in the ground untapped. This is termed as the nanotechnology-enabled 'information intensive' extraction processes. One of the recovery processes involves steam injection in the fields to mobilize oil. In one study, Paul (2001) reported use of fibre-optic thermometers with ultra-miniaturized flow sensors to track the flow of oil and steam in the recovery process. They demonstrated that with this type of monitoring, less injected steam was required and they were able to target it more effectively as well. This adaptation allowed energy savings close to 20% for the recovery process.

Conclusion

Nanomaterials and nanotechnology can play a major role in thermochemical as well as biochemical routes for biofuel production. Nanotechnology can improve all the steps for the biochemical pathways such as crude biomass pretreatment, conversion, bioenergy product separation and extraction and storage. As regards the thermochemical process, nanotechnology can improve the efficiency by providing novel nanomaterials for catalysis, immobilization matrices, product separation/recovery and storage of bioenergy products. For the hydrogen-based processes, nanotechnology provides novel nanophotocatalysts and higher-efficiency materials for microbial fuel cell components such as electrodes, membranes and electrolytes. Novel electrospun nanomaterials offer unique advantages for a variety of aspects such as matrices for hydrogen storage, electrode materials and catalysts. Since nanomaterials offer high surface areas with respect to the traditionally used bulk materials, nano-enabled technologies allow higher efficiencies while allowing resource conservation, making them highly cost-effective and sustainable. Due to the nascent nature of this field, a large amount of research is required to optimize procedures. Various forms of carbon-based nanomaterials have great potential for numerous applications for this field. At present, the efforts are dedicated to nanotubes and nanofibres. The other morphologies such as nanoribbons or plated should be studied as well. Intense efforts should be dedicated for development of new nanomaterials and processes for fast, non-invasive characterization of the materials, especially for the catalyst components in these processes. Inspite of the infancy of this field of technology and the apparent shortcomings, there is no denying the potential and advantages of this approach. The technology is definitely a highly promising solution for the bioenergy field.

References

Ajayan, P., Schadles, L. and Braun, P. (eds) (2003) *Nanocomposite Science and Technology*. Wiley-VCH.
Amalvy, I., Unali, G.-F., Li,Y., Granger-Bevan, S., Armes, A., Binks, B., Rodrigues, J. and Whitby, C. (2004)Synthesis of sterically stabilized polystyrene latex particles using cationic block copolymers and macromonomers and their application as stimulus-responsive particulate emulsifiers for oil-in-water emulsions. *Langmuir* 20, 14345–14354.
Ashok, G., Barry, D.M., Sharon, M.K., Nicholas, C.P., Olaf, J.R. and Peter, J.H. (2009) Optical spectroscopic methods for probing the conformational stability of immobilised enzymes. *Chemical Physic Chem* 10(9–10), 1492–1499.
Ansari, S.A. and Husain, Q. (2011) Potential applications of enzymes immobilised on/in nano materials.*Biotechnology Advances* 21, 223–204.

Ashok, G., Barry, D.M., Sharon, M.K., Nicholas, C.P., Olaf, J.R. and Peter, J.H. (2009) Optical spectroscopic methods for probing the conformational stability of immobilised enzymes. *Chemical Physic Chem* 10(9–10), 1492–1499.

Asuri, P., Karajanagi, S.S., Yang, H.C., Yim, T.J., Kane, R.S. and Dordick, J.S. (2006) Increasing protein stability through control of the nanoscale environment. *Langmuir* 22(13), 5833–5836.

Atsumi, S. and Liao, J.C. (2008) Metabolic engineering for advanced biofuels production from *Escherichia coli. Current Opinions in Biotechnology* 19(5), 414–419.

Atsumi, S., Cann, A.F., Connor, M.R., Shen, C.R., Smith, K.M., Brynildsen, M.P., Chou, K.J.Y., Hanai, T. and Liao, J.C. (2008a) Metabolic engineering of *Escherichia coli* for 1-butanol production. *Metabolic Engineering* 10(6), 305–311.

Atsumi, S., Hanai, T. and Liao, J.C. (2008b) Non-fermentative pathways for synthesis of branched-chain higher alcohols as biofuels. *Nature* 451, 86–89.

Atsumi, S., Wu, T.Y., Eckl, E.M., Hawkins, S.D., Buelter, T. and Liao, J.C. (2010) Engineering the isobutanol biosynthetic pathway in *Escherichia coli* by comparison of three aldehyde reductase/alcohol dehydrogenase genes. *Applied Microbiology and Biotechnology* 85(3), 651–657.

Bahnemann, D. (2004) Photocatalytic water treatment: solar energy applications. *Solar Energy* 77, 445–459.

Besteman, K., Lee, J.O., Wiertz, F.G.M., Heering, H.A. and Dekker, C. (2003). Enzyme-coated carbon nanotubes as single-molecule biosensors. *Nano Letters* 3, 727–730.

Binks, B.P. (2002) Particles as surfactants – similarities and differences. *Current Opinion in Colloid & Interface Science* 7, 21–41.

Bowen, T.C., Wyss, J.C., Noble, R.D. and Falconer, J.L. (2004) Measurements of diffusion through a zeolite membrane using isotopic-transient evaporation. *Microporous Mesoporous Materials* 71(1–3), 199–210.

Boz, N., Degirmenbasi, N. and Kalyon, D.M. (2009) Conversion of biomass to fuel: transesterification of vegetable oil to biodiesel using KF loaded nano–[Gamma]–Al2O3 as catalyst. *Applied Catalysis: B Environmental* 89, 590–596.

Bramucci, M., Nagarajan, V., Sedkova, N. and Singh, M. (2008) Solvent tolerant microorganisms and methods of isolation. US Patent 7659104.

Branagan, D., Breitsameter, M., Meacham, B. and Belashchenko, V. (2005) High-performance nanoscale composite coatings for boiler applications. *Journal of Thermal Spray Technology* 14,196.

Brannon-Peppas, L. and Blanchette, J.O. (2004) Nanoparticle and targeted systems for cancer therapy. *Advances in Drug Delivery Reviews* 56(11), 1649–1659.

Cann, A.F. and Liao, J.C. (2008) Production of 2-methyl-1-butanol in engineered *Escherichia coli. Applied Microbiology Biotechnology* 81(1), 89–98.

Cardona Alzate, C.A. and Sanchez Toro, O.J. (2006) Energy consumption analysis of integrated flowsheets for production of fuel ethanol from lignocellulosic biomass. *Energy* 31(13), 2111–2123.

Carp, O., Huisman, C.L. and Reller, A. (2004) Photoinduced reactivity of titanium dioxide. *Progress in Solid State Chemistry* 32, 33–177.

Carton, A., Benito, G.G., Rey, J.A. and de la Fuente, M. (1998) Selection of adsorbents to be used in an ethanol fermentation process –adsorption isotherms and kinetics. *Bioresource Technology* 66(1), 75–78.

Chawla, V., Sidhu, B., Puri, D. and Prakash, S. (2008) State of art: Plasma sprayed nanostructured coatings: A review. *Materials Forum*, 32.

Chen, G., Xu, C., Mao, Z., Li, Y., Zhu, J., Ci, L., Wei, B., Liang, J. and Wu, D.(2000) Preparation and characterisation of platinum-based electrocatalysts on multiwalled carbon nanotubes for proton exchange membrane fuel cells. *Chinese Science Bulletin* 45, 134.

Chen, H., Zhang, Y. and Ding, C. (2002) Tribological properties of nanostructured zirconia.*Wear* 253 (7), 885–888.

Choi, S., Fu, Y., Ahn, Y., Jo, S. and Manthiram, A. (2008) Nafion-impregnated electrospun polyvinylidene fluoride composite membranes for direct methanol fuel cells. *Journal of Power Sources* 180, 167.

Clauwaert, P., Aelterman, T., Pham, L., De Schamphelaire, Carballa,M., Rabaey, K. and Verstraete, W. (2008) Minimizing Losses in Bio-Electrochemical Systems: The Road to Applications. *Applied Microbiology and Biotechnology* 79(6), 901–913

Crossley, S., Faria, J., Shen, M. and Resasco, D. (2010) Solid nanoparticles that catalyze biofuel upgrade reactions at the water/oil interface. *Science* 327(5961), 68.

Crumbliss, A.L., Perine, S.C., Stonehuerner, J., Tubergen, K.R., Zhao, J. and Henkens,R.W. (1992)Colloidal gold as a biocompatible immobilisation matrix suitable for the fabrication of enzyme electrodes by electrodeposition. *Biotechnology Bioengineering* 40, 483–490.

Cruz, J.C., Pfromm, P.H., Tomich, J.M. and Rezac, M.E. (2010) Conformational changes and catalytic competency of hydrolases adsorbing on fumed silica nanoparticles: I. Tertiary structure. *Colloids and Surfaces B: Biointerfaces* 79(1), 97–104.

Dai, H., Hafner, J.H., Rinzler, A.G., Colbert, D.T. and Smalley, R. (1996) Nanotubes as nanoprobes in scanning probe microscopy. *Nature* 384, 147.

Davis, P.B. and Cooper, M.J. (2007) Vectors for airway gene delivery. *American Association of Pharmaceutical Scientists Journal* 9(1), E11–E17.

Deitzel, J., Kleinmeyer, J., Harris, D. and Beck Tan, N. (2001) The effect of processing variables on the morphology of electrospun nanofibers and textiles. *Polymer* 42, 261.

Department of Energy in US Hydrogen Storage Technologies Roadmap (2005) Available at: http://www1.eere.energy.gov/vehiclesandfuels/pdfs/program/hydrogen_storage_roadmap.pdf.

Dillon, A.C. and Heben, M.J. (2001) Hydrogen storage using carbon adsorbents: past, present and future. *Applied Physics A* 72, 133–142.

Dillon, A.C., Jones, K.M., Bekkedahl, T.A., Kinag, C.H., Bethune, D.S. and Heben, M.J. (1997) Hydrogen storage in carbon nanotubes. *Nature* 386, 377–379.

Dinsmore, A., Hsu, M., Nikolaides, M., Marquez, M., Bausch, A. and Weitz, D. (2002) Colloidosomes: Selectively Permeable Capsules Composed of Colloidal Particles. *Science* 298(5595), 1006.

Dorado, M.P., Ballesteros, E., Lopez, F.J. and Mittelbach, M. (2004) Optimisation of alkali-catalyzed transesterification of *Brassica carinata* oil for biodiesel production. *Energy Fuels* 18, 77–83.

Dosbaeva, G., Veldhuis, S., Yamamoto, K., Wilkinson, D., Beake, B., Jenkins, N., Elfizy, A. and Fox-Rabinovich, G. (2010) Oxide scales formation in nano-crystalline TiAlCrSiYN PVD coatings at elevated temperature. *International Journal of Refractory Metals and Hard Materials* 28(1), 133.

Drbal, L., Westra, K., Boston, P. and Erikson, B. (eds) (1995) *Power Plant Engineering*, 34–37.

Dresselhaus, M.S., Williams, K.A. and Eklund, P.C. (1999) Hydrogen adsorption in carbon materials. *MRS Bulletin* 24, 45–50.

Dujardin, E., Ebbesen, T.W., Hiura, H. and Tanigaki, K. (1994) Capillarity and wetting of carbon nanotubes. *Science* 265, 1850–52.

Ebbesen, T.W., Lezec, H.J., Hiura, H., Bennett, J.W., Ghaemi, H.F. and Thio, T. (1996) Electrical conductivity of individual carbon nanotubes. *Nature* 382, 54–59.

Encinar, J.M., Gonzalez, J.F., Rodriguez, J.J. and Tejedor, A. (2002) Biodiesel fuels from vegetable oils: transesterification of *Cynara cardunculus* L. oils with ethanol. *Energy Fuels* 16, 443–445.

Fecht, H.J. (1996) In: Edelstein, A.S. and Cammarata, R.C. (eds) *Nanomaterials: Synthesis, Properties, and Applications*. Institute of Physics, Philadelphia, Pennsylvania, USA.

Fischer, N.O., McIntosh, C.M., Simard, J.M. and Rotello, V.M. (2002) Inhibition of chymotrypsin through surface binding using nanoparticle-based receptors. *Proceedings of National Academy of Sciences USA* 99(8), 5018–5023.

Fortman, J.L., Chhabra, S., Mukhopadhyay, A., Chou, H., Lee, T.S., Steen, E. and Keasling, J.D. (2008) Biofuel alternatives to ethanol: pumping the microbial well. *Trends in Biotechnology* 26(7), 375–381.

Fridrikh, S., Yu, J., Brenner, M. and Rutledge, G. (2003) Controlling the Fiber Diameter during Electrospinning. *Physics Review Letters* 90, 144502.

Fujishima, A., Rao, T. and Tryk, D. (2000) Titanium dioxide photocatalysis. *Journal of Photochemistry and Photobiology, C* 1.

Gadd, G.E., Blackford, M., Moricca, S., Webb, N., Evans, P.J., Smith, A.M., Jacobsen, G., Leung, S., Day, A. and Hua, Q. (1997) The World's smallest gas cylinders? *Science* 277, 933–936.

Galbe, M., Sassner, P., Wingren, A. and Zacchi, G. (2007) Process engineering economics of bioethanol production. *Advances in Biochemical Engineering Biotechnology* 108, 303–327.

Gandy, D. (2007) *Program on technology innovation: State of knowledge review of nanostructure coatings for boiler tube application*. Technical report, Electric Power Research Institute.

GarcíaMartínez, J., Domínguez, S. and Brugarolas, P. (2007) Ordered circular mesoporosity induced by phospholipids. *Microporous Mesoporous Materials* 100, 63–69.

Gibson, K. (2009) Nanofarming technology harvest biofuel oils without harming algae. News Release, US Department of Energy, Ames Laboratory, Ames, Iowa, USA.

Haas, M.J. (2005) Improving the economics of biodiesel production through the use of low value lipids as feedstocks: vegetable oil soapstock. *Fuel Process Technology* 86, 1087.

Hadjipanayis, G.C. and Siegel, R.W. (eds) (1994) *Nanophase Materials: Synthesis, Properties, Applications*. Kluwer, London.

Hao, D.H., Zhu, H.W., Zhang, X.F., Li, Y.H., Xu, C.L., Mao, Z.Q.*et al*. (2003) Electrochemical hydrogen storage of aligned multi–walled carbon nanotubes. *Chinese ScienceBulletin*48,538–542.

Herrmann, J.M. (1999) Heterogeneous photocatalysis: fundamentals and applications to the removal of various types of aqueous pollutants. *Catalysis Today*53, 115–129.

Hickey, P.J., Juricic, F.P. and Slater, C.S. (1992) The effect of process parameters on the pervaporation of alcohols through organophilic membranes. *Separation Science Technology* 27(7), 843–861.

Hohn, K., Wang, D.H., Pena, L., Ikenberry, M. and Boyle, D. (2009) Acid functionalised nanoparticles for hydrolysis of lignocellulosic feedstocks. *American Society of Agricultural and Biological Engineers* 18, 230–234.

Hohman, M., Shin, M., Rutledge, G. and Brenner, M. (2001) Electrospinning and electrically forced jets. II. Applications. *Physics of Fluids*13, 2221.

Hong, S., Kim, D., Jo, S., Kim, D., Chin, B. and Lee, D. (2007) Graphite nanofibers prepared from catalytic graphitization of electrospun poly(vinylidene fluoride) nanofibres and their hydrogen storage capacity. *Catalysis Today* 120, 413.

Hongfei, J., Guangyu, Z., Bradley, V., Woraphon, K., Darrell, H.R. and Ping, W. (2002) Enzyme-carrying polymeric nanofibers prepared via electrospinning for use as unique biocatalysts. *Biotechnology Progress* 18(5), 1027–1032.

Huang, Z., Zhang, Y., Kotaki, M. and Ramakrishna, M. (2003) A review on polymer nanofibers by electrospinning and their applications in nanocomposites. *Composites Science and Technology* 63, 2223.

Huber, G. and Dumesic, J. (2006) An overview of aqueous-phase catalytic processes for production of hydrogen and alkanes in a biorefinery. *Catalysis Today* 111(1–2), 119.

Iijima, S. (1991) Helical microtubules of graphitic carbon. *Nature* 354, 56–58.

Im, J., Park, S., Kim, T. and Lee, Y. (2009) The metal–carbon–fluorine system for improving hydrogen storage by using metal and fluorine with different levels of electronegativity. *International Journal of Hydrogen Energy* 34, 1423.

Iwasaki, W. (2003) A consideration of the economic efficiency of hydrogen production from biomass. *International Journal of Hydrogen Energy* 28, 939.

Jeevanandam, P. and Klabunde, K.J. (2002) A study on adsorption of surfactant molecules on magnesium oxide nanocrystals prepared by an aerogel route. *Langmuir* 18, 5309–5311.

Jennifer, K. and Peter, V. (2006) Nanotechnology in agriculture and food production. *Project on Emerging Nanotechnologies* 4.

Jia, H., Zhu, G., Vugrinovich, B., Kataphinan, W., Reneker, D.H. and Wang, P. (2002) Enzyme-carrying polymeric nanofibers prepared via electrospinning for use as unique biocatalysts. *Biotechnology Progress* 18, 1027–1032.

Johnson, P.A., Park, H.J. and McConnell, J.T. (2009) *Enzyme Immobilisation on Magnetic Nanoparticles for Cellulose Hydrolysis*. Cited by Khanal,S.K. *et al.* (2010) in: *Bioenergy and Biofuel from Biowastes and Biomass*. Sponsored by Bioenergy and Biofuel Task Committee of the Environmental Council, Environmental and Water Resources Institute (EWRI)of American Society of Civil Engineers, Reston, Virginia: ASCE, 978–0–7844–1089–9, 2010, 505 pp.

Jung, J.H., Han, W.S., Rim, J.A., Lee, S.J., Cho, S.J. and Kim, S.Y. (2006) Hydrogen adsorption in periodic mesoporous organic- and inorganic-silica materials at room temperature. *Chemistry Letters* 35, 32–33.

Kalyanasundaram, D. and Molian, P. (2008) Electrodeposition of nanodiamond particles on aluminium alloy a319 for improved tribological properties. *Micro Nano Letters, IET* 3(4), 110.

Karajanagi, S.S., Yang, H.C., Asuri, P., Sellitto, E., Dordick, J.S. and Kane, R.S. (2006) Protein-assisted solubilisation of single-walled carbon nanotubes. *Langmuir* 22(4), 1392–1395.

Kim, B.C., Nair, S., Kim, J., Kwak, J.H., Grate, J.W., Kim, S.H. and Gu, M.B. (2005) Preparation of biocatalytic nanofibres with high activity and stability via enzyme aggregate coating on polymer nanofibers. *Nanotechnology* 16(7), S382–S388.

Kim, H.J., Kang, B., Kim, M.J., Park, Y.M., Kim, D.K., Lee, J.S. and Lee, K.Y. (2004) Transesterification of vegetable oil to biodiesel using heterogeneous base catalyst. *Catalysis Today* 93–93, 315–320.

Kim, H.J., Kim, S.K., Seo, M.H., Choi, S.M., and Kim, W. B. (2009) Pt and PtRh nanowire electrocatalysts for cyclohexane-fueled polymer electrolyte membrane fuel cell. *Electrochemistry Communications* 11(2), 446–449.

Kim, J. and Grate, J.W. (2003) Single-enzyme nanoparticles armored by a nanometer-scale organic/inorganic network. *Nano Letters* 3(9), 1219–1222.

Kinney, A.J. and Clemente, T.E. (2005)Modifying soybean oil for enhanced performance in biodiesel blends. *Fuel Processing Technology* 86, 1137–1147.

Kisch, H. and Macyk, W. (2002) Visible light photocatalysis by modified titania. *Chemical Physics and Physical Chemistry*3, 399.

Kitano, M., Matsuoka, M., Ueshima, M. and Anpo, M. (2007) Recent developments in titanium oxide based photocatalysts. *Applied Catalysis A*325, 1.

Kondo, A., Murakami, F. and Higashitani, K. (1992) Circular dichroism studies on conformational changes in protein molecules upon adsorption on ultrafinepolystyrene particles. *Biotechnology and Bioengineering* 40, 889–894.

Kramb, J. (2011) Potential Applications of Nanotechnology in Bioenergy. Master's Thesis, Department of Physics, University of Jyvaskyla, Finland.

Kudo, T., Nakamura, Y. and Ruike, A. (2003) Development of rectangular column structured titaniumoxide photocatalysts anchored on silica sheets by a wet process. *Research on Chemical Intermediates* 29(6), 631–639.

Kusdiana, D. and Saka, S. (2004) Effects of water on biodiesel fuel production by supercritical methanol treatment. *Bioresource Technology* 91(3), 289–295.

Laudenslager, M., Scheffer, R.H. and Sigmund, W. (2010) Electrospun materials for energy harvesting, conversion and storage: a review. *Pure and Applied Chemistry* 82, 2137–2156.

Lee, J.S., Han, K.I., Park, S.O., Kim, H.N. and Kim, H. (2004) Performance and impedance under various catalyst layer thicknesses in DMFC. *Electrochimica Acta* 50, 807–810.

Lee, S.M., Jin, L., Kim, J., Han, S., Na, H., Hyeon, T., Koo, Y.M., Kim, J. and Lee, J.H. (2010) Glucosidase coating on polymer nanofibers for improved cellulosic ethanol production. *Bioprocess Biosystems Engineering* 33(1), 141–147.

Li, C., Yoshimoto, M., Fukunaga, K. and Nakao, K. (2007) Characterisation and immobilisation of liposome-bound cellulase for hydrolysis of insoluble cellulose. *Bioresource Technology* 98(7), 1366–1372.

Li, Q., Fang, S., Han, W., Sun, C. and Liang, W. (1997) Synthesis of boron nitride nanotubes from carbon nanotubes by a substitution reaction. *Japanese Journal of Applied Physics* 36, 501–508.

Li, S., Xu, S., Liu, S., Yang, C. and Lu, Q. (2004) Fast pyrolysis of biomass in free-fall reactor for hydrogen-rich gas. *Fuel Process Technology* 85, 1201.

Li, X.F., Hao, X.F., Xu, D., Zhang, G., Zhong, S.L., Na, H. and Wang, D.Y. (2006) Fabrication of sulfonated poly(ether ether ketone) membranes with high proton conductivity. *Journal of Membrane Science* 281, 1.

Lin, J.Y. (2000) Hydrogen storage in nanotubes. *Science* 287, 1929–31.

Lin, V. (2008) *Mesoporous Nanoparticles for Selective Sequestration of Fatty Acids and Fats from Microalgae for Biofuel Applications*. Abstract Papers, American Chemical Society, Washington, DC .

Liu, X.J., Piao, X.L., Wang, Y.J. and Zhu, S.F. (2008) Calcium ethoxide as a solid base catalyst for the transesterification of soybean oil to biodiesel. *Energy Fuels* 22(2), 1313–1317.

Liu, Y., Yu, J., Huang, H., Xu, B., Liu, X., Gao, Y. and Dong, X. (2007) Synthesis and tribological behavior of electroless ni-p-wc nanocomposites coatings.*Surface and Coatings Technology* 201, 16–17, 7246.

Liu, Z., Gao, W. and Li, M. (1999) Cyclic oxidation of sputter-deposited nanocrystalline Fe-Cr-Ni-Al alloy coatings. *Oxidation of Metals* 51, 403.

Liu, Z., Lin, X., Lee, J.Y., Zhang, W., Han, M. and Gan, L.M. (2002) Preparation and characterisation of platinum based electrocatalysts on multiwalled carbon nanotubes for proton exchange membrane fuel cells. *Langmuir* 18, 4054–4060.

Lucena, I.L., Silva, G.F. and Fernandes, F.A.N. (2008) Biodiesel production by esterification of oleic acid with methanol using a water adsorption apparatus. *Industrial Engineering Chemical Research* 47(18), 6885–6889.

Lucia, L.A. and Rojas, O.J. (eds) (2009) *The Nanoscience and Technology of Renewable Biomaterials*. Blackwell Publishing Ltd,West Sussex, UK.

Mandal, D., Bolander, M., Mukhopadhyay, D., Sarkar, G. and Mukharjee, P. (2006) The use of microorganisms for the formation of metal nanoparticles and their applications. *Applied Microbiology and Biotechnology* 69, 485–492.

Mango De Carvalho, L., Tan, A.R. and Gomes, A. (2008) Nanostructured membranes based on sulfonated poly (aryl ether sulfone) and silica for fuel cell applications. *Applied Polymer Science* 110, 1690–1698.

Matsunaga, T. and Kamiya, S. (1987) Use of magnetic particles isolated frommagnetotactic bacteria for enzyme immobilisation. *Applied Microbiology and Biotechnology* 26, 328–332.

Mercerreyes, D., Grande, H., Miguel, O., Marcilla, R. and Cantero, I. (2004) Porous polybenzimidazole membrane doped with phosphoric acid: highly proton conducting solid electrolytes. *Chemical Materials* 16, 604–607.

Mishra, S. (2009) Nano and nanocomposite superhard coatings of silicon carbonitride and titanium diboride by magnetron sputtering. *International Journal of Applied Ceramic Technology* 6(3), 345.

Mitchell, D.T., Lee, S.B., Trofin, L., Li, N., Nevanen, T.K., Suerlund, H. and Martin,C.R. (2002) Smart nanotubes for bioseparations and biocatalysis. *Journal of American Chemical Society* 124, 1186–1187.

Montgomery, C. and Carla, W. (1997) *Environmental Geology*, 5th edn. WCB McGraw Hill.

Moxley, G.Z., Zhu, Z. and Zhang, Y.H. (2008) Efficient sugar release by the cellulose solvent-based lignocellulosic fractionation technology and enzymatic hydrolysis. *Journal of Agricultural and Food Chemistry* 56(17), 7885–7890.

Nagase, Y., Sugimoto, K.,Takamura, Y. and Matsui, K. (1991a) Chemical modification of poly(substituted-acetylene). 6. Introduction of fluoroalkyl group into poly(1-trimethylsilyl-1-propyne) and the improved ethanol permselectivity at pervaporation. *Journal of Applied Polymer Science* 43(7), 1227–1232.

Nagase, Y., Takamura, Y. and Matsui, K. (1991b) Chemical modification of poly(substituted-acetylene). 5. Alkylsilylation of poly(1-trimethylsilyl-1-propyne) and improved liquid separating property at pervaporation. *Journal of Applied Polymer Science* 42(1), 185–190.

Nielsen, D.R., Leonard, E., Yoon, S.H., Tseng, H.C., Yuan, C. and Prather, K.L.J. (2009) Engineering alternative butanol production platforms in heterologous bacteria. *MetabolicEngineering*11(4–5), 262–273.

Nikitin, A., Li, X., Zhang, Z., Ogasawara, H., Dai, H. and Nilsson, A. (2008) Hydrogen storage in carbon nanotubes through the formation of stable C–H bonds. *Nano Letters* 8, 162–167.

Orimo, S., Nakamori, Y., Eliseo, J., Zuttel, A. and Jensen, C. (2003) Complex Hydrides for Hydrogen Storage. *Chemical Reviews* 107, 4111–4124.

Oudshoorn, A., van der Wielen, L.A.M. and Straathof, A.J.J. (2009) Assessment of options for selective 1-butanol recovery from aqueous solution. *Industrial Engineering and Chemical Research* 48(15), 7325–7336.

Pan, H., Li, L., Hu, L. and Cui, X. (2006) Continuous aligned polymer fibers produced by a modified electrospinning method. *Polymer Communications* 47, 490.

Pan, X., Fan, Z., Chen, W., Ding, Y., Luo, H. and Bao, X. (2007) Enhanced ethanol production inside carbon-nanotube reactors containing catalytic particles. *Nature Materials* 6, 507–511.

Pant, D., Van Bogaert, G., Diels, L. and Vanbroekhoven, K. (2010) A review of the substrates used in microbial fuel cells (MFCs) for sustainable energy production.*BioresourceTechnology*101(6), 1533–1543.

Pascal, W., Bemert, L., Engelsttrichen, S. and Strey, R. (2008) Water biofuel microemulsions. Available at: http://strey.uni–koeln.de/333.html?&L=1.

Paul, D. (2001) Presentation at Mackay School of Mines, University of Nevada, Reno.

Pinkerton, F.E., Wicke, B.G., Olk, C.H., Tibbetts, G.G., Meisner, G.P. and Meyer, M.S. (2000) Thermogravimetric measurement of hydrogen absorption in alkali-modified carbon materials. *Journal of Physical Chemistry B*104, 9460–9467.

Planeix, J.M., Coustel, N., Coq, B., Brotons, V., Kumbhar, P.S., Dutartre, R., Geneste, P., Bernier, P. and Ajayan, P.M.J. (1994) Application of carbon nanotubes as supports in heterogeneous catalysis. *Journal of American Chemical Society*116, 7935–39.

Pradhan, B.K., Toba, T., Kyotani, T. and Tomita, A. (1998) Inclusion of crystalline iron oxide nanoparticles in uniform carbon nanotubes prepared by a template carbonisation method. *Chemistry of Materials* 10, 2510–2515.

Pugh, S., McKenna, R., Moolick, R. and Neilson, D.R. (2010) Advances and opportunities at the interface between microbial bioenergy and nanotechnology. *Canadian Journal of Chemical Engineering* 89, 1–12.

Qhobosheane, M., Santra, S., Zhang, P. and Tan, W. (2001) Biochemically functionalised silica nanoparticles. *Analyst* 126(8), 1274–1278.

Rajesh, B., Thampi, K.R., Bonard, J.M. and Viswanathan, B. (2000) Catalyst assisted synthesis of carbon nanotubes using the oxy-acetylene combustion flame method. *Journal of Material Chemistry*10, 1757–1759.

Rajalakshmi, N. and Dhatartreyan, K.S. (2008) Nanotitanium oxide support for proton exchange membrane fuel cells. *International Journal of Hydrogen Energy* 33, 7521–26.

Ramachandrana, S., Haa, J.H. and Kim, D.K. (2007) Hydrogen storage characteristics of metal oxide doped Al-MCM-41 mesoporous materials. *Catalysis Communications* 8, 1934–1938.

Ramakrishna, S., Fujihara, K., Teo, W., Yong, T., Ma, Z. and Ramaseshan, R. (2006) Electrospun nanofibers: solving global issues. *Materials Today* 9, 40.

Rapagna, S., Jand, N. and Foscolo, P.U. (1998) Catalytic gasification of biomass to produce hydrogen rich gas. *International Journal of Hydrogen Energy* 23, 551.

Rege, K., Raravikar, N.R., Kim, D.Y., Schadler, L.S., Ajayan, P.M. and Dordick, J.S. (2003) Enzyme-polymer-single walled carbon nanotube composites as biocatalytic films. *Nano Letters* 3, 829–832.

Reddy, C., Reddy, V., Oshel, R. and Varkade, G. (2006) Room temperature conversion of soybean oil and poultry fat to biodiesel catalyzed by nanocrystalline calcium oxides. *Energy and Fuels* 20 (3), 1310–1314.

Reddy, C., Reddy, V., Oshel, R. and Verkade, J.G. (2009) Conversion of biomass to fuel: transesterification of vegetable oil to biodiesel using KF loaded nano-[Gamma]-A12O3 as catalyst. *Energy Fuel* 20(3), 1310 (S1).

Rosi, N.L., Giljohann, D.A., Thaxton, C.S., Lytton-Jean, A.K.R., Han, M.S. and Mirkin, C.A. (2006) Oligonucleotide-modified gold nanoparticles for intracellular gene regulation. *Science* 312(5776), 1027–1030.

RSC (2005) *Chemical Science Priorities for Sustainable Energy Solutions*.

Rubio-Arroyo, M., Ayona-Argueta, M., Poisot, M. and Ramírez-Galicia, G. (2009) Biofuel Obtained from Transesterification by Combined Catalysis. *Energy & Fuels* 23, 2840–2842.

Ruhl, J., Schmid, A. and Blank, L.M. (2009) Selected *Pseudomonas putida* strains able to grow in the presence of high butanol concentrations. *Applied and Environmental Microbiology* 75(13), 4653–4656.

Saravanan, V., Waijers, D.A., Ziari, M. and Noordermeer, M.A.(2010) Recovery of 1-butanol from aqueous solutions using zeolite ZSM-5 with a high Si/Al ratio; suitability of a column process for industrial applications. *Biochemical Engineering Journal* 49(1), 33–39.

Saville, B., Khavkine, M., Seetharam, G., Marandi, B. and Zou, Y.L. (2004) Characterisation and performance of immobilisedbamylase and cellulase. *Applied Biochemistry and Biotechnology* 113(1), 251–259.

Schechner, P., Kroll, E., Bubis, E., Chervinsky, S. and Zussman, E. (2007) Silver-Plated Electrospun Fibrous Anode for Glucose Alkaline Fuel Cells. *Journal of Electrochemical Society* 154, B942.

Schimmel, H.G., Kearley, G.J., Nijkamp, M.G., Visserl, C.T., Jong de, K.P. and Mulder, F.M. (2003) Hydrogen adsorption in carbon nanostructures: comparison of nanotubes, fibers, and coals. *Chemistry European Journal* 9, 4764–4770.

Schugerl, K. and Hubbuch, J. (2005) Integrated Bioprocesses. *Current Opinion in Microbiology* 8(3), 294–300.

Schwartz, J.A. and Amankwah, K.A.G. (1993) Hydrogen storage systems. In: Howell, D.G. (ed.) *The Future of Energy Gases*. US Government Printing Office, Washington, DC, pp. 725–736.

Serrano, E., Guillermo, R. and Garcia-Martinez, J.(2009) Nanotechnology for sustainable energy. *Renewable and Sustainable Energy Reviews* 13, 2373–2384.

Shah, S., Sharma, S. and Gupta, M.N. (2004) Biodiesel preparation by lipase-catalyzed transesterification of *Jatropha* oil. *Energy Fuels* 18, 154–159.

Sheppard, D.A. and Buckley, C.E.(2008) Hydrogen adsorption on porous silica. *International Journal of Hydrogen Energy* 8, 33,1688–1692.

Sherif, S.A., Barbir, F. and Veziroglu, T.N. (2005) Wind energy and the hydrogen economy – review of the technology. *Solar Energy* 78,647–660.

Shinkai, M., Honda, H. and Kobayashi, T. (1991) Preparation of fine magnetic particles and application for enzyme immobilisation. *Biocatalysis andBiotransformations* 5,61–69.

Sigmund, W., Yuh, J., Park, H., Maneeratana, V., Pyrgiotakis, G., Daga, A., Taylor, J. and Nino, J. (2006) Processing and structure relationships in electrospinning of ceramic fiber systems. *Journal of American Ceramic Society* 89, 395.

Sotiropoulou, S., Vamvakaki, V. and Chaniotakis, N.A. (2005) Stabilisation of enzymes in nanoporous materials for biosensor applications. *Biosensors Bioelectrochemistry* 20(8), 1674–1679.

Stark, I.G. (ed.) (1971) *Biochemical Aspects of Reactions on Solid Supports*. Academic, New York, 71 pp.

Sticklen, M. (2009) Molecular Breeding for Biomass and Biofuels. [PowerPoint slides]. Retrieved from http://bioenergy.msu.edu/presentations

Stoimenov, P.K., Klinger, R.L., Marchin, G.L. and Klabunde, K.J. (2002) Metal oxide nanoparticles as bactericidal agents. *Langmuir* 18, 6679–6686.

Straathof, A.J. (2003) Auxiliary phase guidelines for microbial biotransformations of toxic substrate into toxic product. *Biotechnology Progress* 19(3), 755–762.

Strege, J., Zygarlicke, C., Folkedahl, B. and McCollor, D. (2007) Deactivation in a full-scale coal-fired utility boiler. *Fuel* 87(7), 1341.

Strey, R., Nawrath, A. and Sottman, T. (2004) Microemulsions and use thereof as a fuel. US Patent 7977389.

Subbiah, T., Bhat, G., Tock, R., Parameswaran, S. and Ramkumar, S. (2005) Electrospinning of Nanofibers. *Journal of Applied Polymer Science* 96, 557.

Sun, N. and Klabunde, K.J.J. (1999) Nanocrystal metal oxide-chlorine adducts: selective catalysts for chlorination of alkanes. *Journal of American Chemical Society* 121, 5587–5589.

Sun, Q., Zorin, N.A., Chen, D., Chen, M., Liu, T.X., Miyake, J. and Qian, D.J. (2010) Blodgett films of pyridyldithio-modified multiwalled carbon nanotubes as a support to immobilise hydrogenase. *Langmuir* 26(12), 10259–10265.

Sundaray, B., Subramanian, V., Natarajan, T., Xiang, R., Chang, C. and Fann, W. (2004) Electrospinning of continuous aligned polymer fibers. *Applied Physics Letters* 84, 1222.

Tarleton, E. and Low, J. (2008) Nanofiltration. A technology for selective solute removal from fuels and solvents. *Chemical Engineering Research and Design* 87, 271.

Teng, F., Santhanagopalan, S., Wang, Y. and Meng, D. (2010) *In-situ* hydrothermal synthesis of three-dimensional MnO_2–CNT nanocomposites and their electrochemical properties. *Journal of Alloys and Compounds* 499(2), 259–264.

The Economist (2010) Cleaner Diesel Engines – pouring water on troubled oils. 3 June, 2010, p. 86.

Theron, A., Zussman, E. and Yarin, A. (2001) Electrostatic field-assisted alignment of electrospun nanofibres. *Nanotechnology* 12, 384.

Tsang, S.C., Chen, Y.K., Harris, P.J.F. and Green, M.L.H. (1994) A simple chemical method of opening and filling carbon nanotubes. *Nature* 372, 159–161.

Tummala, S.B., Junne, S.G. and Papoutsakis, E.T. (2003a) Antisense RNA downregulation of Coenzyme A transferase combined with alcohol-aldehyde dehydrogenase overexpression leads to predominantly alcohologenic *Clostridium acetobutylicum* fermentations. *Journal of Bacteriology* 185(12), 3644–3653.

Tummala, S.B., Junne, S.G., Paredes, C.J. and Papoutsakis, E.T. (2003b) Transcriptional analysis of product-concentration driven changes in cellular programs of recombinant *Clostridium acetobutylicum* strains. *Biotechnology and Bioengineering* 84(7), 842–854.

Tummala, S.B., Welker, N.E. and Papoutsakis, E.T. (2003c) Design of antisense RNA constructs for downregulation of the acetone formation pathway of *Clostridium acetobutylicum*. *Journal of Bacteriology* 185(6), 1923–1934.

Vankelecom, I.F.J., Depre, D., Debeukelaer, S. and Uytterhoeven, J.B. (1995) Influence of zeolites in Pdms membranes –pervaporation of water/alcohol mixtures. *Journal of Physical Chemistry* 99(35), 13193–13197.

Vinu, A., Miyahara, M., Mori, T. and Ariga, K. (2006) Carbon nanocage: a large-pore cage-type mesoporous carbon material as an adsorbent for biomolecules. *Journal of Porous Materials* 13(3–4), 379–383.

Wegner, T. and Jones, P. (2007) A fundamental review of the relationship between nanotechnology and lignocellulosic biomass. In: Lucia, L.A. and Rojas, O.J. (eds) *The Nanoscience and Technology of Renewable Biomaterials*. John Wiley & Sons, Chichester, UK, pp. 1–42.

Wang, C., Waje, M., Wang, X., Tang, J.M., Haddon, R.C. and Yan, Y. (2004) Proton exchange fuel cells with carbon nanotube based electrodes. *Nano Letters* 4, 345–348.

Wang, P., Dai, S., Waezsada, S.D., Tsao, A. and Davison, B.H. (2001) Enzyme stabilisation by covalent binding in nanoporous sol-gel glass for nonaqueous biocatalysis. *Biotechnology Bioengineering* 74, 249–255.

Wang, Y. and Hsieh, Y.L. (2008) Immobilisation of lipase enzyme in polyvinyl alcohol (PVA) nanofibrous membranes. *Journal of Membrane Science* 309(1–2), 73–81.

Wang, Y., Jiang, S., Wang, M., Wang, S., Xiao, T. and Strutt, P. (2000) Abrasive wear characteristics of plasma sprayed nanostructured alumina/titania coatings. *Wear* 237(2), 176.

Watanabe, M., Inomata, H. and Arai, K. (2002) Catalytic hydrogen generation from biomass (glucose and cellulose) with ZrO_2 in supercritical water. *Biomass Bioenergy* 22, 405.

Wu, J.Y., Liu, Q.L., Xiong, Y., Zhu, A.M. and Chen, Y. (2009) Molecular simulation of water/alcohol mixtures' adsorption and diffusion in zeolite 4A membranes. *Journal of Physical Chemistry B* 113(13), 4267–4274.

Xiong, C. and Balkus, K. Jr (2005) Fabrication of TiO2 nanofibers from a mesoporous silica film. *Chemistry of Materials* 17, 5136.

Yagi, S., Nakagawa, T., Matsubara, E., Matsubara, S., Ogawa, S. and Tani, H. (2008) Formation of Tin nanoparticles embedded in poly (L-Lactic acid) fiber by electrospinning. *Electrochemical and Solid-State Letters* 11, E25.

Yamamoto, Y., Ito, H. and Kujime, S. (2007) Nano-multilayered CrN/BCN coating for anti-wear and low friction applications. *Surface and Coatings Technology* 201(9-11), 5244. Proceedings of the Fifth Asian-European International Conference on Plasma Surface Engineering – AEPSE 2005.

Yamashita, H., Takeuchi, M. and Anpo, M. (2004) Visible light sensitive photocatalysts. In: *Encyclopedia of Nanoscience and Nanotechnology*. American Scientific Publishers, Stevenson Ranch, California, p. 639.

Yang, D., Lu, B., Zhao, Y. and Jiang, X. (2007) Fabrication of aligned fibrous arrays by magnetic electrospinning. *Advanced Materials* 19, 3702.

Ye, Y., Ahn, C.C., Witham, C., Fultz, B., Liu, J. and Rinzler, A.G. (1999) Hydrogen adsorption and cohesive energy of single-walled carbon nanotubes. *Applied Physics Letters* 74, 2307–2309.

Yim, T.J., Liu, J., Lu, Y., Kane, R.S. and Dordick, J.S. (2005) Highly active and stable DNAzyme-carbon nanotube hybrids. *Journal of American Chemical Society* 127, 12200–12201.

Yuan, Z.Y. and Su, B.L. (2004) Titanium oxide nanotubes, nanofibers and nanowires. *Colloids and Surfaces A* 241, 173.

Yuan, X.Y., Shen, N.X., Sheng, J. and Wei, X. (1999) Immobilisation of cellulase using acrylamide grafted acrylonitrile copolymer membranes. *Journal of Membrane Science* 155(1), 101–106.

Zeltner, W.A. and Tompkin, D.T. (2005) Shedding light on photocatalysis. *ASHRAE Transactions* 111, 532–534.

Zhang, Y., Zhang, Q., Li, Y., Wang, N. and Zhu, J. (2000) Coating of carbon nanotubes with tungsten by physical vapor deposition. *Solid State Communication* 115, 51–55.

Zhang, Y.H., Zhu, Z., Rollin, J. and Sathitsuksanoh, N. (2010) Advances in cellulose solvent- and organic solvent-based lignocellulose fractionation (COSLIF). *Cellulose Solvents: For Analysis, Shaping and Chemical Modification* 365–379.

Zhao, J. and Yang, X. (2003) Photocatalytic oxidation for indoor air purification: a literature review. *Building and Environment* 38, 645–654.

Zhong, Z., Ang, T., Luo, J., Gan, H. and Gedanken, A. (2005) Synthesis of one-dimensional and porous TiO_2 nanostructures by controlled hydrolysis of titanium alkoxide via coupling with an esterification reaction. *Chemistry of Materials* 17, 6814.

Zussman, E., Yarin, A.L., Bazilevsky, A.V., Avrahami, R. and Feldman, M. (2006) Electrospun polyaniline/poly(methyl methacrylate)-derived turbostratic carbon micro-/nanotubes. *Advanced Materials* 18, 348–353.

Züttel, A., Sudan, P., Mauron, P.H., Emmenegger, Ch. and Schlapbach, L. (2002) Hydrogen storage in carbon nanostructures. *International Journal of Hydrogen Energy* 27, 203–212.

Chapter 8

Host Engineering for Biofuel-Tolerant Phenotypes

Becky J. Rutherford and Aindrila Mukhopadhyay

Introduction

Prior to the discovery of recombinant DNA, the early biotechnology industry was successful in producing many important chemicals despite limitations in genetic tools and well-studied host organisms. A pioneering example was Chaim Weizmann's discovery that *Clostridium acetobutylicum* was capable of producing a mixture of acetone, butanol and ethanol (ABE) during fermentation (Jones and Woods, 1986). This ABE process was widely utilized to provide acetone for cordite and gunpowder during World War II, but by the 1940s inexpensive petrochemical routes for production had replaced the ABE fermentation (Jones and Woods, 1986; Lee *et al.*, 2008). Other industrially relevant microorganisms from this era include non-pathogenic species of *Cornyebacterium*, used in the production of glutamic acid for soy sauce (Hermann, 2003), and varieties of *Lactobacillus*, which are used in the production of many dairy products such as cheese and yoghurt (Altermann *et al.*, 2005; Wegmann *et al.*, 2007; Zhu *et al.*, 2009). These microbial production processes were only able to target chemicals as part of the native host metabolism, and further improvements in biotechnology would be necessary to achieve economic microbial production of more novel compounds. Engineering of microbial hosts for large-scale processes has become significantly more sophisticated since the first patent was issued on a microbe in 1972, a *Pseudomonas aeruginosa* strain capable of metabolizing oil (Newell-McGloughlin and Re, 2006). The ability to clone and amplify DNA opened opportunities for expanding the use of microbial hosts through heterologous protein and metabolite production. One of the most famous examples was the use of a recombinant *Escherichia coli* for the production of human insulin, under the brand name Humulin, by Genentech, which was FDA approved in 1982 (Gad, 2008). The diversity of industrial microbes has increased greatly as the potential for their use as 'cell factories' has been realized. Recent biotechnology products have included various high-value chemicals such as anticancer compounds and flavonoids (Ruther *et al.*, 1997; Khosla and Keasling, 2003; Moon *et al.*, 2009; Ajikumar *et al.*, 2010), antimicrobial peptides (Li, 2009) and industrial enzymes (Gupta *et al.*, 2002; Haefner *et al.*, 2005). Recently, issues concerning energy security, global warming and other limitations of petroleum-based chemicals have garnered increased interest in the development of biological systems for the production of fuels and other community chemicals such as plastics. However, since these products have low profit margins, it is vital that the microbes used for these processes be able to grow with minimal nutrients, produce high volumes of the target molecule and be resistant to the chemicals being produced.

Achieving these phenotypes will require work beyond pathway optimization, and will become vital as product titres exceed the native tolerance levels. Most naturally occurring strains are unable to meet all these criteria and will need to be further optimized (Zhang *et al.*, 2009). Because of these limitations, strain improvement strategies have been a focus of both academic and industrial research. Historically, strain improvement has been achieved through serial adaptation or random mutagenesis. These techniques were successful in improving penicillin production over 4000-fold over the original patented strain, and in achieving over 100-fold improvement in avermectin production in *Streptomyces avermitilis* (Demain and Elander, 1999). These strategies are limited in that they rely on random combinatorial events, and it may be difficult to isolate the genetic causes of improved phenotype. It is also difficult to predict how changes in the production process may affect the host organism without a thorough understanding of cellular behaviour on a systems-level.

A modern strategy for host optimization leverages systems biology to characterize responses to metabolic engineering and industrially relevant environmental conditions, and to further develop models that describe the complex networks of reactions occurring. To develop efficient microbial hosts necessary to compete with petrochemical technologies, a cell-wide approach to understanding responses to all the stresses of the biofuel production process will be required. In this chapter, we will be discussing biotechnology for fuel production using recent advances for bacterial systems. For a more historical perspective of this topic, excellent reviews are available from Aristidou and Penttila (2000) and Zaldivar *et al.* (2001). We have chosen to focus on *E. coli* because of its industrial significance and the wide variety of systems biology techniques available for it. *Escherichia coli* is currently used to produce large quantities of chemicals such as plastics (Biebl *et al.*, 1999; Choi and Lee, 1999; Nakamura and Whited, 2003; Burk *et al.*, 2009; Yim *et al.*, 2011), amino acids (Dassler *et al.*, 2000; Bongaerts *et al.*, 2001; Gerigk *et al.*, 2002) and other enzymes. Lin *et al.* and Singh *et al.* (Lin and Tanaka, 2006; Singh and Cu, 2010) provide excellent reviews on other industrial microbial hosts such as *Saccharomyces cerevisiae*.

Characterization of Toxic Fuels and Engineering Tolerance

The economics of next-generation biofuel production in *E. coli* will be severely impacted by the toxicity of these fuels, which can be inhibitory at very low concentrations. These concerns about product toxicity have led to interest in evaluating the systems-level responses to candidate compounds, particularly higher-chain alcohols. Since tolerance phenotypes are complex, and often impossible to recapitulate through the manipulation of a single gene, many studies have focused on global regulators or groups of genes with similar functional annotations. In recent work from Klein-Marcuschamer *et al.* (2009), the α-subunit of RNA polymerase was mutagenized using error-prone PCR, and subsequently used to generate libraries to test for improved tolerance and production of a variety of industrially relevant compounds such as n-butanol, L-tyrosine and hyaluronic acid. Tolerant mutants were found for all tested compounds. In the case of n-butanol, the tolerant mutant was found to have a more rigid membrane and increased rates of proton extrusion.

150   *Microbial Biotechnology: Energy and Environment*

In this study, all three tolerant mutants (n-butanol, L-tyrosine, hyaluronic acid) were able to confer higher L-tyrosine titres. Interestingly, only the n-butanol specific mutant was able to tolerate high concentrations of n-butanol. Theoretically, all global transcriptional regulators could be engineering targets, but it was found that not all were equally good at generating beneficial phenotypes. Sigma factors such as σ^H, σ^D and σ^E showed no significant improvements despite their important roles in several stresses. Combining global regulator mutagenesis with an *E. coli* strain capable of producing higher titres of biofuels may be a future direction for developing interesting phenotypes. Another approach to studying biofuel toxicity and production in *E. coli* couples adaptation during stress with analysis of systems-level data to elucidate which individual genes or functional groups have been perturbed. For example, Goodarzi et al. (2010) used a combined experimental and computational framework to study the genetic basis of adaptation to ethanol stress. As illustrated in Fig. 8.1, their strategy involved serial adaptation of a strain, which contained a

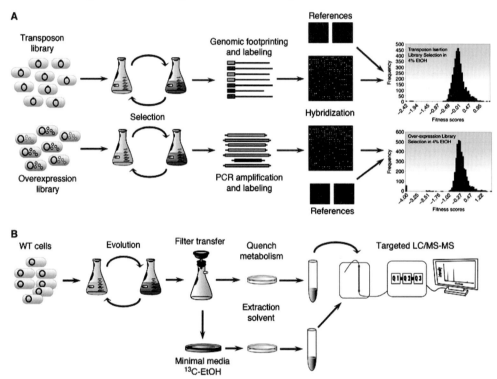

Fig. 8.1. A schematic representation of the overall strategy. (A) Starting from the transposon insertion or over-expression libraries, we enriched for relative increase in ethanol tolerance through several rounds of selection in the presence of ethanol. The changes in the frequency of each mutant in the selected versus unselected samples were then translated into a whole-genome fitness profile. (B) The wild-type strain (MG1655) was grown in minimal media plus glucose and exogenous ethanol to select for increased ethanol tolerance. The resulting tolerant strains were subjected to metabolomic analyses for measuring changes in the steady-state levels of intracellular metabolites. In a parallel experiment, stable-isotope labelling (^{13}C-ethanol) was used to test ethanol assimilation as a possible mechanism for tolerance. (Figure reproduced from Goodarzi et al., 2010 using the Creative Commons Permission.)

comprehensive transposon library and an over-expression library, to increasing concentrations of ethanol (4% and 5.5% v/v).

Following adaptation, affected genes were determined, and fitness scores were calculated based on microarray hybridization signals. Due to the limited benefit of single gene perturbations during ethanol tolerance, few genes passed the authors' statistical tests. Instead of focusing on individual genes, unique gene modules were created combining information from Gene Ontology (GO) annotations (Ashburner *et al.*, 2000), known transcription factor regulons, and stress response pathways. The perturbations measured for these modules were similar to those measured in other published stress studies of ethanol, such as increases in heat-shock stress response, glycine-betaine synthesis, and genes involved in aerobic respiration. After further investigation of these candidates, it was found that the most ethanol-tolerant mutant *ΔfnrΔarcAΔcafA* had a higher level of alcohol dehydrogenase, AdhE. Using labelled ethanol, they were also able to confirm that ethanol was being degraded and assimilated into the TCA cycle in their laboratory-evolved strain. A motivation for understanding microbial stress responses is to engineer increased product titres. In this particular study, *E. coli* was able to tolerate ethanol through degrading it, which is not ideal for a biofuel-producing strain. However, from a microbial perspective, this strategy is a common mechanism to survive in toxic environments, such as *Pseudomonas* strains that are capable of converting toxic phenolic compounds to acetyl-CoA (Allsop *et al.*, 1993). Using this method on a more structurally novel compound may lead to microbial tolerance strategies that are more conducive to producing high titres.

Another study utilizing adaptation and subsequent genomic sequencing found a similar issue when trying to increase biofuel production. An isobutanol-producing *E. coli* strain by Atsumi *et al.* (2010) was adapted to increasing concentrations of isobutanol (4g l^{-1} up to 8 g l^{-1}) using serial transfers. A tolerant strain was isolated, and the genome was sequenced to find which genes were altered during exposure to stress. To elucidate the identity of genes that contributed to the isobutanol-tolerant phenotype, altered genes were repaired and their effect on stress tested. Again, no single gene repair completely eliminated the resistant phenotype, indicating the involvement of several systems. Five key loci were identified that were connected to isobutanol tolerance, *yhbJ*, *acrA*, *marCRAB*, *tnaA* and *gatY*, with inactivation of these sites being beneficial. It is interesting that deletion of *marCRAB* would be beneficial since many other stress studies have indicated up-regulation of these genes during stress (Aono, 1998). To test the effects of a tolerant host on fuel production, these genetic changes were then incorporated into an *E. coli* strain capable of producing isobutanol at titres greater than the toxicity level (20 g l^{-1}). Although the resistant phenotype was observed, the improved ability to cope with stress did not lead to an increase in the amount of isobutanol produced. This may be due to the conditions used to generate the isobutanol-tolerant phenotype. High concentrations of isobutanol were added exogenously to the media, whereas during endogenous production, isobutanol levels increase slowly, though the final concentration in the culture may be equivalent. Often mutants are tolerant only in the specific conditions under which they were selected, so it may not be surprising that the production did not improve significantly. To determine if these genetic changes conferred resistance to other stresses, the authors tested it against exposure to ethanol, n-butanol, 2-methyl-1-butanol and chloramphenicol. The strain was more sensitive to ethanol and chloramphenicol, which suggests that C4-C5 alcohol stress has a distinct stress profile. This specificity of phenotypes mirrors what has been found in other stress studies (Klein-Marcuschamer et al., 2009).

Challenges in Stress Response Studies

Escherichia coli remains one of the most studied microbial production hosts, yet there are aspects of its genetics, metabolism and regulation that remain obscure. As a consequence, attempts to use systems biology tools to engineer beneficial behaviour may be hindered by the lack of knowledge of key systems. Microarrays are a commonly used technique to quantify mRNA changes on a global level during different experimental conditions. As an illustrative example of difficulty interpreting these data, we selected eight *E. coli* microarray datasets from MicrobesOnline that relate to biofuel stress or engineered pathways (Table 8.1). MicrobesOnline is a resource for comparative and functional genomics created by the Virtual Institute for Microbial Stress and Survival (Dehal et al., 2009). These data sets cover a variety of conditions relevant to fuel or metabolite production such as acid stress, ethanol stress (two separate experiments), butanol stress (two separate experiments), isobutanol stress, heat shock and production of threonine. The top 50 up-regulated genes were chosen from each dataset, sorted by the Z scores submitted. Calculation of Z-scores, as done in MicrobesOnline, is described in detail in Mukhopadhyay et al. (2006). We applied no additional score thresholds or cut-offs. A summary of the percentage of proteins of unknown functions among these 50 is provided in Table 8.1.

Table 8.1. Percentage uncharacterized genes in top 50 up-regulated genes.

Experiment	% Uncharacterized in top 50 up-regulated genes
Acid stress	64
Ethanol 1	30
Ethanol 2	32
Butanol 1	20
Butanol 2	48
Isobutanol	32
Heat shock	22
Threonine production	32

The most recent statistics on the characterization of the *E. coli* genome reveals a large percentage of the sequence is still under-annotated. According to EcoCyc v12.5, approximately 2650 (59.3%) of *E. coli* genes have experimental evidence for their functions, and out of the predicted 3359 transcription units, only 940 (28%) have experimental evidence (Keseler et al., 2009). Using these numbers as a baseline, several of the microarray experiments profiled have a high percentage of uncharacterized genes, such as the exposure to low pH for 10 min, which has 64% of the top up-regulated genes annotated as such candidates. It is also interesting that many of these poorly characterized genes appear in several of these datasets despite the difference in strains, types of arrays and time points. For instance, *ycjF* appeared in datasets for butanol, isobutanol and heat stress, and is annotated as a conserved inner membrane protein. An orthologue of the *E. coli* protein YcjF was identified as being expressed during disease in a mouse infection model, with its role being linked to increased survival (Khan and Isaacson, 2002). Another study of the *E. coli* proteome identified it as an inner membrane protein, and that its expression is regulated by TyrR, a dual transcriptional regulator of genes involved in aromatic amino acid biosynthesis and transport (Cornish *et al.*, 1986; Daley *et al.*, 2005). Also identified in the heat shock experiment was *ycjX*, which is predicted to be co-transcribed with *ycjF*. Heat stress and alcohols are both known to disrupt the structure and integrity of the cell membrane, so it is not surprising to see an inner membrane protein being up-regulated.

When applying systems data to design better hosts, it is often difficult to use this information constructively when so little is known about these genes and putative proteins. However, narrowing down the pool of potential engineering candidates via microarrays and other tools has much value. Regardless, experimental assessment of these candidates through over-expression, knockout and mutagenesis is required to reveal if a particular candidate would confer tolerance to these stresses. Another challenge to many systems biology studies is being able to utilize previous studies to draw meaningful conclusions when the approaches (strains, media, time points) differ in slight but significant ways. For example, several studies have recently been published that took a systems-wide approach to studying cellular responses to n-butanol. The major conclusions from most of these studies point to the same systems being perturbed such as respiration, transport functions and induction of stress response systems. For example, Rutherford *et al.* (2010) used cell-wide microarray, proteomics and metabolite assays to find perturbations of the *nuo* and *cyo* operons involved in respiration, significant oxidative stress response, heat shock, envelope stress, metabolite transport and biosynthesis. Specific candidates selected for further follow-up studies were up-regulated genes such as *cpxP*, *evgS*, *sodC*, *yqhD* and *malE*, and down-regulated genes *metE*, *ompF* and *envY*. Another study by Reyes *et al.* (2011) used a genomic library screen to identify genes that were either enriched or depleted during adaptation to n-butanol stress. They identified 11 genes that were enriched in the butanol-tolerant strains, and three beneficial depletions *astE*, *ygiH* and *rph*. Researchers from the National Taiwan University in collaboration with James Liao at UCLA have filed a patent on alcohol-tolerant *E. coli* (Juan *et al.*, 2010), which identified an *ydhF* knockout mutant over-expressing PhoH, and a *potG* knockout mutant over-expressing YqhD. However, it is interesting that despite drawing similar conclusions about high-level cellular perturbations during stress, the individual genes found to confer tolerance are quite different among these studies.

Despite these difficulties in interpreting data, systems-wide studies contain a tremendous quantity of information and can lay the foundation for strain improvement strategies capable of addressing the phenotypic requirements of a production strain beyond pathway manipulation. As production strains with higher yields of advanced biofuels are engineered (11, 41, 49, 55), systems biology studies will be able to expand into the realm of endogenous exposure to these fuels since studying exogenous addition may be providing a crude model of true cellular response (Bond-Watts *et al.*, 2011; Shen *et al.*, 2011). Still to come are studies that focus on other biofuel candidates such as isopentanol, limonene, farnasane and fatty acid esters (Peralta-Yahya and Keasling, 2010). These target compounds may result in cellular burden arising from the expression of multigenic heterologous pathways as well as from the potential toxicity. However, desirable phenotypes for each of these aspects can be obtained using targeted or combinatorial approaches.

Heterologous approaches to engineering tolerance

It is often assumed that the genome of *E. coli* can be manipulated to achieve unlimited phenotypes given enough time for adaptation or mutation, but host engineering efforts may need to look further into nature for novel enzymes and other components. An example of this type of heterologous approach is investigating cellular export systems to provide a direct mechanism for alleviating biofuel toxicity (Dunlop *et al.*, 2011). Different efflux pumps in the hydrophobe/amphiphile (HAE1) family were selected based on their homology to known solvent tolerance pumps, such as the *P. putida* toluene tolerance pump, and were tested against a wide variety of biofuel candidates being sought after as replacements for gasoline, biodiesel and jet fuel. Growth competitions were devised to select for strains harbouring plasmids with the most beneficial pump in the presence of a particular stressor. Several of the most beneficial pumps were from relatively uncharacterized organisms such as *Marinobacter aqueolei* and *Alcanivorax borkumensis*. These organisms have been isolated from oil spills, and may be good sources of solvent tolerance systems. Several of the *P. putidia* pumps were also found to be beneficial in ameliorating solvent toxicity. Interestingly, the native *E. coli* pump AcrAB was found to provide tolerance for several of the compounds tested. This suggests that under the growth conditions used, native regulation may not have been able to respond adequately to the toxicity imposed by the biofuel exposure, though the mechanism was available in the genome. Finally, Dunlop *et al.* (2011) also demonstrated that expression of an effective pump resulted in an improvement in biofuel production.

In addition to stress from the accumulation of biofuels, many inhibitory compounds are produced during biomass saccharification. Since plant biomass is constantly degraded in nature, enzymatic pathways must exist to process these by-products. However, a challenge in utilizing the full range of native biodiversity is the difficulty in culturing and characterizing a majority of soil microbes. Sommer *et al.* (2010) were able to utilize enzymes from soil samples containing many uncultureable microorganisms to engineer tolerance to common biomass inhibitors in *E. coli*. Soil samples were collected from several locations (urban parks, farmland, bogs), and metagenomic DNA was isolated. The authors chose to create four large-insert (50 kb) libraries to test for resistance to hydroquinone, 4-methylcatechol, 4-hydroxy-benzaldehyde, syringaldehyde, 2-furoic acid, furfural and ethanol. Such large inserts were used to allow for expression of multiple genes, which as discussed earlier are often required to confer complex phenotypes. Tolerant phenotypes were found for all tested compounds, but no single library was able to confer tolerance to all inhibitors. Two inhibitory compounds, a lignin derivative syringaldehyde and a sugar dehydration byproduct 2-furoic acid, were chosen for further analysis since these metagenomic fosmids were able to increase the tolerance of *E. coli* by 5.7- and 6.9-fold, respectively. When analysed, the fosmids had little homology to sequences in the NCBI non-redundant database, with the 2-furoic acid winner having 7% homology to a region of the *Pelobacter propionicus* DSM 2379 genome, and syringaldehyde winner having 1% homology to a region of the *Burkholderia ambifaria* AMMD chromosome 2. These were then screened for loss of function, and regions were identified that contained genes that were responsible for the tolerant phenotype. Interestingly, two genes, more than 10 kb apart in the metagenomic fragment, were necessary to confer tolerance to syringaldehyde. If smaller fragments of DNA had been used to create the library, this combination of genes would have been missed. The authors suggest this platform could also be used for selecting for substrate usage, or for production by linking the presence of desired products to a selection marker.

The manipulation of global regulatory systems has been successful in improving phenotypes, and a further refinement of this strategy is to employ non-native regulators. An example is IrrE, a genus-specific global regulator from radiation-resistant bacterium *Deinococcus radiodurans*, which has previously been shown to improve osmotic tolerance, oxidative stress and heat stress in *E. coli* (Pan *et al.*, 2009). It was hypothesized that this regulator may also improve tolerance to industrially relevant fuels such as ethanol, butanol, isobutanol, pentanol and isopentanol since these fuels illicit complex stress signatures that are similar to those improved by expression of IrrE (Chen *et al.*, 2011). Initially IrrE did not help with tolerance, but by mutating it with error-prone PCR, mutants that confer tolerance to each of the aforementioned fuels were obtained. Tolerance to acetate, which is an inhibitory compound generated during most fermentation processes, was also engineered in this manner. The authors hypothesized part of the benefit on IrrE was a reduction in reactive oxygen species, which was confirmed using a fluorescence-based assay. This is one of the first examples of an exogenous regulator being evolved to function beneficially in a new host.

Interesting New Tools for Strain Improvement

Cell-wide mutagenesis and laboratory evolution are popular and powerful techniques for generating diverse phenotypes, but it is a challenging task to differentiate between beneficial and neutral mutations when studying the genomes of the resulting strain. Goodarzi *et al.* (2009) developed an array-based discovery of adaptive mutations (ADAM) tool to address this issue. ADAM requires a library of selectable markers, a mechanism for transferring those markers to an evolved strain, and a method to measure the frequency of insertions (Fig. 8.2). More specifically, the authors infected a strain containing a Kan^R transposon library with P1vir lysate, and transferred this into another strain with a phenotype of interest. Next, the population is divided, and grown in selective conditions (antibiotic, stress, carbon source, etc.) and non-selective conditions. The mRNA from these populations is then labelled and hybridized to arrays to determine the genetic differences. If genes necessary for the phenotype were disrupted, then those loci are depleted after growth on selective media, whereas neutral mutations will remain. The major advantage of this technique is determination of beneficial genes or mutations can be determined without whole genome resequencing, and is specific for genes related to the phenotype of interest.

For proof of principle of the ADAM strategy, an *E. coli* strain containing a chloramphenicol (cm) marker was used as the 'evolved' strain with the beneficial phenotype. After growth in media containing chloramphenicol, the transposon insertions in loci close to *lacZ* were depleted. To quantify the differences in genetic changes, a depletion score was defined as the ratio between marker frequencies in the non-selective conditions to the selective conditions. ADAM's efficiency was also tested on more complex phenotypes, such as growth on asparagine as illustrated in Fig. 8.2, and exposure to ethanol. In the first case, three regions were found to be highly depleted, *sstT*, *ansA* and *lrp*, and further experiments showed all three were necessary for growth on asparagine. In the case of ethanol adaptation, four adaptive mutations were found that were also all required for the tolerant phenotype. ADAM could also be used in other bacteria using other generalized transducing phages, although the array technology may not be commercially available.

Generating vast quantities of genetic mutations in a scalable, tunable and high-throughput manner is a goal of many studies striving to create diverse phenotypes. Current methods of targeting and screening single genes are laborious, and can be difficult to target selected regions of the genome. A system to help address these issues is Multiplex Automated Genome Engineering (MAGE), a method of continually evolving cell populations to generate targeted combinatorial mutations (Wang *et al.*, 2009). A cycle of MAGE consists of growing cells to mid-log phase, inducing λ red ssDNA binding protein β, transforming the cells with oligos targeted to the area of interest, and recovering before further rounds. Allelic replacement is achieved by directing oligos to the lagging strand of the replication fork during DNA replication. New genetic modifications (mismatches, deletions, or insertions) were achieved in greater than 30% of the population in about 2 h. Efficiency in incorporation of genetic changes depends on the amount of homology the oligos have with the target regions, which is an important design consideration. The diversity of the genetic changes is also dependent on the degeneracy of the targeted oligos.

To demonstrate MAGE's potential for metabolic engineering use, the 1-deoxy-D-xylulose-5-phosphate (DXP) pathway, which produces lycopene, was optimized by designing oligos to manipulate the expression of genes previously determined to be important in production. In total, 24 genes were chosen for optimization. For 20 of those genes, 90-mer oligos were designed with degenerate ribosome binding sites (RBS) to bring the native RBS closer to the Shine-Dalgarno canonical sequence. The other four candidate genes were targeted for deletion with oligos that created nonsense mutations in the gene's open reading frame.

After 35 cycles of MAGE, it was estimated that nearly 15 billion genetic variants were generated. Screening of mutant strains was accomplished by isolating intense red colonies, indicative of lycopene production. After 3 days, mutated strains were isolated that had five-fold improved production of lycopene. This platform was also used to increase the incorporation of non-natural amino acids into native proteins, and to construct strains resistant to multiple viruses. The use of MAGE for host engineering efforts may be limited by the available screening techniques for the resulting mutant population.

Conclusions

Systems biology approaches have limitations that must be acknowledged when developing hypotheses for strain engineering. Often, differentially expressed genes or proteins are focused upon as the key mechanisms required for stress response, but this may not always be the case. For example, the efflux pump AcrAB-TolC is up-regulated in response to a variety of solvent-like stresses including butanol stress (Rutherford *et al.*, 2010). However, follow-up on this observation via an expression strain for this pump does not provide a tolerant strain for butanol exposure (Dunlop *et al.*, 2011). The availability of resources such as the KEIO and ASKA collections and the promoter libraries in *E. coli* enables efficient follow-up on hypotheses before a full-scale strain optimization can be conducted (Baba *et al.*, 2006; Kitagawa *et al.*, 2005). Another important limitation to most systems studies is that it does not take into account the behaviour of subpopulations in the culture. It is well documented, especially during stress, that microbial populations exhibit heterogeneity. As a

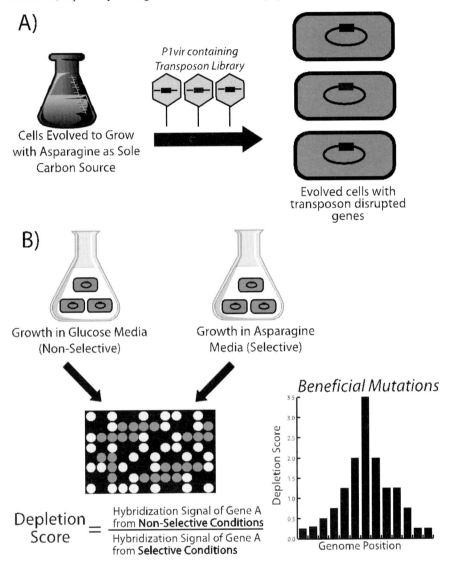

Fig. 8.2. *E. coli* cells were evolved to grow on asparagine as a sole carbon source. These evolved cells were then infected with P1 vir lysate from a KanR transposon library in the parental background. After growth in either selective or non-selective media, a hybridization-based genetic footprinting approach was used to determine the frequency of transposon insertion in each locus. Insertions in genomic regions that were beneficial for the evolved phenotype will appear to be depleted compared with the control.

result, the responses measured are averaged over multiple populations. Further, there is a danger of associating a response from one subpopulation with the entire culture. Strain engineering based on genes whose response originated from a subpopulation with a phenotype contradictory to the process being optimized would lead to an ineffectual effort. Emerging methods that allow cell-wide studies at a single cell resolution may provide ways to avoid such scenarios (Ottesen et al., 2006). Significant improvements in high throughput methods to generate and examine strain libraries may also facilitate identification of effective gene targets and minimize the time lost from following candidates that provide no benefit.

As strain engineering benefits from the exponential increase in resources and research efforts, a final optimized bacterial strain may be considerably altered from the starting strain. Changes in native genes via gene deletions and altered expression levels, incorporation of heterologous mechanisms to counter not only the burden of metabolite production but also other parameters associated with the process such as complex carbon source and industrial scale cultivation, are likely to result in highly specialized fuel production microbes. With recent technological improvements for sequencing as well as gene synthesis, entire synthetic genomes can be envisioned in the near future. The recent demonstration of the chemical synthesis of an entire genome (Gibson et al., 2010) and its transformation into a related but different species is the first step in this direction.

Acknowledgements

We thank Dr Zain Y. Dossani and Dr Kevin George for their input on the manuscript. This work was funded by the US Department of Energy's Joint BioEnergy Institute (http//www.jbei.org) supported by the US Department of Energy, Office of Science, Office of Biological and Environmental Research, and through contract DE-AC02-05CH11231 between Lawrence Berkeley National Laboratory and the US Department of Energy.

References

Ajikumar, P.K., Xiao, W.H., Tyo, K.E., Wang, Y., Simeon, F., Leonard, E., Mucha, O., Phon, T.H., Pfeifer, B. and Stephanopoulos, G. (2010) Isoprenoid pathway optimization for Taxol precursor overproduction in *Escherichia coli*. *Science* 330, 70–74.

Allsop, P.J., Chisti, Y., Mooyoung, M. and Sullivan, G.R. (1993) Dynamics of phenol degradation by *Pseudomonas putida*. *Biotechnology and Bioengineering* 41, 572–580.

Altermann, E., Russell, W.M., Azcarate-Peril, M.A., Barrangou, R., Buck, B.L., McAuliffe, O., Souther, N., Dobson, A., Duong, T., Callanan, M., Lick, S., Hamrick, A., Cano, R. and Klaenhammer, T.R. (2005) Complete genome sequence of the probiotic lactic acid bacterium *Lactobacillus acidophilus* NCFM. *Proceedings of the National Academy of Sciences of the United States of America* 102, 3906–3912.

Aono, R. (1998) Improvement of organic solvent tolerance level of *Escherichia coli* by overexpression of stress-responsive genes. *Extremophiles* 2, 239–248.

Aristidou, A. and Penttila, M. (2000) Metabolic engineering applications to renewable resource utilization. *Current Opinion in Biotechnology* 11, 187–198.

Ashburner, M., Ball, C.A., Blake, J.A., Botstein, D., Butler, H., Cherry, J.M., Davis, A.P., Dolinski, K., Dwight, S.S., Eppig, J.T., Harris, M.A., Hill, D.P., Issel-Tarver, L., Kasarskis, A., Lewis, S., Matese, J.C., Richardson, J.E., Ringwald, M., Rubin, G.M., Sherlock, G. and Gene Ontology C (2000) Gene Ontology, tool for the unification of biology. *Nature Genetics* 25, 25–29.

Atsumi, S., Wu, T.Y., Machado, I.M., Huang, W.C., Chen, P.Y., Pellegrini, M. and Liao, J.C. (2010) Evolution, genomic analysis, and reconstruction of isobutanol tolerance in *Escherichia coli*. *Molecular Systems Biology* 6, 449.

Baba, T., Ara, T., Hasegawa, M., Takai, Y., Okumura, Y., Baba, M., Datsenko, K.A., Tomita, M., Wanner, B.L. and Mori, H. (2006) Construction of *Escherichia coli* K-12 in-frame, single-gene knockout mutants, the Keio collection. *Molecular Systems Biology* 2.

Biebl, H., Menzel, K., Zeng, A.P. and Deckwer, W.D. (1999) Microbial production of 1,3-propanediol. *Applied Microbiology and Biotechnology* 52, 289–297.

Bond-Watts, B.B., Bellerose, R.J. and Chang, M.C.Y. (2011) Enzyme mechanism as a kinetic control element for designing synthetic biofuel pathways. *Nature Chemical Biology* 7, 222–227.

Bongaerts, J., Kramer, M., Muller, U., Raeven, L. and Wubbolts, M. (2001) Metabolic engineering for microbial production of aromatic amino acids and derived compounds. *Metabolic Engineering* 3, 289–300.

Burk, M.J., Burgard, A.P., Osterhout, R.E. and Sun, J. (2009) Microorganisms for the production of 1,4-butanediol. Office USPaT (ed). Genomatica, Inc., USA.

Chen, T., Wang, J., Yang, R., Li, J., Lin, M. and Lin, Z. (2011) Laboratory-evolved mutants of an exogenous global regulator, IrrE from *Deinococcus radiodurans*, enhance stress tolerances of *Escherichia coli*. *PLoS ONE* 6, e16228.

Choi, J.I. and Lee, S.Y. (1999) High-level production of poly(3-hydroxybutyrate-co-3-hydroxyvalerate) by fed-batch culture of recombinant *Escherichia coli*. *Applied and Environmental Microbiology* 65, 4363–4368.

Cornish, E.C., Argyropoulos, V.P., Pittard, J. and Davidson, B.E. (1986) Structure of the *Escherichia coli* K12 regulatory gene *tyrR*. Nucleotide sequence and sites of initiation of transcription and translation. *Journal of Biological Chemistry* 261, 403–410.

Daley, D.O., Rapp, M., Granseth, E., Melen, K., Drew, D. and von Heijne, G. (2005) Global topology analysis of the *Escherichia coli* inner membrane proteome. *Science* 308, 1321–1323.

Dassler, T., Maier, T., Winterhalter, C. and Bock, A. (2000) Identification of a major facilitator protein from *Escherichia coli* involved in efflux of metabolites of the cysteine pathway. *Molecular Microbiology* 36, 1101–1112.

Dehal, P.S., Joachimiak, M.P., Price, M.N., Bates, J.T., Baumohl, J.K., Chivian, D., Friedland, G.D., Huang, K.H., Keller, K., Novichkov, P.S., Dubchak, I.L., Alm, E.J. and Arkin, A.P. (2009) MicrobesOnline, an integrated portal for comparative and functional genomics. *Nucleic Acids Research* 38 (Database issue), D396–D400

Demain, A.L. and Elander, R.P. (1999) The beta-lactam antibiotics, past, present, and future. *Antonie Van Leeuwenhoek International Journal of General and Molecular Microbiology* 75, 5–19.

Dunlop, M.J., Dossani, Z.Y., Szmidt, H.L., Chu, H.C., Lee, T.S., Keasling, J.D., Hadi, M.Z. and Mukhopadhyay, A. (2011) Engineering microbial biofuel tolerance and export using efflux pumps. *Molecular Systems Biology* 7.

Gad, S.C. (2008) *Pharmaceutical Manufacturing Handbook, Production and Processes*, Vol. 10. Wiley-Interscience.

Gerigk, M., Bujnicki, R., Ganpo-Nkwenkwa, E., Bongaerts, J., Sprenger, G. and Takors, R. (2002) Process control for enhanced L-phenylalanine production using different recombinant *Escherichia coli* strains. *Biotechnology and Bioengineering* 80, 746–754.

Gibson, D.G., Glass, J.I., Lartigue, C., Noskov, V.N., Chuang, R.Y., Algire, M.A., Benders, G.A., Montague, M.G., Ma, L., Moodie, M.M., Merryman, C., Vashee, S., Krishnakumar, R., Assad-Garcia, N., Andrews-Pfannkoch, C., Denisova, E.A., Young, L., Qi, Z.Q., Segall-Shapiro, T.H., Calvey, C.H., Parmar, P.P., Hutchison, C.A. III, Smith, H.O. and Venter, J.C. (2010) Creation of a bacterial cell controlled by a chemically synthesized genome. *Science* 329, 52–56.

Goodarzi, H., Hottes, A.K. and Tavazoie, S. (2009) Global discovery of adaptive mutations. *Nature Methods* 6, 581–583.

Goodarzi, H., Bennett, B.D., Amini, S., Reaves, M.L., Hottes, A.K., Rabinowitz, J.D. and Tavazoie, S. (2010) Regulatory and metabolic rewiring during laboratory evolution of ethanol tolerance in *E. coli*. *Molecular Systems Biology* 6, 378.

Gupta, R., Beg, Q.K. and Lorenz, P. (2002) Bacterial alkaline proteases, molecular approaches and industrial applications. *Applied Microbiology and Biotechnology* 59, 15–32.

Haefner, S., Knietsch, A., Scholten, E., Braun, J., Lohscheidt, M. and Zelder, O. (2005) Biotechnological production and applications of phytases. *Applied Microbiology and Biotechnology* 68, 588–597.

Hermann, T. (2003) Industrial production of amino acids by coryneform bacteria. *Journal of Biotechnology* 104, 155–172.

Jones, D.T. and Woods, D.R. (1986) Acetone-butanol fermentation revisited. *Microbiological Reviews* 50, 484–524.

Juan, H.-F., Mori, H., Chang, H.-Y., Huang, H.-C., Huang, T.-C. and Liao, J.C. (2010) Alcohol tolerant *Escherichia coli* and methods of preparation thereof. Office USPaT (ed). National Taiwan University, USA.

Keseler, I.M., Bonavides-Martinez, C., Collado-Vides, J., Gama-Castro, S., Gunsalus, R.P., Johnson, D.A., Krummenacker, M., Nolan, L.M., Paley, S., Paulsen, I.T., Peralta-Gil, M., Santos-Zavaleta, A., Shearer, A.G. and Karp, P.D. (2009) EcoCyc, a comprehensive view of *Escherichia coli* biology. *Nucleic Acids Research* 37, D464–D470.

Khan, M.A. and Isaacson, R.E. (2002) Identification of *Escherichia coli* genes that are specifically expressed in a murine model of septicemic infection. *Infection Immunology* 70, 3404–3412.

Khosla, C. and Keasling, J.D. (2003) Timeline – metabolic engineering for drug discovery and development. *Nature Reviews Drug Discovery* 2, 1019–1025.

Kitagawa, M., Ara, T., Arifuzzaman, M., Ioka-Nakamichi, T., Inamoto, E., Toyonaga, H. and Mori, H. (2005) Complete set of ORF clones of *Escherichia coli* ASKA library (A complete Set of *E. coli* K-12 ORF archive), Unique resources for biological research. *DNA Research* 12, 291–299.

Klein-Marcuschamer, D., Santos, C.N.S., Yu, H.M. and Stephanopoulos, G. (2009) Mutagenesis of the bacterial RNA polymerase alpha subunit for improvement of complex phenotypes. *Applied and Environmental Microbiology* 75, 2705–2711.

Lee, S.Y., Park, J.H., Jang, S.H., Nielsen, L.K., Kim, J. and Jung, K.S. (2008) Fermentative butanol production by clostridia. *Biotechnology and Bioengineering* 101, 209–228.

Li, Y.F. (2009) Carrier proteins for fusion expression of antimicrobial peptides in *Escherichia coli*. *Biotechnology and Applied Biochemistry* 54, 1–9.

Lin, Y. and Tanaka, S. (2006) Ethanol fermentation from biomass resources, current state and prospects. *Applied Microbiology and Biotechnology* 69, 627–642.

Moon, T.S., Yoon, S.H., Lanza, A.M., Roy-Mayhew, J.D. and Prather, K.L.J. (2009) Production of glucaric acid from a synthetic pathway in recombinant *Escherichia coli*. *Applied and Environmental Microbiology* 75, 589–595.

Mukhopadhyay, A., He, Z.L., Alm, E.J., Arkin, A.P., Baidoo, E.E., Borglin, S.C., Chen, W.Q., Hazen, T.C., He, Q., Holman, H.Y., Huang, K., Huang, R., Joyner, D.C., Katz, N., Keller, M., Oeller, P., Redding, A., Sun, J., Wall, J., Wei, J., Yang, Z.M., Yen, H.C., Zhou, J.Z. and Keasling, J.D. (2006) Salt stress in *Desulfovibrio vulgaris* Hildenborough, an integrated genomics approach. *Journal of Bacteriology* 188, 4068–4078.

Nakamura, C.E. and Whited, G.M. (2003) Metabolic engineering for the microbial production of 1,3-propanediol. *Current Opinion in Biotechnology* 14, 454–459.

Newell-McGloughlin, M. and Re, E.B. (2006) *The Evolution of Biotechnology, From Natufians to Nanotechnology.* Springer, Dordrecht, the Netherlands.

Ottesen, E.A., Hong, J.W., Quake, S.R. and Leadbetter, J.R. (2006) Microfluidic digital PCR enables multigene analysis of individual environmental bacteria. *Science* 314, 1464–1467.

Pan, J., Wang, J., Zhou, Z.F., Yan, Y.L., Zhang, W., Lu, W., Ping, S., Dai, Q.L., Yuan, M.L., Feng, B., Hou, X.G., Zhang, Y., Ma, R., Liu, T., Feng, L., Wang, L., Chen, M. and Lin, M. (2009) IrrE, a global regulator of extreme radiation resistance in *Deinococcus radiodurans*, enhances salt tolerance in *Escherichia coli* and *Brassica napus*. *Plos ONE* 4, 9.

Peralta-Yahya, P.P. and Keasling, J.D. (2010) Advanced biofuel production in microbes. *Biotechnology Journal* 5, 147–162.

Reyes, L.H., Almario, M.P. et al. (2011) Genomic library screens for genes involved in n-butanol tolerance in *Escherichia coli*. *Plos One* 6(3): e17678.

Ruther, A., Misawa, N., Boger, P. and Sandmann, G. (1997) Production of zeaxanthin in *Escherichia coli* transformed with different carotenogenic plasmids. *Applied Microbiology and Biotechnology* 48, 162–167.

Rutherford, B.J., Dahl, R., Price, R., Szmidt, H.L., Benke, P.I., Mukhopadhyay, A. and Keasling, J.D. (2010) Functional genomic study of exogenous *n*-butanol stress in *Escherichia coli*. *Applied Environmental Microbiology* 76, 1935–1945.

Shen, C.R., Lan, E.I., Dekishima, Y., Baez, A., Cho, K.M. and Liao, J.C. (2011) Driving forces enable high-titer anaerobic 1-butanol synthesis in *Escherichia coli*. *Applied Environmental Microbiology* 77, 2905–2915.

Singh, J. and Cu, S. (2010) Commercialization potential of microalgae for biofuels production. *Renewable & Sustainable Energy Reviews* 14, 2596–2610.

Sommer, M.O., Church, G.M. and Dantas, G. (2010) A functional metagenomic approach for expanding the synthetic biology toolbox for biomass conversion. *Molecular Systems Biology* 6, 360.

Wang, H.H., Isaacs, F.J., Carr, P.A., Sun, Z.Z., Xu, G., Forest, C.R. and Church, G.M. (2009) Programming cells by multiplex genome engineering and accelerated evolution. *Nature* 460, 894–898.

Wegmann, U., O'Connell-Motherwy, M., Zomer, A., Buist, G., Shearman, C., Canchaya, C., Ventura, M., Goesmann, A., Gasson, M.J., Kuipers, O.P., van Sinderen, D. and Kok, J. (2007) Complete genome sequence of the prototype lactic acid bacterium *Lactococcus lactis* subsp *cremoris* MG1363. *Journal of Bacteriology* 189, 3256–3270.

Yim, H., Haselbeck, R. *et al.* (2011) Metabolic engineering of *Escherichia coli* for direct production of 1,4-butanediol. *Nature Chemical Biology* 7, 445–452.

Zaldivar, J., Nielsen, J. and Olsson, L. (2001) Fuel ethanol production from lignocellulose, a challenge for metabolic engineering and process integration. *Applied Microbiology and Biotechnology* 56, 17–34.

Zhang, Y.P., Zhu, Y. and Li, Y. (2009) The importance of engineering physiological functionality into microbes. *Trends in Biotechnology* 27, 664–672.

Zhu, Y., Zhang, Y.P. and Li, Y. (2009) Understanding the industrial application potential of lactic acid bacteria through genomics. *Applied Microbiology and Biotechnology* 83, 597–610.

Chapter 9

Microbial Fuel Cells: Electricity Generation from Organic Wastes by Microbes

Kun Guo, Daniel J. Hassett and Tingyue Gu

Introduction

Microbial fuel cells (MFCs) are bioreactors that convert chemical energy stored in the bonds of organic chemicals into electricity through biocatalysis of microorganisms (Potter, 1911; Cohen, 1931; Davis and Yarbrough, 1962; Moon et al., 2006). The schematic of a typical MFC is shown in Fig. 9.1; the anodic and cathodic chambers are separated by a proton exchange membrane (PEM) (Wilkinson, 2000; Gil et al., 2003) that allows transport of protons while blocking oxygen and other compounds. Microbes in the anodic chamber degrade organic matter and produce electrons, protons and carbon dioxide (CO_2). Electrons and protons produced by microbes are then transported to the cathodic chamber via external circuit and a proton exchange membrane (PEM), respectively. In the cathodic chamber, protons and electrons react with oxygen to form water. Because the terminal electron acceptor (i.e. oxygen) is kept away from the anodic chamber, electrons are allowed to pass through the external load to generate electricity (Du et al., 2007).

Using acetate as example for fuel, the two electrode half-reactions and the overall redox reaction are shown below:

$$\text{Anodic reaction: } CH_3COOH + 2H_2O \xrightarrow{microbe} 2CO_2 + 8H^+ + 8e^- \quad (9.1)$$

$$\text{Cathodic reaction: } 8H^+ + 8e^- + 2O_2 \longrightarrow 4H_2O \quad (9.2)$$

$$\text{Overall reaction: } CH_3COOH + 2O_2 \xrightarrow{microbe} 2H_2O + 2CO_2 \quad (9.3)$$

Reaction 9.3 shows that acetate (substrate) is oxidized to produce CO_2 and water. Theoretically, the anodic potential (E_{An}) is around −0.300 V (versus NHE) while the cathode potential (E_{Cat}) is about 0.805 V. Therefore, the maximum cell potential is 1.105V (Logan et al., 2007). However, in practice, the cathode potential with oxygen as the terminal electron acceptor is much less than the theoretical maximum due to the overpotential of cathodic reaction. Typically, the open circuit potential is about 0.4 V for an

MFC with an air cathode, while the working potential is only around 0.25 V even when prohibitively expensive Pt is used as the cathodic catalyst (Liu and Logan, 2004).

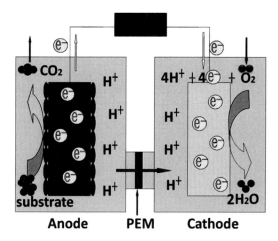

Fig. 9.1. Schematic diagram of a typical two-chamber MFC.

Compared with traditional technologies used for energy generation from organic matter, MFCs hold many inherent advantages and have much wider applications.

1. MFCs have a wide range of substrates, such as carbohydrates, proteins, lipids and even the organic matters in wastewater (Pant *et al.*, 2010).
2. MFCs possess a high-energy transformation efficiency, since it converts the chemical energy stored in substrates into electricity directly.
3. Single-chamber MFCs do not require energy input for aeration, which lowers operational costs (Rabaey and Verstraete, 2005).
4. MFCs have great potentials for widespread applications such as wastewater treatment, biological oxygen demand (BOD) sensors, bioremediation, hydrogen production and electricity generation (Logan and Regan, 2006a).

MFC History

Over a century ago, Potter was the first to demonstrate that electrical current can be generated during the degradation of organic substrates by bacteria or yeast (Potter, 1911). Two decades later, Cohen confirmed Potter's results and produced an overall voltage of 35 V at a current of 0.2 mA using a stacked bacterial fuel cell (Cohen, 1931). These publications are generally considered the first reported cases of MFCs, but they did not generate much interest since the current density and power output were very small.

It was not until the 1960s that the idea of microbial electricity generation was picked up again as a potential method to convert human wastes into electricity during long space flights (Canfield *et al.*, 1963). It was realized that the complicated underlying bioelectrochemical processes in MFCs' operations require systematic and long-term research efforts (Cohn, 1963; Lewis, 1966). Before MFC research could take off, the rapid advances in other energy technologies (e.g. photovoltaics) forced MFC research to the back burner (Schroder, 2007). Another milestone of MFCs is the discovery in 1980s that externally added electron mediators could greatly enhance MFC current density and power output (Delaney *et al.*, 1984; Roller *et al.*, 1984). These exogenous electron mediators include dyes and metallorganics such as neutral red (NR), methylene blue (MB), thionine, meldola's blue (MelB), 2-hydroxy-1,4-naphthoquinone (HNQ) and Fe(III)EDTA (Vega and Fernandez, 1987; Allen and Bennetto, 1993; Park and Zeikus, 2000; Tokuji and Kenji, 2003; Ieropoulos *et al.*, 2005b). Unfortunately, these synthetic compounds usually tend to be cytotoxic, unstable and expensive, thus limiting their applications beyond laboratory tests (Du *et al.*, 2007).

The latest and most remarkable breakthrough of MFCs was made near the end of the 20th century when some microbes were found to be capable to transfer electrons directly to the anode (Kim *et al.*, 1999). Examples of this kind of bioelectrochemical active bacterial species are *Shewanella putrefaciens* (Kim *et al.*, 2002), *Geobacter sulfurreducens* (Bond and Lovley, 2003), *Rhodoferax ferrireducens* (Chaudhuri and Lovley, 2003) and *Geobacter metallireducens* (Min *et al.*, 2005a). These microbes could form a biofilm on an anode and transfer the electrons from organic carbon oxidation to the anode via the cell membrane or special conductive pili (also known as nanowires). Since mediators are not needed in this kind of MFC, the operational cost can be reduced and the concern for environmental pollution caused by artificial mediators is eliminated (Ieropoulos *et al.*, 2005b). Furthermore, this form of MFC is operationally more stable and yields a high coulombic efficiency (Du *et al.*, 2007). Hence, mediator-less MFCs are more suitable for wastewater treatment and power generation.

In the past decade, rapid progress has been made in MFC research and the number of publications in this area has increased exponentially. Most of the studies focused on MFC design, exoelectrogenic bacteria and cost-effective electrode materials (Logan and Regan, 2006a; Logan *et al.*, 2006; Rozendal *et al.*, 2008a). The power densities of MFCs have increased from below 0.1 to 6860 mWm^{-2} over the past decade (Kim *et al.*, 1999; Fan *et al.*, 2008). The increase will eventually be limited by the sustained electron transfer rate achieved by the bacteria used in MFCs. In a practical MFC, a mixture of metabolically versatile organisms is typically present. Researchers have realized that understanding the roles of different bacteria in a synergistic biofilm consortium and the electron transfer mechanisms by key bacteria is needed to improve MFC power output (Logan and Regan, 2006b).

Electricity-producing Bacteria and their Electron Transfer Mechanisms

Electricity-producing bacterial communities

Theoretically, most anaerobes (or facultative anaerobes) have the potential to be the biocatalyst in MFCs if final electron acceptors such as oxygen, nitrate and sulfate are absent in the culture and proper electron shuttles are present. In recent years, there has been a significant increase in the open literature on many kinds of electrochemically active microbes that are capable of producing electricity (Logan, 2008). A variety of bacteria can produce a modicum of electricity in an MFC if a mediator is used to shuttle the electrons between the anodic surface and the cells, while many other bacteria have been found to transfer electrons from fuel (substrate) oxidation to a working electrode without a mediator. There are several reviews that summarized the microbial species used in MFCs (Chang *et al.*, 2006; Du *et al.*, 2007; Logan, 2009). A list of microbial species used in MFCs is shown in Table 9.1 with information compiled from Chang *et al.* (2006), Du *et al.* (2007) and Logan (2009).

Mechanisms of Extracellular Electron Transfer

Electrochemically active bacteria (EAB) are able to transfer electrons to an electrode from the inside of the cell via two different extracellular electron transfer mechanisms (Schroder, 2007; Rozendal *et al.*, 2008a). The first mechanism is mediated electron transfer (MET), which depends on redox cycling of mediators between the microbes and the electrode. The mediators can either be naturally present chemicals such as humic acids and some sulfur species (Stams*et al.* 2006), or chemicals secreted by the microbes such as quinones (Newman and Kolter, 2000) or phenazines (Rabaey *et al.*, 2005). The second mechanism is direct electron transfer (DET), which depends on the direct contact between microbes and the electrode surface. Some EAB use a series of membrane redox proteins (i.e. cytochromes) to transfer the electrons (Lovley, 2006), while some other EAB utilize electrically conductive pili on the surface of their cells to transfer electrons (Reguera *et al.*, 2005; Gorby *et al.*, 2006).

Mediated electron transfer

Mediators are required for electron transfer by most bacteria to transport electrons from inside cells to the outside, since their cell membranes consist of non-conductive lipids, peptidoglycans and lipopolysaccharides (Davis and Higson, 2007; Du *et al.*, 2007). Mediators in an oxidized state are reduced in the cytoplasm or periplasm by absorbing electrons released by enzyme-catalysed organic carbon oxidation inside the bacteria. The reduced mediators diffuse to an anodic surface and donate the electrons to the anode and in the mean time become oxidized again (Schroder, 2007). The oxidized mediators are now ready to repeat the same process. This cyclic process is depicted in Fig. 9.2A. Figure 9.2B shows the other scheme for mediated electron transfer in which the mediators do not cross the cell membrane. They are reduced when they encounter electron transport proteins (such as cytochrome) on the outer cell membrane instead.

Judging from their sources, mediators can be divided into two groups, namely the exogenous (artificial) mediators and endogenous mediators. In nature, some microorganisms may use externally available (exogenous) electron shuttling compounds such ashumic acids or metal chelators (Stams *et al.*, 2006). In several reports, artificial redox mediators such as neutral red (Park and Zeikus, 1999), thionin (Choi *et al.*, 2003) and methyl viologen (Aulenta *et al.*, 2007) were added into MFC reactors. These mediators were able to enhance the electric current generation, but a big disadvantage of the use of exogenous redox mediators is the need for repeated addition, which is costly, because they are usually unstable. Furthermore, these artificial mediators may pose environmental concerns due to their toxicity (Schroder, 2007). Thus, the application of artificial mediators

Table 9.1. Microbes used in MFCs.

Microbe	Comment	Reference
Mediator-needed electricity-producing bacteria		
Proteus mirabilis	Thionine as mediator	Thurston et al., 1985
Erwinia dissolven	Ferric chelate complex as mediators	Vega and Fernandez, 1987
Lactobacillus plantarum	Ferric chelate complex as mediators	Vega and Fernandez, 1987
Streptococcus lactis	Ferric chelate complex as mediators	Vega and Fernandez, 1987
Desulfovibrio desulfuricans	Sulfate/sulfide as mediator	Park et al., 1997
Actinobacillus succinogenes	Neutral red or thionine as electron mediator	Park and Zeikus, 1999
Gluconobacter oxydans	Mediator (HNQ, resazurin or thionine) needed	Lee et al., 2002
Escherichia coli	Mediators such as methylene blue needed	Schroder et al., 2003
Proteus mirabilis	Thionin as mediator	Choi et al., 2003
Pseudomonas aeruginosa	Pyocyanin and phenazine-1-carboxamide as mediator	Rabaey et al., 2004
Klebsiella pneumoniae	HNQ as mediator	Rhoads et al., 2005
Shewanella oneidensis	Anthraquinone-2,6-disulfonate (AQDS) as mediator	Ringeisen et al., 2006
Mediator-less electricity-producing bacteria		
Shewanella putrefaciens IR-1	A dissimilatory metal-reducing bacterium	Kim et al., 1999
Desulfuromonas acetoxidans	Deltaproteobacteria identified from a sediment MFC	Bond et al., 2002
Geobacter metallireducens	Shown to generate electricity in a poised potential system	Bond et al., 2002
Geobacter sulfurreducens	generated current without poised electrode	Bond and Lovley, 2003
Rhodoferax ferrireducens	Betaproteobacteria used glucose as substrate	Claudhuri and Lovley, 2003
Aeromonas hydrophila	Deltaproteobacteria	Pham et al., 2003
Desulfobulbus propionicus	Deltaproteobacteria	Holmes et al., 2004a
Escherichia coli	Found to produce current after a long acclimation time	Zhang et al., 2006
Shewanella oneidensis DSP10	Achieved a high power density 2 W m^{-2}	Ringeisen et al., 2007
S. oneidensis MR-1	Various mutants identified that increase current or lose the ability for current generation	Bretschger et al., 2007
Pichia anomala	Current generation by yeast (kingdom Fungi).	Prasad et al., 2007
Rhodopseudomonas palustris DX-1	Produced high power densities of 2.72 Wm^{-2}	Xing et al., 2008
Ochrobactrumanthropi YZ-1	An opportunistic pathogen, e.g. P. aeruginosa	Zuo et al., 2008
Desulfovibrio desulfuricans	Reduced sulphate when growing on lactate	Zhao et al., 2008
Acidiphilium sp. 3.2 Sup5	Power production at low pH	Borole et al., 2008
Klebsiella pneumoniae L17	Produced current without a mediator for the first time	Zhang et al., 2008

Thermincola sp. strain JR	Phylum Firmicutes	Wrighton *et al.*, 2008
Geopsychrobacter electrodiphilus	Psychrotolerant Deltaproteobacteria	Holmes *et al.*, 2004b

is rather limited and is gradually being abandoned. Fortunately, in some systems exogenous mediators are not needed since some bacteria possess the ability to secrete mediators by themselves. These mediators typically come from reversibly reducible compounds (secondary metabolites) and oxidizable metabolites (primary metabolites) (Rabaey and Verstraete, 2005). Examples for such secondary metabolites are pyocyanin, 2-amino-3-carboxy-1, 4-naphthoquinone and ACNQ (Hernandez and Newman, 2001; Rabaey *et al.*, 2004), which are able to shuttle electrons to an electrode. The primary metabolites that are involved in the electron transfer include H_2 and H_2S, which are used by *Escherichia coli* K12 (Schroder *et al.*, 2003) and *Sulfurospirillum deleyianum* (Straub and Schink, 2004) as mediators, respectively.

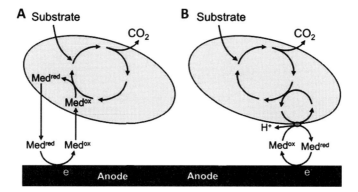

Fig. 9.2. Simplified schematic illustration of MET. (A) Shuttling via cytoplasmic or periplasmic redox couples; (B) shuttling via outer cell membrane electron transport proteins. (Figure redrawn with modifications after Schroder, 2007.)

Direct electron transfer

Certain microbes seem to also possess the ability to transport electrons out of the cell via the cell membrane or special conductive pili rather than mediators. This form of electron transfer approach is known as DET (Schroder, 2007). Since the outer layers of most bacteria are non-conductive, a series of membrane-bound electron transport proteins are required for DET, such as *c*-type cytochromes and haem proteins (Du *et al.*, 2007). Two proposed DET pathways are illustrated in Fig. 9.3. One is through the membrane-bound electron transport proteins (Fig. 9.3A), and the other is through the pili (nanowires) that are connected to the membrane-bound electron transport proteins (Fig. 9.3B).

In Fig. 3A, the DET via outer membrane-bound cytochromes requires that the bacterial cells adhere to the surface of the anode directly (Schroder, 2007). Consequently, only the bacteria in the first monolayer of the sessile cells directly on the anode surface contribute to the current generation in the MFC (Lovley, 2006).

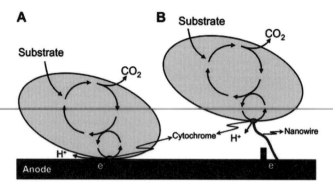

Fig. 9.3. Simplified schematic illustration of DET: (A) electron transfer via membrane-bound cytochromes; (B) electron transfer via nanowires. (Reproduced by permission of the PCCP Owner Societies from Schroder, 2007.)

The MFC current output thus depends on the cell density in this bacterial monolayer. Thus, the power densities of these MFCs are usually very limited. For example, the maximum current densities for MFCs based on *Shewanella putrefaciens* (Kim et al., 2002), *Rhodoferax ferrireducens* (Chaudhuri and Lovley, 2003) and *G. sulfurreducens* (Bond and Lovley, 2003) were as low as 0.6, 3 and 6.5 μAcm^{-2}, respectively.

Figure 9.3B shows the simplified mechanism of DET via electronically conducting molecular pili. The pili are connected to the membrane-bound cytochromes that transfer the electrons from the inside to the outside of a cell. It has been reported that some bacteria (e.g. *Geobacter* and *Shewanella*) strains can utilize their electronically conducting molecular pili as nanowires to transfer electrons (Reguera et al., 2005; Gorby et al., 2006). The nanowires may allow the microorganisms to utilize more space around the anode and develop thicker electrochemically active biofilms beyond the first monolayer of sessile cells on the anode, thereby increasing the functional bacteria density and boosting the current generation (Schroder, 2007). For instance, Reguera et al. (2006) reported that the nanowires of *G. sulfurreducens* may network multilayers of sessile cells in the biofilm to transfer electrons efficiently and thus increase electricity production by more than ten-fold.

Enrichment of Electricity-producing Bacteria

Currently, electricity-producing bacteria are generally isolated from anaerobic sludge using an MFC reactor as a selection tool. This method was initially proposed based on the observation that the metabolism of *S. putrefaciens* was stimulated by the presence of an MFC anode when electron acceptors were not present (Hyun et al., 1998). Kim et al. (2004) described a process to enrich microbes for MFCs. They used sludge collected from a maize-processing wastewater treatment plant as inoculum for the anodic chamber of an MFC and fed it with wastewater from the same source. Initially the MFC generated a current of 20 μA with an external load of 10 Ω. When the anode solution was replaced with a more nutritious wastewater from a difference source, the current increased with concomitant COD (chemical oxygen demand) reduction. After repeated replenishments of the wastewater the current output eventually reached 1.2 mA (Chang *et al*., 2006). Figure 9.4 shows the process of enrichment and selection of electricity-producing bacteria.

Figure 9.4A shows that in the initial stage of enrichment, there are only planktonic cells. No electricity is generated because there are no sessile cells on the anode to transfer electrons to the anode or mediators secreted by these organisms. When sessile cells start to appear on the anode, electron transfer starts (Fig. 9.4B). It is believed that electrode reducing is an energy-conserving microbial respiration process (Bond *et al*., 2002), and the electron-donating bacteria can be enriched during MFC operation (Chang *et al*., 2006). When electrons are removed from the cytoplasm, beneficial organic carbon oxidation proceeds forward. This means that with no other electron acceptors in the culture medium in an MFC operation, the electricity-producing bacteria tend to attach to the anodic surface to donate the electrons to the anode. During the enrichment process, the current output of the MFC increases gradually. After the enrichment, a variety of electricity-producing bacteria form a biofilm on the anodic surface and the current output of the MFC reaches a maximum value and becomes stabilized (Fig. 9.4C).

Nutrients can be introduced to facilitate the enrichment process in the MFC. Copiotrophic cultures can be enriched with artificial wastewater containing acetate, propionate or even glucose and glutamate, and oligotrophic cultures with artificial wastewater or river water with added nutrients (Kang *et al*., 2004; Phung *et al*., 2004). Introducing fermentable substrates promoted a more diverse microbial population in the MFC than non-fermentable substrates such as acetate. Electrochemically inactive bacteria may also be promoted together with EAB in the same biofilm consortium. Their metabolism may play a critical role in assisting EAB during electricity production (Chang *et al*., 2006).

Genetically Engineered 'Super-bugs' to Reduce or Eliminate Electron Transfer Resistance: Roles of Cytochromes, Polysaccharide and Type IV Pili

Harnessing electrogenic properties

The study of specific factors in bacteria that are critical for optimal electrogenesis has really only just begun in the past few years. Harnessing maximal electrogenic properties in bacteria requires organisms that are genetically tractable and possessing properties that include: (i) the ability to metabolize a wide variety of carbon-rich substrates; (ii) the ability to form surface-associated communities on anodic surfaces known as biofilms; and (iii) the capacity to synthesize and secrete mediators. Below is a brief synopsis of three factors that are known to be involved in optimal MFC current generation. Future advances in this area present an exciting possibility for using genetically engineered super-bugs to maximize MFC power output beyond any other existing MFC optimization methods could achieve.

Cytochromes

The first genome-wide screen to date for the assessment of the role of different bacteria gene products in electrogenesis was performed in *G. sulfurreducens* (Kim *et al*., 2008). Bacteria lacking OmcF, an outer membrane *c*-type cytochrome, possessed significantly

170   *Microbial Biotechnology: Energy and Environment*

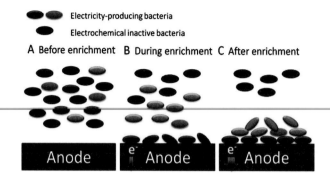

Fig. 9.4. Enrichment of electricity-producing bacteria. (Figure redrawn with modifications after Chang *et al.*, 2006.)

reduced overall current production (~0.3 relative to 0.75 mA over 200 h). The reduced current generated by the *omcF* mutant was not dependent upon iron reduction, as an *omcB* mutant was shown to generate near wild-type current. DNA microarray analysis conducted on the *omcF* mutant versus wild-type bacteria revealed decreased transcription of genes known to be upregulated when bacteria were on the anodic surface. Among these were genes encoding metal efflux and/or type I secretion proteins, OmcS and OmcE, and outer membrane cytochromes, and several hypothetical proteins. More recently, Krushkal and co-workers showed that an *omcB* mutant, lacking the ability to reduce soluble or insoluble iron, was able to adapt and grow on soluble iron with time. It was then speculated that multiple regulatory elements played a role in the adaptive behaviour (Krushkal *et al.*, 2009).

Polysaccharide

In *S. oneidensis* MR-1, Kouzuma and co-workers (Kouzuma *et al.*, 2010) used a transposon mutagenesis approach to identify a mutant in the *SO3177* gene, encoding a putative formyl tranferase involved in cell surface polysaccharide biosynthesis. This mutant possessed a cell surface that was more hydrophobic than wild-type organisms, an ability to adhere better to graphite felt anodes, and a 50% better current generation capacity. Similar to *Geobacter* species, the extended respiratory chain of *S. oneidensis* MR-1 is also involved in the process of dissimilatory iron-reduction. The outer membrane cytochromes MtrC and MtrF were found to be powerful reductases of chelated ferric iron, birnessite, and a carbon anode in an MFC (Bucking *et al.*, 2010).

Type IV pili: involvement in attachment to anodic surfaces and electron conduction

A major means by which bacteria can conduct electrons to anodic surfaces is via the synthesis of type IV pili, which emanate from the surface of many bacterial genera, including many of those listed in Table 9.1. These 'nanowires' have been demonstrated and characterized on electrogenic *Shewanella* and *Geobacter* species (Reguera *et al.*, 2005, 2006).

Conduction atomic force microscopy has clearly established that these type IV pili are highly conductive under modest voltage biases. The conducting pili of *Shewanella* and

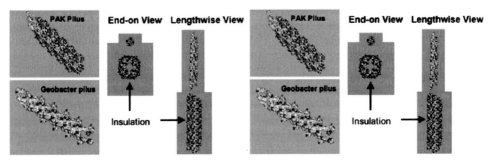

Fig. 9.5. Fundamental 3-dimensional structure of the pilus 'nanowire' of *P. aeruginosa* strain PAK and *Geobacter* sp., indicating the insulating properties of this unique appendage (Randall T. Irvin, U. of Alberta, personal communication with DJH).

Geobacter are 3-start helical assemblages of a short type IV pilin that display a high homology to the *N*-terminal region of the classical type IV pilin of *P. aeruginosa* (Fig. 9.5).Recently, researchers have established that the *P. Aeruginosa* pilin structural protein possesses a binding domain that mediates direct contact and exceptionally strong contact with stainless steel (Giltner *et al.*, 2006).The pilin binding domain de-localizes electrons from the stainless steel and effectively increases the electron work function (EWF) of the stainless steel surface substantially while a truncated monomeric form of pilin bound to the surface has a substantially lower EWF (i.e. it readily releases electrons when exposed to a voltage bias).This establishes that the *P. aeruginosa*pilin structural protein, like those of *Shewanella* and *Geobacter*, is a functional 'semi-conductor'.

In *G. sulfurreducens*, the conductive pili are essential for Fe(III) oxide reduction and optimal current production in MFCs (Reguera *et al.*, 2005, 2006). In fact, the bacteria that utilize insoluble iron or manganese specifically synthesize both flagella and type IV pili so that they may undergo a motile process toward these electron sources known as chemotaxis (Childers *et al.*, 2002). PilR is a recently discovered transcription factor that is required for the genetic activation of pilA, encoding the pilus structural subunit. Surprisingly, PilR is required for both soluble and insoluble iron reduction (Juarez *et al.*, 2009). More recently, PilR DNA-binding motifs were assessed in a genome-wide screen. Using Pattern Location software, Krushkal and co-workers probed the *G. sulfurreducens* genome for candidate PilR-binding domains and identified 523 putative sequence elements (Krushkal *et al.*, 2010). These included 328 category I tandem repeat patterns 5'-[(N)$_{4-6}$STGTC]r-3' and 5'-[(N)$_{5-7}$TGTC]r-3', with r = 2 or 3. Many of these motifs resided immediately upstream of genes involved in pilus and flagellum, secretion, and cell wall synthesis.

What genetic modifications should we engineer to create better electrogens?

There are many methods to create a far better electrogenic bacterium.

1. To mutate the organism to generate more polar, conductive pili, such as pilT mutations in *P. aeruginosa* (Fig. 9.6). It is well known that both pili and flagella are required for biofilm formation in electrogenic *P. aeruginosa*. Figure 9.7 demonstrates that both the type IV pilus and the single flagellum are required for optimal biofilm formation in *P. aeruginosa* (Yoon *et al.*, 2002).

2. Construct mutants that are fully capable of forming biofilms, yet unable to disperse from the biofilm.
3. Limit production of polysaccharides so that nutrient or feedstock flow to electrogenic organisms would be maximized to the bacteria and not impaired matrix components such as polysaccharide DNA, lipid or protein.
4. Restrict anaerobic respiratory metabolism in denitrifying bacteria such that electrons will only flow to the anode and not to another anaerobic terminal electron acceptor such as nitrate, sulfate or sulfur.
5. Control cell division, as rapid cell growth could clog the anodic compartment.
6. Increase the ability of certain organisms to generate mediators. Thus, such bacteria would not only generate electricity via biofilm (e.g. pilus) mediate conductivity, but also that facilitated through mediators.
7. Entertain the possibility of cloning in an uncoupling protein (e.g. the UCP class in eukaryotes) that would increase the rate of oxidation of substrates, thereby increasing electron flow to the anode.

These are testable hypotheses currently being investigated in our research collaborations.

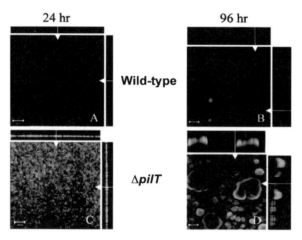

Fig. 9.6. Wild-type *P. aeruginosa* 24 and 96 h biofilms versus those of an isogenic *pilT* mutant (bottom panels). Note the dramatically more robust biofilms formed by the *pilT* mutant. (Reprinted from Chiang and Burrows, 2003 with permission from American Society for Microbiology.)

MFC Reactor Designs

As mentioned above, a typical MFC has an anode and a cathode separated by a proton exchange membrane to prevent direct oxidation of organic carbons in the anodic chamber. In fact, any reactor that can provide the anode an anaerobic environment and the cathode an aerobic one as well as a pathway for charge exchange is possible for MFC operations. For MFCs with a biocathode, the cathodic chamber can either be aerobic or anaerobic depending on whether aerobes or anaerobes are used in the chamber (Lefebvre et al., 2008). Various MFCs were designed for different purposes.

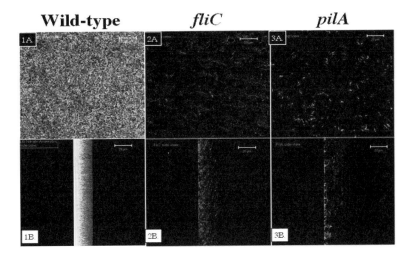

Fig. 9.7. Confocal microscopic examination of biofilm formation by *P. aeruginosa* wild-type, flagellum-deficient (fliC), and pilus-deficient (pilA) bacteria grown under anaerobic conditions. A, top view; B, saggital view 22 mm into the biofilm. (From Yoon *et al.*, 2002, with permission from Elsevier.)

Two-chamber MFCs

Many different configurations are possible for two-chamber MFCs. The most commonly used design is an H-shaped two-chamber MFC (Fig. 9.8A) (Logan *et al.*, 2006). However, these MFCs typically have a high internal resistance because of the long distance between the two electrodes and small membrane area, hence limiting power density output. They are often used to study various MFC parameters, such as new substrates, electrode materials, membranes or microbial communities, or for MFC-based sensors.

In order to reduce the internal resistance, MFCs with a larger membrane area and a shorter electrode distance, such as cube-type MFCs and flat-type MFCs, reduced the internal resistance substantially, thereby increasing the power generation. Kim and co-workers constructed a cube-type MFC with two plexiglas cylindrical chambers, each 2 cm long and 3 cm in diameter separated by a membrane (Fig. 9.8B). The reactor was used to study the internal resistances for different membranes (Kim .*et al.*, 2007b). Min and Logan (2004) created a flat-plate MFC (Fig. 9.8C) for continuous wastewater treatment. A carbon cloth that was hot pressed to a Nafion PEM formed the cathode that was in direct contact with a carbon paper anode. The smallest MFC for power generation reported so far was built by Ringeisen *et al.* (2006), which had a diameter of about 2 cm. It had a high volume power density. This type of miniature MFC reactor may be used as power source for autonomous sensors in remote areas for long-term operations. Recently, Hou *et al.* (2009) described a micro-array of 24 MFCs through microfabrication for the purpose of fast screening of electrogenic microbes.

The above two-chamber MFCs, however, are not attractive in scaling up because of their complexity and high costs. He and co-workers used a tubular upflow MFC that was

174 *Microbial Biotechnology: Energy and Environment*

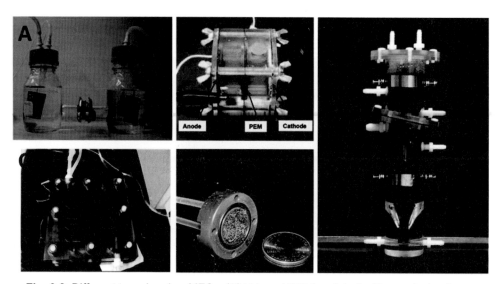

Fig. 9.8. Different two-chamber MFCs. (A) H-type MFC (reprinted with permission from Logan and Regan, 2006a. Copyright 2006 American Chemical Society). (B) Cube-type MFC (reprinted with permission from Kim *et al.*, 2007b. Copyright 2007 American Chemical Society). (C) Flat-type MFC (reprinted with permission from Min and Logan, 2004. Copyright 2004 American Chemical Society). (D) Miniature MFC (reprinted from Ringeisen *et al.*, 2008. US Government publication). (E) Tubular upflow MFC (reprinted with permission from He *et al.*, 2005. Copyright 2005 American Chemical Society).

made from plexiglas (He *et al.*, 2005). The cathode chamber (9 cm tall, 250 cm^3 in wet volume) was placed above the anode chamber (20 cm tall, 520 cm^3 in wet volume), and both of them were packed with reticulated vitreous carbon (RVC) (Fig. 9.8E). A cation exchange membrane (CEM) was used to partition the two chambers with a 15° angle to prevent accumulation of gas bubbles. The reactor had a power density of 170 mWm^{-2} membrane area when fed a sucrose solution in anode and ferricyanide solution in cathode, with coulombic efficiencies (C_E) ranging from 0.7% to 8.1%. The low efficiencies were likely due to the 84 Ω internal resistance.

Single-chamber MFCs

Researchers have proved that it is not essential to have a cathode filled with liquid catholyte or in a dedicated chamber when oxygen or air is used at the cathode. Therefore, the cathode can be in direct contact with oxygen or air with or without a membrane (Logan *et al.*, 2006). This type of MFC was referred to as a single-chamber MFC. Single-chamber MFCs have many inherent advantages over two-chamber MFCs, such as simpler designs, cost savings, and no need for aeration in the cathode chamber. Figure 9.9 shows the schematic of a typical single-chamber MFC.

Liu and Logan (2004) designed a cube-type air-cathode reactor consisting of an anode and a cathode placed on the opposite sides in a plastic cylindrical chamber (28 ml) (Fig. 9.10A). The anode was a carbon paper without wet proofing, and the cathode was a carbon electrode/PEM assembly. A standalone rigid carbon paper without PEM was also used as a cathode for the reaction. Without the membrane, the power density was higher (494 versus

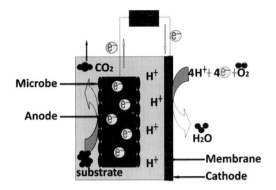

Fig. 9.9. Schematic of a typical single-chamber MFC.

232 mWm^{-2}). Logan *et al.* (2007) showed a side-arm MFC, which is similar to the classic H-shaped dual-chamber MFC, but without the cathode bottle (Fig. 9.10C). It reached a maximum power density of 1430 mW m^{-2} with glucose as fuel.

Among single-chamber MFC designs, tubular MFC systems are gaining popularity again because of their scalable characteristics. Open-air cathodes can be put either outside or inside the anode chamber. Liu and co-workers constructed a horizontal tubular single-chamber reactor (Liu *et al.*, 2004). Its cathode was in the centre of an acrylic tube (Fig. 9.10B). The anode consisted of eight graphite rods surrounding the cathode. The cathode was a carbon cloth hot-pressed on a Nafion membrane. The MFC generated a power density of 26 mW m^{-2}(C_E < 12%) with 80% OCD removal (Liu *et al.*, 2004). You *et al.* (2007) reported a tubular MFC configuration with an outer cathode (Pt-coated carbon cloth) wrapped around an anode (carbon granules) (Fig. 9.10D). The anode chamber was a 3cm diameter, 13.5cm high and 20mm thick cylindrical plexiglas tube with numerous small holes (2.0 mm in diameter) on the wall for proton transport. This reactor produced a maximum power density of 50.2 Wm^{-3} with an internal resistance of 28 Ω (You *et al.*, 2007). Flexible materials are often used for the air cathode of tubular MFCs so that the tubular MFCs are easily scaled up. However, with increasing volume, mechanical strength of the cathode material becomes a concern, because hydrostatic pressure on the cathode may cause its deformation and water leakage.

Stacked MFCs

A single MFC has a low voltage and current density. Stacked MFCs either in series or parallel or their combination seems to be a simple and easily used means to solve this problem. Only a few studies in the literature reported stacked MFCs. These studies demonstrated that voltage or current output was enhanced by using stacked MFCs (Aelterman *et al.*, 2006), but voltage reversal remains a large obstacle (Oh and Logan, 2007).

Aelterman and co-workers (Aelterman *et al.*, 2006) designed a six-cell stack and tested the performance of the stack using acetate as the substrate and ferricyanide as the catholyte (Fig. 9.11A). The individual MFC cells were separated by rubber sheets and connected using copper wires. The anode and cathode chambers each had a volume of 156 ml (60 ml liquid volume). Graphite rods were inserted into the beds of graphite granules. In the

176   *Microbial Biotechnology: Energy and Environment*

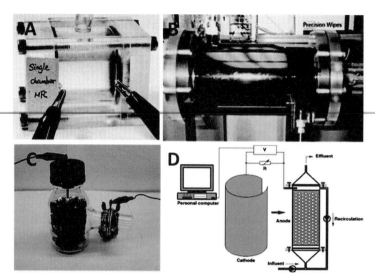

Fig. 9.10. Different single-chamber MFC designs. (A) Cube-type MFC (reprinted with permission from Liu and Logan, 2004. Copyright 2004 American Chemical Society). (B) Horizontal tube-type MFC (reprinted with permission from Liu *et al.*, 2004. Copyright 2004 American Chemical Society). (C) Side-arm bottle MFC (reprinted with permission from Logan *et al.*, 2007. Copyright 2007 American Chemical Society). (D) Upflow-type MFC (reprinted from You *et al.*, 2007, with permission from Elsevier).

parallel mode, the MFC had a power density of 59 Wm^{-3} compared with 51 $W\ m^{-3}$ in series based on a total volume of 1.9 l. The average C_E for the cells was 12% when cells were arranged in series, but it increased to 78% in parallel. Voltages from the cells were found to be unequal when cells were connected in series. Some cells even yielded negative voltages that adversely affected power generation (Aelterman *et al.*, 2006). Oh and Logan (2007) showed that voltage reversal could happen for a cell due to substrate depletion. Logan (2008) suggested that the voltage reversal was due to fluctuations in biological systems. Diodes can be used to prevent voltage reversal.

Figure 9.11B shows a bipolar-plate type of stacked MFC consisting of five cells tested by Shin *et al.* (2006). The reactor used thionin as a mediator for electron transfer by *Proteus vulgaris* grown on glucose. It had a power density of 1300 mWm^{-2} using ferricyanide as catholyte compared to 230 mWm^{-2} with pure oxygen gas. Liu *et al.* (2008) used a novel design consisting of stacked MFCs bridged internally through an extra cation exchange membrane (Fig. 9.11C). The MFC stack, assembled from two single MFCs, doubled voltage output and halved the optimal internal resistance. The COD removal rate increased from 32.4% to 54.5%. The performance improvement could be attributed to the smaller internal resistance and enhanced cation transfer. Their investigation of a half-cell study further confirmed the important role of the extra CEM. This study proved that it was beneficial in terms of increased voltage output and reduced cell resistance when the anode and cathode were sandwiched between two CEMs (Liu *et al.*, 2008).

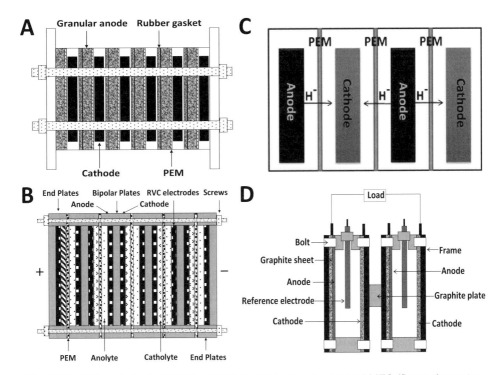

Fig. 9.11. Different stacked MFCs. (A) Six individual units stacked MFC (figure drawn to illustrate a photo in Aelterman *et al.*, 2006). (B) Bipolar plate MFC (redrawn with modifications after Shin *et al.*, 2006). (C) MFC stack bridged internally through an extra CEM (redrawn with modifications after Liu *et al.*, 2008). (D) Two-chamber stack (redrawn with modifications after Oh and Logan, 2007).

Materials Used in Constructing MFCs

Anode materials

In an MFC, the anode is where electricity-producing bacteria form a biofilm and it is the receptor of electrons. Therefore, an ideal anode material should be inexpensive, highly conductive with a large specific surface area, non-corrosive and resistant to fouling (Logan, 2008). A variety of materials have been used as anodic electrodes, including carbon materials, graphite materials, conductive polymers and metals. Table 9.2 lists the commonly used electrode materials and their properties.

In order to boost the performance of an anode, several treatment methods have been reported, such as ammonia gas treatment, electrochemical treatment, heat treatment and addition of mediators. Cheng and Logan (2007) treated carbon cloth using 5% NH_3 gas in a helium carrier gas at 700°C for 60 min. This reduced the start-up time of MFC by 50% and increased power density from 1640 mWm^{-2} to 1970 mWm^{-2}. Their analysis suggested that this could be attributed to the increase of the positive surface charge of the cloth from 0.38

Table 9.2. Commonly used electrode materials in MFCs.

Materials	Advantages	Disadvantages
Carbon paper	High conductivity	Brittle, low specific surface area, expensive
Carbon cloth	High conductivity, flexible, high specific surface area	Expensive
Reticulated vitreous carbon	High conductivity, high porosity, large specific surface area	Brittle
Graphite rod	High conductivity, defined surface area	Low specific surface area, expensive
Graphite felt	High conductivity, high porosity, large specific surface area, flexible	Low strength
Graphite granules bed	Low cost, high porosity, high surface area	High contact resistance
Graphite fibre brush	High conductivity, high porosity, large specific surface area, flexible	Expensive
Conductive polymers	Large surface area, flexible	Low conductivity
Stainless steel	High conductivity, low cost	Poor bacteria attachment, low power production

to 3.99 meqm^{-2}. Wang and co-workers found that heat treatment of carbon mesh at 450°C for 30 min decreased atomic O/C ratio and removed contaminants that interfered with charge transfer, thereby enhancing the performance of anode (Wang et al., 2009b). Park and Zeikus (2002) bound neutral red (NR) to a woven graphite electrode and increased power to 9.1 mWm^{-2} compared to only 0.02 mWm^{-2} without NR, using a pure culture of *S. putrefaciens* and lactate as fuel. Tang and co-workers electrochemically treated graphite felt with a constant current density (30 mAcm^{-2} based on the projected area of anode) for 12 h, and this produced a current of 1.13 mA that was 39.5% higher than untreated anodes. This enhancement came from the newly created carboxyl groups that have a strong hydrogen bonding with the peptide bonds in the cytochromes of the microbes (Tang et al., 2011).

Cathode materials

The same aforementioned anode materials can also be used for cathodes. If ferricyanide or MnO_2 is used as the final electron acceptors, no catalyst is needed for cathodic reactions (Logan, 2008). But if oxygen is used as the final electron acceptor, a catalyst must be employed because graphite and carbon materials are poor catalysts for oxygen reduction (e.g. cathodic reaction). Platinum or a metal plated with platinum is typically used as the cathode catalyst due to its excellent catalytic ability, but platinum's high cost makes its large-scale applications such as wastewater treatment prohibitive. Therefore, a series of metals and their complexes have been investigated as replacements for platinum in the cathode in MFCs, such as Fe(III) (Park and Zeikus 2002, 2003; terHeijne et al., 2006), cobalt complexes (Zhao et al., 2005; Cheng et al., 2006), manganese oxide (Mao et al., 2003; Rhoads et al., 2005), lead dioxide (Morris et al., 2007) and manganese dioxide (Li et al., 2010). Some of these non-precious catalysts could produce a power density at a level comparable to those achieved with a Pt-based cathode, but the longevity of such materials is not well studied.

In order to avoid using expensive metal catalysts, bacteria have been used as biocatalysts in the cathode chamber. This is known as a biocathode (He and Angenent 2006; Huang et al., 2011). A biocathode can be either aerobic or anaerobic. Aerobic biocathodes rely on aerobes to catalyse the oxidation of transition-metal compounds, such as Mn(II) or Fe(II) by oxygen. Anaerobic biocathodes utilize anaerobes or facultative anaerobes capable of reducing non-oxygen oxidants such as nitrate, nitrite, sulfate, iron, manganese, selenate, arsenate, urinate, fumarate and CO_2 (Lefebvre et al., 2008). These non-oxygen oxidants serve as terminal electron acceptors. They have much lower standard potentials than oxygen, meaning their MFC voltage output will likely be much lower.

Membranes

The majority of MFC designs require the separation of the anode and the cathode compartments by a membrane. The purpose of the membrane is to keep anode and cathode solutions separated while allowing ion transfer. This prevents direct oxidation of organic matters that happens when oxidants cross into the anode chamber from the cathode chamber. Nafion is the most commonly used CEM to allow the passage for ion exchange while partitioning the anode chamber and the cathode chamber. Besides CEM, AEM, bipolar membrane (BPM), charge mosaic membrane (CMM), ultrafiltration membrane (UFM) may play the role of transporting ions through the membrane to maintain electro-neutrality. Fig. 9.12 shows the mechanisms of charge transport of four different types of ion exchange membranes.

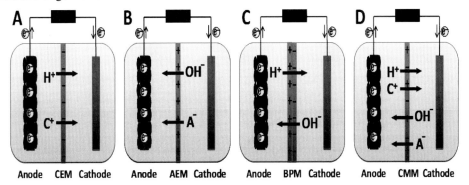

Fig. 9.12. Mechanisms of charge transport for four different types of ion exchange membranes (C^+: Na^+, K^+, NH_4^+, Mg^{2+}; A^-: Cl^-, HCO_3^-, $H_2PO_4^-$, CH_3COO^-) (redrawn with modifications after Rozendal et al., 2008b).

Laboratory MFCs often use membranes to partition the anode chamber and the cathode chamber. A major disadvantage of using membranes is their high costs and fouling. Membranes also present significant internal resistances that lead to reduced power production (Logan, 2008). There is a growing consensus among MFC researchers that it is not essential to use membranes in MFCs to separate the biological anode from the cathode reactions. Salt bridge (Min et al., 2005a), porcelain septum (Park and Zeikus, 2003), microporous filter (Biffinger et al., 2007) and physical barriers (Jang et al., 2004) are alternative ion exchange systems. In fact, it was observed that the anode could even be kept anaerobic due to diffusional resistance between the anode and the cathode without any physical partitioning (Bond et al., 2002). For example, in a sediment MFC, nothing but the aqueous solution separates the anode from the cathode. The distance between the two electrodes serves as the barrier for oxygen diffusion although it is not 100% efficient. It is likely that the first successful large-scale implementation of MFCs will be membrane-less because a membrane is too costly for large-scale applications and it is also easily fouled.

MFC Applications

Power supply

MFCs convert the chemical energy in the chemical bonds of organic compounds directly into electricity instead of producing low-grade heat from their direct oxidation that is limited by the Carnot cycle thermal efficiency. Therefore, just like chemical fuel cells, MFCs possess much higher conversion efficiency (>70%) (Du et al., 2007). However, so far, the power produced by MFCs is still too low to be useful in most applications, and it is likely that MFCs will not contribute to the power grid even with major advances in the future. After all, MFCs do not use high density fuels such as pure hydrogen used in a hydrogen fuel cell. But MFCs are especially suitable for use as power supplies in small telemetry systems and wireless sensors that are not energy intensive in remote area (Ieropoulos et al., 2005a; Shantaram et al., 2005).

Ieropoulos et al. (2005) developed the robot EcoBot-II and used MFCs as the onboard energy supply. EcoBot-II was able to perform sensing, communication and actuation using flies as substrate. An MFC consisting of a sacrificial anode combined with the reduction of biomineralized manganese oxides was used to power electrochemical sensors and small telemetry systems to transmit the acquired data to remote receivers. To ensure power supply, a capacitor was used to store the energy produced by the MFC, and energy stored was used in short bursts when needed (Shantaram et al., 2005).A Benthic Unattended Generator (BUG) MFC was also tested to power a data buoy that monitored air temperature, pressure and humidity, as well as water temperature. The system radioed data every 5 min to a shore-based receiver (Tender et al., 2008). MFCs for a long-term space flight such as a mission to Mars are attractive since they can generate electricity while treating organic wastes onboard a spacecraft. It is conceivable in the future that a miniature MFC inside a human body fuelled by the nutrients inside the body can be used to power an implanted medical device for long-term uses (Chiao et al., 2007).

Wastewater treatment

MFCs are regarded as a promising future technology for wastewater treatment for the several reasons. MFCs are able to harvest energy from organic matters and treat wastewaters at the same time. The power generated by MFCs in wastewater treatment helps offset the power input needed in the aerobic treatment stage (Du et al., 2007). Furthermore, MFCs can reduce solids by 50–90% (Holzman, 2005), because MFCs break down organic molecules such as acetate, propionate and butyrate into CO_2 and H_2O. A wide range of wastewater can be treated by MFCs, if proper electricity-producing bacteria were to be enriched on the anode. MFCs can treat many different types of wastewater rich in organic matters, including sanitary wastes, swine and other animal wastewater, and food processing wastewater(Suzuki *et al*., 1978; Liu *et al*., 2004; Min *et al*., 2005b; Oh and Logan, 2005; Zuo *et al*., 2006). Unlike the expensive pure hydrogen used in a hydrogen fuel cell, these wastes are practically free fuels for MFCs. Finally, MFCs with biocathodes use wastewater that contains oxidants other than oxygen as terminal electron acceptors, allowing treatment of two different wastewater streams at the same time with concomitant electricity generation (Virdis *et al*., 2010; Huang *et al*., 2011). MFCs also hold promise in some niche applications such as reduced power demand in treating black or grey wastewaters at military camps where fuel costs are up to ten times higher than normal (such as some US military camps in current war-torn Afghanistan.)

The current densities generated by laboratory MFCs almost approach levels required for practical applications. However, so far, those MFCs with high power density required for practical applications are typically operated on small scales, varying from just a few millilitres to several litres (Rozendal *et al*., 2008a). The performance of MFCs with real organic wastewater was also far lower than artificial wastewater (Pant *et al*., 2010). Because of the complexity of MFC operations, MFC scale-up for wastewater treatment is not straightforward. More work is required using real wastewater streams. It is also imperative to reduce the capital costs for large-scale MFC devices if they are to be used for wastewater treatment (Rozendal *et al*., 2008a).

Biohydrogen production

MFCs can be modified for hydrogen production by adding a small voltage to the system and eliminating oxygen from the cathode, and the MFC-like reactor is referred to as a microbial electrolysis cell (MEC) (Liu *et al*., 2005; Rozendal and Buisman, 2005). Figure 9.13 shows the fundamental principle of MEC. In an MEC, exoelectrogens in an anodic biofilm oxidize organic matters, generating electrons, protons and CO_2. Electrons flow through an external circuit to reach the cathode and protons diffuse through a PEM to the cathodic chamber. To overcome a thermodynamic barrier, an external voltage is applied to produce hydrogen at the cathode. Typically 0.3V or larger is needed when acetate is used as fuel. This voltage is much less than the voltage required for direct water electrolysis, which is around 1.8–2.0 V (Rozendal *et al*., 2006a). If a more energetic organic carbon was to be used, such as lactate that has a more negative standard reduction potential than acetate, the external voltage requirement would be substantially less or even eliminated.

Since both the anodic chamber and the cathodic chamber are anaerobic in an MEC, a membrane (e.g. PEM) is not necessary for the system. Laboratory experiments have demonstrated that eliminating the membrane reduces the cost and lowers the internal resistance, avoids the pH gradient across the membrane, and increases hydrogen production rate (Call and Logan, 2008; Hu *et al*., 2008). Hence, single-chamber MECs become more attractive for scale-up. One drawback in the absence of a membrane is the hydrogen consumption by methanogens, which reduces the hydrogen production and purity. Apart from this, hydrogen reoxidization by exoelectrogens also causes energy losses (Clauwaert and Verstraete, 2009; Wang *et al*., 2009a).

While MECs were invented only a few years ago, performances of MECs have improved significantly in terms of hydrogen production rate, conversion efficiency of substrate and energy recovery (Logan *et al.*, 2008). MECs are able to convert a wide range

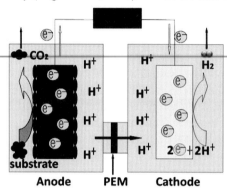

Fig. 9.13. Schematic diagram of a typical two-chamber MEC.

of low-grade organic matters into hydrogen (Ditzig *et al.*, 2007), and the energy input is many times smaller that of water electrolysis. Therefore, MEC is a promising new technology for renewable and sustainable biohydrogen production. However, further research work is needed to reduce construction costs, enhance hydrogen production rate and inhibit methane generation in MECs.

Challenges and Prospects

While MFCs hold great potentials for various applications, major challenges remain for MFCs to be practical. The power densities of MFCs must be augmented because they are too low for most envisioned applications. One key bottleneck for MFCs is the electron transfer rate limitation in microbial biofilms. A breakthrough may be achieved by fundamental observations, or through transposon (or other) mutagenesis or hypothesis-driven genetically engineering new super-bugs that possess large number of pili per cell and/or secrete efficient electron mediators at sufficiently high local concentrations. These may be achieved by several microbes working together in a synergistic biofilm consortium, where one organism provides a by-product of normal metabolism that another organism can use to continue an electrogenic cycle of events. Conceivably, such a breakthrough can greatly reduce or eliminate the electron transfer bottleneck. Using the powerful traits of the voraciously metabolic bacteria *P. aeruginosa*, a number of viable strategies were outlined earlier to improve electron transfer. Using such a virulent organisms, and coupled with other bacteria that could offer synergistic properties to the MFC, exciting new breakthroughs in MFC research are ahead.

Another bottleneck in MFCs is the high internal resistance, resulting from slow proton diffusion from anode to cathode and slow oxygen reduction kinetics at the cathode (Kim, *et al.*, 2007a). In MFCs, catalysed cathodic reaction consumes more protons at the cathode than the membrane can deliver. This causes an undesirable pH gradient across the membrane and increases the internal resistance. Proton transfer in an aqueous solution is seriously hampered by other cations, such as Na^+, K^+, NH_4^+ and Mg^{2+}, whose concentrations are typically 10^5 times greater than that of protons at neutral pH (Rozendal et al., 2006b; Zhao et al., 2006). Development of a proton-specific membrane may help solve this problem (Kim, B.H. et al., 2007). Another effective approach is through optimized reactor design, such as shortening the distance between the anode and the cathode (Logan et al., 2006; Du et al., 2007). Using platinum as the catalyst for oxygen reduction could reduce the cathodic overpotential, achieving a maximum current three to four times higher (Pham et al., 2004). However, the cost of using platinum is prohibitive for large-scale applications such as wastewater treatment.

The construction cost of MFCs must be reduced. Currently, expensive proton exchange membranes (such as Nafion) and precious metals (such as platinum) are generally used in laboratory-scale MFCs. They account for more than 80% of the construction cost. The high cost of these materials makes MFCs uncompetitive against energy production from wind and solar and biofuels. MFCs also need to compete with inexpensive methane digesters that are popularly used in some third-world countries to produce methane for household cooking and lighting (Verstraete et al., 2005). If expensive laboratory materials are used, the cost of a large-scale MFC will be orders of magnitude higher than a conventional wastewater treatment system (Rozendal et al., 2008a). Low-cost materials are desired. Researchers have demonstrated that Nafion membranes can be substituted with relatively inexpensive ion exchange membranes (Kim et al., 2007b) and some designs do not even require membranes (Liu and Logan, 2004). To avoid using expensive platinum for catalysis of oxygen reduction, various metals and their complexes have been investigated as materials for cathodes in MFCs, such as Fe(III) (Park and Zeikus, 2002, 2003; TerHeijne et al., 2006), cobalt complexes (Zhao et al., 2005; Cheng et al., 2006) and manganese oxide (Mao et al., 2003; Rhoads et al., 2005) have also been studied. These materials cost much less than platinum although they tend to be less effective (Kim et al., 2007a). Zhao et al. (2005) claimed that Fe(II) phthalocyanine and cobalt tetramethoxyphenylporphyrin catalysed oxygen reduction in some tests with comparable performance as platinum. Biocathodes do not need expensive metal catalysts, because they use biocatalysis to reduce non-oxygen oxidant. This presents a different approach to the reduction of MFC costs. Biocathodes also have the added ability to process a second wastewater stream in the cathode chamber (Huang et al., 2011).

There is still a lack of understanding of electricity-producing bacteria's electron transfer mechanisms, especially in microbial communities involving electrochemically inactive bacteria. Researchers found that in MFCs with electrochemically active iron-reducing bacteria such as *Shewanella* and *Geobacter* species, there were more non iron-reducing bacteria in the biofilm communities (Logan and Regan, 2006b). These electrochemically inactive bacteria enhanced electricity generation by the electrochemically active bacteria because MFCs with pure cultures generated two to three orders of magnitude less power (Rabaey et al., 2004). It is possible that some of the electrochemically inactive bacteria secreted electron mediators that accelerated electron transfer by the electrochemically active bacteria. It is also possible that there are still undiscovered electrochemical mechanisms in synergistic biofilm communities that can be exploited to improve MFC performance. With a better understanding of how biofilm communities function electrochemically, new super-bug communities can be genetically engineered to improve MFC performances beyond what any conventional approaches could achieve.

References

Aelterman, P., Rabaey, K., Pham, T.H., Boon, N. and Verstraete, W. (2006) Continuous electricity generation at high voltages and currents using stacked microbial fuel cells. *Environmental Science and Technology* 40, 3388–3394.

Allen, R.M. and Bennetto, H.P. (1993) Microbial fuel cells: electricity production from carbohydrates. *Applied Biochemistry and Biotechnology* 39–40, 27–40.

Aulenta, F., Catervi, A., Majone, M., Panero, S., Reale, P. and Rossetti, S. (2007) Electron transfer from a solid-state electrode assisted by methyl viologen sustains efficient microbial reductive dechlorination of TCE.*Environmental Science and Technology* 41, 2554–2559.

Biffinger, J.C., Ray, R., Little, B. and Ringeisen, B.R. (2007) Diversifying biological fuel cell designs by use of nanoporous filters. *Environmental Science and Technology* 41,1444–1449.

Bond, D.R. and Lovley, D.R. (2003) Electricity production by *Geobacter sulfurreducens* attached to electrodes. *Applied and Environmental Microbiology* 69, 1548–1555.

Bond, D.R., Holmes, D.E., Tender, L.M. and Lovley, D.R. (2002) Electrode-reducing microorganisms that harvest energy from marine sediments. *Science* 295, 483–485.

Borole, A.P., O'Neill, H., Tsouris, C. and Cesar, S. (2008) A microbial fuel cell operating at low pH using the acidophile *Acidiphilium cryptum*. *Biotechnology Letters* 30, 1367–1372.

Bretschger, O., Obraztsova, A., Sturm, C.A., Chang, I.S., Gorby, Y.A., Reed, S.B., Culley, D.E., Reardon, C.L., Barua, S., Romine, M.F., Zhou, J., Beliaev, A.S., Bouhenni, R., Saffarini, D., Mansfeld, F., Kim, B.H., Fredrickson, J.K. and Nealson, K.H. (2007) Current production and metal oxide reduction by *Shewanella oneidensis* MR-1 wild type and mutants. *Applied and Environmental Microbiology* 73, 7003–7012.

Bucking, C., Popp, F., Kerzenmacher, S. and Gescher, J. (2010) Involvement and specificity of *Shewanella oneidensis* outer membrane cytochromes in the reduction of soluble and solid-phase terminal electron acceptors. *FEMS Microbiology Letters* 306, 144–151.

Call, D. and Logan, B.E. (2008) Hydrogen production in a single chamber microbial electrolysis cell lacking a membrane. *Environmental Science and Technology* 42, 3401–3406.

Canfield, J.H., Goldner, B.H. and Lutwack, R. (1963) NASA Technical report, Magna Corporation, Anaheim, California, USA, 63 pp.

Chang, I.S., Moon, H., Bretschger, O., Jang, J.K., Park, H.I., Nealson, K.H. and Kim, B.H. (2006) Electrochemically active bacteria (EAB) and mediator-less microbial fuel cells. *Journal of Microbiology and Biotechnology* 16, 63–177.

Chaudhuri, S.K. and Lovley, D.R. (2003) Electricity generation by direct oxidation of glucose in mediatorless microbial fuel cells. *Nature Biotechnology* 21, 1229–1232.

Cheng, S. and Logan, B.E. (2007) Ammonia treatment of carbon cloth anodes to enhance power generation of microbial fuel cells. *Electrochemical Communication* 9, 492–496.

Cheng, S., Liu, H. and Logan, B.E. (2006) Power densities using different cathode catalysts (Pt and CoTMPP) and polymer binders (Nafion and PTFE) in single chamber microbial fuel cells. *Environmental Science and Technology* 40, 364–369.

Chiang, P. and Burrows, L.L. (2003) Biofilm formation by hyperpiliated mutants of *Pseudomonas aeruginosa*. *Journal of Bacteriology* 185, 2374–2378.

Chiao, M., Lin, L. and Lam, K. (2007) Implantable, miniatured microbial fuel cell. US Patent 7160637B2.

Childers, S.E., Ciufo, S. and Lovley, D.R. (2002) *Geobacter metallireducens* accesses insoluble Fe(III) oxide by chemotaxis. *Nature* 416, 767–769.

Choi, Y., Kim, N., Kim, S. and Jung, S. (2003) Dynamic behaviors of redox mediators within the hydrophobic layers as an important factor for effective microbial fuel cell operation. *Bulletin of the Korean Chemical Society* 24, 437–440.

Clauwaert, P. and Verstraete, W. (2009) Methanogenesis in membraneless microbial electrolysis cells. *Applied Microbiology and Biotechnology* 82, 829–836.

Cohen, B. (1931) The bacterial culture as an electrical half-cell.*Journal of Bacteriology* 21, 18–19.

Cohn, E.M. (1963) Perspectives on biochemical electricity. *Developments in Industrial Microbiology* 4,53–58.

Davis, F. and Higson, S.P.J. (2007) Biofuel cells – recent advances and applications. *Biosensor and Bioelectronics* 22, 1224–1235.

Davis, J.B. and Yarbrough, H.M.Jr (1962) Preliminary experiments on a microbial fuel cell. *Science* 137, 615–616.

Delaney, G.M., Bennetto, H.P., Mason, J.R., Roller, S.D., Stirling, J.L. and Thurston, C.F. (1984) Electron-transfer coupling in microbial fuel cells. 2. performance of fuel cells containing selected microorganism–mediator–substrate combinations. *Journal of Chemical Technology and Biotechnology* 34, 13–27.

Ditzig, J., Liu, H. and Logan, B.E. (2007) Production of hydrogen from domestic wastewater using a bioelectrochemically assisted microbial reactor (BEAMR). *International Journal of Hydrogen Energy* 32, 2296–2304.

Du, Z., Li, H. and Gu, T. (2007) A state of the art review on microbial fuel cells: a promising technology for wastewater treatment and bioenergy. *Biotechnology Advances* 25, 464–482.

Fan, Y., Sharbrough, E. and Liu, H. (2008) Quantification of the internal resistance distribution of microbial fuel cells. *Environmental Science and Technology* 42, 8101–8107.

Gil, G.C., Chang, I.S., Kim, B.H., Kim, M., Jang, J.Y., Park, H.S. and Kim, H.J. (2003) Operational parameters affecting the performance of a mediatorless microbial fuel cell. *Biosensor and Bioelectronics* 18, 327–334.

Giltner, C.L., van Schaik, E.J., Audette, G.F., Kao, D., Hodges, R.S., Hassett, D.J. and Irvin, R.T. (2006) The *Pseudomonas aeruginosa* type IV pilin receptor binding domain functions as an adhesion for both biotic and abiotic surfaces. *Molecular Microbiology* 59, 1083–1096.

Gorby, Y.A., Yanina, S., McLean, J.S., Rosso, K.M., Moyles, D., Dohnalkova, A., Beveridge, T.J., Chang, I.S., Kim, B.H., Kim, K.S., Culley, D.E., Reed, S.B., Romine, M.F., Saffarini, D.A., Hill, E.A., Shi, L., Elias, D.A., Kennedy, D.W., Pinchuk, G., Watanabe, K., Ishii, S., Logan, B., Nealson, K.H. and Fredrickson, J.K. (2006) Electrically conductive bacterial nanowires produced by *Shewanella oneidensis* strain MR-1 and other microorganisms. *Proceedings of the National Academy of Sciences* 103, 11358–11363.

He, Z. and Angenent, L.T. (2006) Application of bacterial biocathodes in microbial fuel cells. *Electroanalysis* 18, 2009–2015.

He, Z., Minteer, S.D. and Angenent, L.T. (2005) Electricity generation from artificial wastewater using an upflow microbial fuel cell. *Environmental Science and Technology* 39, 5262–5267.

Hernandez, M.E. and Newman, D.K. (2001) Extracellular electron transfer. *Cellular and Molecular Life Sciences* 58, 1562–1571

Holmes, D.E., Bond, D.R., O'Neil, R.A., Reimers, C.E., Tender, L.R. and Lovley, D.R. (2004a) Microbial communities associated with electrodes harvesting electricity from a variety of aquatic sediments. *Microbial Ecology* 48, 178–190.

Holmes, D.E., Nicoll, J.S., Bond, D.R. and Lovley, D.R. (2004b) Potential role of a novel psychrotolerant member of the family Geobacteraceae, *Geopsychrobacter electrodiphilus* gen. nov., sp. nov., in electricity production by a marine sediment fuel cell. *Applied and Environmental Microbiology* 70, 6023–6030.

Holzman, D.C. (2005) Microbe power. *Environmental Health Perspectives* 113(11), 754–757.

Hou, H., Li, N., Cho, Y., Figueiredo, P.D. and Han, A. (2009) Microfabricated microbial fuel cell arrays reveal electrochemically active microbes. *PLoS ONE* 4, e6570.

Hu, H., Fan, Y. and Liu, H. (2008) Hydrogen production using single-chamber membrane-free microbial electrolysis cells. *Water Research* 42, 4172–4178.

Huang, L., Regan, J.M. and Quan, X. (2011) Electron transfer mechanisms, new applications, and performance of biocathode microbial fuel cells. *Bioresource Technology* 102, 316–323.

Hyun, M.S., Kim, H.J. and Kim, B.H. (1998) Use of a fuel cell to enrich electrochemically active Fe(III)-reducing bacteria. *98th General Meeting of American Society for Microbiology*, Atlanta, USA, pp. 309–309.

Ieropoulos, I., Melhuish, C., Greenman, J. and Horsfield, I. (2005a) EcoBot-II: an artificial agent with a natural metabolism. *International Journal of Advanced Robotic Systems* 2, 295–300.

Ieropoulos, I.A., Greenman, J., Melhuish, C. and Hart, J. (2005b) Comparative study of three types of microbial fuel cell. *Enzyme and Microbial Technology* 37, 238–245.

Jang, J.K., Pham, T.H., Chang, I.S., Kang, K.H., Moon, H., Cho, K.S. and Kim, B.H. (2004) Construction and operation of a novel mediator- and membrane-less microbial fuel cell. *Process Biochemistry* 39, 1007–1012.

Juarez, K., Kim, B.C., Nevin, K., Olvera, L., Reguera, G., Lovley, D.R. and Methe, B.A. (2009) PilR, a transcriptional regulator for pilin and other genes required for Fe(III) reduction in *Geobacter sulfurreducens*. *Journal of Molecular Microbiology and Biotechnology* 16, 146–158.

Kang, K.H., Jang, J.K., Lee, J.Y., Moon, H., Chang, I.S., Kim, J.M. and Kim, B.H. (2004) A low BOD sensor using a microbial fuel cell. *Journal of KSEE* 26, 58–63.

Kim, B.C., Postier, B.L., Didonato, R.J., Chaudhuri, S.K., Nevin, K.P. and Lovley, D.R. (2008) Insights into genes involved in electricity generation in *Geobacter sulfurreducens* via whole genome microarray analysis of the OmcF-deficient mutant. *Bioelectrochemistry* 73, 70–75.

Kim, B.H., Kim, H.J., Hyun, M.S. and Park, D.H. (1999) Direct electrode reaction of Fe(III)-reducing bacterium, *Shewanella putrifaciens*. *Journal of Microbiology and Biotechnology* 9, 127–131.

Kim, B.H., Park, H.S., Kim, H.J., Kim, G.T., Chang, I.S., Lee, J. and Phung, N.T. (2004) Enrichment of microbial community generating electricity using a fuel-cell-type electrochemical cell. *Applied Microbiology and Biotechnology* 63, 672–681.

Kim, B.H., Chang, I.S. and Gadd, G.M. (2007a) Challenges in microbial fuel cell development and operation. *Applied Microbiology and Biotechnology* 76, 485–494.

Kim, H.J., Park, H.S., Hyun, M.S., Chang, I.S, Kim, M. and Kim, B.H. (2002) A mediatorless microbial fuel cell using a metal reducing bacterium, *Shewanella putrifaciens*. *Enzyme and Microbial Technology* 30, 145–152.

Kim, J.R., Cheng, S., Oh, S.E. and Logan, B.E. (2007b) Power generation using different cation, anion and ultrafiltration membranes in microbial fuel cells. *Environmental Science and Technology* 41, 1004–1009.

Kouzuma, A., Meng, X.Y., Kimura, N., Hashimoto, K. and Watanabe, K. (2010) Disruption of the putative cell surface polysaccharide biosynthesis gene SO3177 in *Shewanella oneidensis* MR-1 enhances adhesion to electrodes and current generation in microbial fuel cells. *Applied and Environmental Microbiology* 76, 4151–4157.

Krushkal, J., Juarez, K., Barbe, J.F., Qu, Y., Andrade, A., Puljic, M., Adkins, R.M., Lovley, D.R. and Ueki, T. (2010) Genome-wide survey for PilR recognition sites of the metal-reducing prokaryote *Geobacter sulfurreducens*. *Gene* 469, 31–44.

Krushkal, J., Leang, C., Barbe, J.F., Qu, Y., Yan, B., Puljic, M., Adkins, R.M. and Lovley, D.R. (2009) Diversity of promoter elements in a *Geobacter sulfurreducens* mutant adapted to disruption in electron transfer. *Functional and Integrative Genomics* 9, 15–25.

Lee, S.A., Choi, Y., Jung, S. and Kim, S. (2002) Effect of initial carbon sources on the electrochemical detection of glucose by *Gluconobacter oxydans*. *Bioelectrochemistry* 57, 173–178.

Lefebvre, O., Al–Mamun, A. and Ng, H.Y. (2008) A microbial fuel cell equipped with a biocathode for organic removal and denitrification. *Water Science and Technology* 58, 881–885.

Lewis, K. (1966) Symposium on bioelectrochemistry of microorganisms IV biochemical fuel cells. *Bacteriology Reviews* 30, 101–113.

Li, X., Hu, B., Sui, S., Lei, Y. and Li, B. (2010) Manganese dioxide as a new cathode catalyst in microbial fuel cells. *Journal of Power Sources* 195, 2586–2591.

Liu, H. and Logan, B.E. (2004) Electricity generation using an air-cathode single chamber microbial fuel cell in the presence and absence of a proton exchange membrane. *Environmental Science and Technology* 38, 4040–4046.

Liu, H., Ramnarayanan, R. and Logan, B.E. (2004) Production of electricity during wastewater treatment using a single chamber microbial fuel cell. *Environmental Science and Technology* 38, 2281–2285.

Liu, H., Grot, S. and Logan, B.E. (2005) Electrochemically assisted microbial production of hydrogen from acetate. *Environmental Science and Technology* 39, 4317–4320.

Liu, Z., Liu, J., Zhang, S. and Su, Z. (2008) A novel configuration of microbial fuel cell stack bridged internally through an extra cation exchange membrane. *Biotechnology Letters* 30, 1017–23.

Logan, B.E. (2008) *Microbial Fuel Cell*. John Wiley & Sons, New Jersey.

Logan, B.E. (2009) Exoelectrogenic bacteria that power microbial fuel cells. *Nature Reviews Microbiology* 7, 375–381.

Logan, B.E. and Regan J.M. (2006a) Microbial fuel cells–challenges and applications. *Environmental Science and Technology* 40, 5172–5180.

Logan, B.E. and Regan, J.M. (2006b) Electricity-producing bacterial communities in microbial fuel cells. *Trends in Microbiology* 14, 512–518.

Logan, B.E., Hamelers, B., Rozendal, R., Schröder, U., Keller, J., Freguia, S., Aelterman, P., Verstraete, W. and Rabaey, K. (2006) Microbial fuel cells: methodology and technology. *Environmental Science and Technology* 40, 5181–5192.

Logan, B.E., Cheng, S., Watson, V. and Estadt, G. (2007) Graphite fiber brush anodes for increased power production in air-cathode microbial fuel cells. *Environmental Science and Technology* 41, 3341–3346.

Logan, B.E., Call, D., Cheng, S., Hamelers, H.V.M., Sleutels, T.H., Jeremiasse, A.W. and Rozendal, R.A. (2008) Microbial electrolysis cells for high yield hydrogen gas production from organic matter. *Environmental Science and Technology* 42, 8630–8640.

Lovley, D.R. (2006) Bug juice: harvesting electricity with microorganisms. *Nature Reviews Microbiology* 4, 497–508.

Mao, L., Zhang, D., Sotomura, T., Nakatsu, K., Koshiba, N. and Ohsaka, T. (2003) Mechanistic study of the reduction of oxygen in air electrode with manganese oxides as electrocatalysts. *Electrochimica Acta* 48, 1015–1021.

Min, B. and Logan, B.E. (2004) Continuous electricity generation from domestic wastewater and organic substrates in a flat plate microbial fuel cell. *Environmental Science and Technology* 38(21), 5809–5814.

Min, B., Cheng, S. and Logan, B.E. (2005a) Electricity generation using membrane and salt bridge microbial fuel cells. *Water Research* 39, 1675–1686.

Min, B., Kim, J.R., Oh, S.E., Regan, J.M. and Logan, B.E. (2005b) Electricity generation from swine wastewater using microbial fuel cells. *Water Research* 39, 4961–4968.

Moon, H., Chang, I.S. and Kim, B.H. (2006) Continuous electricity production from artificial wastewater using a mediator-less microbial fuel cell. *Bioresource Technology* 97, 621–627.

Morris, J.M., Jin, S., Wang, J., Zhu, C. and Urynowicz, M.A. (2007) Lead dioxide as an alternative catalyst to platinum in microbial fuel cells. *Electrochemistry Communications* 9, 1730–1734.

Newman, D.K. and Kolter, R. (2000) A role for excreted quinones in extracellular electron transfer. *Nature* 405, 94–97.

Oh, S.E. and Logan, B.E. (2005) Hydrogen and electricity production from a food processing wastewater using fermentation and microbial fuel cell technologies. *Water Research* 39, 4673–4682.

Oh, S.E. and Logan, B.E. (2007) Voltage reversal during microbial fuel cell stack operation. *Journal of Power Source* 167, 11–17.

Pant, D., Bogaert, G.V., Diels, L. and Vanbroekhoven, K. (2010) A review of the substrates used in microbial fuel cells (MFCs) for sustainable energy production. *Bioresource Technology* 101, 1533–1543.

Park, D.H. and Zeikus, J.G. (1999) Utilization of electrically reduced neutral red by *Actinobacillus succinogenes*: physiological function of neutral red in membrane-driven fumarate reduction and energy conservation. *Journal of Bacteriology* 181, 2403–2410.

Park, D.H. and Zeikus, J.G. (2000) Electricity generation in microbial fuel cells using neutral red as an electronophore. *Applied and Environmental Microbiology* 66,1292–1297.

Park, D.H. and Zeikus, J.G. (2002) Impact of electrode composition on electricity generation in a single-compartment fuel cell using *Shewanella putrefucians*. *Applied Microbiology and Biotechnology* 59, 58–61.

Park, D.H. and Zeikus, J.G. (2003) Improved fuel cell and electrode designs for producing electricity from microbial degradation. *Biotechnology and Bioengineering* 81, 348–355.

Park, D.H., Kim, B.H., Moore, B., Hill, H.A.O., Song, M.K. and Rhee, H.W. (1997) Electrode reaction of *Desulfovibriodesulfuricans* modified with organic conductive compounds. *Biotechnology Techniques* 11, 145–58.

Pham, C.A., Jung, S.J., Phung, N.T., Lee, J., Chang, I.S., Kim, B.H., Yi, H. and Chun, J. (2003) A novel electrochemically active and Fe(III)-reducing bacterium phylogenetically related to *Aeromonas hydrophila*, isolated from a microbial fuel cell. *FEMS Microbiology Letters* 223,129–134.

Pham, T.H., Jang, J.K., Chang, I.S. and Kim, B.H. (2004) Improvement of cathode reaction of a mediator-less microbial fuel cell. *Journal of Microbiology and Biotechnology* 14, 324–329.

Phung, N.T., Lee, J.K., Kang, H., Chang, I.S., Gadd, G.M. and Kim, B.H. (2004) Analysis of microbial diversity in oligotrophic microbial fuel cell using 16S rDNA analyses. *FEMS Microbiology Letters* 233, 77–82.

Potter, M.C. (1911) Electrical effects accompanying the decomposition of organic compounds. *Proceedings of the Royal Society of London* 84, 260–276.

Prasad, D., Arun, S., Murugesan, M., Padmanaban, S., Satyanarayanan, R.S., Berchmans, S. and Yegnaraman, V. (2007) Direct electron transfer with yeast cells and construction of a mediatorless microbial fuel cell. *Biosensors and Bioelectronic* 22, 2604–2610.

Rabaey, K. and Verstraete, W. (2005) Microbial fuel cells: novel biotechnology for energy generation. *Trends in Biotechnology* 23, 291–298.

Rabaey, K., Boon, N., Siciliano, S.D., Verhaege, M. and Verstraete, W. (2004) Biofuel cells select for microbial consortia that self-mediate electron transfer. *Applied and Environmental Microbiology* 70, 5373–5382.

Rabaey, K., Boon, N., Höfte, M. and Verstraete, W. (2005) Microbial phenazine production enhances electron transfer in biofuel cells. *Environmental Science and Technology* 39, 3401–3408.

Reguera, G., McCarthy, K.D., Mehta, T., Nicoll, J.S., Tuominen, M.T. and Lovley, D.R. (2005) Extracellular electron transfer via microbial nanowires. *Nature* 435, 1098–1101.

Reguera, G., Nevin, K.P., Nicoll, J.S., Covalla, S.F., Woodard, T.L. and Lovley, D.R. (2006) Biofilm and nanowire production leads to increased current in *Geobacter sulfurreducens* fuel cells. *Applied and Environmental Microbiology* 72, 7345–7348.

Rhoads, A., Beyenal, H. and Lewandowshi, Z. (2005) Microbial fuel cell using anaerobic respiration as an anodic reaction and biomineralized manganese as a cathodic reactant. *Environmental Science and Technology* 39, 4666–4671.

Ringeisen, B.R., Henderson, E., Wu, P.K., Pietron, J., Ray, R., Little, B., Biffinger, J.C. and Jones-Meehan, J.M. (2006) High power density from a miniature microbial fuel cell using *Shewanella oneidensis* DSP10. *Environmental Science and Technology* 40, 2629–2634.

Ringeisen, B.R., Ray, R. and Little, B. (2007) A miniature microbial fuel cell operating with an aerobic anode chamber. *Journal of Power Sources* 165, 591–597.

Ringeisen, B.R., Biffinger, J.C., Pietron, J., Ray, R. and Little, B. (2008) Aerobic miniature microbial fFuel cells. *Chemical/Biochemical Research* 141–142.

Roller, S.D., Bennetto, H.P., Delaney, G.M., Mason, J.R., Stirling, J.L. and Thurston, C.F. (1984) Electron-transfer coupling in microbial fuel cells: 1. comparison of redox-mediator reduction rates and respiratory rates of bacteria. *Journal of Chemical Technology and Biotechnology* 34, 3–12.

Rozendal, R.A. and Buisman, C.J.N. (2005) Process for producing hydrogen. Patent WO2005005981.

Rozendal, R.A., Hamelers, H.V.M., Euverink, G.J.W., Metz, S.J. and Buisman, C.J.N. (2006a) Principle and perspectives of hydrogen production through biocatalyzed electrolysis. *International Journal of Hydrogen Energy* 31, 1632–1640.

Rozendal, R.A., Hamelers, H.V.M. and Buisman, C.J.N. (2006b) Effects of membrane cation transport on pH and microbial fuel cell performance. *Environmental Science and Technology* 40, 5206–5211.

Rozendal, R.A., Hamelers, H.V.M., Rabaey, K., Keller, J. and Buisman, C.J.N. (2008a) Towards practical implementation of bioelectrochemical wastewater treatment. *Trends in Biotechnology* 26, 450–459.

Rozendal, R.A., Sleutels, T.H.J.A., Hamelers, H.V.M. and Buisman, C.J.N. (2008b) Effect of the type of ion exchange membrane on performance, ion transport, and pH in biocatalyzed electrolysis of wastewater. *Water Science and Technology* 57, 1757–1762.

Schroder, U. (2007) Anodic electron transfer mechanisms in microbial fuel cells and their energy efficiency. *Physical Chemistry Chemical Physics* 9, 2619–2629.

Schroder, U., Niessen, J. and Scholz, F. (2003) A generation of microbial fuel cells with current outputs boosted by more than one order of magnitude. *Angewandte Chemie International Edition* 42, 2880–2883.

Shantaram, A., Beyenal, H., Veluchamy, R.R.A. and Lewandowski, Z. (2005) Wireless sensors powered by microbial fuel cells. *Environmental Science and Technology* 39, 5037–5042.

Shin, S.H., Choi, Y., Na, S.H., Jung, S. and Kim, S. (2006) Development of bipolor plate stack type microbial fuel cells. *Bulletin of the Korean Chemical Society* 27, 281–285.

Stams, A.J., de Bok, F.A., Plugge, C.M., van Eekert, M.H., Dolfing, J. and Schraa, G. (2006) Exocellular electron transfer in anaerobic microbial communities. *Environmental Microbiology* 8, 371–382.

Straub, K.L. and Schink, B. (2004) Ferrihydrite-dependent growth of *Sulfurospirillum deleyianum* through electron transfer via sulfur cycling. *Applied and Environmental Microbiology* 10, 5744–5749.

Suzuki, S., Karube, I. and Matsunaga, T. (1978) Application of a biochemical fuel cell to wastewater. *Biotechnology and Bioengineering Symposium* 8, 501–511.

Tang, X., Guo, K., Li, H., Du, Z. and Tian, J. (2011) Electrochemical treatment of graphite to enhance electron transfer from bacteria to electrodes. *Bioresource Technology* 102, 3558–3560.

Tender, L.M., Gray, S.A., Groveman, E., Lowy, D.A., Kauffman, P., Melhado, J., Tyce, R.C., Flynn, D., Petrecca, R. and Dobarro, J. (2008) The first demonstration of a microbial fuel cell as a viable power supply: powering a meteorological buoy. *Journal of Power Sources* 179, 571–575.

Ter Heijne, A., Hamelers, H.V.M., De Wilde, V., Rozendal, R.R. and Buisman, C.J.N. (2006) Ferric iron reduction as an alternative for platinum-based cathodes in microbial fuel cells. *Environmental Science and Technology* 40, 5200–5205.

Thurston, C.F., Bennetto, H.P., Delaney, G.M., Mason, J.R., Roller, S.D. and Stirling, J.L. (1985) Glucose metabolism in a microbial fuel cell. Stoichiometry of product formation in a thionine-mediated *Proteus vulgaris* fuel cell and its relation to coulombic yields. *Journal of General Microbiology* 131, 1393–1401.

Tokuji, I. and Kenji, K. (2003) Vioelectrocatalyses-based application of quinoproteins and quinprotein-containing bacterial cells in biosensors and biofuel cells. *Biochimica et Biophysica Acta* 1647, 121–126.

Vega, C.A. and Fernandez, I. (1987) Mediating effect of ferric chelate compounds in microbial fuel cells with *Lactobacillus plantarum*, *Streptococcus lactis*, and *Erwinia dissolvens*. *Bioelectrochemistry and Bioenergetics* 17, 217–222.

Verstraete, W., Morgan–Sagastume, F., Aiyuk, S., Rabaey, K., Waweru, M. and Lissens, G. (2005) Anaerobic digestion as a core technology in sustainable management of organic matter. *Water Science and Technology* 52, 59–66.

Virdis, B., Rabaey, K., Rozendal, R.A., Yuan, Z.G. and Keller, J. (2010) Simultaneous nitrification, denitrification and carbon removal in microbial fuel cells. *Water Research* 44, 2970–2980.

Wang, A., Liu, W., Cheng, S., Xing, D., Zhou, J. and Logan, B.E. (2009a) Source of methane and methods to control its formation in single chamber microbial electrolysis cells. *International Journal of Hydrogen Energy* 34, 3653–3658.

Wang, X., Cheng, S., Feng, Y., Merrill, M.D., Saito, T. and Logan, B.E. (2009b) Use of carbon mesh anodes and the effect of different pretreatment methods on power production in microbial fuel cells. *Environmental Science and Technology* 43, 6870–6874.

Wilkinson, S. (2000) 'Gastrobots' – benefits and challenges of microbial fuel cells in food powered robot applications. *Autonomous Robots* 9, 99–111.

Wrighton, K.C., Agbo, P., Warnecke, F., Weber, K.A., Brodie, E.L., DeSantis, T.Z., Hugenholtz, P., Andersen, G.L. and Coates, J.D. (2008) A novel ecological role of the Firmicutes identified in thermophilic microbial fuel cells. *The ISME Journal* 2, 1146–1156.

Xing, D., Zuo, Y., Cheng, S., Regan, J.M. and Logan, B.E. (2008) Electricity generation by *Rhodopseudomonas palustris* DX-1. *Environmental Science and Technology* 42, 4146–4151.

Yoon, S.S., Hennigan, R.F., Hilliard, G.M., Ochsner, U.A., Parvatiyar, K., Kamani, M.C., Allen, H.L., DeKievit, T.R., Gardner, P.R., Schwab, U. *et al.* (2002) *Pseudomonas aeruginosa* anaerobic respiration in biofilms: relationships to cystic fibrosis pathogenesis. *Developmental Cell* 3, 593–603.

You, S., Zhao, Q., Zhang, J., Jiang, J., Wan, C., Du, M. and Zhao, S. (2007) A graphite-granule membrane-less tubular air-cathode microbial fuel cell for power generation under continuously operational conditions. *Journal of Power Source* 173, 172–177.

Zhang, L., Zhou, S., Zhuang, L., Li, W., Zhang, J., Lu, N. and Deng, L. (2008) Microbial fuel cell based on *Klebsiella pneumoniae* biofilm. *Electrochemistry Communications* 10, 1641–1643.

Zhang, T., Cui, C., Chen, S., Ai, X., Yang, H., Shen, P. and Peng, Z.(2006) A novel mediatorless microbial fuel cell based on direct biocatalysis of *Escherichia coli*. *Chemical Communications* 2006, 2257–2259.

Zhao, F., Hamisch, F., Schroder, U., Scholz, F., Bogdanoff, P. and Henmann, I. (2005) Application of pyrolysed iron phthalocyanine and CoTMPP based oxygen reduction catalysts as cathode materials in microbial fuel cells. *Electrochemistry Communications* 7, 1405–1410.

Zhao, F., Harnisch, F., Schroder, U., Scholz, F., Bogdanoff, P. and Herrmann, I. (2006) Challenges and constraints of using oxygen cathodes in microbial fuel cells. *Environmental Science and Technology* 40, 5193–5199.

Zhao, F., Rahunen, N., Varcoe, J.R., Chandra, A., Avignone-Rossa, C., Thumser, A.E. and Slade, R.C. (2008) Activated carbon cloth as anode for sulfate removal in a microbial fuel cell. *Environmental Science and Technology* 42, 4971–4976.

Zuo, Y., Maness, P.C. and Logan, B.E. (2006) Electricity production from steamexploded corn stover biomass. *Energy and Fuels* 20, 1716–1721.

Zuo, Y., Xing, D., Regan, J.M. and Logan, B.E. (2008) Isolation of the exoelectrogenic bacterium *Ochrobactrum anthropi* YZ–1 by using a U-tube microbial fuel cell. *Applied and Environmental Microbiology* 74, 3130–3137.

Chapter 10

Integration of Anaerobic Digestion and Oil Accumulation: Bioenergy Production and Pollutants Removal

Mi Yan, Jianguo Zhang and Bo Hu

Introduction

Anaerobic digestion (AD) is an industrial process to decompose organic materials by microorganisms in an enclosed vessel with no oxygen. It has been widely commercialized worldwide in the past few decades to treat waste water while harvesting biogas as an energy source. However, excess nutrients, particularly nitrogen and phosphorus, remain in AD effluents, and further treatment is needed before they can be discharged to the environment. Cells growing on the nutrients of AD effluents could address this need. An oleaginous cell cultivation step, following AD treatment, can remove the remaining organics, ammonia and phosphate, while the cell biomass can be harvested for the bioenergy/lipid production. Several options exist for the most appropriate oleaginous microbial species in the cultivation step. Recently, oleaginous microalgae have been attracting research attention, due to their high content of oil in certain stressed conditions and their capability to assimilate ammonia and phosphate during the cell growth. The majority of microalgae cells in AD effluents still grow in heterotrophic conditions, because rich organic nutrients and high turbidity inhibit the microalgae to grow in sunlight and CO_2. Dilution of the AD effluent is probably needed since highly concentrated waste water may inhibit microalgae growth. There are also possibilities to integrate AD with oil accumulation processes via other oleaginous species, such as red yeast and filamentous fungi, in order to remove the remaining pollutants and produce biofuel. This chapter reviews the current literature on the integrated process development to combine the AD and microbial oil accumulation, and defines the benefits and opportunities of biofuel production by using AD effluents. Critical benefits and limitations of current process developments are also summarized and discussed in order to provide future directions for research and development in this field.

Anaerobic Digestion

Anaerobic digestion (AD) is a simple process that can greatly reduce the amount of organic matter that might otherwise be landfilled or burned in an incinerator, with potentially serious environmental concerns. Compared to the traditional energy-consuming aerobic wastewater treatment, the use of AD is advantageous and suitable to most types of organic wastes such as agricultural manure, waste paper, grass clippings, municipal waste, and food and fruit/vegetable processing waste. End products of the AD process are mainly methane, carbon dioxide and AD liquid effluents that contain stabilized organic matter and other nutrients. The benefits of AD include substantial odour reduction, production of a renewable energy source (biogas), reduction of greenhouse gas (GHG) emissions, potential pathogen reduction, production of less sludge for disposal, and enhanced nutrient management (Verstraete and Vandevivere, 1999). Of all the current bioenergy options, AD is one of only a few processes already in massive industrial scale that also may generate profits without governmental tax credit.

Different groups of microorganisms work together in a food chain to degrade organic materials and produce methane as the final product. Briefly, insoluble organic materials (e.g. polysaccharide sugar, protein and fat) are hydrolysed to become soluble materials such as simple sugar, amino acid and long-chain fatty acid. In a process called acidogenesis, acidogenic bacteria degrade soluble organics to produce volatile fatty acid (VFA) and hydrogen. Then, acetogenic bacteria produce acetate from VFA and solvents by acetogenesis. A group of acetogenic bacteria can synthesize acetate from hydrogen and CO_2, referred to as homoacetogenesis. Finally, methanogens use acetate or hydrogen to produce methane as the final product. There are also other steps involved in AD; for example, sulfate-reducing bacteria generate hydrogen sulfide during the anaerobic degradation (Jeong et al., 2005). Chemical oxygen demand (COD) will be significantly reduced in the AD process; and organic nitrogen and phosphorus in the protein are converted to inorganic matters such as ammonia and phosphate, and released in the effluent solution.

With capabilities of odour reduction, pathogen reduction and water pollution control, AD is a favourable solution for waste water and animal waste management. However, AD has its limitations. The ordinary AD process is time consuming because the acidogenic bacteria are not efficient in generating hydrolysis enzymes, causing the hydrolysis and pretreatment as the rate-controlling step, and the methanogens grow extremely slowly. In addition, AD has limited capability to remove all the pollutants from the wastes. For instance, with AD treatment of dairy manure, between 40% and 60% of COD can remain, mainly as fine fibres and lignins, even after digestion. Meanwhile, total N and P remain constant, although AD converts organic N and P to ammonia and phosphate. The AD effluents usually need further treatment before they can be discharged to the receiving water. Besides the low growth rate of the anaerobic microorganisms, especially methanogens, AD is not very stable. If AD is subjected to disturbances, due to biomass washout, toxic substances, or shock loading, it may take a long time for the AD system to return to the normal operating conditions (Voolapalli and Stuckey, 1998). High-rate anaerobic bioreactors, such as upflow anaerobic sludge blanket (UASB) (Casserly and Erijman, 2003), anaerobic filter (Reyes et al., 1999), anaerobic membrane bioreactor (Berube et al., 2006) and sequential batch reactor (Dearman et al., 2006), are designed to maintain relatively high sludge retention times (SRTs) in order to increase the overall efficiency of anaerobic degradation and alleviate the damage caused by stress conditions, such as those mentioned above. Increasing reaction temperature is also proposed to increase the overall efficiency of AD. Compared to mesophilic digestion (20–45°C), thermophilic anaerobic digestion (above 45°C) has additional benefits including a high degree of waste stabilization, more thorough destruction of viral and bacterial pathogens, and improved post-treatment sludge dewatering (Lo et al., 1985). However, these designs may increase the capital investment and maintenance cost of AD.

In general, high concentrations of ammonia and phosphate are released in the effluent together with magnesium and potassium, which can support the growth of other microorganisms. Due to recent increases in oil price and energy demand, waste management researchers are focusing not only on environmental protection, but also on efficient energy production. Besides the methane gas, the biomass and microbial lipids produced from oleaginous cell cultures on AD effluents can serve as another potential energy source for possible biodiesel production.

Integration of Anaerobic Digestion with Oil Accumulation Process

Agricultural effluent and municipal waste water after AD cannot be disposed directly because of their high nutrient level (Levine et al., 2011). In contrast, these waste waters can be considered as a cost-effective candidate of raw materials for biodiesel production (Siddiquee and Rohani, 2011). Cultivation of microalgae, fungi, or yeast can be integrated with AD systems to reduce the remaining COD, phosphate and ammonia. Microalgae were also studied to absorb metal ion, waste pharmaceutical chemicals and dye into their cell biomass in order to remove the pollutants from waste water once their cell biomass was stabilized and harvested. A typical example is the recently developed high rate algae pond (HRAP) in the tertiary wastewater treatment facility (El Hamouri, 2009). HRAP functions behind a two-step upflow anaerobic reactor (pre-treatment) and was followed by one maturation pond (MP) for polishing. The HRAP was revealed to have no activity for removing the COD from the waste water; however, it removed 85% of total N and 63% of total P. Nitrogen removal was discovered due to the assimilation of microalgae for their growth, and denitrification did not play any role in removing the nitrogen in this process. Phosphorus removal in this process was attributed to chemical precipitation and biological assimilation (around 50% each). In removing ammonia, the HRAP is superior to the traditional bacterial nitrification–denitrification process, which requires the assimilation of extra-organic carbon as a carbon source. Phosphorus removal by microalgae is largely thought to be due to its uptake for normal growth, as an essential element required for making cellular constituents such as phospholipids, nucleotides and nucleic acids. Under certain conditions microalgae can be triggered to uptake more phosphorus than is necessary for survival, in the form of polyphosphate (Powell et al., 2011). Phosphorus removal by luxury uptake (amount of P uptake more than growth required) was confirmed to occur in the microalgae growing in the wastewater treatment facility and further research is needed in this prosperous field about the detailed mechanism and applications.

On the other hand, many microalgae and fungi species can accumulate high lipid content (e.g. *Chlorella protothecoides*, *Aspergillus oryzae*, or *Mortierella isabellina*). Microbial lipid accumulation as a means to provide alternative oil resources and for biodiesel production is garnering increased attention due to its high production efficiency and lower demand of agricultural land (Miao and Wu, 2006; Xu et al., 2006; Chisti, 2007; Li et al., 2007). Microalgae oil accumulation provides an especially ideal situation for biofuel generation because most algae species can utilize sunlight and CO_2 for oil synthesis, considered promising for commercial trials and academic research. Based on a recent estimation (Chisti, 2007, 2008), growing microalgae as a fuel-producing crop would require only around 2–3% of existing US crop hectarage to meet 50% of present transportation fuel needs. This is compared to 800% of existing US crop area under maize and 350% in soybeans. While this estimation may be optimistic, it is important to note that the process will only be feasible if the microalgae cultivation is integrated with the wastewater treatment, where the cost of water and nutrients can be compromised. With the microbial oil accumulation process integrated with wastewater treatment, nutrient removal can be achieved and renewable feedstock of biofuel can be harvested. Therefore, the integration of these two processes is a win–win strategy.

Cultivation of microalgae in the waste water is still primarily at the laboratory scale, and fairly limited in commercial applications (Shilton et al., 2008). There is a list of literatures that discuss the nutrient recovery from swine and dairy manures. In a recent study about algae growth in manure waste, total nitrogen was reduced by 75.7–82.5%, and total phosphorus was also found to receive a significant reduction (62.5–74.7%) for all samples (Wang et al., 2010a). These results are similar to high nutrient removal observed by Woertz (2007). However, the nutrient level does affect the lipid content in algae biomass. It is also revealed that the oleaginous algae, which have lipid content around 22.7–27.1% under optimum growth conditions, increased their lipid content to around 44.6–45.7% under nutrient-stressed conditions, indicating that low nutrient level is beneficial for the lipid accumulation (Hu et al., 2008). Using waste water to supply nutrients for algal growth minimizes the application of synthetic fertilizer, provides on-site nutrient removal and reduces greenhouse gas emissions. Following is a brief summary of recent research activities on integrated microalgae cultivation by using anaerobic digestion effluents.

The integrated system is mostly evaluated by using flasks or batch cultivation. In most cases, fast growth rate microalgae species, such as *Chlorella* sp., were selected as the model strain. Dilution of AD effluents is needed due to barriers such as ammonia inhibition and high pH. Three microalgae species (*Chlorella* sp., *Scenedesmus obliquus*, *Phormidium bohneri*) were reported to grow in digested swine manure effluent at different diluted manure medium in 1 l flasks for 12 days cultivation with light. NH_4^+ was completely removed, together with 90% phosphorus and 60–90% COD, depending on the species (Delanoue and Basseres, 1989). *Chlorella* was reported as the best candidate for the batch treatment as *Chlorella vulgaris* production can be achieved in a short time by feeding digested effluent once (Kumar et al., 2010). The effects of dilution ratio on cultivation of green algae *Chlorella* sp. for fatty acid production was described in detail (Wang et al., 2010a). A reverse linear relationship ($R^2 = 0.982$) was found between the average specific growth rate in the first 7 cultivation days and the initial turbidity. This algae cultivation removed ammonia, total nitrogen, total phosphorus and COD by 100%, 75.7–82.5%, 62.5–74.7% and 27.4–38.4%, respectively, and it was reported that CO_2 provided carbon into microalgae biomass. Although the obtained microalgae had 13.7% of lipids in their dry cell biomass, this is still considered as a successful case of integrated biofuel production on AD effluents due to the significant pollution reduction. The process of nitrogen and phosphorus removal process by microalgae was assumed to be a catalytic reaction. *Chlorella vulgaris* was used as model to measure kinetic coefficients (Aslan and Kapdan, 2006). Park described the kinetic coefficients of three microalgae species (*Microcystis aeruginosa*, *C.*

vulgaris and *Euglena gracilis*) growing on the effluents of either anaerobically digested or aerobically digested animal wastewater. The results showed that the microalgae growth was actually affected by the concentration of organic matter, instead of the dilution rates of the AD effluents (Park *et al.*, 2009).

Other microalgae species were also chosen as model strains to study the integration of AD with microalgae cultivation. *Neochloris oleoabundans*, green algae, grew in both synthetic media and anaerobic digester effluents. These algae cells assimilated 90–95% of the initial nitrate and ammonium after 6 days and yielded 10–30% fatty acid methyl esters on a dry weight basis. Their C18:1 content in microbial lipids steadily increased, while polyunsaturated fatty acid content decreased after nitrogen was depleted. High quality biodiesel was produced from digested manure, while accomplishing wastewater treatment (Levine *et al.*, 2011). Using *Scenedesmus* sp. as a model species, concentrated algal cell suspensions growing on slightly diluted aerated manure is a promising way of simultaneously treating pig waste and producing good quality biomass. With this method, nutrient removal was up to 3.3 mmol l^{-1} day^{-1} (59 mg l^{-1} day^{-1}) for N-NH$_4^+$ (Martin *et al.*, 1985). *Spirulina platensis* removed 26.5–30.0% COD of waste water in the medium when AD effluent was integrated with *Spirulina platensis* cultivation. The maximum removal of total phosphorus (41.6%) was with 8.5% of waste water. This supports the feasibility of nutrient removal and biodiesel production by microalgae (Mezzomo *et al.*, 2010).

A semi-continuous cultivation was set up based on Wang's previous work. A higher organic carbon source in undigested manure aids mixotrophic microalgae growth, therefore a shorter hydraulic retention time (HRT) is needed for nitrogen removal. Algae in digested manure could harvest more CO$_2$ (1.68 mg CO^2 mg^{-1} dry weight (DW) (Wang *et al.*, 2010b). The effect of HRT of a continuous system on removal efficiencies was stated as an exponential relationship. COD, BOD and orthophosphate removal rate decreased with increased effluent concentration while the inorganic nitrogen removal rate increased with an increasing effluent substrate concentration (Travieso *et al.*, 2004).

There are a few onsite microalgae cultivations reported in the references. In year-long high-rate algal ponds in Spain, the average COD and TKN removal efficiencies were 76±11% and 88± 6%, respectively. The productivity of microalgae was 21 to 28 g m^{-3} day^{-1}. COD removal shifted slightly higher between different seasons. It was recommended that high organic carbon is prevailing with high microalgae diversities in late spring and summer (de Godos *et al.*, 2009). In Mexico, 31 species of microalgae were tested over 25 days at different phases in a wastewater plant. Different algae were abundant in different phases because of culture conditions and nutrient source. From this work, we can clearly obtain evidence for the suitability of microalgae in wastewater treatment (Bernal *et al.*, 2008). Microalgae cultivation was also integrated with fish farm effluent to remove nutrients because of EU regulations. It was found that 80–100% ammonia and 41–100% phosphorus were removed depending on species. The integrated systems of microalgae cultivation and fish cultivation were recommended again due to their promising results (Borges *et al.*, 2005).

Microalgae Species Developed to Accumulate Lipids

Lipid accumulation with microalgae is a heavily researched topic to provide alternative oil resources for biodiesel production (Roessler *et al.*, 1994). Databases of microalgae collections exist in the USA at the University of Texas at Austin, as well as other universities such as University of Minnesota, University of Hawaii, and some federal agencies. Using the fast growth rate and high lipid content as searching criteria, a few microalgae species, which may have some industrial potential, and their performances are listed in Table 10.1.

Table 10.1. Some microalgae species with fast growth rate (from UTEX culture collection of algae).

Microalgae	Culture (trophic)			Substrates	Growth rate	Lipid Content (%)
	Photo	Mixo	Hetero			
Chlorella protothecoides	X		X	glucose, acetate/CO$_2$	3.74 g l^{-1} – 144 h	55.2
Chlorella vulgaris	X	X	X	glucose, acetate, lactate/CO$_2$	0.098 h^{-1}	
Crypthecodinium cohnii			X	glucose/CO$_2$	40g l^{-1} – 60–90 h	15–30
Scenedesmus obliquus	X		X	glucose/CO$_2$	Double in 14 h after adaptation	14–22
Chlamydomonas reinhardtii	X	X	X	acetate/CO$_2$	Exponential during the first 20 h	21
Micractinium pusillum		X	X	CO$_2$	0.94 g l^{-1} – 24 h	
Euglena gracilis			X	CO$_2$		14–20
Schizochytrium sp.				glycerol/CO$_2$	130–140 h (28°C)	55
Spirulina platensis	X		X	glucose/CO$_2$	0.008 h^{-1}	
Botryococcus braunii			X	CO$_2$	low growth rate	20–86
Dunaliella salina	X	X		CO$_2$		~70

Unlike microalgae species widely seen in wastewater treatment facilities with relatively large (>70 um) cells such as *Coelastrum* and *Spirulina*, oleaginous algae species mostly belong to species approaching bacterial dimensions (<30 μm), such as *Scenedesmus*, *Dunaliella* and *Chlorella* (Brennan and Owende, 2010). It is challenging to have ideal strains that are both suitable to grow in the AD effluents, which feature high N and P concentration, and to efficiently remove nutrients while maintaining a high level of lipid accumulation. Novel microalgae screening is an efficient method because the right growth condition provides the microalgae with potential ability for biofuel production and robust substrate utilization. A consortium of 15 native algae isolates showed higher than 96% nutrient removal in the treated waste water. About 63.9% algal oil obtained can be converted into biodiesel (Chinnasamy *et al.*, 2010). Two marine microalgal isolates (*Phaeodactylum tricornutum*, *Oscillatoria* sp.) out of 102 microalgae species from a sewage outfall site in St Andrews Bay, Scotland were cultured on corrugated raceways (2.5 m long, 0.2 m wide) to remove nutrients and purify water; and their capability to remove pollutants remained unaltered during the diurnal cycle (Craggs et al., 1997). Further research is needed in this specific area, targeting oleaginous microalgae screening from different wastewater treatment processes in order to obtain ideal strains.

Growth Mode of Microalgae in the AD Effluents

Microalgae can rapidly accumulate lipid, which fits the industrial needs, with either autotrophic growth or heterotrophic growth mode. For the autotrophic growth mode, the microalgae assimilate the CO_2 from the atmosphere as their carbon source, and sunlight in most cases is their energy source. The heterotrophic growth of microalgae cells is to use organic carbon such as glucose, acetate, or glycerol in the dark. Past studies for large-scale cultivation of algae relied on open-pond systems, which made it difficult to successfully cultivate algae due to the high downstream processing cost. Open-pond cultures are only commercialized to produce some value-added health food supplements such as feed and reagents (Chisti, 2007). Photobioreactors are developed to achieve higher productivity and to maintain monoculture of algae; however, the illuminated areal, volumetric productivity and cost of production in these enclosed photobioreactors are not better than those achievable in open-pond cultures despite higher biomass concentration and better control of culture parameters (Lee, 2001). Alternatively, cultivating microalgae in heterotrophic mode in sterilized fermenters has achieved some commercial success due to several superior characteristics of this process. Like yeast and fungi, heterotrophic algae can accumulate biomass and lipids using organic carbon as their source instead of CO_2 and sunlight. Compared with phototrophic algae, the heterotrophic growth process has the advantages of no light limitation, a high degree of process control, higher productivity and low costs for biomass harvesting (Barclay et al., 1994). Most heterotrophically cultured algae have greater than ten times the biomass concentration, while the lipid productivity is also significantly higher than the theoretical data for autotrophic cultivation. For certain algae strains, it was suggested the heterotrophically cultured cells exhibited better capability for biomass and lipid production. Miao and Wu (2006) reported that the oil content of heterotrophically cultured *Chlorella potothecoides* was approximately four times greater than that in the corresponding autotrophic culture. Liu et al. (2010) demonstrated that the heterotrophically cultured cells of *Chlorella zofingiensis* showed 411% and 900% increase in dry cell weight and lipid yield, respectively, compared to autotrophically cultured cells.

Although few studies touched on the growth mode of microalgae culturing in the AD effluents, the high turbidity actually blocks the access of sunlight to microalgae cells in many cases, and they either have to grow heterotrophically or their growth is generally inhibited inside the pond. AD effluents either have to be diluted, filtered, or pretreated with chemical additives to remove or dilute the fine particles and growth inhibitors (Levine et al., 2011). This is also an area for microalgae screening to obtain microalgae species that can survive in high turbidity conditions. Wang (Wang et al., 2010a) cultured a wild strain found in the local fresh water in the digested dairy manure waste water and conducted an elemental analysis on their culture of *Chlorella* sp. They found that algae used the organic carbon in the digested dairy manure only as part of their carbon source, while CO_2 was also assimilated through the process of autotrophic photosynthesis (Wang et al., 2010a). Similar results were reported that mixotrophic *Chlorella vulgaris* produced higher biomass concentrations than those under heterotrophic and autotrophic conditions (Liang *et al.*, 2009). This is not widely seen in many microalgae species because they cannot utilize light in the presence of carbon sources. *Chlorella protothecoides*, for instance, is a typical example of this amphitrophic organism, which term is used to describe organisms able to live either auto- or heterotrophically; for this organism, the addition of sunlight did not increase the growth rate at all (Becker, 1994). These results show the potential advantages to the mixotrophic microalgae species in dealing with high turbidity substrates.

Stress Conditions for the Microalgae Cell Growth and Oil Accumulation

Culture stability and contaminant mitigation is the primary concern for microbial oil accumulation, especially when the cells are cultured in large unprotected ponds or raceways. In general, high lipid accumulation only occurs under several stressed conditions. Nitrogen deprivation or limitation was commonly thought as the default stress condition to induce the elevated lipid content (Illman *et al.*, 2000; Takagi *et al.*, 2000). Oleaginous cells still assimilate the carbon sources and convert to lipid as energy storage if the nitrogen limitation restrains their growth. Some species, especially *C. protothecoides*, tend to accumulate higher lipid content in the heterotrophic growth mode than the autotrophic mode (Xu *et al.*, 2006). This behaviour is primarily due to the increased carbon:nitrogen ratio that the heterotrophic growth mode can reach. Other factors, such as low temperature (Renaud *et al.*, 2002), high salt concentration (Takagi *et al.*, 2000) and high iron concentration (Liu *et al.*, 2008), have also been shown to be stress conditions for higher lipid production. In order to reach higher lipid content in algae required for efficient biodiesel production, nitrogen limitation is usually induced to stimulate the oil accumulation. But the nitrogen limitation also induces a strong decrease in growth rate (Droop, 1983). Consequently the increase in the lipid content is generally not compensated and eventually productivity is decreased (Rodolfi *et al.*, 2009). The requirement for stress conditions to repress the cell growth brings serious issues for this process to be commercialized, due to the contamination or invasion from foreign species if the growth of oleaginous microalgae is inhibited. The oil accumulation feature also generates some dilemma in the integration of microalgae cultivation with anaerobic digestion, because the AD effluents generally have high concentrations of nitrogen and phosphorus while their carbon sources are difficult to further digest and are quantitatively limited.

Sterilization of AD Effluents and Mixed Culture of Microalgae and Bacteria

Data are generally lacking on the bacteria–algae interactions on the microalgae culturing with AD effluents, although many research efforts directly applied the effluents with microalgae cultivation without sterilization. Early work by the Aquatic Species Program (1978–1996) revealed that polycultures of endogenous, unicellular algae on waste water only produced microalgae biomass with very low oil content (5–15% lipids on a dry weight basis) (Sheehan *et al.*, 1998). Mixed periphyton cultures grown on raw and digested animal manure were even less suitable for biodiesel production, containing only 0.6–1.5% lipids

(Levine et al., 2011). However, mixed culture of microalgae and aerobic bacteria can remove nutrients more efficiently (Jacome-Pilco et al., 2009), and this integrated system was confirmed in 16 l laboratory ponds (Travieso et al., 2006). Oxygen produced by algae was required by acclimatized bacteria to degrade hazardous pollutants such as polycyclic aromatic hydrocarbons, phenolics and organic solvents. The integration system is stable when cultivation condition was under proper control (microalgae concentration, nutrients concentration, light density). The plant growth-promoting bacteria (PGPB) used to increase the growth and nutrient removal capacity of microalgae so far belong to the genus *Azospirillum* and are widely used as an inoculant to promote the growth and yield of numerous crop plants, mainly by affecting hormonal metabolism and mineral absorption of the plants (de-Bashan and Bashan, 2010; Tang et al., 2010). However, it is still not determined whether polycultures of microalgae in waste water or monocultures of dedicated oleaginous microalgae are better for high lipid content. Complete removal of bacteria from the AD effluents is difficult, as a result of spore production by many *Clostridia* species from the anaerobic digestion process.

Immobilization of Microalgae and Self-aggregation

Immobilization of microalgae is a common strategy to increase the nutrient removal from AD effluents, due to the relative easy harvest of microalgae cells. In addition, immobilizing microalgae-promoting bacteria is used to facilitate microalgae growth. Numerous review articles have been published in this field. For instance, Mallick (2002) reviewed the immobilized microalgae cultivations for their effectiveness in removing nitrogen, phosphorus and metals, as well as their potential commercial application of microalgae cultivation in different reactors. Immobilization affects the growth rate and morphology of microalgae, leading to different metabolism and cell productivity. The common issue of microalgae immobilization is cell leaking, even though the issue is not significant in many cases (Cordoba and Hernandez, 1995). The influence of the recirculation in the efficiency of the immobilized microalgae process was also evaluated for 3 months. A recirculation factor of 10:1 was able to remove a considerable amount of nitrogen and phosphorus by immobilized microalgae (Cordoba and Hernandez, 1995).

Besides the commonly studied immobilization methods by using external materials, such as covalent coupling, affinity immobilization, adsorption, confinement in liquid–liquid emulsion, capture behind semi-permeable membrane and entrapment in polymers (Mallick, 2002), many filamentous microorganisms tend to aggregate and grow as pellets/granules. They are spherical or ellipsoidal masses of hyphae with variable internal structure, ranging from loosely packed hyphae, forming 'fluffy' pellets, to tightly packed, compact, dense granules (Hu and Chen, 2007, 2008; Hu et al., 2009; Xia et al., 2011). This self-immobilization approach appears to be the most promising option to achieve both a high-quality treated effluent in terms of total suspended solids and economically recovering algal biomass for biofuel use (Uduman et al., 2010). It will also be more environmentally sound than current procedures, which may need chemical addition. Many of the algal species in the wastewater treatment processes often form large colonies (50–200 μm), and their cell aggregation can be achieved through nitrogen limitation and CO_2 addition. Several cyanobacteria are filamentous species, especially in the conditions lacking nitrogen sources (Park et al., 2011). However, most of these microalgae species are not oleaginous species.

New oleaginous microalgae with filamentous features or methods to enable oleaginous microalgae aggregation during their cultivation are strategically and urgently needed.

Cost-effective Harvesting

How to harvest microalgae cells is dependent on the characteristics of the microalgae, such as size and density. Currently available harvest approaches, including flocculation, flotation, centrifugal sedimentation and filtration, all have limitations for effective, cost-efficient production of biofuel. For instance, flotation methods, based on the trapping of algae cells using dispersed micro-air bubbles, is very limited in its technical and economic viability. Most conventional and economical separation methods, such as filtration and gravitational sedimentation, are widely applied in wastewater treatment facilities to harvest relatively large (>70 µm) microalgae such as *Coelastrum* and *Spirulina*. However, they cannot be used to harvest algae species approaching bacterial dimensions (<30 µm) like *Scenedesmus*, *Dunaliella* and *Chlorella* (Brennan and Owende, 2010), to which most oleaginous microalgae species belong. Centrifugation is a method widely used to recover microalgae biomass, especially small-sized algae cells; however, its application is restricted to algae cultures for high-value metabolites due to intensive energy needs and high equipment maintenance requirements. While flocculation is used to harvest small-sized microalgae cells, it is a preparatory step to aggregate the microalgae cells and increase the particle size so that other harvesting methods such as filtration, centrifugation or gravity sedimentation can be applied (Molina Grima *et al.*, 2003). Several flocculants have been developed to facilitate the aggregation of microalgae cells, including multivalent metal salts like ferric chloride ($FeCl_3$), aluminium sulfate ($Al_2(SO_4)_3$) and ferric sulfate ($Fe_2(SO_4)_3$), and organic polymers such as Chitosan (Li *et al.*, 2008). Chemical flocculation can be reliably used to remove small algae cells from pond water by forming large (1–5 mm) sized flocs (Sharma *et al.*, 2006). However, the chemical reactions are highly sensitive to pH and the high doses of flocculants required produce large amounts of sludge and may leave a residue in the treated effluent. In summary, most technologies including chemical and mechanical methods greatly increase operational costs for algal production and are only economically feasible for production of high-value products (Park *et al.*, 2011).

Besides traditional methods mentioned above, there are several new technology developments in this field. DOE-ARPA-E recently funded a research project for Algae Venture Systems (AVS) to develop a Harvesting, Dewatering and Drying (AVS-HDD) technology by using the principles of liquid adhesion and capillary action to pull water out of dilute microalgae solutions. Attached algal culture systems have also been developed for growing microalgae on the surface of polystyrene foam to simplify the cell harvest (Wilkie and Mulbry, 2002; Johnson and Wen, 2010).

A recent life-cycle assessment (LCA) of algal biodiesel production from *C. vulgaris* indicated that drying and hexane extraction accounted for up to 90% of the total process energy (Campbell *et al.*, 2011). Cost-effective harvesting and lipid extraction are perhaps the most critical needs in the area of microalgal lipids. A biodiesel production process that obviates biomass drying and organic solvent use for oil extraction could lead to significant energy and cost savings (Levine *et al.*, 2010), while the development is still in process.

Utilization of Algae and Biofuel Products from Microbial Lipids

The algal biomass produced and harvested could be converted through various pathways to biofuels, for example anaerobic digestion to biogas, transesterification of lipids to produce biodiesel, fermentation of carbohydrate to bioethanol, and high temperature conversion to bio-crude oil. Microalgae contain varying proportions of proteins, lipids, carbohydrates, nucleic acids, pigments and vitamins. Several types of fuel can be extracted from algae, and it attracts scientists to study how to utilize the algae biomass efficiently for a long time. Research on the utilization of algae for biofuel production started more than 50 years ago. Among the various utilization routes, biodiesel is of particular interest because it can easily and quickly be used as vehicle fuel. Lipids can be converted to biodiesel via several conversion technologies, and transesterification to produce FAME is the most widely used. Transesterification is a reversible reaction, in which an ester is transformed into another through interchange of the alkoxy moiety. The chemical and enzymatic transesterification are well applied to the commercial production due to their high conversion efficiency and low cost. Most of the biodiesel that is currently made uses soybean oil, methanol and an alkaline catalyst, while lipid content in algae can also be extracted and used as substrate for biodiesel production. Recent interest in using oleaginous microalgae as a non-edible biodiesel feedstock has grown considerably (Levine et al., 2010). Considering the generally low content of lipids for the microalgae cells cultured in the AD effluents or waste water, it is still arguable whether it will be economically feasible to extract lipids from microalgae cells and convert them to biodiesel. Pyrolysis and gasification are proposed to directly convert raw microalgae biomass to liquid fuel.

Integration of Anaerobic Digestion with Cultivation of Cyanobacteria, Fungi and Yeast

Besides ordinary microalgae species that have been reported as the biofuel producer from agricultural effluent, blue-green algae (cyanobacteria) have also been introduced to remove ammonia and other nutrients. After 7 days of microalgae cultivation in effluents from an anaerobic dairy lagoon, 99% nitrogen was removed with maximum removal rate at 24 mg l^{-1} day^{-1} (Lincoln et al., 1996). However, cyanobacteria generally do not accumulate triglycerides, and gene modification to enable this feature is currently under initial investigation.

Oleaginous yeasts and fungi have been examined and found to be of great importance in biofuel production because of their percentage in lipid accumulation and also for their fatty acids, which are similar to those found in vegetable oil. Species of oleaginous fungi and yeast are listed in Table 10.2.

The cultivation of oleaginous yeast and fungi on AD effluents is rarely researched, although numerous references report the utilization of yeast and fungal strains for removing nutrients from AD effluents and other types of waste water (More et al., 2010). Thanh and Simard (1973) reported that 17 fungal species removed phosphate (84.1%), total nitrogen (68.1%) and COD (39.3%); and the cell biomass from the cultivation effluent reached 451.2 mg l^{-1}. Recently, the potential of pollutant removal by filamentous fungi was summarized (More et al., 2010). *Trichothecium roseum* was the most efficient strain in phosphate removal (97.5%), while *Epicoccum nigrum, Geotrichum candidum* and *Trichoderma* sp. showed largest removal of total nitrogen (86.8%) and COD (72.3%). A yeast species, *Candida geotrichum*, can reduce the COD (91–95%) using dilute effluent solution. Decreased phosphate and ammonia concentration also was observed after the yeast-screening procedure for reducing pollutant materials (Arnold *et al.*, 2000).

Table 10.2. Lipid production on different substrates by yeasts and fungi.

Yeast strain	Substrate	Lipid content (%)	Reference
Yeast			
Y. lipolytica	Industrial glycerol	42	Papanikolaou and Aggelis (2002)
Y. lipolytica	Glycerol and stearin	30	Papanikolaou and Aggelis (2002)
C. curvatus	Glycerol	25	Meesters *et al.* (1996)
C. curvatus	Glucose	53	Hassan *et al.* (1996)
A. cuvatum	Glucose	45.6	Hassan *et al.* (1993)
L. starkeyi	Glucose and xylose (2:1)	61.5	Zhao *et al.* (2010)
R. glutinis	Sugarcane molasses	39.2	Alvarez *et al.* (1992)
R. glutinis	Glucose	66	Johnson *et al.* (1992)
R. toruloides and S. fibuliger	Starch	36.5	Dostalek (1986)
T. fermentans	Mannose	50.4	Huang *et al.* (2009)
T. fermentans	Cellobiose	65.6	Huang *et al.* (2009)
T. fermentans	Galactose	59.0	Huang *et al.* (2009)
T. fermentans	Dilute sulfuric acid pretreated rice straw hydrolysate	40.1	Huang *et al.* (2009)
Fungi			
M. circinelloides	Acetic acid	28	Du Preez *et al.* (1995)
R. oryzae	Glucose	57	Baldwin (1986)
A. niger	Glycerol	41	Andre *et al.* (2010)
A. fischeri	–	53	Ratledge (1997)
A. terreus	–	57	Ratledge (1997)
C. globosum	–	54	Ratledge (1997)
M. isabellina	Glucose	50–55	Papanikolaou *et al.* (2004)
M. circinelloides	Acetic acid	34.4	Du Preez *et al.* (1995)
M. vinacea	–	66	Ratledge (1984)

Activated sludge obtained from aerobic treatment of AD effluents is reported to contain lipids that can be extracted for biofuel production. The oleaginous microorganisms were converting waste nutrients in undiluted AD effluent to biofuel in this process, without external organic supplements (Yousuf *et al.*, 2010). Fungal species were found in the sludge treatment (More *et al.*, 2010; Revellame *et al.*, 2010). Undoubtedly, low final lipid yield is inevitable because the majority of microorganisms in the sludge are not oil-producing species, but the yield can be improved through introduction of specific oleaginous organisms into the sludge process. In a study using yeast (*Rhodotorula glutinis*) for COD degradation and lipid production, 25 g l^{-1} of biomass, 20% of lipid content and 45% of COD degradation were obtained, respectively (Xue *et al.*, 2008). When using starch waste water as substrate, cultivation in a 5 l fermenter yielded more than 60 g l^{-1} biomass with 30% (w/w) lipid content after 60 h (Xue *et al.*, 2008). A pilot-scale in a 300 l fermenter yielded 40 g l^{-1} biomass and 35% lipid content with 80% COD degradation after only 30–40 h cultivation (Xue *et al.*, 2010).

Significance and Problems

Oleaginous microbial cultivation coupled with anaerobic digestion seems to be a promising approach to degrade organic pollutants while generating biofuel; however, the overall process is still in its infancy and significant research efforts are needed. Microalgae cultivation has been proven both in academia and industry to effectively reduce COD and nutrient level of anaerobic digestion effluents, especially agricultural manure and municipal waste water. High content of microbial lipids can also be potentially accumulated via microalgae cultivation, although this research is generally still in the laboratory setting and using synthetic defined culture medium. There are several technical barriers that seriously inhibit the integration of these two processes and the outcome is generally economically unfeasible so far. One of the keys to this integration is the novel microalgae strains that can both efficiently remove pollutants while at the same time accumulate high content of oil in their cell biomass. The ideal organism would be filamentous so that the cell harvest can be dramatically simplified. This situation can also be applied to fungi and yeast. The exploration of new species provides a possible solution, although this type of research is highly unpredictable. Microorganism screening, heterogeneous gene expression, protein engineering by mutagenesis, molecular evolution, genetic manipulation and control of carbon fluxes are also tools to overcome these barriers.

References

Alvarez, R.M., Rodriguez, B., Romano, J.M., Diaz, A.O., Gomez, E., Miro, D., Navarro, L., Saura, G. and Garcia, J.L. (1992) Lipid accumulation in *Rhodotorula glutinis* on sugar cane molasses in single-stage continuous culture. *World Journal of Microbiology and Biotechnology* 8(2), 214–215.

Andre, A., Diamantopoulou, P., Philippoussis, A., Sarris, D., Komaitis, M. and Papanikolaou, S. (2010) Biotechnological conversions of bio-diesel derived waste glycerol into added-value compounds by higher fungi, production of biomass, single cell oil and oxalic acid. *Industrial Crops and Products* 31(2), 407–416.

Arnold, J.L., Knapp, J.S. and Johnson, C.L. (2000) The use of yeasts to reduce the polluting potential of silage effluent. *Water Research* 34(15), 3699–3708.

Aslan, S. and Kapdan, I.K. (2006) Batch kinetics of nitrogen and phosphorus removal from synthetic wastewater by algae. *Ecological Engineering* 28(1), 64–70.

Baldwin, A. R. (ed.) (1986) *Proceedings World Conference on Emerging Technologies in the Fats and Oils Industry.* American Oil Chemists' Society, 326 pp.

Barclay, W., Meager, K. and Abril, J. (1994) Heterotrophic production of long chain omega-3 fatty acids utilizing algae and algae-like microorganisms. *Journal of Applied Phycology* 6(2), 123–129.

Becker, E.W. (1994) *Microalgae, Biotechnology and Microbiology.* Cambridge University Press, Cambridge, UK.

Bernal, C.B., Vazquez, G., Quintal, I.B. and Bussy, A.L. (2008) Microalgal dynamics in batch reactors for municipal wastewater treatment containing dairy sewage water. *Water Air And Soil Pollution* 190(1–4), 259–270.

Berube, P.R., Hall, E.R. and Sutton, P.M. (2006) Parameters governing permeate flux in an anaerobic membrane bioreactor treating low-strength municipal wastewaters, a literature review. *Water Environment Research* 78(8), 887–896.

Borges, M.T., Silva, P., Moreira, L. and Soares, R. (2005) Integration of consumer-targeted microalgal production with marine fish effluent biofiltration – a strategy for mariculture sustainability. *Journal of Applied Phycology* 17(3), 187–197.

Brennan, L. and Owende, P. (2010) Biofuels from microalgae – a review of technologies for production, processing, and extractions of biofuels and co-products. *Renewable and Sustainable Energy Reviews* 14(2), 557–577.

Campbell, P.K., Beer, T. and Batten, D. (2011) Life cycle assessment of biodiesel production from microalgae in ponds. *Bioresource Technology* 102(1), 50–56.

Casserly, C. and Erijman, L. (2003) Molecular monitoring of microbial diversity in an UASB reactor. *International Biodeterioration and Biodegradation* 52(1), 7–12.

Chinnasamy, S., Bhatnagar, A., Hunt, R.W. and Das, K.C. (2010) Microalgae cultivation in a wastewater dominated by carpet mill effluents for biofuel applications. *Bioresource Technology* 101(9), 3097–3105.

Chisti, Y. (2007) Biodiesel from microalgae. *Biotechnology Advances* 25(3), 294–306.

Chisti, Y. (2008) Biodiesel from microalgae beats bioethanol. *Trends in Biotechnology* 26(3), 126–131.

Cordoba, L.T. and Hernandez, E.S. (1995) Final treatment for cattle manure using immobilized microalgae. 2. Influence of the recirculation. *Resources Conservation and Recycling* 13(3–4), 177–182.

Craggs, R.J., McAuley, P.J. and Smith, V.J. (1997) Wastewater nutrient removal by marine microalgae grown on a corrugated raceway. *Water Research* 31(7), 1701–1707.

de-Bashan, L.E. and Bashan, Y. (2010) Immobilized microalgae for removing pollutants, review of practical aspects. *Bioresource Technology* 101(6), 1611–1627.

de Godos, I., Blanco, S., Garcia-Encina, P.A., Becares, E. and Munoz, R. (2009) Long-term operation of high rate algal ponds for the bioremediation of piggery wastewaters at high loading rates. *Bioresource Technology* 100(19), 4332–4339.

Dearman, B., Marschner, P. and Bentham, R.H. (2006) Methane production and microbial community structure in single-stage batch and sequential batch systems anaerobically co-digesting food waste and biosolids. *Appl. Microbiol. Biotechnol.* 69(5), 589–596.

Delanoue, J. and Basseres, A. (1989) Biotreatment of anaerobically digested swine manure with microalgae. *Biological Wastes* 29(1), 17–31.

Dostalek, M. (1986) Production of lipid from starch by a nitrogen-controlled mixed culture of *Saccharomycopsis fibuliger* and *Rhodosporidium toruloides*. *Appl. Microbiol. Biotechnol.* 24(1), 19–23.

Droop, M. R. (1983). 25 years of algal growth-kinetics – a personal view. *Botanica Marina* 26(3), 99–112.

Du Preez, J.C., Immelman, M., Kock, J.L.F. and Kilian, S.G. (1995) Production of gamma-linolenic acid by *Mucor circinelloides* and *Mucor rouxii* with acetic acid as carbon substrate. *Biotechnology Letters* 17(9), 933–938.

El Hamouri, B. (2009) Rethinking natural, extensive systems for tertiary treatment purposes. The high-rate algae pond as an example. *Desalination and Water Treatment* 4(1–3), 128–134.

Hassan, M., Blanc, P.J., Granger, L.M., Pareilleux, A. and Goma, G. (1993) Lipid production by an unsaturated fatty acid auxotroph of the oleaginous yeast *Apiotrichum curvatum* grown in single-stage continuous culture. *Appl. Microbiol. Biotechnol.* 40(4), 483–488.

Hassan, M., Blanc, P.J., Granger, L.-M., Pareilleux, A. and Goma, G. (1996) Influence of nitrogen and iron limitations on lipid production by *Cryptococcus curvatus* grown in batch and fed-batch culture. *Process Biochemistry (Oxford)* 31(4), 355–361.

Hu, B. and Chen, S.L. (2007) Pretreatment of methanogenic granules for immobilized hydrogen fermentation. *International Journal of Hydrogen Energy* 32(15), 3266–3273.

Hu, B. and Chen, S.L. (2008) Biological hydrogen production using chloroform-treated methanogenic granules. *Applied Biochemistry and Biotechnology* 148(1–3), 83–95.

Hu, B., Zhou, X., Forney, L. and Chen, S.L. (2009) Changes in microbial community composition following treatment of methanogenic granules with chloroform. *Environmental Progress and Sustainable Energy* 28(1), 60–71.

Hu, Q., Sommerfeld, M., Jarvis, E., Ghirardi, M., Posewitz, M., Seibert, M. and Darzins, A. (2008) Microalgal triacylglycerols as feedstocks for biofuel production, perspectives and advances. *Plant Journal* 54(4), 621–639.

Huang, C., Zong, M.-H., Wu, H. and Liu, Q.-P. (2009) Microbial oil production from rice straw hydrolysate by *Trichosporon fermentans*. *Bioresource Technology* 100(19), 4535–4538.

Illman, A.M., Scragg, A.H. and Shales, S.W. (2000) Increase in *Chlorella* strains calorific values when grown in low nitrogen medium. *Enzyme and Microbial Technology* 27(8), 631–635.

Jacome-Pilco, C.R., Cristiani-Urbina, E., Flores-Cotera, L.B., Velasco-Garcia, R., Ponce-Noyola, T. and Canizares-Villanueva, R.O. (2009) Continuous Cr(VI) removal by *Scenedesmus incrassatulus* in an airlift photobioreactor. *Bioresource Technology* 100(8), 2388–2391.

Jeong, H.S., Suh, C.W., Lim, J.L., Lee, S.H. and Shin, H.S. (2005) Analysis and application of ADM1 for anaerobic methane production. *Bioprocess and Biosystems Engineering* 27(2), 81–89.

Johnson, M.B. and Wen, Z.Y. (2010) Development of an attached microalgal growth system for biofuel production. *Appl. Microbiol. Biotechnol.* 85(3), 525–534.

Johnson, V., Singh, M., Saini, V.S., Sista, V.R. and Yadav, N.K. (1992) Effect of pH on lipid accumulation by an oleaginous yeast, *Rhodotorula glutinis* IIP-30. *World Journal of Microbiology and Biotechnology* 8(4), 382–384.

Kumar, M.S., Miao, Z.H.H. and Wyatt, S.K. (2010) Influence of nutrient loads, feeding frequency and inoculum source on growth of *Chlorella vulgaris* in digested piggery effluent culture medium. *Bioresource Technology* 101(15), 6012–6018.

Lee, Y.K. (2001) Microalgal mass culture systems and methods, their limitation and potential. *Journal of Applied Phycology* 13(4), 307–315.

Levine, R.B., Pinnarat, T. and Savage, P.E. (2010) Biodiesel production from wet algal biomass through *in situ* lipid hydrolysis and supercritical transesterification. *Energy and Fuels* 24, 5235–5243.

Levine, R.B., Costanza-Robinson, M.S. and Spatafora, G.A. (2011) *Neochloris oleoabundans* grown on anaerobically digested dairy manure for concomitant nutrient removal and biodiesel feedstock production. *Biomass and Bioenergy* 35(1), 40–49.

Li, X.F., Xu, H. and Wu, Q.Y. (2007) Large-scale biodiesel production from microalga *Chlorella protothecoides* through heterotropic cultivation in bioreactors. *Biotechnology and Bioengineering* 98, 764–771.

Li, Y., Horsman, M., Wu, N., Lan, C.Q. and Dubois-Calero, N. (2008) Biofuels from microalgae. *Biotechnology Progress* 24(4), 815–820.

Liang, Y.N., Sarkany, N. and Cui, Y. (2009) Biomass and lipid productivities of *Chlorella vulgaris* under autotrophic, heterotrophic and mixotrophic growth conditions. *Biotechnology Letters* 31(7), 1043–1049.

Lincoln, E.P., Wilkie, A.C. and French, B.T. (1996) Cyanobacterial process for renovating dairy wastewater. *Biomass and Bioenergy* 10(1), 63–68.

Liu, J., Huang, J., Sun, Z., Zhong, Y., Jiang, Y. and Chen, F. (2010) Differential lipid and fatty acid profiles of photoautotrophic and heterotrophic *Chlorella zofingiensis*, assessment of algal oils for biodiesel production. *Bioresource Technology* 102(1), 106–110.

Liu, Z.Y., Wang, G.C. and Zhou, B.C. (2008) Effect of iron on growth and lipid accumulation in *Chlorella vulgaris*. *Bioresource Technology* 99(11), 4717–4722.

Lo, K.V., Liao, P.H. and March, A.C. (1985) Thermophilic anaerobic digestion of screened dairy manure. *Biomass* 6(4), 301–315.

Mallick, N. (2002) Biotechnological potential of immobilized algae for wastewater N, P and metal removal, a review. *Biometals* 15(4), 377–390.

Martin, C., Delanoue, J. and Picard, G. (1985) Intensive cultivation of fresh-water microalgae on aerated pig manure. *Biomass* 7(4), 245–259.

Meesters, P.A.E.P., Huijberts, G.N.M. and Eggink, G. (1996) High cell density cultivation of the lipid accumulation yeast *Cryptococcus curvatus* using glycerol as a carbon source. *Applied Microbiology and Biotechnology* 45(5), 575–579.

Mezzomo, N., Saggiorato, A.G., Siebert, R., Tatsch, P.O., Lago, M.C., Hemkemeier, M., Costa, J.A.V., Bertolin, T.E. and Colla, L.M. (2010) Cultivation of microalgae *Spirulina platensis* (*Arthrospira platensis*) from biological treatment of swine wastewater. *Ciencia E Tecnologia De Alimentos* 30(1), 173–178.

Miao, X. and Wu, Q. (2006) Biodiesel production from heterotrophic microalgal oil. *Bioresource Technology* 97(6), 841–846.

Molina Grima, E., Belarbi, E.H., Acién Fernández, F.G., Robles Medina, A. and Chisti, Y. (2003) Recovery of microalgal biomass and metabolites, process options and economics. *Biotechnology Advances* 20(7–8), 491–515.

More, T.T., Yan, S., Tyagi, R.D. and Surampalli, R.Y. (2010) Potential use of filamentous fungi for wastewater sludge treatment. *Bioresource Technology* 101(20), 7691–7700.

Papanikolaou, S. and Aggelis, G. (2002) Lipid production by *Yarrowia lipolytica* growing on industrial glycerol in a single-stage continuous culture. *Bioresource Technology* 82(1), 43–49.

Papanikolaou, S., Komaitis, M. and Aggelis, G. (2004) Single cell oil (SCO) production by *Mortierella isabellina* grown on high-sugar content media. *Bioresource Technology* 95(3), 287–291.

Park, J.B.K., Craggs, R.J. and Shilton, A.N. (2011) Wastewater treatment high rate algal ponds for biofuel production. *Bioresource Technology* 102(1), 35–42.

Park, K.Y., Lim, B.R. and Lee, K. (2009) Growth of microalgae in diluted process water of the animal wastewater treatment plant. *Water Science and Technology* 59(11), 2111–2116.

Powell, N., Shilton, A., Pratt, S. and Chisti, Y. (2011) Luxury uptake of phosphorus by microalgae in full-scale waste stabilisation ponds. *Water Science and Technology* 63(4), 704–709.

Ratledge, C. (1984) Microbial oils and fats – an overview. *AOCS Monograph* 11 (Biotechnol. Oils Fats Ind.), 119–127.

Ratledge, C. (1997) *Microbial Lipids.* Wiley–VCH Verlag GmbH.

Renaud, S.M., Thinh, L.V., Lambrinidis, G. and Parry, D.L. (2002) Effect of temperature on growth, chemical composition and fatty acid composition of tropical Australian microalgae grown in batch cultures. *Aquaculture* 211(1–4), 195–214.

Revellame, E., Hernandez, R., French, W., Holmes, W. and Alley, E. (2010) Biodiesel from activated sludge through *in situ* transesterification. *Journal of Chemical Technology and Biotechnology* 85(5), 614–620.

Reyes, O., Sanchez, E., Rovirosa, N., Borja, R., Cruz, M., Colmenarejo, M.F., Escobedo, R., Ruiz, M., Rodriguez, X. and Correa, O. (1999) Low-strength wastewater treatment by a multistage anaerobic filter packed with waste tyre rubber. *Bioresource Technology* 70(1), 55–60.

Rodolfi, L., Zittelli, G.C., Bassi, N., Padovani, G., Biondi, N., Bonini, G. and Tredici, M.R. (2009) Microalgae for oil, strain selection, induction of lipid synthesis and outdoor mass cultivation in a low-cost photobioreactor. *Biotechnology and Bioengineering* 102(1), 100–112.

Roessler, P.G., Brown, L.M., Dunahay, T.G., Heacox, D.A., Jarvis, E.E., Schneider, J.C., Talbot, S.G. and Zeiler, K.G. (1994) Genetic-engineering approaches for enhanced production of biodiesel fuel from microalgae. *Enzymatic Conversion of Biomass for Fuels Production* 566, 255–270.

Sharma, B.R., Dhuldhoya, N.C. and Merchant, U.C. (2006) Flocculants – an ecofriendly approach. *Journal of Polymers and the Environment* 14(2), 195–202.

Sheehan, J., Dunnahay, T., Benemann, J. and Roessler, P. (1998) A look back at the US Department of Energy's Aquatic Species Program – biodiesel from algae. *National Renewable Energy Laboratory*.

Shilton, A.N., Mara, D.D., Craggs, R. and Powell, N. (2008) Solar-powered aeration and disinfection, anaerobic co-digestion, biological CO_2 scrubbing and biofuel production, the energy and carbon management opportunities of waste stabilisation ponds. *Water Science and Technology* 58(1), 253–258.

Siddiquee, M.N. and Rohani, S. (2011) Lipid extraction and biodiesel production from municipal sewage sludges, a review. *Renewable and Sustainable Energy Reviews* 15(2), 1067–1072.

Takagi, M., Watanabe, K., Yamaberi, K. and Yoshida, T. (2000) Limited feeding of potassium nitrate for intracellular lipid and triglyceride accumulation of *Nannochloris* sp UTEX LB1999. *Appl. Microbiol. Biotechnol.* 54(1), 112–117.

Tang, X., He, L.Y., Tao, X.Q., Dang, Z., Guo, C.L., Lu, G.N. andYi, X.Y. (2010) Construction of an artificial microalgal–bacterial consortium that efficiently degrades crude oil. *Journal of Hazardous Materials* 181(1–3), 1158–1162.

Thanh, N. and Simard, R. (1973) Biological treatment of domestic sewage by fungi. *Mycopathologia* 51(2), 223–232.

Travieso, L., Sanchez, E., Borja, R., Benitez, F., Leon, M. and Colmenarejo, M.F. (2004) Evaluation of a laboratory and full-scale microalgae pond for tertiary treatment of piggery wastes. *Environmental Technology* 25(5), 565–576.

Travieso, L., Benitez, F., Sanchez, E., Borja, R. and Colmenarejo, M.F. (2006) Production of biomass (algae–bacteria) by using a mixture of settled swine and sewage as substrate. *Journal of Environmental Science and Health Part A – Toxic/Hazardous Substances and Environmental Engineering* 41(3), 415–429.

Uduman, N., Qi, Y., Danquah, M.K., Forde, G.M. and Hoadley, A. (2010) Dewatering of microalgal cultures, a major bottleneck to algae-based fuels. *Journal of Renewable and Sustainable Energy* 2(1).

Verstraete, W. and Vandevivere, P. (1999) New and broader applications of anaerobic digestion. *Critical Reviews in Environmental Science and Technology* 29(2), 151–173.

Voolapalli, R.K. and Stuckey, D.C. (1998) Stability enhancement of anaerobic digestion through membrane gas extraction under organic shock loads. *Journal of Chemical Technology and Biotechnology* 73(2), 153–161.

Wang, L., Li, Y.C., Chen, P., Min, M., Chen, Y.F., Zhu, J. and Ruan, R.R. (2010a) Anaerobic digested dairy manure as a nutrient supplement for cultivation of oil-rich green microalgae *Chlorella* sp. *Bioresource Technology* 101(8), 2623–2628.

Wang, L.A., Wang, Y.K., Chen, P. and Ruan, R. (2010b) Semi-continuous cultivation of *Chlorella vulgaris* for treating undigested and digested dairy manures. *Applied Biochemistry and Biotechnology* 162(8), 2324–2332.

Wilkie, A.C. and Mulbry, W.W. (2002) Recovery of dairy manure nutrients by benthic freshwater algae. *Bioresource Technology* 84(1), 81–91.

Woertz, I.C. (2007) Lipid productivity of algae grown on dairy wastewater as a possible feedstock for biodiesel. Masters thesis, California Polytechnic University, USA, pp. 57–58.

Xia, C., Zhang, J., Zhang, W. and Hu, B. (2011) A new cultivation method for bioenergy production – cell pelletization and lipid accumulation by *Mucor circinelloides*. *Biotechnology for Biofuels* 4, 15.

Xu, H., Miao, X. and Wu, Q. (2006) High quality biodiesel production from a microalga *Chlorella protothecoides* by heterotrophic growth in fermenters. *Journal of Biotechnology* 126(4), 499–507.

Xue, F.Y., Miao, J.X., Zhang, X., Luo, H. and Tan, T.W. (2008) Studies on lipid production by *Rhodotorula glutinis* fermentation using monosodium glutamate wastewater as culture medium. *Bioresource Technology* 99(13), 5923–5927.

Xue, F.Y., Gao, B., Zhu, Y.Q., Zhang, X., Feng, W. and Tan, T.W. (2010) Pilot-scale production of microbial lipid using starch wastewater as raw material. *Bioresource Technology* 101(15), 6092–6095.

Yousuf, A., Sannino, F., Addorisio, V. and Pirozzi, D. (2010) Microbial conversion of olive oil mill wastewaters into lipids suitable for biodiesel production. *Journal of Agricultural and Food Chemistry* 58(15), 8630–8635.

Zhao, X., Hu, C., Wu, S., Shen, H. and Zhao, Z. (2010) Lipid production by *Rhodosporidium toruloides* Y4 using different substrate feeding strategies. *Journal of Industrial Microbiology and Biotechnology* 38(5), 627–632.

Chapter 11

Biohydrogen Generation Through Solid Phase Anaerobic Digestion from Organic Solid Waste

S. Jayalakshmi

Introduction

In the last decade, there has been a remarkable progression in global warming, which imposes threats such as abnormal weather and increased sea levels. This global warming is assumed to be caused by carbon dioxide exhausted by burning fossil fuels such as oil and coal. Therefore, an alternative energy resource to fossil fuel is needed in the future. Hydrogen is considered an alternative energy resource to fossil fuel and provides clean energy for fuel cells that efficiently generate electricity. In addition, hydrogen is advantageous as a renewable energy source and produces no greenhouse gas. Thus, many motor companies have rapidly developed fuel-cell vehicles in recent years. On the other hand, the recycling of organic waste is an emergency requirement because of the stringency of the waste landfill capacity and increase in the amount of organic waste. Anaerobic digestion (AD) is an appropriate technique for the treatment of organic solid waste before final disposal and it is employed worldwide as the oldest and most important process for waste stabilization (Brummeler *et al.*, 1992).

Solid Phase Anaerobic Digestion/Dry Digestion

AD processes can be classified according to the total solid (TS) content of the sludge in the reactor. Low solids (LS) systems contain less than 10% TS, medium solids (MS) contain about 15–20%, and high solids (HS) contain 22–40% solids (Techobanoglous *et al.*, 1993).

High solid anaerobic digestion

High solid anaerobic digestion (HSAD) is a new application of the proven and conventional low-solids anaerobic digestion technology used throughout the world (Rglesia *et al.*, 1998). The HSAD process can be applied to many different waste feedstocks including paper and packaging waste, food processing waste, agricultural waste, sewage and industrial sludge, yard or green waste and municipal solid waste. The HSAD process is specifically designed to process solid organic waste (Alvarez *et al.*, 2000). It can easily accommodate different combinations of solid and liquid, industrial or municipal waste. Blends of rapidly degrading feedstocks, such as fat, oil and grease, and slower-degrading materials, like paper and yard waste, make superior feedstocks for the HSAD process. Blended feedstocks provide consistency of composition with improved process control and higher conversion rates.

The benefits of HSAD systems are (Hoffman, 2000): (i) HSAD residues are suitable for composting; (ii) HSAD does not require special techniques such as slurry pumps, mixers, shredders and liquid manure injectors for distribution; (iii) process energy demand for operating the machinery is lower because of the reduced reactor size; (iv) improved process stability and reliability –problems such as foam formation and sedimentation will not occur; (vi) reduced odour emission because there is no sludge involved; and (vii) reduced nutrient runoff during storage of sludge because there is no liquid in the sludge.

Biohydrogen

The anaerobic digestion of the organic fraction of municipal solid waste (OFSWM) for the generation of methane has received increased interest in the last 15 years (Cecchi et al., 1989; Varaldo et al., 1997, 2002). Yet, even the use of methane as a fuel could be debatable, due to the generation of CO_2 known to contribute to the greenhouse effect (Dickinson and Cicerone, 1986). This leads to the study of alternate ecofriendly, high efficiency future energy, i.e. hydrogen. Hydrogen is a clean fuel, since its combustion with oxygen does not generate polluting emissions. According to energy experts, hydrogen is safe, versatile and has high energy content, high utilization efficiency and is the best option for transport applications (Veziroglu and Barbir, 1995; Das and Veziroglu, 2001).

Much recent interest has been expressed in the biological generation of hydrogen from organic waste materials by dark fermentation, due to its potential importance in our economy (Yokoi et al., 2001; Logan et al., 2002; Yu et al., 2002; Zhang et al., 2003). Biological hydrogen generation shares many common features with methanogenic anaerobic digestion, especially the relative ease with which the two gaseous products can be separated from the treated waste. The mixed communities involved in both bioprocesses share some common elements but with one important difference: successful biological hydrogen generation requires inhibition of hydrogen-utilizing microorganisms, such as homoacetogens and methanogens (Klass, 1998). Inhibition is commonly accomplished by heat treatment of the inoculum to kill all microorganisms except for spore-forming fermenting bacteria (for example, species from the families Clostridiaceae, Streptococcaceae, Sporolactobacillaceae, Lachnospiraceae and Thermoanaerobacteriaceae) (Ueno et al., 2001; Fang and Liu, 2002; Fang et al., 2002). Other methods that have been used include the operation of reactors at high dilution rates (Zhang et al., 2003) or low pH (Oh et al., 2003). Conceptually, important efforts are those that prevent consumption of hydrogen by methanogens.

It was found that two types of fermentation usually occur, namely butyric acid fermentation at pH >6 (Cohen et al., 1984) and ethanol fermentation at pH <4.5 (Ren et al., 1997). According to most of the studies of acetogenic fermentation, the major products of the process are butyric acid (Shin et al., 2004). Ethanol-type fermentation was identified, in which liquid products mostly consisted of ethanol and acetic acid.

Research in hydrogen generation has been a long-term focus worldwide (Hawkes et al., 2002). The Hawaii Natural Energy Institute at the University of Hawaii has carried outresearch and development on biological hydrogen generation since the early 1990s (Bolton, 1996). They made some significant contributions to the advancement of photobiological hydrogen generation. It was the first project that attempted to develop an indirect biophotolysis process. The concept was to use nitrogen-limited microalgae mass cultures to maximize CO_2 fixation into carbohydrates and then to generate hydrogen in both dark and light.

Hydrogen has been called the 'most alternative' of the alternative fuels: if it is made by electrolysis of water using electricity from a non-polluting source such as wind or solar power, then no pollutants of any kind are generated by burning it in an internal combustion engine except for trace amounts of nitrogen oxides, and if it is used in a fuel cell then even these disappear (Gustavo et al., 2008). Using hydrogen as the 'battery' to store energy from a non-polluting, renewable source would result in a truly unlimited supply of clean fuel (Larminie and Dicks, 2000). The advantage of using hydrogen to store energy rather than a battery pack is that a hydrogen tank can be refilled in minutes rather than recharged in hours, and it takes less space and weight to store enough hydrogen to drive a given distance on a single refuelling than it does to carry enough battery capacity to go the same distance on a single recharging.

Hydrogen gas generated from organic waste could be sold, used as a heating fuel, as an off- or on-site vehicle fuel, or could be used to make electricity. Of these options, resale of the hydrogen gas makes the most economical sense. The value of hydrogen as a heating fuel can be calculated on the basis of its heat content and the cost of other fuels. For example, if H_2 were compared on the basis of the cost of equal heating content of methane, the heating value of hydrogen and methane is 141,790 $kJkg^{-1}$ and 55,530 $kJkg^{-1}$, respectively (Hansel and Lindblad, 1998).

Methods of Biohydrogen Generation

There are two main approaches to microbial hydrogen generation, photochemical and fermentation. The first uses photosynthetic microorganisms such as algae and photosynthetic bacteria (Ike et al., 1997; Melis and Happe, 2001). The second approach is by fermentative hydrogen-generating microorganisms such as facultative anaerobes and obligate anaerobes (Joyner and Winter, 1997; Nandi and Sengupta, 1998). The various technologies used for biohydrogen generation are: (i) direct biophotolysis; (ii) indirect biophotolysis; (iii) photo-fermentation; (iv) hydrogen synthesis via water-gas shift reaction; and (v) anaerobic digestion.

Direct biophotolysis

Direct biophotolysis of hydrogen generation is a biological process using the microalgae photosynthetic system (Lee et al., 2002) to convert solar energy into chemical energy in the form of hydrogen (Eqn 11.1).

$$2H_2O \xrightarrow{light\ energy} 2H_2 + O_2 \quad (11.1)$$

Indirect biophotolysis

According to Gaudernack (1998), the concept of indirect biophotolysis involves the following four steps: (i) biomass generation by photosynthesis; (ii) biomass concentration; (iii) aerobic dark fermentation yielding 4 $molH_2 mol^{-1}$ glucose in the algae cell, along with 2 mol of acetates; and (iv) conversion of 2 mol of acetates into hydrogen. In a typical indirect biophotolysis, Cyanobacteria are used to generate hydrogen via the reactions represented in Eqns 11.2 and 11.3 (Levin et al., 2004):

$$12H_2O + 6CO_2 \xrightarrow{\text{light energy}} C_6H_{12}O_6 + 6O_2 \qquad (11.2)$$

$$C_6H_{12}O_6 + 12H_2O \xrightarrow{\text{light energy}} 12H_2 + 6CO_2 \qquad (11.3)$$

Photo-fermentation

Photosynthetic bacteria have the capacity to generate hydrogen through the action of their nitrogenase using solar energy and organic acids or biomass (Masukawa et al., 2002). This process is known as photo-fermentation. In recent years, some attempts have been made for hydrogen generation from industrial and agricultural waste for effective waste management (Kondo et al., 2002; Masukawa et al., 2002). Hydrogen can be generated by photo-fermentation of various types of wastes. However, these processes have three main drawbacks: (i) use of nitrogenase enzyme with high-energy demand; (ii) low solar energy conversion efficiency; and (iii) demand for elaborate anaerobic photo-bioreactors covering large areas (Fedorov et al., 1998). Hence, at the present time, the photo-fermentation process is not a competitive method for hydrogen generation. Equation 11.4 represents the hydrogen generation by photo-fermentation.

$$C_6H_{12}O_6 + 12H_2O \xrightarrow{\text{light energy}} 12H_2 + 6CO_2 \qquad (11.4)$$

Hydrogen synthesis via water–gas shift reaction

Some photoheterotrophic bacteria, such as *Rhodospirillum rubrum*, can survive in the dark by using CO as the sole carbon source to generate ATP by coupling the oxidation of CO to the reduction of H^+ to H_2 (Kerby et al., 1995). Equation 11.5 represents the hydrogen generation by water–gas shift reaction.

$$CO(g) + H_2O(l) \rightarrow CO_2(g) + H_2(g) \qquad (11.5)$$

The biological water–gas shift reaction for hydrogen generation is still under laboratory scale and only a few trials have been reported. The common objectives of these works were to identify suitable microorganisms that had high CO uptake and to estimate the hydrogen generation rate.

Anaerobic digestion

A promising method to generate biological hydrogen is anaerobic digestion (AD). Hydrogen is generated in the acidogenic and acetogenic phases of fermentation. The substrate is decomposed into hydrogen, carbon dioxide and organic acids by acetogenic bacteria. The metabolites are mainly acetic and butyric acid (Shin et al., 2004). The maximum hydrogen generation rate of 4 $molH_2\,mol^{-1}$ glucose can be achieved when acetic acid (CH_3COOH) is the end product (eqn11.6). Only 2 $molH_2\,mol^{-1}$ glucose can be generated with butyric acid ($CH_3(CH_2)_2COOH$) as an end product (eqn11.7) (Hawkes *et al.*, 2002).

$$C_6H_{12}O_6 + 2H_2O \rightarrow 2CH_3COOH + 2CO_2 + 4H_2 \quad (11.6)$$

$$C_6H_{12}O_6 \rightarrow CH_3(CH_2)_2COOH + 2CO_2 + 2H_2 \quad (11.7)$$

If mixed cultures are used for biological hydrogen generation, acetic and butyric acids will always be generated. In this case, a maximum generation of 2.5 mol $H_2 mol^{-1}$ glucose can be expected (eqn11.8) (Hallenbeck and Benemann, 2002):

$$4C_6H_{12}O_6 + 2H_2O \rightarrow 2CH_3COOH + 3CH_3(CH_2)_2COOH + 8CO_2 + 10H_2 \quad (11.8)$$

During the anaerobic fermentation of glucose, lactic acid ($CH_3CHOHCOOH$) is a potential fermentation product as well. This kind of metabolism is undesirable since no hydrogen will be generated (eqn11.9). Furthermore, lactic acid bacteria such as *Lactobacillus paracasei* or *Enterococcus durans* could produce toxic or inhibitory intermediates for the hydrogen-generating bacteria (Noike and Mizuno, 2000).

$$C_6H_{12}O_6 \rightarrow 2CH_3CHOHCOOH \quad (11.9)$$

Thus, the highest theoretical yields of H_2 are associated with acetate as the fermentation end-product. In practice, however, high H_2 yields are associated with a mixture of acetate and butyrate fermentation products, and low H_2 yields are associated with propionate and reduced end-products (alcohols, lactic acid) (Ren *et al.*, 1997).

Fermentation reactions can be operated at mesophilic (25–40°C), thermophilic (40–65°C), extreme thermophilic (65–80°C), or hyperthermophilic (>80°C) temperatures (Ahring and Ibrahim, 2001). Bacteria known to generate hydrogen include species of *Enterobacter, Bacillus* and *Clostridium*. Carbohydrates are the preferred substrate for hydrogen fermentation (Das and Veziroglu, 2001). Glucose, isomers of hexose, or polymers in the form of starch or cellulose, yield different amounts of H_2 per mole of glucose, depending on the fermentation pathway and end-product(s). *Clostridium pasteurianum*, *C. butyricum* and *C. beijerinkii* are high H_2 producers, while *C. propionicum* is a poor H_2 producer (Hawkes *et al.*, 2002).

Advantages of Biohydrogen Generation by Anaerobic Digestion

Advantages of H_2 generation by anaerobic fermentation (HAF) are that many fermentative bacteria are capable of producing around a 100 times higher hydrogen generation rate (Francou and Vignais, 1984) and H_2 is generated throughout day and night at a constant rate since it does not depend on the energy provided by sunlight. HAF generation has been investigated using pure cultures in sterile conditions and undefined mixed cultures in non-sterile conditions (Liang *et al.*, 2002). Moreover, some authors have shown that this valuable fuel can be generated from OFMSW and industrial wastes (Sparling *et al.*, 1997; Lay *et al.*, 1999, 2003; Okamoto *et al.*, 2000; Jayalakshmi and Sukumaran, 2009). Some advantages of using mixed culture over pure culture in HAF are lower operational costs, operational control based on differential kinetics of microbial subgroup, and feasibility to use septic organic wastes as substrate (Kotsopoulos *et al.*, 2006).

Limitations in hydrogen generation by anaerobic digestion

The long-term process feasibility of continuous and semi-continuous processes is yet to be fully demonstrated. Most HAF-related processes rely on the disruption of the hydrogen uptake of methanogenic archaea, since the latter are recognized to be the most significant hydrogen-consuming microbial group in anaerobic consortia (Morvan et al., 1996; Weijma et al., 2002). The suppression of hydrogen-consuming microbial subgroups such as methanogens remains as a major challenge (Oh et al., 2003). Several approaches have been attempted and reported for achieving this goal, e.g. the use of chemical inhibitors such as acetylene and bromoethane sulfonic acid (Sparling et al., 1997), heat shock pre-treatment of inocula (Lay et al., 1999, 2003; Okamoto et al., 2000; Logan et al., 2002) and keeping the pH of the cultures in the acidic range (5.8–6.5) (Lin and Chang, 1999; Fang and Liu, 2002). Heat shock pre-treatments rely on the killing or thermal suppression of methanogenic archaea and non-sporulating eubacteria, whereas the culture is enriched in sporulating, hydrogen-generating bacteria such as *Clostridia*.

Operating bioreactors at low hydrogen partial pressure, perhaps by sparging with nitrogen gas to strip hydrogen from the solution as fast as it is generated (Oh et al., 2003; Hussy et al., 2005), accomplishes both efforts simultaneously. Unfortunately, optimization of biohydrogen generation focuses on a relatively small fraction of the total hydrogen equivalents that are present in wastewater. For example, optimization of hydrogen generation from hexose, at best results in the generation of 4 mol H_2 mol^{-1} hexose, because 2 mol acetate are also formed; complete oxidation to carbon dioxide and hydrogen, however, would generate 12 mol H_2 mol^{-1} hexose. Actual yields are even lower than the 4 mol of hydrogen that are theoretically possible, typically ranging from 1 to 2.5 molH_2 mol^{-1} hexose. When butyric acid is generated as a major fermentation product, only 2 mol H_2 can be generated. Hydrogen yield is even lower when more reduced organic compounds, such as lactic acid, propionic acid and ethanol, are generated as fermentation products, because these represent end-products of metabolic pathways that bypass the major hydrogen-generating reaction in carbohydrate fermentations (Hallenbeck, 2005).

Limitations for hydrogen generation means that, even under optimized conditions, one cannot expect to recover more than 15% of hydrogen in a high-carbohydrate waste, thus, it is not surprising that several research groups are considering implementing two-step processes, involving biohydrogen generation followed by methanogenic anaerobic digestion to increase the energy yield of the overall process (Ginkel et al., 2001; Logan et al., 2002). As described previously, methanogenic anaerobic digestion is a mature, reliable technology that has been demonstrated in thousands of full-scale facilities worldwide. Catalytic conversion of methane to hydrogen gas is also a well-developed and reliable process (Witt and Schmidt, 1996). Therefore, direct biological generation of hydrogen through AD appears to be restricted to a pre-treatment step in a larger bioenergy or biochemical production concept. Another anticipated disadvantage of large-scale hydrogen generation that needs to be addressed during scale-up is the escape of hydrogen through large plastic enclosures and thin metal sheets that might occur due to the high diffusivity of hydrogen (Kraemer and Bagley, 2005).

Biohydrogen Synthesis Rate

A comparison of H_2 generation rates reported for several biohydrogen systems presented in Table 11.1 reveals the wide range of H_2 synthesis by different biohydrogen systems.

Light-dependent biohydrogen system (direct photolysis, indirect photolysis and photo-fermentation) have rates of H_2 synthesis mostly below 1 $lH_2 l^{-1} h^{-1}$. Dark fermentation systems generate H_2 at rates above 1 $lH_2 l^{-1} h^{-1}$. The rates of H_2 synthesis by an undefined consortium of thermophilic *Clostridium* and by the extreme thermophilic *Caldicellulosiruptor saccharolyticus* (Niel *et al.*, 2002) are very similar (28 and 29 $lH_2 l^{-1} h^{-1}$, respectively).

A pure strain of mesophilic *Clostridium* (Taguchi *et al.*, 1996) demonstrated a higher rate of H_2 synthesis with xylose as a substrate (72 $lH_2 l^{-1} h^{-1}$), and two AD systems that utilized undefined consortia of mesophilic bacteria (Lay *et al.*, 2003) had impressively higher rates of H_2 synthesis (221 and 414 l $H_2 l^{-1} h^{-1}$, respectively).

Table 11.1. Comparison of H_2 synthesis rates by different technologies.

S.No.	Bio H_2 technology	H_2 synthesis rate $lH_2 l^{-1} h^{-1}$	Reference
1.	Direct photolysis	0.24	Francou and Vignasis, 1984
2.	Indirect photolysis	1.21	Taguchi *et al.*, 1996
3.	Photo-fermentation	0.55	Melis, 2002
4.	Water–gas shift reaction	328	Zhu *et al.*, 2002
5.	Darkfermentations		
	(i) Mesophilic, pure strain[a]	72	Ueno *et al.*, 1996
	(ii) Mesophilic, undefined[b]	221	Jouanneau *et al.*, 1984
	(iii) Mesophilic, mixed	414	Morvan *et al.*, 1996
	(iv) Thermophilic, mixed	28	Lindblad *et al.*, 2002
	(v) Extreme thermophilic, pure strain[c]	29	Kondo *et al.*, 2002

[a] *Clostridium* species; [b] A consortium of unknown microorganisms culture from a natural substrate and selected by the bioreactor culture conditions; [c] *Cakdicellulosiruptor saccharolyticus*

Feedstocks for biohydrogen generation through anaerobic digestion

The conversion of carbohydrates to hydrogen and organic acids is preferred because it yields the highest amount of hydrogen per mole of substrate. These carbohydrates can be monosaccharides but may also be polymers such as starch, cellulose or xylan. Besides carbohydrates, formate and peptides have also been studied as substrates for AD. Amino acids can also be oxidized to hydrogen by certain strains (Ravot *et al.*, 1995). With the large number of hydrogen-generating microbial species such as *Bacillus*, *Enterobacter* and *Clostridium* etc., found that carbohydrates are more suitable feedstock for AD hydrogen fermentation. Proteins, peptides and amino acids are probably less suitable for AD, whereas biopolymers like lipids will be unsuited.

The great potential of dark hydrogen fermentation, i.e. the vast range of potential organic substrates, has also been recognized by other workers in the field (Ginkel *et al.*, 2001; Cheng *et al.*, 2006). Noike and Mizuno (2000) and Yu *et al.* (2002) refer to several forms of organic waste streams ranging from solid wastes like rice straw to waste water from a sugar factory and a rice winery, which have been successfully used for AD hydrogen generation. Besides the organic substrates, CO in syngas (a mixture of (mainly) CO and H_2) has been used as feedstock for biological H_2 generation. A wide range of anaerobic microorganisms such as *Enterobacter* and *Clostridium* are capable of CO oxidation with concomitant H_2 generation in a biological variant of the water–gas shift reaction (Cheng *et al.*, 2006). These organisms could serve as a biological alternative for chemical catalysts to remove CO from H_2-rich gases, and generates H_2 linked to CO oxidation. CO uptake has been shown to occur at very high rates. The process could be used for fuel gas conditioning and upgrading, both by CO removal and H_2 generation. A technological challenge is to enhance CO mass transfer, which is the rate-limiting step in the process.

Even though there have been reports on AD hydrogen fermentation using solid organic waste, this phenomenon has not yet been established efficiently. Besides this contrast, which is probably due to different species being involved, there is the even more basic discussion concerning the configuration of the feedstock on the molecular level. Pre-treatment of biomass may be needed from the process technological point of view for improving rheological properties (Cheng *et al.*, 2006). Apart from the more easily degradable feedstocks such as starch and cellulose, the main components of future feedstocks will, most probably, to a large extent be derived from lignocellulosic raw materials (Shin *et al.*, 2004). Lignocellulose is a biopolymer consisting of tightly bound lignin, cellulose and hemicellulose. Whereas cellulose and hemicellulose can be feedstocks for hydrogen fermentation, lignin is not degraded under anaerobic conditions. Moreover, lignin strongly hampers the utilization of cellulose and hemicellulose because: (i) the bonding in lignocellulose resists mobilization; and (ii) chemically degraded lignin is often inhibitory to microbial growth. Major waste materials, which can be used for hydrogen generation, are (Kapdan and Karg, 2006): (i) agricultural or food industry wastes; (ii) carbohydrate-rich industrial wastes; and (iii) waste sludge from wastewater treatment plants.

Species Involved in Hydrogen Generation

Biohydrogen can be generated by strict and facultative anaerobes (clostridia, micrococci, methanobacteria, enterobacteria, etc.) and photosynthetic bacteria (Nandi and Sengupta, 1998) as reported in Table 11.2. The organisms belonging to genus *Clostridium* such as *C. buytricum* (Yokoi *et al.*, 2001), *C. thermolacticum* (Collet *et al.*, 2004), *C. pasteurianum* (Lin and Lay, 2004), *C. paraputrificum* M-21 (Evvyernie *et al.,* 2001) and *C. bifermentants* (Wang *et al.*, 2003) are obligate anaerobes and spore-forming organisms. Clostridia species generate hydrogen during the exponential growth phase. In batch growth of clostridia the metabolism shifts from a hydrogen/acid generation phase to a solvent generation phase, when the population reaches to the stationary growth phase. Investigations on microbial diversity of a mesophilic hydrogen-generating sludge indicated the presence of clostridia species as 64.6% (Fang *et al.*, 2002). The dominant culture of clostridia can be easily obtained by heat treatment of biological sludge. The spores formed at high temperatures can be activated when required environmental conditions are provided for hydrogen generation.

The species of the family Enterobactericeae have the ability to metabolize glucose by mixed acid or 2-3-butanediol fermentation.In both patterns, CO_2 and H_2 are generated from formic acid in addition to ethanol and the 2-3-butanediol. Hydrogen generation capacity of anaerobic facultative bacterial culture *Enterobacter aerogenes* has been widely studied. *Enterobacter cloacae* ITT-BY 08 generated 2.2 $molH_2 mol^{-1}$ glucose. *Aeromonas* spp.,

Pseudomonas spp. and *Vibrio* spp. were also identified for hydrogen generation (Nandi and Sengupta, 1998). Anaerobic cultures such as *Actinomyces* spp., *Porphyromonas* spp. as well as *Clostridium* spp. have been detected in anaerobic granular sludge. The hydrogen yield varied between 1 and 1.2 mmolmol^{-1} glucose when the cultures were cultivated under anaerobic conditions (Oh *et al.*, 2003). Hydrogen generation by *Thermotogales* spp. and *Bacillus* spp. were detected under mesophilic acidogenic conditions (Shin *et al.*, 2004).

Hydrogen generation capacity of some anaerobic thermophilic organisms belonging to the genus *Thermoanaerobacterium* has also been investigated (Ueno *et al.*, 2001; Zhang *et al.*, 2003;Shin *et al.*, 2004). Shin *et al.* (2004) reported *T. hermosaccharolyticum* and *Desulfotomaculumgeothermicum* strains generating hydrogen in thermophilic acidogenic condition. A hyperthermophilic archeon, *Thermococcus kodakaraensis* KOD1 with 85°C optimum growth temperature was isolated from a geothermal spring in Japan and identified as a hydrogen-generating bacteria. *Clostridium thermolacticum* can generate hydrogen from lactose at 58°C. Recently, a hydrogen-generating bacterial strain *Klebsiella oxytoca* HP1was isolated from hot springs with maximal hydrogen generation rate at 35°C.

Methods to enhance biohydrogen generation

There are several methods to improve the yield of H_2 in AD. They are: (i) gas separation; (ii) nitrogen/argon gas sparging; (iii) heat treatment of inoculums; (iv) acid treatment of inoculums; (v) sludge granulation; and (vi) genetic modification of H_2-generating microorganism.

Gas separation

Hydrogen build-up is inhibiting to the process. The effect of increased H_2 partial pressure may limit the use of higher feed strengths. For maximum hydrogen yield H_2 should be removed as it is generated (Hawkes *et al.*, 2002). Vacuum can also be applied to the headspace to reduce the H_2 partial pressure. Increased agitator speed may also lower dissolved H_2 concentration.

A hollow fibre/silicone rubber membrane effectively reduced biogas partial pressure in a dark-fermentation system, resulting in 10% improvement in the rate of H_2 generation and 15% increase in H_2 yield (Liang *et al.*, 2002), and a non-porous, synthetic polyvinyltrimethylsilane (PVTMS) membrane was used for generation of high-purity H_2 from three different H_2-generating bioreactor systems.

Silicone rubber membrane can be used to separate biogas from the liquid medium in the hydrogen fermentation reactor. The hollow fibre membrane made with silicone rubber can facilitate the transfer of biogas in bulk solution and reduce the partial pressure of biogas in the culture liquid, enhancing hydrogen generation based on glucose in a hydrogen fermentation reactor.

The pH is an extremely important factor for continuous H_2 synthesis. As H_2 concentrations increase, H_2 synthesis decreases and the metabolic activity shifts to pathways that synthesize more reduced substrates. The concentration of CO_2 also affects the rate of synthesis and final yield of H_2. Cells synthesize succinate and formate using CO_2, pyruvate, and reduced nicotinamide adenine dinucleotide hydrogen (NADH) via the hexose monophosphate pathway (Das and Veziroglu, 2001). This pathway competes with reactions in which H_2 is synthesized by NADH-dependent hydrogenases. Efficient removal of CO_2

Table 11.2. Hydrogen generation rates and yield coefficients from pure and complex substrates under batch semi-continuous and continuous operation.

Sl No	System	Inoculum	Substrate[a]	Volumetric H_2 generation rate ($|H_2|_{culture}^{-1} h^{-1}$)[f]	H_2 yield	Culture conditions[a] HRT (h), Load, pH, Temperature (°C), H_2 in biogas (%v/v)
1	Batch	Clostridium butyricum CGS5	Sucrose (20 g COD l^{-1})	28	2.78 l H_2 l^{-1} sucrose	–, –, 5.5–6.0[c], 37, 64
2		Clostridium saccharoperbutyl acetonicum ATCC 27021	Crude cheese whey (ca.41.4 g lactose l^{-1})	32	2.7 l H_2 l^{-1} lactose	–, –, 6.0[d], 30, NR[b]
3		Escherichia coli strains	Glucose (4 g l^{-1})	NR[b]	~2 l H_2 l^{-1} glucose	–, –, 7.0, 37, NR
4		Escherichia coli strains	Formic acid (25 mM)	40,339	1 l H_2 l^{-1} formate	–, –, 6.5[d], 37, NR
5		Defined consortium (1:1:1, and separately tested): Enterobacter cloacae IIT-BT 08, Citrobacterfreundii IIT-BT L139, Bacilluscoagulans IIT-BT S1	Glucose (10 g l^{-1})	NR	41.23 ml H_2 g^{-1} COD$_{removed}$	–, –, 6.0[d], 37, NR
6	Batch	Mesophilic bacterium HN001	Starch (20 g l^{-1})	202	2 l H_2 l^{-1} glucose	–, –, 6.0[c], 37, NR
7		Aerobic and anaerobic sludges, soil and lake sediment (acid and heat conditioned)	Glucose (20 g l^{-1})	NR	1.4 l H_2 l^{-1} glucose	–, –, 6.0[c], 35, NR
8		Aerobic sludge (heat conditioned)	Glucose (2 g l^{-1})	NR	2.0 l H_2 l^{-1} glucose	–, –, 6.2[d], 30, 87.4
9		Soil (heat conditioned)	Organic matter present in four carbohydrate-rich wastewaters	621	100 ml H_2 g^{-1} COD$_{removed}$	–, –, 6.1[d], 23, 60
10		Anaerobic sludge (acid treatment and acclimated in a CSTR)	Sucrose (20 g COD l^{-1})	328	1.74 l H_2 l^{-1} sucrose	–, –, 6.1[d], 40, 45
11		Anaerobic sludge (heat conditioned)	Glucose (10 g l^{-1})	93 l H_2 g$_{vss}^{-1}$ l$_{culture}^{-1}$ h^{-1}	1.75 l H_2 l^{-1} glucose	–, –, 6.0[d], 37, 40
12	Batch	Anaerobic sludge (acid treatment)	Glucose (~21.3 g l^{-1})	17–30	0.8–1.0 l H_2 l^{-1} hexose	–, –, 5.7[c], 34.5, 59
13		Microflora from a	Wheat straw	9.2 l	9 l H_2 g$_{TVS}^{-1}$	–, –, 7.0, 36, 52

No.	Reactor	Inoculum	Substrate	H₂ rate	H₂ yield	Other parameters
		cowdung^d compost (heat treatment)	wastes (25 gl^{-1})	$H_2 g^{-1} TVS|_{culture} h^{-1}$		
14		Anaerobic sludge (heat treated)	Sucrose (10 gl^{-1})	27.4	1.9 lH$_2$l^{-1} sucrose	–, –, 5.5^c, 35, NR
15		Anaerobic sludge (heat treated)	Sucrose (24.8 gl^{-1})	68.4	3.4 lH$_2$l^{-1} sucrose	–, –, 5.5^c, 34.8, 64
16		Anaerobic sludge (heat treated)	Glucose (3.76 gl^{-1})	30.8	1.0 lH$_2$l^{-1} glucose	–, –, 6.2^d, 30, 66
17		Anaerobic sludge (heat treated)	Glucose (2.82 gl^{-1})	NR	0.968 lH$_2$l^{-1} glucose	–, –, 6.2^d, 25, 57–72
18	Batch	Microflora from soil (heat shocked)	Glucose, sucrose, molasses, lactate, potato starch, cellulose (each: 4 g CODl^{-1})	NR	0.92 lH$_2$l^{-1} glucose, 1.8 lH$_2$l^{-1} sucrose, 0.59 lH$_2$l^{-1} potato starch^e 0.01 l H$_2$l^{-1} lactate, 0.003 l H$_2$l^{-1} cellulose^e	–, –, 6.0^d, 26, 62 H$_2$ mol^{-1} sucrose,
19	Fed batch	Mixed culture	OFMSW-Semisolid substrate	0.53 lH$_2$g$_{VSdestroyed}^{-1}$	NR^b	504, 11 g$_{VS}$ kg$_{wmr}^{-1}$day^{-1}, 6.4, 55, 58
20		Anaerobic POME sludge Windrow yard waste compost	POME (2.5% w/v) Glucose (2 gl^{-1})	60.9 25.5	NR 1.75 l H$_2$l^{-1}glucose	24, NR, 5.5, 60, 66 76, NR, 5.4, 55, NR
21	CSTR	Mixed culture	Sucrose (20 g CODl^{-1})	58	3.5 lH$_2$l^{-1}sucrose	12, NR, 6.8, 35, 45.9
22	CSTR	Mixed culture	Sucrose (40 gl^{-1})	68.4	1.15 lH$_2$l^{-1} hexose	12, 80 gl^{-1}day^{-1}, 5.2, 35, 60
23		Mixed culture immobilized in silicone gel	Sucrose (30 g CODl^{-1})	2095	3.86 l H$_2$l^{-1} sucrose	0.5, NR, 6.5, 40, 44
24		Mixed culture	Xylose (20 g CODl^{-1})	17.1	1.1 lH$_2$l^{-1} xylose	12, NR, 7.1, 35, 32
25		Mixed culture	Broken kitchen wastes (10 kg CODm^{-3}day^{-1}) and corn starch (10 kg CODm^{-3}day^{-1})	5.8	NR	96, NR, 5.3–5.6, 35, NR
26		Mixed culture	Glucose (15 g CODl^{-1})	45.25	1.93 lH$_2$l^{-1} glucose	4.5, 80 g CODl^{-1}day^{-1}, 5.5, 37, 67
27		Dewatered and thickened sludge	Glucose (4 g CODl^{-1})	11.87	1.9 lH$_2$l^{-1} glucose	10, NR, 5.5, 35,67
28	CSTR	Mixed culture	Sucrose (20 g CODl^{-1})	53.35	3.6 l H$_2$l^{-1} sucrose	12, NR, 5.5, 35, 50
29		Mixed culture	Organic wastewater (4000 mg CODl^{-1})	16.96	NR	12, NR, 4.4, 8 kgCODm^{-3}day^{-1}, 30
30		Sewage sludge	Sucrose (20 g CODl^{-1})	179.89	3.43 lH$_2$l^{-1} sucrose	12, NR, 6.8, 35, 50.9

#	Reactor	Culture	Substrate	Yield	H$_2$ production rate	Other params
31		Mixed culture	Sucrose and sugarbeet	17.61	1.9 H$_2$l^{-1} hexose	15, 16 kg sugarm^{-3}day^{-1}, 5.2, 32, NR
32		Mixed culture	Glucose (15 gl^{-1})	0.115 g H$_2$COD/COD$_{Feed}$	1.38 lH$_2$l^{-1} hexose	10, NR, 5.5, 35,45
33		C. thermolacticum (DSM 2910)	Lactose (10 gl^{-1})	8.82	2.1–3 l H$_2$l^{-1} lactose	17.2, NR, 7.0, 58, 55
34		Seed sludge	Molasses (3000 mg CODl^{-1})	89.37 lH$_2$kg^{-1} COD removed	NR	11.4, 27.98 kg CODm^{-3} reactorday^{-1}, 4.5, 35, 45
35		Mixed culture	Glucose (10 gl^{-1})	7.55	2.47 l H$_2$l^{-1} glucose	26.7, NR, 4.8–5.5, 70, NR
36		Mixed culture	Sucrose rich wastewater	20.28	1.61 lH$_2$l^{-1} glucose	12, NR, 7, 39, NR
37	UASB	Mixed culture	Citric acid waste water (18 kg CODl^{-1})	4.21	0.84 lH$_2$l^{-1} hexose	12, 38.4 kg COD m^{-3}day^{-1}, 7, 35, NR
39		Mixed culture	Sucrose (20 g CODl^{-1})	38.65	1.5 l H$_2$l^{-1} sucrose	8, 175 mmol sucrosel^{-1}day^{-1}, 6.7, 35, 42.4
40		Mixed culture	Glucose (7.7 gl^{-1})	62.93	1.7 lH$_2$l^{-1} glucose	2, NR, 6.4, 55, 36.8
			Glucose (1.3 gl^{-1})	64.98	0.7 lH$_2$l^{-1} glucose	2, NR, 4.4, 35,29.4
41	CSTR and UASB	Mixed culture	Starch (10 gl^{-1}) and xylose (1:1 w/w)	15.39 / 16.28 / 8.69	NR	32.9, NR, 7, 35, 68 / 6.7, NR, 7, 35, 68 / 20.5, NR, 7, 35, 68
42	CSTR, UASB	Mixed culture	Glucose (6.86 gl^{-1})	128.25	1.6 lH$_2$l^{-1} glucose	12, NR, 5.5, 60, 48
43	UFBR	Clostridium acetobutylicum (ATCC 824)	Glucose (10.5 gl^{-1})	30.44	0.9 lH$_2$l^{-1} glucose	0.035, 8.3 gl^{-1}h^{-1}, 4.9, 30, 74
44	TBR	Mixed culture	Glucose (2 gl^{-1})	NR	2.48 lH$_2$l^{-1} glucose	0.5, 96 kgm^{-3}day^{-1}, 7.7, 30, NR
45	CIGSB	Mixed culture	Sucrose (17.8 gl^{-1})	1.02	3.88 lH$_2$l^{-1} sucrose	0.5, NR, 6.7, 40, 42
46	PBR	Cow dung	POME (5–60 g CODl^{-1})	0.42 l biogasg^{-1}COD destroyed	NR	3–7, NR, 5, NR, 53–56
47	UACF	Cow dung	Jackfruit peel (22.5 g$_{VS}$l^{-1})	0.72 l biogasg^{-1}Versusd estroyed	NR	288, NR, 5, NR, 56
48	MBR	Mixed culture	Glucose (10 gl^{-1})	244.2	1.1 lH$_2$l^{-1} glucose	0.79, NR, 5.5, 37, 70
49	FBR DTFBR	Mixed culture	Sucrose (20 g CODl^{-1})	171.9 / 325.69	2.10 l H$_2$l^{-1} sucrose / 1.22 l H$_2$l^{-1} sucrose	2, NR, 6.9, 40, 40 / 0.5, NR, 7, 40,35

[a] When optimization trials were carried out, optimum values are reported
[b] NR: not reported
[c] Controlled value
[d] Initial, not controlled
[e] Starch, cellulose: $((C_6H_{10}O_5)n)$
[f] In some cases unit conversions were made according to the conditions reported by the authors.

CIGSB, carrier induced granular sludge bed; COD, chemical oxygen demand; CSTR, continuous stirred tank reactor; DTFBR, draft tube fluidized bed reactor; FBR, fluidized bed bioreactor; kg$_{wmr}$, kilograms of wet mass in the reactor; MBR, membrane bioreactor; OFMSW, organic fraction of municipal solid wastes; PBR, packed bed reactor; POME, palm oil mill effluent; TBR, trickling biofilter; TVS, total volatile solids; UACF, up-flow anaerobic contact filter; UASB, up-flow anaerobic sludge blanket; UFBR, up-flow fixed bed reactor; VS, volatile solids; VSS, volatile suspended solids. Source: Vazquez *et al.* (2008)

from the fermentation system would reduce competition for NADH and thus result in increased H_2 synthesis.

Nitrogen/argon gas sparging

Sparging with N_2 or argon (Ar) gas in the headspace of anaerobic digesters has maintained strict anaerobic condition in the reactor, which increased the hydrogen generation. Too much sparging, however, dilutes the H_2 and creates a serious problem with respect to separation of the H_2 from the sparging gas.

Heat treatment of inoculum

One of the difficulties associated with hydrogen generation using mixed communities in continuous flow systems is the coexistence of hydrogen-consuming microorganisms, such as methanogens. Several studies have adopted heat treatment for the inoculum used as seed in the reactors as a method to inactivate or eliminate these microorganisms (Ginkel *et al.*, 2001).

Acid treatment of inoculum

Acid treatment may be used to control methanogens present in the inoculum for controlling the methanogenic activity. The acidic pre-treatment involved decreasing the pH of the sludge or granule solution to 3.0 using 0.1N HCl solutions for 24 h and a readjustment of pH back to the operating pH by 0.1N NaOH solution (Khanal *et al.*, 2004).

Sludge granulation

The granular sludge formation played a pivotal role in the dramatic increase of hydrogen generation rate. Sludge granulation enabled a high-cell density fermentation system that attained a high overall hydrogen generation; it would also be interesting to know whether the specific activity can be maintained, as mass transfer in the culture become less efficient when cell concentration gets too high.

Genetic modification of hydrogen-generating microorganisms

H_2 generation may be enhanced through genetic modification of H_2-generating bacteria. Hydrogen-generating strains of bacteria can be genetically modified in several ways to increase H_2 synthesis, including over-expression of cellulases, hemicellulases and lignases that can maximize substrate (glucose) availability; elimination of uptake hydrogenases; over-expression of H_2 evolving hydrogenases that have themselves been modified to be oxygen tolerant; and elimination of metabolic pathways that compete for reducing equivalents required for H_2 synthesis.

Conclusion

Both light-dependent (direct photolysis, indirect photolysis and photo-fermentation) and AD biohydrogen systems are under intense investigation to find ways to improve both the rates of H_2 generation and the ultimate yield of H_2. Hydrogen generation by direct photolysis using green algae is currently limited by: (i) solar conversion efficiency of the photosynthetic apparatus; (ii) H_2 synthesis process (i.e. the need to separate the processes of H_2O oxidation from H_2 synthesis); and (iii) bioreactor design and cost. A number of approaches to improve H_2 generation by green algae are currently under investigation. These include genetic engineering of light-gathering antennae, optimization of light input into photo-bioreactors, and improvements to the two-phase H_2 generation systems used with green algae.

Hydrogen generation via AD can be improved by screening for pure strains possessing highly active hydrogen-evolving enzymes (hydrogenases), in combination with high heterocyst formation. Genetic modification of strains to eliminate uptake hydrogen and increase levels of bidirectional hydrogenase activity may yield significant increases in H_2 generation. Optimization of cultivation conditions such as pH, temperature and nutrient content, as well as maintaining low partial pressures of H_2 (10–6 bars) and CO_2 will contribute to increased H_2 generation.

AD systems also appear to have the great potential to be developed as practical biohydrogen systems. For biological hydrogen generation, a broad spectrum of substrates can be used. However, very few studies are undertaken using the waste materials such as food industry waste, agricultural waste and wastewater treatment-plant sludge. Kitchen waste, whose disposal is a major environmental problem, has potential to be used for hydrogen generation through AD.

Fermentative H_2 generation is considered to be a complex process and needs optimization with respect to the type of inoculum and pre-treatment, substrate nature and composition, co-substrate addition, fermentation pH, fermentation period, etc. prior to up-scaling. Inoculum selection and its pre-treatment is one of the important aspects which have a vital role in selecting the requisite bacteria for efficient H_2 generation. Several types of pre-treatment procedures (heat treatment, chemical treatment, pH treatment, etc.) were reported in the literature for a variety of inocula. Additionally, system operating conditions (operating pH, short fermentation period and sludge retention time) will also have significant effect on H_2 evolution.

References

Ahring, B.K. and Ibrahim, A.A.Z. (2001) Effect of temperature increase from 55 to 65ºC on performance and microbial population dynamics of an anaerobic reactor treating cattle manure. *Water Research*35(10), 2446–2452.

Alvarez, M.J., Mace, S. and Llabres, P. (2000) Anaerobic digestion of organic wastes: an overview of research achievements and perspectives. *Bioresource Technology* 74, 3–16.

Bolton, J.R. (1996) Solar photo production of hydrogen.*Solar Energy*57, 37–50.

Brummeler,E.T., Aarnink, M.M.J. and Koster, I.W. (1992) Dry anaerobic digestion of solid organic waste in a biocel reactor at pilot-scale plants. *Water Science and Technology* 25(7), 301–310.

Cecchi, F., Marcomini, A., Pavan, P., Fazzini, G. and Mata Alvarez, J. (1989) Mesophilic digestion of the refuse organic fraction sorted by plant performance and kinetic. *Waste Management and Research* 22, 33–41.

Cheng, S.S., Li, S.L., Kuo, S.C., Lin, J.S., Lee, Z.K. and Wang, Y.H. (2006) A feasibility study of biohydrogenation from kitchen waste fermentation. In: *Proceedings of the 16th World Hydrogen Energy Conference*, Lyon, France, pp. 56–60.

Cohen, A., Van Gemert, J.M., Zoetemeyer, R.J. and Breure, A.M. (1984) Main characteristics and stoichiometric aspects of acidogenesis of soluble carbohydrate containing wastewater. *Process Biochemistry* 19, 228–237.

Collet, C., Adler, N., Schwitzguebel, J.P. and Peringer, P. (2004) Hydrogen production by *Clostridium thermolacticum* during continuous fermentation of lactose. *International Journal of Hydrogen Energy* 29, 1479–1485.

Das, D. and Veziroglu, N.T. (2001) Hydrogen production by biological processes: a survey of literature. *International Journal of Hydrogen Energy* 26, 13–28.

Dickinson, R.E. and Cicerone, R.J. (1986) Future global warming from atmospheric trace gases. *Nature* 319, 109–135.

Evvyernie, D., Morimoto, K., Karita, S., Kimura, T., Sakka, K. and Ohmiya, K. (2001) Conversion of waste to hydrogen gas by *Clostridium paraputrificum* M-21. *Journal of Bioscience and Bioengineering* 91, 339–343.

Fang, H.H.P. and Liu, H. (2002) Effect of pH on hydrogen production from glucose by a mixed culture. *Bioresource Technology* 31, 958–968.

Fang, H.H.P., Zhang, T. and Liu, H. (2002) Microbial diversity of mesophilic hydrogen producing sludge. *Applied Microbiology and Biotechnology* 58, 112–118.

Fedorov, A.S., Tsygankov, A.A., Rao, K.K. and Hall, D.O. (1998) Hydrogen photo production by *Rhodobacter sphaeroides* immobilised on polyurethane foam. *Biotechnology Letters* 20, 1007–1009.

Francou, N. and Vignasis, P.M. (1984) Hydrogen production by *Rhodopseudomonas capsulata* cells entrapped in carrageenan beads. *Biotechnology Letters* 6, 639–644.

Gaudernack, B. (1998) Photo production of hydrogen. In: Elam, C.C. (ed.) *IEA Agreement on the Production and Utilization of Hydrogen Annual Report*. National Renewable Energy Laboratory, Golden, Colorado, pp. 57–63.

Ginkel, S.V., Sung, S. and Lay, J.J. (2001) Biohydrogen production as a function of pH and substrate concentration. *Environmental Science and Technology* 35, 4726–4730.

Gustavo, D.V., Arriaga, S., Alatriste-Mondragon, F., de Leon-Rodriguez, A., Rosales-Colunga, L.M. and Razo-Flores, E. (2008) Fermentative biohydrogen production: trends and perspectives. *Reviews in Environmental Science and Biotechnology* 7, 27–45.

Hallenbeck, P.C. (2005) Fundamentals of the fermentative production of hydrogen. *Water Science and Technology* 52(1–2), 21–29.

Hallenbeck, P.C. and Benemann, J.R. (2002) Biological hydrogen production: fundamentals and limiting processes. *International Journal of Hydrogen Energy* 27, 1185–1194.

Hansel, A. and Lindblad, P. (1998) Mini-review: toward optimization of cyanobacteria as biotechnologically relevant producers of molecular hydrogen – a clean energy source. *Applied and Environmental Microbiology* 50, 153–160.

Hawkes, F.R., Dinsdale, R., Hawkes, D.L. and Hussy, I. (2002) Sustainable fermentative biohydrogen: challenges for process optimization. *International Journal of Hydrogen Energy* 27, 1339–1347.

Hoffman, M. (2000) Dry fermentation: development status and prospects. *Landtechnik* 56, 410–413.

Hussy, I., Hawkes, F.R., Dinsdale, R. and Hawkes, D.L. (2005) Continuous fermentative hydrogen production from sucrose and sugar beet. *International Journal of Hydrogen Energy* 30(5), 471–483.

Ike, A., Toda, N., Tsuji, N., Hirata, K. and Miyamoto, K. (1997) Hydrogen photo production from CO_2 fixing micro algal biomass: application of halotolerant photosynthetic bacteria. *Journal of Fermentation and Bioengineering* 84, 606–669.

Jayalakshmi, S.K.J. and Sukumaran, V. (2009) Bio hydrogen generation from kitchen waste in an inclined plug flow reactor. *International Journal of Hydrogen Energy* 34, 8854–8858.

Jouanneau, Y., Lebecque, S. and Vignais, P.M. (1984) Ammonia and light effect on nitrogenase activity in nitrogen-limited continuous cultures of *Rhodopseudomonas capsilata*: role of glutamate synthetase. *Archives of Microbiology* 119, 326–331.

Joyner, A.E. and Winter, W.T. (1997) Studies on some characteristics of hydrogen production by cell-free extracts of rumen anaerobic bacteria. *Canadian Journal of Microbiology* 23, 346–563.

Kapdan, I.K. and Fikret Karg (2006) Bio-hydrogen production from waste materials. *Enzyme and Microbial Technology* 38, 569–582.

Kerby, R.L., Ludden, P.W. and Robert, G.P. (1995) Carbon monoxide dependent growth of *Rhodospirillum rubrum*. *Journal of Bacteriology* 177, 2241–2244.

Khanal, S.K., Chen, W.H., Li, L. and Sung, S. (2004) Biological hydrogen production: Effects of pH and intermediate products. *International Journal of Hydrogen Energy* 29, 1120–1123.

Klass, D.L. (1998) *Biomass for Renewable Energy, Fuels and Chemicals*. Academic Press, USA, pp. 28–37.

Kondo, T., Arawaka, M., Wakayama, T. and Miyake, J. (2002) Hydrogen production by combining two types of photosynthetic bacteria with different characteristics. *International Journal of Hydrogen Energy* 27, 1303–1308.

Kotsopoulos, T.A., Zeng, R.J. and Angelidaki, I. (2006) Biohydrogen production in granular up-flow anaerobic sludge blanket (UASB) reactors with mixed cultures under hyper-thermophilic temperature (78°C). *Biotechnology and Bioengineering* 94 (2), 296–302.

Kraemer, J.T. and Bagley, D.M. (2005) Continuous fermentative hydrogen production using a two-phase reactor system with recycle. *Environmental Science and Technology* 39(10), 3819–3825.

Larminie, J. and Dicks, A. (2000) *Fuel Cell Systems Explained*. Wiley Publishers, New York, pp. 1–5.

Lay, J.J., Lee, Y.J. and Noike, T. (1999) Feasibility of biological hydrogen production from organic fraction of municipal solid waste. *Water Research* 33, 2579–2586.

Lay, J.J., Fan, K.S., Chang, J. and Ku, C.H. (2003) Influence of chemical nature of organic wastes and their conversion to hydrogen by heat-shock digested sludge. *International Journal of Hydrogen Energy* 28, 1361–1367.

Lee, C.M., Chen, P.C., Wang, C.C. and Tung, Y.C. (2002) Photohydrogen production using purple non-sulfur bacteria with hydrogen fermentation reactor effluent. *International Journal of Hydrogen Energy* 27, 1308–1314.

Levin, D.B., Pitt, L. and Love, M. (2004) Biohydrogen production: prospects and limitations to practical application. *International Journal of Hydrogen Energy* 29, 173–185.

Liang, T.M., Cheng, S.S. and Wu, K.L. (2002) Behavioural study on hydrogen fermentation reactor installed with silicone rubber memberane. *International Journal of Hydrogen Energy* 27, 1157–1165.

Lin, C.Y. and Chang, R.C. (1999) Hydrogen production during the anaerobic acidogenic conversion of glucose. *Journal of Chemical Technology and Biotechnology* 74, 498–500.

Lin, C.Y. and Lay, C.H. (2004) Carbon/nitrogen ratio effect on fermentative hydrogen production by mixed microflora. *International Journal of Hydrogen Energy* 29, 41–45.

Lindblad, P., Christensson, K., Lindberg, P., Federov, A., Pinto, F. and Tsygankov, A. (2002) Photoproduction of H_2 by wild type *Aabaena* PCC 1720 and a hydrogen uptake deficient mutant: from laboratory to outdoor culture.*International Journal of Hydrogen Energy* 27, 1271–1281.

Logan, B.E., Oh, S.E., Kim, I.N.S. and Van Ginkel, S. (2002) Biological hydrogen production measured in batch anaerobic respirometers. *Environmental Science and Technology* 2530–2535.

Masukawa, H., Mochimaru, M. and Sakurai, H. (2002) Hydrogenases and photobiological hydrogen production utilizing nitrogenase system in cyanobacteria. *International Journal of Hydrogen Energy* 27(11–12), 1471–1474.

Melis, T. (2002) Green algae production: Process, challenges, and prospects. *International Journal of Hydrogen Energy* 27, 1217–1228.

Melis, A. and Happe, T. (2001) Hydrogen production green algae as a source of energy. *Plant Physiology* 127, 740–748.

Morvan, B., Bonnemoy, F., Fonty, G. and Gouet, P. (1996) Quantitative determination of H_2-utilizing acetogenic and sulfate-reducing bacteria and methanogenic archaea from digestive of different mammals. *Current Microbiology* 32, 129–133.

Nandi, R. and Sengupta, S. (1998) Microbial production of hydrogen and overview. *Critical Reviews in Microbiology* 24, 61–84.

Niel, V.E.W.J., Claassen, P.A.M. and Stams, A.J.M. (2002)Substrate and product inhibition of hydrogen production by the extreme thermophile *Caldicellulosiruptor saccharolyticus*. *Biotechnology and Bioengineering* 81, 255–262.

Noike, T. and Mizuno, O. (2000) Hydrogen fermentation of organic municipal waste. *Water Science and Technology* 42, 155–162.

Oh, S.E., Ginkel, S.V. and Logan, B.E. (2003) The relative effectiveness of pH control and heat treatment for enhancing biohydrogen gas production. *Environmental Science and Technology* 37, 5186–5190.

Okamoto, M., Miyahara, O., Mizuno, O. and Noike, T. (2000) Biological hydrogen potential of material characteristic of the organic fraction of municipal solid waste. *Water Science and Technology* 41, 25–30.

Ravot, G., Magot, M., Fardeau, M.L., Patel, B.K.C., Prensier, G., Egan, A., Garcia, J.L. and Ollivier, B. (1995) *Thermotoga elfii* sp. a novel thermophilic bacterium from an african oil-producing well. *International Journal of Systematic Bacteriology* 45, 308–314.

Ren, N., Wang, B. and Huang, J. (1997) Ethanol type fermentation of carbohydrate wastewater in a high rate acidogenic reactor. *Biotechnology and Bioengineering* 54, 428–433.

Rglesia, J., Castrillon, L., Maranon, E. and Sastre, H. (1998) Solid-state anaerobic digestion of unsorted municipal solid waste in a pilot-plant scale digester. *Bioresource Technology* 63, 29–35.

Shin, H.S., Youn, J.H. and Kim, S.H. (2004) Hydrogen production from food waste in anaerobic mesophilic and thermophilic acidogenesis. *International Journal of Hydrogen Energy* 29, 1355–1363.

Sparling, R., Risbey, D. and Poggi-Vvaraldo, H.M. (1997) Hydrogen production from inhibited anaerobic compost.*International Journal of Hydrogen Energy* 22, 563–566.

Taguchi, F., Yamada, K., Hasegawa, K., Saito-Taki, T. and Hara, K. (1996) Continuous hydrogen production by *Clostridium* sp. No. 2 from cellulose hydrolysis in an aqueous two-phase system.*Journal of Fermentation and Bioengineering* 82, 80–83.

Techobanoglous, G., Theisen, H. and Vigi, S. (1993) *Integrated Solid Waste Management*. McGraw-Hill, Singapore.

Ueno, Y., Otauka, S. and Morimoto, M. (1996) Hydrogen production from industrial wastewater by anaerobic microflora in chemostat culture. *Journal of Fermentation and Bioengineering* 82, 194–197.

Ueno, Y., Haruta, S., Ishii, M. and Igarashi, Y. (2001) Characterization of a microorganism isolated from the effluent of hydrogen fermentation by microflora. *Journal of Bioscience and Bioengineering* 92, 397–400.

Varaldo, H.M., Rodriguez-Vazquez, R., Fernandez-Villagomez, G. and Esporza, F. (1997) Inhibition of mesophilic solid substrate anaerobic digestion by ammonia nitrogen. *Applied Microbiology and Biotechnology* 47, 284–291.

Varaldo, H.M., Gomez-Cisneros, E., Rodriguez-Vazquez, R., Trejo-Espino, J. and Rinderknecht-Seijas, N. (2002) Unsuitability of anaerobic compost from solid substrate anaerobic digestion as soil amendment. In: *Microbiology of Composting*. Springer, Heidelberg, Germany, pp. 287–297.

Vazquez, G.D., Arriaga, S., Alatriste-Mondragon, F., de Leon-Rodriguez, A., Rosales-Colunga, L.M. and Razo-Flores, E. (2008) Fermentative biohydrogen production: trends and perspectives. *Reviews in Environmental Science and Biotechnology* 7, 27–45.

Veziroglu, T.N. and Barbir, F. (1995) Transportation fuel-hydrogen. *Energy Technology and Environment* 4, 2712–2730.

Wang, C.C., Chang, C.W., Chu, C.P., Lee, D.J., Chang, B.V. and Liao, C.S. (2003) Producing hydrogen from wastewater sludge by *Clostridum bifermentans*. *Journal of Biotechnology* 102, 83–92.

Wejma, J., Gubbels, F., Hulshoff Pol, L.W., Stams, A.J.M., Lens, P. and Lettinga, G. (2002) Competition for H_2 between sulfate reducers methanogens and homo-acetogens in a gas-lift reactor. *Water Science and Technology* 45, 75–80.

Witt, P.M. and Schmidt, L.D. (1996) Effect of flow rate on the partial oxidation of methane and ethane. *Journal of Catalysis* 163, 465–475.

Yokoi, H., Saitsu, A.S., Uchida, H., Hirose, J., Hayashi, S. and Takasaki, Y. (2001) Microbial hydrogen production from sweet potato starch residue. *Journal of Bioscience and Bioengineering* 91, 58–63.

Yu, H., Zhu, Z., Hu, W. and Zhang, H. (2002) Hydrogen production from rice winery wastewater in an upflow anaerobic reactor by using mixed anaerobic cultures. *International Journal of Hydrogen Energy* 27, 1359–1365.

Zhang, T., Liu, H. and Fang, H.H.P. (2003) Biohydrogen production from starch in wastewater under thermophilic conditions. *Journal of Environmental Management* 69, 149–156.

Zhu, H., Ueda, S., Asada, Y. and Miyake, J. (2002) Hydrogen production as a novel process of wastewater treatment-studies on tofu wastewater with entrapped *R. sphaetoides* and mutagenes. *International Journal of Hydrogen Energy* 27, 1349–1358.

Chapter 12

Algae: A Novel Biomass Feedstock for Biofuels

Senthil Chinnasamy, Polur Hanumantha Rao, Sailendra Bhaskar, Ramasamy Rengasamy and Manjinder Singh

Introduction

Once considered a taxonomist's delight, algae have now begun to fascinate technologists, due to their usein various applications, which include food, feed and nutraceuticals, recently more so in areas such as bioenergy and the environment. They represent a large group of genetically diverse, heterogeneous photosynthetic organisms belonging to different phylogenetic groups and evolutionary lineages, with approximately 30,000 known species. Algae are defined as primitive plants (thallophytes) and lack well-defined structures such as roots, shoots, leaves, seeds and fruits (Lee, 1980). They can be microscopic or macroscopic, prokaryotic or eukaryotic, unicellular (coccoid, palmelloid, colonial and filamentous) or multicellular, motile or non-motile, attached or free-living, terrestrial or aquatic (marine or freshwater) and aerial or sub-aerial. Ranging from microscopic picoplanktons to the complex, giant kelps, these living fossils form the bulk of the primary producers of the aquatic ecosystem and they are responsible for nearly one-third of the Earth's total carbon fixation. They also generate nearly 70% of our planet's total atmospheric oxygen. Algae are widespread in water and soil habitats, at different geographical latitudes and altitudes, and occur in waters with different degrees of salinity, pH and temperature. A majority of these are microalgae that survive in a range of hostile environments including permafrost zones, hot springs, saltine and soda lakes, aphotic areas, desiccating conditions, hyperbaric environments, anaerobic niches and they can be even endolithic. They exhibit a wide range of reproductive strategies including simple asexual cell division, vegetative reproduction and complex forms of sexual reproduction. Macroalgae, most of which grow in shallow marine environments, are commonly referred to as seaweeds. Algae find numerous applications in food, feed, agriculture, aquaculture, cosmetic, pharmaceutical and nutraceutical industries; they offer us an array of value-added chemicals, such as pigments (e.g. phycobilins, carotenoids, astaxanthin), agar, alginates, carrageenan, antioxidants and polyunsaturated fatty acids (PUFAs). Considering the above, mass cultivation of both microalgae and macroalgae becomes essential, in which temperature, light, nutrients, dissolved oxygen and carbon dioxide (CO_2), and type of the reactor play major roles in improving the biomass production.

Algae for Alternative Fuels and Bioenergy

©CAB International 2012. *Microbial Biotechnology: Energy and Environment*
(ed R. Arora)

In the present-day scenario, rising prices of fossil fuels, concern for energy security and climate change are the major drivers that have led to growing worldwide interests in alternative fuels and bioenergy. The problem of reducing CO_2 and greenhouse gas (GHG) emissions is compounded by the prediction that the global population will increase from 6.6 billion in 2008 to 9.2 billion by 2050 and the resultant increase in fuel use will be further exacerbated by the increasing energy demands of the rapidly expanding economies of China and India. Of the clean energy technologies being developed, almost all target the electricity market (e.g. photovoltaic, solar thermal, geothermal, wind and wave power), which currently accounts for only 33% of global energy consumption. However, to secure future fuel supplies (66% of global energy), biofuels represent the only viable option (Stephens *et al.*, 2010). Biofuels are renewable fuels derived from biological feedstocks, and include both liquid forms such as bioethanol (gasoline-equivalent) or biodiesel (diesel-equivalent), and gaseous forms such as biogas, syngas and biohydrogen (Koh and Ghazoul, 2008). Recently, global biofuel production has tripled from 18 billion l in 2000 to about 60.5 billion l in 2007 with the USA and Brazil contributing to 75% of total production (Jegannathan *et al.*, 2009). First-generation biofuels such as vegetable oils are those obtained from oilseed crops such as soybean, groundnut and sunflower. Second-generation biofuels comprise bio-oil and lignocellulosic ethanol from cheap and abundantnon-food plant waste biomass, and third-generation biofuels such as bioethanol and biodiesel are from microbial sources such as algae and genetically modified organisms (GMOs). First-generation biofuels are not viable because of limited feedstock availability and food versus fuel issues. Although the second-generation biofuels are environment-friendly and do not compete with food, their production will continue to face major constraints in commercial deployment. Third-generation biofuels are gaining prominence because of the use of microbes. In the process of diversifying the second- and third-generation biofuels feedstock resource base, algae seem to lead the race, for various reasons. Of late, algae are definitely gaining momentum because algal biofuels are an appealing choice due to the rapid growth rate of these organisms, high lipid content, comparatively low land usage and high CO_2 absorption and uptake rate (Schenk *et al.*, 2008). Extensive research has been conducted to investigate the utilization of microalgae as an energy feedstock, with applications being developed for the production of biodiesel, bioethanol, biomethane and biohydrogen (Huntley and Redalje, 2007).

Algae Versus Terrestrial Crops

Algae can create clean renewable fuels, remediate wastewater and still produce high-value biochemicals, and thus they have a clear advantage over terrestrial plants for use in biofuel applications. The renaissance of algae in biofuel applications after a gap of nearly two decades apparently shows the possibility of overshadowing the use of higher plants for biofuel production.Higher plants require large areas of land, huge amounts of freshwater, will not thrive in harsh conditions and pose an insidious threat to food crops thus resulting in food versus fuel competition. Moreover, the seeds from which oil is usually extracted are seasonal, thus making year-round production of fuel difficult. In contrast, unlike other biofuel feedstocks (i.e. maize and soybean), algae do not compete with existing food crops, can grow on marginal lands not suitable for conventional agriculture, can be produced in extremely high volumes and can grow in wastewater and seawater (saline and hypersaline environments). Unlike plants, they do not require herbicides or pesticides (Rodolfi *et al.*,

2009). Indeed, seaweeds can be cultivated in shallow sea-waters and microalgae can be cultivated in raceway ponds or photobioreactors, thus making the upstream processing of algae-based biofuels more competitive than that of traditional plant-based biofuels. Proponents of algal bioenergy production systems argue that from an efficiency perspective, algae can meet global demand more easily than field crop-based bioenergy production systems (Gerber *et al.*, 2009). A widely stated claim is that microalgae are capable of producing 30 times more oil per unit area of land than terrestrial oilseed crops (Sheehan *et al.*, 1998). A comparison of annual productivity of oil from microalgae with that of oilseed crops is given in Table 12.1.

According to energy yield calculations by Chisti (2008a), the total energy yield from the microalgal culture, obtained from a raceway system, with a biomass productivity of 0.025 kg m^{-2} d^{-1} is 1444 GJ ha^{-1} year^{-1}, whereas the total energy output of sugarcane with a cane productivity of approximately 75 t ha^{-1} is 163.9 GJ ha^{-1} year^{-1}, or merely 11% of the total energy output from algae. This 6- to 12-fold energy yield advantage of microalgae over terrestrial plants can be achieved because they are efficient solar energy converters, thrive across a greater range of light environments and are full canopy absorbers having superior light capture efficiencies (Dismukes *et al.*, 2008). Microalgae are buoyant aquatic microcells that do not require structural biopolymers essential for the growth of higher plants in terrestrial environments. The resulting absence of lignin in algal biomass simplifies the processing and improves conversion efficiency through bypassing the most costly and inefficient steps for conversion of plant-derived cellulose to ethanol (Lynd *et al.*, 2008). To summarize, biofuel production via microalgal farming offers many advantages over biofuel production from land crops: (i) the high growth rate of microalgae makes them the ideal candidate to satisfy the massive demand for biofuels using limited land resources without causing potential biomass deficit; (ii) microalgae consume less water than land crops; (iii) the tolerance of microalgae to high CO_2 content in gas streams allows high-efficiency CO_2 mitigation; (iv) nitrous oxide release in flue gases could be minimized when microalgae are used for biofuel production; and (v) microalgal farming could be potentially more cost-effective than conventional farming (Li *et al.*, 2008).

Requirements for Raising Algal Biomass Feedstock – a Comparison with Crop-based Feedstock

Water

Microalgae can utilize low-quality water, such as agricultural runoff or municipal, industrial or agricultural wastewaters, as a source of water for the growth medium as well as a source of major secondary and micronutrients (Becker, 1994). Water footprint analysis during microalgal biodiesel production has shown that 3726 kg of water is required to generate 1 kg of microalgal biodiesel if freshwater is used without recycling, whereas recycling reduces the water usage by 84% (Yang *et al.*, 2011). Furthermore, water requirement reduces significantlyif algal production can be designed to use wastewater or seawater. In fact, using seawater/wastewater as culture medium decreases 90% of water requirement. These positive sustainability indices of algae have attracted substantial research investments from federal governments of various countries, private investors and energy industries to developtechnology for the productionof next-generation biofuel feedstock (Subhadra and Edwards, 2010). In the case of crop-based biodiesel, soybean and jatropha crops require 13,676 and 19,924 kg of water, respectively, for the production of 1 l of biodiesel (Gerbens-Leenes *et al.*, 2009). The total water footprint of biodiesel derived fromterrestrial cropsis very muchhigher than that of algae biodiesel; moreover, water recycling and growing higher plants in seawater/wastewater are almost impossible options.Hence, microalgae appear a better bet in sustainable use of valuable water resources.

Nutrients

For optimal growth, algalcultures must be provided with nutrients in adequate amounts. These include several major nutrients such as carbon, nitrogen, phosphorus, potassium, sulfur and a number of trace elements,and vitamins. The advantage with microalgae is that they can grow autotrophically, heterotrophically or mixotrophically and most microalgae can grow in more than one mode of nutrition. Under heterotrophic or mixotrophic conditions, some microalgal species can metabolize a variety of organic carbon sources such as sugars, molasses and acetic acid as well as organic compounds present in the wastewater(Becker, 1994; Bhatnagar *et al.*, 2011). Based on the nutrient requirement of particular microalga, the growth media, including seawater and wastewater containing heavy organic loads such as sewage, can be selected. In addition, microalgae possess a greater ability to fix CO_2 than higher plants (Tredici, 2010). Therefore, to attain maximum growth and productivity in algal systems, CO_2 supplementation becomes necessary. Usual sources of CO_2 for microalgae include atmospheric CO_2, CO_2 from industrial exhaust gases (e.g. flue gas and flaring gas) and CO_2 in the form of soluble carbonates (e.g. $NaHCO_3$ and Na_2CO_3). Some cyanobacteria also possess the ability to fix atmospheric nitrogen and hence are capable of growing in media devoid of nitrogen. Thus, the varied nutrient uptake potential of algae allow them to grow in the presence of organic and inorganic pollutants, which is lacking in higher plants.

Land/area

When increased demand for food and energy combines, pressure on land conversion is aggravated, leading to further climate change, which in turn may affect the productivity and availability of land, thereby creating a potentially vicious cycle resulting in the challenges related to energy–food–environment (Harvey and Pilgrim, 2011). Considering the areal productivity, for instance, if oil palm, a high-yielding oil crop, is grown for biofuel production, 24% of the total arable land of the USA will need to be dedicated to its cultivation to meet 50% of the transport fuel needs of the USA. Obviously, oil crops cannot contribute to replacing petroleum-derived liquid fuels in a significant manner in the foreseeable future. This scenario changes dramatically in the case of microalgae, as between 1% and 3% of the total US cropping area would be sufficient for producing algal biomass that satisfies 50% of the transportation fuel needs of the USA; the comparison of microalgae with other oil-producing crops in the context of land usage is shown in Table 12.1 (Chisti, 2007).

Weather

The chemical composition of microalgae is influenced by environmental factors including

Table 12.1. Comparison of various biodiesel feedstocks (Source: Chisti, 2007).

Source	Oil yield (lha^{-1})	Land area needed in Mha[a]
Maize	172	1540
Soybean	446	594
Canola	1190	223
Jatropha	1892	140
Coconut	2689	99
Oil palm	5950	45
Microalgae (with 70% oil)	136,900	2
Microalgae (with 30% oil)	58,700	4.5

[a]For meeting 50% of the total transport fuel needs of the USA

temperature and light (Oliveira et al., 1999). Temperature is the most crucial factor that regulates cellular, morphological and physiological responses of microalgae; higher temperatures generally accelerate the metabolic rates of microalgae (Munoz and Guieysse, 2006). Thus, algal cultivation can be best carried out in non-arable land of tropical regions where the average annual temperature is in the range of 30°C and the solar radiation is between 6000 and 8000 MJ m^{-2} year^{-1}. The relationship between temperature and chemical composition varies from species to species. However, in the study conducted by Oliveira et al. (1999), high growth temperature has been related to a significant decrease in protein content, together with increases in lipids and carbohydrates. Microalgae can survive extreme temperature variations even though the growth rate decreases with either of the temperature extremes. In contrast, higher plants are not sturdy enough to withstand extreme temperature fluctuations. Similarly, optimal light conditions can enhance the growth rate of microalgae. This can be accomplished by controlling the depth or thickness of the culture, mixing rate, dilution rate for continuous systems and culture density (Molina et al., 2001). However, this sort of manipulation of cultivation conditions with respect to weather parameters is not possible for higher plants.

Manpower

Microalgal cultivation for raising algal biomass feedstock requires very little manpower. Allprocesses can be automated including harvesting, where mechanical and chemical means are used. However, in the case of terrestrial plants and even in seaweeds, considerable manpower is required for harvesting the oil-containing material (Singh et al., 2011). In addition, manpower involvement is higherin the case of land-based crops with respect to fertilizer and pesticide application and irrigation, while these operations are either not required in microalgal cultivation or if necessary, requires much lower manpower input.

AlgaeBiomass Production and its Utilization

Both macroalgal and microalgal biomass can be efficiently used for biofuel applications. For algae biofuels, generating huge amounts of algal biomass remains a major challenge.Consistent efforts are being made worldwide to achieve the ideal combination of algae and growing conditions.In a large study on the feasibility of methane production from macroalgal biomass (Chynoweth et al., 1993), base case scenarios assumed yields of 11 dry t ha^{-1} year^{-1}, based on data from commercial growers. For optimized cultivation systems, yield of 45 dry t ha^{-1} year^{-1} was assumed. In some highly controlled environments, such yields were actually obtained, however, at costs precluding scale-up and commercialization (Chynoweth *et al.*, 1993). Similarly, research is being carried out to cultivate microalgal species with maximum lipid contents in order to make the conversion of microalgal biomass to biofuels more profitable. With the right algal species and right conditions in place, 1 t of wet algal biomass can yield about 200 l of oil (Singh and Gu, 2010).

The gross chemical composition of microalgae and macroalgae varies from species to species, and in particular, this composition of microalgae can be manipulated easily according to the requirements. This is possible as the composition of microalgae largely depends on environmental factors such as light intensity, temperature and nutrient availability (Becker, 1994; Rodolfi *et al.*, 2009). In general, microalgae contain varying proportions of carbohydrates, proteins, lipids, nucleic acids, pigments and vitamins (Becker, 1994). Proteins and lipids are found in higher quantities, with proteins ranging from 10% to 60% (w/w) and lipids from 2% to 70% (w/w) (Becker, 1994; Spolaore *et al.*, 2006). Among green algae, species including *Botryococcus braunii*, *Chlamydomonas reinhardtii*, *Dunaliella salina* and various *Chlorella* species can contain over 50% (w/w) of lipid, much of which is secreted into the cell wall (Metzger and Largeau, 2005). Such a high content of lipid renders microalgae a suitable choice for use in the production of biodiesel. Carbohydrates in microalgae in the form of starch, glucose, sugars and other polysaccharides are present in concentrations ranging from 5% to 50% (w/w) (Spolaore *et al.*, 2006). In addition, microalgae contain highly valuable substances such as pigments, long-chain PUFAs, including eicosapentaenoic acid and docosahexaenoic acid, and vitamins such as A, B_1, B_2, B_6, B_{12}, C, E, nicotinic acid, biotin, folic acid and pantothenic acid (Becker, 1994; Spolaore *et al.*, 2006). Based on the chemical composition of the algal biomass feedstock, various biofuel products such as biodiesel, bioethanol, biomethane, biohydrogen and thermochemical conversion products such as bio-oil, biocrude and syngas can be obtained.

Since the 1990s, the production of biodiesel from algae has been an area of considerable interest (Hu *et al.*, 2008). Algae use photosynthesis to convert solar energy into chemical energy. They store this energy in the form of oils, carbohydrates and proteins, and the oil can be converted to biodiesel. The more efficient a particular plant is at converting solar energy into chemical energy, the better it is from a biodiesel perspective, and algae are among the most photosynthetically efficient plants on earth (Demirbas and Demirbas, 2011). Algal oil can be converted into biodiesel using a technique calledtransesterification, wherein the reaction time is much shorter compared toenzymatic and immobilized whole cell processes. Transesterification is a chemical reaction between triglycerides and alcohol in the presence of a low-cost catalyst such as KOH to produce fatty acid methyl esters (FAME) that are termed as biodiesel (Sharma and Singh, 2009).

$$\text{Triglyceride} + 3 \text{ methanol} \xrightarrow{\text{catalyst}} \text{Glycerine} + 3 \text{ FAME (biodiesel)} \quad (12.1)$$

Although the production of biodiesel remains the ultimate objective for algae technologists, it is possible to economically co-produceother biofuelsfrom the algal biomass. Anaerobic digestion of lipid-extracted algal biomass by methanogenic bacteria leads to the production of methane. Biogas production through anaerobic digestion results in a net reduction in greenhouse gas (GHG) emissions, because methane would otherwise be released into the atmosphere, provoking a greenhouse effect that is 21-fold that of CO_2 (Fredriksson et al., 2006). Chisti (2008b) discussed the recovery of energy from microalgae residues after biodiesel production, highlighting its potential to meet most of the energy demands of the preceding processes. He theoretically estimated that an average heating value of 9360 MJ t^{-1} of microalgae residues was recoverable as methane. Furthermore, co-digesting the microalgae residues with glycerol obtained during transesterification of algal oils was observed to augment the methane yields by 5–8% compared with the anaerobic digestion of the algal residues alone (Ehimen et al., 2009).

Lipid-extracted algal biomass can also be used for producing ethanol, which is currently produced mainly from maize in the Americas. Bioethanol has a future as an alternative fuel, but it is extremely important to make sure that its production is not obstructed by raw material constraints (Harun et al., 2010). In this context, microalgae have a very short harvesting cycle (1–10 days) compared with other feedstocks (harvested once or twice a year) and thus algae provide enough supplies to meet ethanol production demands (Schenk et al., 2008). Besides, algae have higher photon conversion efficiency and can synthesize and accumulate large quantities of carbohydrate biomass for bioethanol production (Subhadra and Edwards, 2010). Hon-Nami (2006) has reported the fermentation of *Chlamydomonas perigranulata* to produce ethanol, butanediol, acetic acid and CO_2, thus showing the potential of algal biomass for the production of multiple products. Harun et al. (2010) have reported that the lipid-extracted microalgae yielded 60% higher ethanol concentration compared with the dried/intact non-lipid extracted microalgae, thus implying the significance of using the spent biomass for ethanol production.

Apart from the above-mentioned chemical and biological processes for generating energy, thermochemical conversion is another attractive option. Although thermochemical conversion technologies of algal biomass are certainly not the pick of the options at present, the fact is that combustion is responsible for over 97% of the world's bioenergy production (Balat, 2009). Major thermochemical conversion processes are pyrolysis, gasification and thermochemical liquefaction, which result in the production of bio-oil, biocrude and syngas.

Another potentially attractive fuel option from microalgae is biohydrogen. In an increasingly carbon-constrained world, H_2 is anticipated to become a more important clean fuel in the future. Microalgal H_2 production, particularly within a biorefinery model, has considerable potential to contribute to sustainable H_2 supply. Moreover, as H_2 is a volatile product that can be readily collected from the culture, it can be an excellent component of the biorefinery (Kruse and Hankamer, 2010). However, the current status of H_2 production is not encouraging as shown by Laurinavichene et al. (2008). In their study, continuous H_2 production in microalgae by immobilized non-motile *Chlamydomonas reinhardtii* cultures was 6–8 ml H_2 day^{-1}, which is very low for commercial-scale production. As of today, biological hydrogen production is the most challenging area of biotechnology with respect to both technical knowhow and commercialization possibilities. The future of biological hydrogen production depends on strain improvement through genetically engineered microorganisms.

Aviation biofuel is rapidly becoming a major focus of aviation industry as it is one of the primary means by which the industry can reduce its carbon footprint. Also, it is clear that if petrocrude-based fuels become unavailable in the future, aviation industry has no option but to use biofuels derived from renewable feedstocks such as algae. After a multi-year technical review from aircraft makers, engine manufacturers and oil companies, biofuels were approved for commercial use in July 2011. Since then, multiple airlines have begun the use of biofuels on commercial flights. The focus of the industry is on second-and third-generation sustainable jetfuels. Algae as feedstock for production of aviation biofuel are expected to be a promising option as they do not compete with food needs and have a high sustainability index.

The best approach to using algal biomass as a sustainable feedstock lies in the maximum utilization of the biomass within an algal biorefinery, which is a new concept. Bennett (2008) points out that as feedstock, algae could fit into most of the integrated biorefinery designs as its primary components such as oils, carbohydrates and proteins can be converted into various biofuels such as biocrude, jetfuels, biodiesel, bioethanol and biomethane. Therefore, an integrated biorefinery may be a more profitable option and biorefining approaches depicted in Fig. 12.1 can be employed to generate energy and value-added chemicals, to treat wastewater, to utilize CO_2 and to avoid carbon emissions. It drastically reduces biofuel production costs, as there are a variety of outputs and sources of revenue from the main product and co-products.

Algal Cultivation Systems

Mass cultivation of algae is commonly carried out in open ponds and closed photo-bioreactors. Open pond production systems for microalgal cultivation have been in use since the 1950s (Borowitzka, 1999) and raceway ponds are the most commonly used cultivation system (Jiménez *et al.*, 2003). However, later, closed photo-bioreactor technologies were designed to overcome some of the major problems, such as contamination, associated with the open-pond production systems. Some of the types of closed systems include flat-plate, tubular and column photo-bioreactors. Flat-plate photo-bioreactors have space constraints and tubular photo-bioreactorshave design limitations. Column photo-bioreactors are the most efficient, which provide efficient mixing, the highest volumetric mass transfer rates and the best controllable growth conditions (Eriksen, 2008). In general, open ponds are considered advantageous compared to photo-bioreactors because they are lower cost, easier to construct and operate, easy to clean up after cultivation and good for mass cultivation of microalgae (Ugwu *et al.*, 2008). Moreover, cultivation in open ponds finds favour because of escalating equipment costs and lack of reliable scale-up technologies for closed systems. Recently, a two-step hybrid cultivation process, i.e. a combination of photo-bioreactor and open-pond cultivation, has been suggested. The first step aids rapid growth of algae in the photo-bioreactor, while the second step ensures mass cultivation in open ponds (Bruton *et al.*, 2009). Attached algal cultivation systems are also receiving attention of late for wastewater treatment.Harvesting of biomass in attached growth systems such as algal turf scrubbers is much easier compared to suspended cultivation systems, which include open ponds and photo-bioreactors (Adey *et al.*, 1993). However, algal turf scrubbers are suitable only for growing consortiaof filamentous algae, which tend to be poor lipid producers.

Approaches to Enhance Biomass Productivity and Biofuel Production

Bioprospecting is one option where phycologists look for natural algal strains with faster growth rates for enhancing biomass production. Genetic manipulation can enhance growth rate and lipid and carbohydrate content in algae. Thus, achieving faster growth rate with increased biomass, and lipid and carbohydrate production should be the approach for the

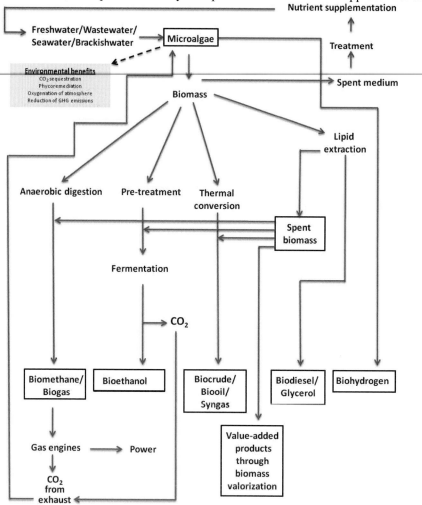

Fig. 12.1. Flow diagram for an integrated algal biorefinery with multiple product streams.

future. Ideal reactor design, appropriate nutrient concentrations for maximizing the growth rate, appropriate inoculum volumes, standardization of harvesting volumes and cycles, and optimum level of agitation are very important in algae mass cultivationto improve biomass productivity. In particular, agitation of the culture media plays an important role such that it not only avoids sedimentation, photo-inhibition, nutrient limitation and thermal stratification but also increases light-conversion efficiency, thereby resulting in a higher rate of photosynthesis. Mixing has its positive effect, particularly at sunrise and sunset hours, when the production rate may be 2.5 times more than that for non-mixed culturesin the early morning when incident light is low (Bosca et al., 1991). Optimization of CO_2 supply can also result in increased production of biomass. CO_2-rich flue gases from thermal power plants can be used as ideal carbon source for the cultivation of algae. Hunt et al.

(2011) advocated the use of biostimulants such as naphthalene acetic acid (NAA) to enhance the biomass productivity of microalgae. Therefore, research is required towards creating innovative approaches, optimizingculture conditions/parameters to improve the growth rate/biomass productivity.

Biofuel Production Coupled with Phycoremediation

Increasing awareness about our environment and growing demand for energy and food are forcing us to explore the feasibility of treatment and utilization of wastewater for various applications. Within this context, treatment of wastewater and CO_2-rich flue gas using microalgae (commonly called phycoremediation) is particularly attractive becausealgaescavenge pollutants, convert solar energy into useful renewable biomass, remove nitrogen and phosphorus from the wastewater, and improve the dissolved O_2 content (De La Noue and De Pauw, 1988). Likewise, growing algae in wastewater/seawater/brackish water for the generation of biomass feedstock is an interesting option. Efforts should be made for successful cultivation of microalgae in marine environments and commercialization of the same, as algae will notcompetewith food crops for land and water. Thus, the combination of the three roles of microalgae viz. CO_2 cycling, wastewater treatment, and cultivation in wastewater, seawater and brackish water for biofuel production has the potential to maximize the impact of sustainable microalgae biofuel production systems (Guzzon et al., 2008). The use of algae to treat wastewater has been investigated for over 50 years, with one of the first descriptions of this application being reported by Oswald and Gotaas (1957). The use of microalgae for the treatment of municipal wastewater has been a subject of research and development for several decades (Oswald, 1963). According to Clarens et al. (2010), most of the environmental burdens associated with algae would be offset if wastewater were used as a nutrient source. They concluded that algae biofuel production using freshwater and fertilizers would consume more energy, have higher GHG emissions and use more water than biofuel production from land-based crops such as switchgrass, canola and maize. In contrast, algal production from wastewater offers a far more attractive proposition from an environmental viewpoint, and their modelling of algae-based biofuels has shown that algae biofuel production coupled to growing algae in sewage and agricultural wastewaters improved the life cycle burden of these biofuels and proved to be more environmentally beneficial than terrestrial plant-based biofuels. Therefore, the ideal approach would be to produce algal biomass using wastewater as a nutrient source. Furthermore, phycoremediation, apart from generating the biomass, has the following advantages: (i) CO_2 cycling; (ii) oxygenation of the atmosphere; (iii) reduction in treatment cost; and (iv) energy conservation. The industry, which employs phycoremediation for the treatment of their wastewaters, not only can use the generated algal biomass for various applications, but can also claim significant carbon and energy credits.

In India, Sivasubramanian et al. (2009) have implemented a large-scale continuous phycoremediation plant at an alginate industry for treating the acidic effluent generated at the facility situated at Ranipet, Tamil Nadu, which has been in operation since 2006 (Fig. 12.2). This plant treats the effluent with a pH range of 1.4–1.8 and the daily volume generated being approximately 30 Kl. In their study, the pH of the treated effluent remained around 7 and very minimal sludge formation was observed with efficient utilization of generated biomass. Apart from this, microalgae are also efficient in the treatment of wastewaters containing toxic compounds, e.g. heavy metal-containing effluents. Microalgae with good adsorption capacities are used for this purposeand the sequestered heavy metals can be re-extracted from the biomass and the biomass can be utilized for biofuel applications. In anearlierstudy by one of the authors, *Chlorella vulgaris* showed a lead adsorption capability of 30.62 mg g^{-1} of DW, which is much higher than the value obtained from studies of plant-based adsorbents (Hanumantha Rao *et al.*, 2011). Chinnasamy *et al*. (2010a, b) reported cultivation of native algae consortia in treated and untreated carpet industry wastewater for the removal of nutrients such as nitrogen and phosphorus and production of renewable biomass. Thus, phycoremediation has tremendous potential in mitigatingpollution while making available feedstock for biofuel production in the future.

Transgenic Algae for Biofuel Applications – a Possible Option

The application of genetic engineering to improve biofuel production in microalgae is still in its infancy. In future, genetic and metabolic engineering of algae are going toplay an important role in improving the economics of production of algae biofuels (Roessler *et al.*, 1994; Dunahay *et al.*, 1996). Microalgae can be genetically engineered to potentially increase photosynthetic efficiency, biomass yield, growth rate, lipid and carbohydrate productivity; improve temperature tolerance; eliminate the light saturation phenomenon; and reduce photo-inhibition and photo-oxidation (Chisti, 2007). However, the 'laboratory to farm' transition will not be easy as it has to overcome issues such as social acceptance, political lobbying and compliance with regulatory norms. In addition, issues such as genetic cross-contamination and use of closed photo-bioreactors instead of open ponds need to be thoroughly evaluated before implementation.

Figure 12.2. Large-scale phycoremediation plant, which has been in operation since 2006, for treating the highly acidic effluent from an alginate industry (photo courtesy of Prof. V Sivasubramanian, Vivekananda Institute of Algal Technology, Chennai, India, and Mr Rajasekaran, SNAP Natural & Alginate Products (P) Limited, Ranipet, Tamil Nadu, India).

Challenges and Problems in Cultivation, Harvesting and Processing of Algae

Cultivation of microalgae in open ponds poses a real challenge as the process is subject to contamination with other weed algae, and algal predators such as zooplanktons, protozoa and insects. These problems can be prevented by cultivating algae in closed photo-bioreactors, but again, this is not an economically viable option as of today, as the capital investment for closed photo-bioreactors exceed US$100 m^{-2} compared to US$10 m^{-2} for open ponds. Another key challenge in microalgal cultivation is the use of strains that can maintain a high growth rate in addition to a high metabolic rate, thus leading to significant lipid and carbohydrate yields. This major challenge can be duly addressed through extensive bioprospecting or target-oriented genetic engineering, both of which are now starting to appear as promising approaches (Kumar *et al.*, 2010). The next major hurdle comes in the form of harvesting the algal biomass. Conventional processes used to harvest microalgae include concentration through centrifugation, foam fractionation, chemical flocculation, electroflocculation, membrane filtration and ultrasonic separation. Harvesting costs may contribute to approximately 20–30% of the total cost of algal biomass and the above methods would be viable only if the biomass harvested is used for extracting high-value products such as nutraceuticals (Molina Grima *et al.*, 2003). Microalgae are typically small with a diameter of 3–30 µm and the culture broths may be quite dilute at less than 0.5 g solids l^{-1}. Thus, large volumes of culture media must be handled to harvest algal biomass, which makes harvestingan energy-intensive process. Due to the small size of algal cells and relatively low biomass concentration in the cultivation medium, harvesting, dewatering and lipid extraction from microalgal biomass are considered challenging issues for the commercial-scale production of algae biofuels.

Processing of the harvested algal biomass is also a major problem. To extract oil, the algal biomass needs to be dried and the process is again energy intensive. Normal drying methods, which require energy, including spray drying, drum drying, flash drying and solar drying, are not considered efficient. The use of harvested biomass directly for fermentation is also not feasible due to the lack of easily fermentable sugars such as glucose or sucrose in the algal biomass. Hence, pretreatment of the algal biomass has to be carried out to break down the polysaccharides into individual monomer constituents, which again adds to the cost (Singh *et al.*, 2011).

Therefore, to overcome the above-mentioned bottlenecks,research efforts are warranted to:(i) cultivate algae with higher cell densities without contamination; (ii)develop cheap and energy-efficient harvesting techniques; (iii) develop a direct fermentation process of algal biomass without any pre-treatment; (iv) develop an efficient low-cost lipid extraction technique that can use wet algal biomass as feedstock; and (v) develop an energy-efficient hydrothermal liquefaction process to produce biocrude from wet algal biomass.

Environmental Impact

Algae biofuels are environmentallyfriendly due to their non-toxic and biodegradable characteristics and they also are devoid of GHG emissions. Replacement of fossil fuels with algae biofuels will result in GHG abatement and in turn will protect the environment by preventing climate change. However, this will not be the case with first-generation biofuels because a large-scale expansion of these biofuels could cause the release of GHG emissions to the atmosphere through land-use change as farmers might clear existing forests to meet increased crop demand to supply food and feedstock for biofuels (Timilsina and Shrestha, 2011).

Economics and Technical Feasibility

As of today, the per litre production cost of algal biodiesel is US$8.80, whereas that of crop-based biodiesel is US$1.17 (Lee, 2011). Even though the cost of microalgal biodiesel is higher, the increased oil productivity of microalgae per hectare (i.e. more than ten times than that of oil palm) is definitely advantageous over crop-based biodiesel.

Efficient and viable algal production systems provide consistently low-cost, uninterrupted and year-round feedstock supply. Integrated algal biorefineries focus on simultaneous production of multiple products with contrasting market values to ensure economic sustainability. For low-value biofuel production systems, use of fatter algae with 60% oil content can result inreduction of the size and footprint of algae biofuel production systems, leading to significant capital and operating cost reductions and savings (Singh and Gu, 2010).

Sustainability of AlgaeBiofuels

Continued use of fossil fuels is now widely recognized as unsustainable because of depleting supplies and the contribution of these fuels to the accumulation of CO_2 in the environment. Algae biofuels are considered sustainable as: they are carbon neutral; do not affect the quality and quantity of available natural resources such as water and soil; do not compete with food production; and do not affect biodiversity (Lora *et al.*, 2011). Algae biofuels are renewable, and hence can result in environmental and economic sustainability. Although algae biofuels are promoted as environmentallyfriendly, the carbon footprint of algae biofuels has to be thoroughly studied before it is claimed as a sustainable fuel. With existing technologies not in a position to commercially produce biofuels at a price competitive to fossil fuels, one of the main challenges facing the algae biofuels sector is to attain high productivity while reducing capital and operating costs.The profit or return on investment is comparatively low for algae biofuel-based farming, which is another constraint from an investment perspective on using algae biomass solely as an energy feedstock (Subhadra, 2010). Therefore, to make the technology sustainable, research on integrated approaches to produce multiple products is necessary.

Presently, most carbon capture and sequestration discussions are about geological storage of CO_2. Even if this is proved safe, the biggest difficulty with this approach is ocean acidification and release of CO_2 from the sequestration sites into the atmosphere. According to Packer (2009), carbon capture for biofuels is mitigating only in that it reduces new fossil reserves being released, by recycling carbon from the atmosphere.

Conclusion

It isimportant to devise and adopt policies to promote renewable energy sources, which are capable of sequestering atmospheric CO_2 to maintain environmental and economic sustainability. The production and utilization of algal biomass as a novel feedstock is considereda sustainable and eco-friendly approach for renewable biofuel production. The first- and second-generation biofuels such as biodiesel and bioethanol derived from plantsgrown on arable lands will adversely impactfood prices. The third-generation biofuels from algae grown on non-arable lands and wastewater/seawater/brackish water seem to be the best solutionto the food versus fuel conflict. The utilization of nutrients from wastewater and the recycling of atmospheric CO_2 into the food chain have made algae the candidate of choice for the production of renewable biomass. Although the prospects for algal biomass as a feedstock for the biofuel industry are promising, there are still substantial challenges in algal cultivation, harvesting and processing to make production of biofuels from algae economically viable. The high investments being made in research, development and commercialization of algae biofuels byover 150 companies worldwide (Deng *et al.*, 2009) clearly indicate that this technology is promising and could well meet global energy needs in future.

References

Adey, W., Luckett, C. and Jensen, K. (1993) Phosphorus removal from natural waters using controlled algal production. *Restorative Ecology* 1, 29–39.
Balat, M. (2009) New biofuel production technologies. *Energy Education Science and Technology Part A*22,147–161.
Becker, E.W.(1994) *Microalgae Biotechnology and Microbiology*. Cambridge University Press, New York.
Bennett, S. (2008) Algal biorefineries. Cleantech magazine.Available at: http://www.cleantechinvestor.com/portal/biofuels/1758-algal-biorefinery.html
Bhatnagar, A., Chinnasamy, S., Singh, M. and Das, K.C. (2011) Renewable biomass production by mixotrophic algae in the presence of various carbon sources and wastewaters. *Applied Energy* 88, 3425–3431.
Borowitzka, M.A. (1999) Commercial production of microalgae: ponds, tanks, tubes and fermenters. *Journal ofBiotechnology* 70, 313–321.
Bosca, C., Dauta, A. and Marvalin, O. (1991) Intensive outdoor algal cultures: how mixing enhances the photosynthetic production rate. *Bioresource Technology* 38, 185–188.
Bruton, T., Lyons, H., Lerat, Y., Stanley, M. and BoRasmussen, M. (2009) A review of the potential of marine algae as a source of biofuel in Ireland. Report for Sustainable Energy Ireland. Available at: http://www.seambiotic.com/uploads/algae%20report%2004%202009.pdf
Chinnasamy, S., Bhatnagar, A., Hunt, R.W. and Das, K.C. (2010a) Microalgae cultivation in a wastewater dominated by carpet mill effluents for biofuel applications. *Bioresource Technology* 101, 3097–3105.
Chinnasamy, S., Bhatnagar, A., Claxton, R. and Das, K.C. (2010b) Biomass and bioenergy production potential of microalgae consortium in open and closed bioreactors using untreated carpet industry effluent as growth medium. *Bioresource Technology* 101, 6751–6760.
Chisti, Y. (2007) Biodiesel from microalgae. *Biotechnology Advances* 25, 294–306.
Chisti, Y. (2008a) Response to Reijnders: do biofuels from microalgae beat biofuels from terrestrial plants? *Trends in Biotechnology* 26, 351–352.
Chisti, Y. (2008b) Biodiesel from microalgae beats bioethanol. *Trends in Biotechnology* 26, 126–131.
Chynoweth, D.P., Turick, C.E., Owens, J.M., Jerger, D.E. and Peck, M.W. (1993) Biochemical methane potential of biomass and waste feedstocks. *Biomass and Bioenergy* 5, 95–111.
Clarens, A.F., Resurreccion, E.P., White, M.A. and Colosi, L.M. (2010) Environmental life cycle comparison of algae to other bioenergy feedstocks. *Environmental Science and Technology* 44,1813–1819.
De La Noue, J. and De Pauw, N. (1988) The potential of microalgal biotechnology: A review of production and uses of microalgae. *Biotechnology Advances* 6, 725–770.
Demirbas, A. and Demirbas, M.F. (2011) Importance of algae oil as a source of biodiesel. *Energy Conversion and Management* 52, 163–170.

Deng, X., Li, Y. and Fei, X. (2009) Microalgae: a promising feedstock for biodiesel. *African Journal of Microbiology Research* 3, 1008–1014.

Dismukes, G.C., Carrieri, D., Bennette, N., Ananyev, G.M. and Posewitz, M.C. (2008) Aquatic phototrophs: efficient alternatives to land-based crops for biofuels. *Current Opinion in Biotechnology* 19, 235–240.

Dunahay, T.G., Jarvis, E.E., Dais, S.S. and Roessler, P.G. (1996) Manipulation of microalgal lipid production using genetic engineering. *Applied Biochemistry and Biotechnology* 57–58, 223–231.

Ehimen, E.A., Connaughton, S., Sun, Z. and Carrington, C.G. (2009) Energy recovery from lipid extracted transesterified and glycerol co-digested microalgae biomass. *Global Change Biology Bioenergy* 1, 371–381.

Eriksen, N. (2008) The technology of microalgal culturing. *Biotechnology Letters* 30, 1525–1536.

Fredriksson, H., Baky, A., Bernesson, S., Nordberg, Å., Norén, O. and Hansson, P.A. (2006) Use of on-farm produced biofuels on organic farms – evaluation of energy balances and environmental loads for three possible fuels. *Agricultural Systems* 89, 184–203.

Gerbens-Leenes, P.W., Hoekstra, A.Y. and van der Meer, T.H. (2009) The water footprint of bioenergy. *Proceedings of the National Academy of Sciences USA* 106, 10219–10223.

Gerber, N., van Eckert, M. and Breuer, T. (2009) The impacts of biofuel production on food prices: a review. University of Bonn, Centre for Development Research (ZEF) Discussion Paper No. 127. Available at: http://ssrn.com/abstract=1402643

Guzzon, A., Bohn, A., Diociaiuti, M. and Albertano, P. (2008) Cultured phototrophic biofilms for phosphorus removal in wastewater treatment. *Water Research* 42, 4357–4367.

HanumanthaRao, P., Ranjith, K.R., Raghavan, B.G., Subramanian, V.V. and Sivasubramanian, V. (2011) Is phycovolatilization of heavy metals a probable (or possible) physiological phenomenon? An *in situ* pilot-scale study at a leather-processing chemical industry. *Water Environmental Research* 83(4), 291–297.

Harun, R., Danquah, M.K. and Forde, G.M. (2010) Microalgal biomass as a fermentation feedstock for bioethanol production. *Journal of Chemical Technology and Biotechnology* 8, 199–203.

Harvey, M. and Pilgrim, S. (2011) The new competition for land: food, energy, and climate change. *Food Policy* 36, S40–51.

Hon-Nami, K. (2006) A unique feature of hydrogen recovery in endogenous starch-to-alcohol fermentation of the marine microalga, *Chlamydomonas perigranulata*. *Applied Biochemistry and Biotechnology* 131, 808–828.

Hu, Q., Sommerfeld, M., Jarvis, E., Ghirardi, M., Posewitz, M., Seibert, M. and Darzins, A. (2008) Microalgaltriacylglycerols as feedstocks for biofuel production: perspectives and advances. *Plant Journal* 54, 621–639.

Hunt, R.W., Chinnasamy, S. and Das, K.C. (2011) The effect of naphthalene acetic acid on biomass productivity and chlorophyll content of green algae, coccolithophore, diatom and cyanobacterium cultures. *Applied Biochemistry and Biotechnology* 164(8), 1350–1365.

Huntley, M. and Redalje, D.G. (2007) CO_2 mitigation and renewable oil from photosynthetic microbes: a new appraisal. *Mitigation and Adaptation Strategies for Global Change* 12, 573–608.

Jegannathan, K.R., Chan, E.-S. and Ravindra, P. (2009) Harnessing biofuels: a global renaissance in energy production? *Renewable and Sustainable Energy Reviews* 13, 2163–2168.

Jiménez, C., Cossío, B.R., Labella, D. and Xavier Niell, F. (2003) The feasibility of industrial production of *Spirulina* (Arthrospira) in southern Spain. *Aquaculture* 217, 179–190.

Koh, L.P. and Ghazoul, J. (2008) Biofuels, biodiversity, and people: understanding the conflicts and finding opportunities. *Biological Conservation* 141, 2450–2460.

Kruse, O. and Hankamer, B. (2010) Microalgal hydrogen production. *Current Opinion in Biotechnology* 21, 238–243.

Kumar, A., Ergas, S., Yuan, X., Sahu, A., Zhang, Q., Dewulf, J., Malcata, F.X. and Langenhove, H. (2010) Enhanced CO_2 fixation and biofuels production via microalgae: recent developments and future directions. *Trends in Biotechnology* 28, 371–380.

Laurinavichene, T.V., Kosourov, S.N., Ghirardi, M.L., Seibert, M. and Tsygankov, A.A. (2008) Prolongation of H_2 photoproduction by immobilized, sulfur-limited *Chlamydomonas reinhardtii* cultures. *Journal of Biotechnology* 134, 275–277.

Lee, D.H. (2011) Algal biodiesel economy and competition among bio-fuels. *Bioresource Technology* 102, 43–49.

Lee, R.E. (1980) *Phycology*. Cambridge University Press, New York.

Li, Y., Horsman, M., Wu, N., Lan, C.Q. and Dubois-Calero, N. (2008) Biofuels from microalgae. *Biotechnology Progress* 24, 815–820.

Lora, E.E.S., Palacio, J.C.E., Rocha, M.H., Renó, M.L.G., Venturini, O.J. and Olmo, O.A. (2011) Issues to consider, existing tools and constraints in biofuels sustainability assessments. *Energy* 36, 2097–2110.

Lynd, L.R., Laser, M.S., Bransby, D., Dale, B.E., Davison, B., Hamilton, R., Himmel, M., Keller, M., McMillan, J.D., Sheehan, J. and Wyman, C.E. (2008) How biotech can transform biofuels. *Nature Biotechnology* 26, 169–172.

Metzger, P. and Largeau, C. (2005) *Botryococcus braunii*: a rich source for hydrocarbons and related ether lipids. *Applied Microbiology and Biotechnology* 66, 486–496.

Molina, E., Fernández, J., Acién, F.G. and Chisti, Y. (2001) Tubular photobioreactors design for algal cultures. *Journal of Biotechnology* 92, 113–131.

MolinaGrima, E., Belarbi, E.-H., Fernández, F.G., Medina, A.R. and Chisti, Y. (2003) Recovery of microalgal biomass and metabolites: process options and economics. *Biotechnology Advances* 20, 491–515.

Munoz, R. and Guieysse, B. (2006) Algal-bacterial processes for the treatment of hazardous contaminants: a review. *Water Research* 40, 2799–2815.

Oliveira, M.A.S., Monteiro, M.P., Robbs, P.G. and Leite, S.G. (1999) Growth and chemical composition of *Spirulina maxima* and *Spirulinaplatensis* biomass at different temperatures. *Aquaculture International* 7, 261–275.

Oswald, W.J. (1963) High rate ponds in waste disposal. *Developments in Industrial Microbiology* 4, 112–119.

Oswald, W.J. and Gotaas, H.B. (1957) Photosynthesis in sewage treatment. *Transactions of American Society of Civil Engineers* 122, 73–105.

Packer, M. (2009) Algal capture of carbon dioxide; biomass generation as a tool for greenhouse gas mitigation with reference to New Zealand energy strategy and policy. *Energy Policy* 37, 3428–3437.

Rodolfi, L., Zittelli, G.C., Bassi, N., Padovani, G., Biondi, N., Bonini, G. and Tredici, M.R. (2009) Microalgae for oil: strain selection, induction of lipid synthesis and outdoor mass cultivation in a low-cost photobioreactor. *Biotechnology and Bioengineering* 102, 100–112.

Roessler, P.G., Brown, L.M., Dunahay, T.G., Heacox, D.A., Jarvis, E.E., Schneider, J.C., Talbot, S.G. and Zeiler, K.G. (1994) Genetic engineering approaches for enhanced production of biodiesel fuel from microalgae. *ACS Symposium Series* 566, 255–270.

Schenk, P.M., Thomas-Hall, S.R., Stephens, E., Marx, U.C., Mussgnug, J.H., Posten, C., Kruse, O. and Hankamer, B. (2008) Second generation biofuels: high-efficiency microalgae for biodiesel production. *Bioenergy Research* 1, 20–43.

Sharma, Y.C. and Singh, B. (2009) Development of biodiesel: current scenario. *Renewable and Sustainable Energy Reviews* 13, 1646–1651.

Sheehan, J., Dunahay, T., Benemann, J. and Roessler, P. (1998) *A look back at the US Department of Energy's Aquatic Species Program – Biodiesel from Algae*. National Renewable Energy Laboratory, Golden, Colorado, Report NREL/TP-580-24190.

Singh, A., Nigam, P.S. and Murphy, J.D. (2011) Mechanism and challenges in commercialization of algal biofuels. *Bioresource Technology* 102, 26–34.

Singh, J. and Gu, S. (2010) Commercialization potential of microalgae for biofuels production. *Renewable and Sustainable Energy Reviews* 14, 2596–2610.

Sivasubramanian, V., Subramanian, V.V., Raghavan, B.G. and Ranjithkumar, R. (2009) Large scale phycoremediation of acidic effluent from an alginate industry. *Science Asia* 35, 220–226.

Spolaore, P., Joannis-Cassan, C., Duran, E. and Isambert, A. (2006) Commercial applications of microalgae. *Journal ofBioscience and Bioengineering* 101, 87–96.

Stephens, E., Ross, I.L., Mussgnug, J.H., Wagner, L.D., Borowitzka, M.A., Posten, C., Kruse, O. and Hankamer, B. (2010) Future prospects of microalgal biofuels production systems. *Trends in Plant Science* 15, 554–564.

Subhadra, B. (2010) Sustainability of algal biofuel production using integrated renewable energy park (IREP) and algal biorefinery approach. *Energy Policy* 38, 5892–5901.

Subhadra, B. and Edwards, M. (2010) Algal biofuel production using integrated renewable energy park approach in United States. *Energy Policy* 38, 4897–4902.

Timilsina, G.R. and Shrestha, A. (2011) How much hope should we have for biofuels? *Energy* 36, 2055–2069.

Tredici, M.R. (2010) Photobiology of microalgae mass cultures: understanding the tools for the next green revolution. *Biofuels* 1, 143–162.

Ugwu, C.U., Aoyagi, H. and Uchiyama, H. (2008) Photobioreactors for mass cultivation of algae. *Bioresource Technology* 99, 4021–4028.

Yang, J., Xu, M., Zhang, X., Hu, Q., Sommerfeld, M. and Chen, Y. (2011) Life-cycle analysis on biodiesel production from microalgae: water footprint and nutrients balance. *Bioresource Technology* 102, 159–165.

Chapter 13

Biofuel from Microalgae: Myth versus Reality

Jubilee Purkayastha, Hemanta Kumar Gogoi, Lokendra Singh and Vijay Veer

Introduction

The world's economic scenario is in a rapidly changing mood, where fulfilment of energy requirement has been recognized as a key issue for socio-economic and national development. In this burning point of high fuel prices, biofuels are seen as a more attractive option. Therefore, research into biofuels could now be more cost-effective. Being the sixth largest power consumer in the world, oil shortage has been a real influencing factor on Indian economic and societal development. In this scenario, it is of strategic importance to find a new source of sustainable energy to replace conventional fossil fuel. Biomass energy, especially biodiesel, has been becoming the world's largest contributor towards renewable power needs, providing substantial environmental and economic benefits. Biomass material can be found in almost every corner of the world, making it a theoretically inexhaustible renewable energy source with almost unlimited potential. Biomass not only includes intentionally grown and harvested crops but also other products such as biodegradable wastes, forest residues such as dead trees, branches and tree stumps, yard clippings, wood chips, garbage as well as plant or animal matter used for production of fibres or chemicals. An almost untapped source of biomass energy, especially biodiesel, is the microalgal biomass, although the idea of using microalgae as a source of fuel is not new (Chisti, 1980–1981), but only now it is being taken seriously because of the increasing price of petroleum and also due to the budding concern about burning fossil fuels and global warming (Gavrilescu and Chisti, 2005). Microalgal biomass has the potential to grow rapidly and yields are far higher than any other currently used feedstock. It has the possibility of a much higher and much more efficient energy yield per unit. However, comparatively little research and development is currently being put into microalgal energy research.

Why Microalgal Biodiesel?

Fossil fuels were generated hundreds of millions of years ago, and are depleting at an ever faster rate due to increased energy demand (Veziroglu and Sahin, 2008). India contains only 0.7% of the world's reserves for oil (Oil and Gas Journal, 2005). Fossil fuel dependence without reserves means depending on other countries' reserves for supply. Moreover, the impact of the by-products from fossil fuel combustion are dangerous, causing air pollution, global warming, water/land pollution, oil spills and adverse health effects (Veziroglu and Sahin, 2008). The main products of fossil-fuel burning are carbon dioxide (CO_2), nitrous oxides and hydrocarbons, including mixture of gases commonly referred to as 'greenhouse gases' resulting in an imbalance in the environment contributing to climate change. Among the gases emitted by fossil fuels, CO_2 is one of the most significant, output of which has increased by 25% since the beginning of fossil fuel burning and is predicted to increase by 1.9% each year. Biodiesel is a fossil diesel substitute fuel that is manufactured from living sources such as vegetable oils, waste oils, animal fats or microalgal lipids. However, the primary source of all the biofuel is plants (including microalgae), because green plants produce oils from sunlight and air; on the other hand animal fats are produced when the animal consumes plant oils and other fats, which are also renewable. Biodiesel is an established fuel and the technology for producing and using biodiesel has been known for a long time (Knothe *et al*., 1997; Fukuda *et al*., 2001). Biodiesel is produced currently from plant and animal oils, mainly from soybeans, canola oil, animal fat, palm oil, maize oil, waste cooking oil (Kulkarni and Dalai, 2006) and jatropha oil (Barnwal and Sharma, 2005).

Biodiesel is much more rewarding than petroleum-based diesel and ethanol fuel: (i) it is energy efficient and can be used in existing diesel engines with little or no modification to the engine; (ii) biodiesel is renewable and may be continually obtained from plants, animals and microalgae if cultivated; and (iii) biodiesel is environment friendly. Biodiesel contains practically no sulfur or aromatics, and the use of biodiesel in a conventional diesel engine results in substantial reduction of unburned hydrocarbons, CO_2 and particulate matter. Biodiesel has advantages over other biofuels such as ethanol, which yields 25% more energy than the energy invested in its production, whereas biodiesel yields 93% more. If compared to ethanol, biodiesel releases just 1.0%, 8.3% and 13% of the agricultural nitrogen, phosphorus and pesticide pollutants, respectively, per net energy gain; greenhouse gas emissions are reduced 12% by the use of ethanol and 41% by biodiesel and biodiesel also releases less air pollutants per net energy gain than ethanol (Hill *et al*., 2006).

Microalgae are living solar driven units that convert CO_2 to potential biofuels, foods, feeds and high-value bioactives (Melis, 2002; Lorenz and Cysewski, 2003; Metzger and Largeau, 2005; Singh *et al*., 2005; Walter *et al*., 2005; Spolaore *et al*., 2006). Microalgae can provide several different types of renewable biofuels such as: methane by anaerobic digestion of the algal biomass (Spolaore *et al*., 2006); biodiesel from microalgal oil (Roessler *et al*., 1994; Sawayama *et al*., 1995; Banerjee *et al*., 2002; Gavrilescu and Chisti, 2005); and also biohydrogen (Fedorov *et al*., 2005; Kapdan and Kargi, 2006). It is possible to find species best suited to local environments or specific growth characteristics as different microalgal species can adapt themselves to a variety of environmental conditions, which is not possible with other biodiesel feedstocks (Mata *et al*., 2010).

However, to date there appears to be no commercial scale biodiesel production unit from microalgae. This may change in the near future as several companies along with research organizations are attempting commercial microalgal biodiesel promotion.

Present Perspective for Promoting Microalgal Fuel

Getting microalgae to produce lipids is like a dream come true by using underutilized resources. Put microalgae in waste water unfit for other uses, expose them to the sun in areas unsuitable for growing crops – unlike terrestrial plants, they do not require rainfall or good soil – and finally feed them exhaust gas that threatens the world climate and in return get algal biomass with high oil content. Warm climate and seashore areas are unsuitable for traditional agriculture, but highly suitable for microalgae growth. They naturally store energy as oil and also use the lipids to regulate their buoyancy. Growth rates are far higher than terrestrial plants and oil production is as much as 50 times that per hectare of oilseed crops.

From 1978 to 1996, the US Department of Energy Office of Fuel Development funded a programme to develop renewable transportation fuels from algae. The main focus of the programme, known as the Aquatic Species Program (ASP), was the production of biodiesel from high lipid content algae grown in ponds utilizing waste CO_2 from coal-fired plants. During this programme, the high cost of algae production remained an obstacle. An Outdoor Test Facility was established in Roswell, New Mexico to carry out an engineering design assessment (Weissman and Tillett, 1989) and two 0.1 ha ponds were operated to understand microalgal growth rates in mass culture, CO_2 utilization and pond design parameters (Weissman and Tillett, 1992), along with study of lipid extraction and conversion to biodiesel (Nagle et al., 1988; Nagle and Lemke, 1990). Even DNA from microalgae with potential for fuel production has been analysed (Jarvis et al., 1992) and genetic transformation protocols for introducing foreign genes into microalgae (Dunahay, 1993). An important enzyme in lipid metabolism, acetyl-CoA carboxylase, was purified (Roessler, 1990) and subsequently this key gene was cloned and sequenced (Roessler and Ohlrogge, 1993).

Military jet fuels are also very costly, with the difficulty of transportation adding to the cost. But hydroprocessing of microalgal oil to jet fuel using strategically located refining capacity around the world may ease the problem.

The National Renewable Energy Laboratory (NREL), having extensive experience in cultivating and manipulating microalgae, is taking initiatives using hydroprocessing to catalytically remove impurities or reduce molecular weight for biodiesel production from microscopic algae. They are planning for production of a kerosene-like fuel very similar to petroleum-derived commercial and military jet fuels or into a fuel designed for multipurpose military use. By manipulating nutrients and other growth conditions and by selecting and genetically engineering strains to increase oil production, NREL researchers were able to attain quite high lipid production levels.

In partnership with oil refiners, NREL is now looking to reestablish its microalgal oil research, with a particular view towards jet fuel production. The programme was discontinued in 1996, when diesel was less costly than the current cost of both diesel and jet fuel. Microalgal screening and genetic engineering technologies have also made significant advancement during this period. The Aquatic Species Project initially involved a large effort to collect algal species from natural habitats, followed by an extensive screening process to determine lipid production and mass culture capability (Barclay et al., 1987). This led to the establishment of an extensive microalgal culture collection with a large gene pool for additional study and manipulation.

Recently, Xu et al. (2006) and Miao and Wu (2006) have reported production of high quality diesel from a microalga *Chlorella protothecoides* by heterotrophic growth in fermenters. The technique of metabolic controlling used by Xu et al. (2006) resulted in rise of lipid content in microalgae to 55%. They used corn powder hydrolysate (instead of conventional glucose) as the organic carbon source in the culture medium in fermenters. This procedure not only reduced the production cost of algae but also increased the net biomass production. Miao and Wu (2006) reported that best processing conditions for transesterification of algal oil with methanol was 100% catalyst quantity (based on oil weight) with 56:1 molar ratio of methanol to oil at 30°C. A recent paper from Michael Briggs at the UNH Biodiesel Group offers estimates for the realistic replacement of all vehicular fuel with biodiesel by utilizing algae that has a greater than 55% natural oil content, which he suggests can be grown on algae ponds at wastewater treatment plants. In 2006, Aquaflow Bionomic Corporation from Marlborough, New Zealand announced that it has produced its first sample of biodiesel fuel made from algae found in sewage ponds. Unlike previous attempts, the algae were naturally grown in pond discharge from the Marlborough District Council's sewage treatment works.

Oil content in microalgae can reach 75% by weight of dry biomass but may be associated with low productivities (e.g. *Botryococcus braunii*). Most common algae (*Chlorella*, *Crypthecodinium*, *Cylindrotheca*, *Dunaliella*, *Isochrysis*, *Nannochloris*, *Nannochloropsis*, *Neochloris*, *Nitzschia*, *Phaeodactylum*, *Porphyridium*, *Schizochytrium*, *Tetraselmis*) have oil levels between 20 and 50% but higher productivities can be reached.

The selection of the most adequate species needs to take into account many factors, such as the ability of microalgae to grow under natural environmental condition or under specific environmental conditions; they should have a desirable fatty acid composition as this can have a significant effect on the characteristics of biodiesel produced. For example, Thomas *et al.* (1984) analysed the fatty acid compositions of seven freshwater microalgae species and reported that the relative intensity of other individual fatty acid chains is species specific, e.g. C16:4 and C18:4 in *Ankistrodesmus* sp., C18:4 and C22:6 in *Isochrysis* sp., C16:2, C16:3 and C20:5 in *Nannochloris* sp., C16:2, C16:3 and C20:5 in *Nitzschia* sp.

In India, the production of algae to harvest oil for biodiesel has not been undertaken on a commercial scale, but attempts have been made to conduct feasibility studies to arrive at the yield estimate. Ravishankar and his group from the Central Food Technological Research Institute, Mysore (Ambati *et al.*, 2010) have done excellent work on the isolation and characterization of *B. braunii* (Tripathi *et al.*, 2001; Dayananda *et al.*, 2005, 2006). Recently, Rengasamy and his team from the University of Madras (Ashokkumar and Rengasamy, 2012) have successfully accomplished cultivation of *B. braunii* in open raceway ponds (Sivasubramanian, 2009). Kaur *et al.* (2009) from the Defence Research Laboratory have done extensive work on algal diversity exploration for biodiesel production.

The crude oil consumption of India is about 113 million metric tonnes per annum (MMTPA), out of which only one-third is produced in the country, putting a heavy load on the Indian economy. Indian Defence forces are major consumers of petroleum products for their vehicles, ships and aircraft. Therefore, investment in R&D on biofuel production technology from microalgae by the Defence Research and Development Organization (DRDO) makes strong economic sense.

Culture of Microlagae: Mass-Cultivation Systems

Most microalgae are strictly photosynthetic and are usually photo-autotrophs. However, many algae are capable of growing in darkness and of using organic carbons (such as glucose or acetate) as energy and carbon sources and are termed heterotrophic. Heterotrophic microalgal culture is not justifiable for biodiesel production due to high capital and operational costs. And hence, to be realistic, algal-biofuel production must rely on photo-autotrophic algal growth. The requirements of photo-autotrophic microalgae include a light source, CO_2, water (water temperature 15–30°C), and inorganic salts as a source of nitrogen, phosphorus, iron, and sometimes silicon (Grobbelaar, 2004). Nutrients are provided during daytime for algal reproduction and in order to prevent settling, and algal cells are continuously mixed (Molina Grima et al., 1999). Up to one-quarter of algal biomass produced can be lost through respiration during the night (Chisti, 2007).

At present open ponds are used as a series of 'raceways' (Terry and Raymond, 1985) and enclosed sophisticated photobioreactors (Molina Grima et al., 1999; Sánchez Mirón et al., 1999; Tredici, 1999) for algal biofuel production.

Open ponds

Open ponds are designed in a raceway configuration and are shallow ponds usually about 30 cm deep and are the simplest and cheapest system for mass cultivation of microalgae. These are normally made of concrete or even sometimes dug into the soil and lined with a plastic to prevent liquid soaking into the soil. A paddlewheel is used to circulate and mix the algal cells and nutrients. These are operated in a continuous mode, where nutrients are added in front of the paddlewheel, allowed to circulate in the pond and algal biomass is harvested behind the paddlewheel. The paddlewheel operates all the time to prevent sedimentation.

Raceways are less expensive than photo-bioreactors, because they cost less to build and operate. Although raceways are low-cost, this culture system has its intrinsic disadvantages. They often experience a lot of water loss due to evaporation, open ponds do not allow microalgae to use CO_2 as efficiently, optimal culture conditions are difficult to maintain in open ponds, and recovering the biomass from such a dilute culture is expensive (Molina Grima et al., 1999). All these factors lead to low biomass productivity compared with photo-bioreactors (Chisti, 2007). Productivity is also affected by contamination with unwanted algae and other microbes. Feasibility of raceway ponds for production of microalgal biomass for making biodiesel has been extensively evaluated by the United States Department of Energy (Sheehan et al., 1998).

Closed photo-bioreactors

Photo-bioreactors are an excellent system, permitting culture of single microalgal species for prolonged durations and have been successfully evaluated (Molina Grima et al., 1999; Tredici, 1999; Pulz, 2001; Carvalho et al., 2006). These systems are made of transparent materials and are illuminated by natural light if placed outside or by artificial illumination if kept inside. These systems overcome the contamination and evaporation problems associated with open ponds (Molina Grima et al., 1999).

Moreover, the biomass productivity of photo-bioreactors is about 13 times more than that of a raceway pond and harvest of biomass from photobioreactors is less expensive as microalgal biomass from a photo-bioreactor is more concentrated than from raceway ponds (Chisti, 2007).

The commonly used photo-bioreactor is a tubular, having an array of clear transparent tubes up to 10 cm in diameter; the media broth is circulated through a pump to the tubes from the reservoirs and a portion of the algae is harvested after it passes through the solar collection tubes. Another type is a helical-tubular photo-bioreactor, where a mechanical pump or an airlift pump maintains a highly turbulent flow within the reactor, which prevents the algal biomass from settling down (Chisti, 2007). However, this system is also not free from disadvantages. The photosynthesis process generates oxygen in this closed system and the culture must be returned to a degassing zone from time to time. Temperature must be regulated in the photo-bioreactor. Also, the use of CO_2 for enhanced biomass production results in an increase in pH, which must be taken care of. The reactors are difficult to scale up, light limitation is another important issue and attachment of cells to the tube walls adds to the problem.

Harvesting of Microalgae

Microalgae harvesting, i.e. the concentration of diluted algae suspension until a thick algae paste is obtained, is one of the major challenges in algae biodiesel initiatives. The drawbacks are that microalgae mass cultures are dilute (500–1000 mg l^{-1} on a dry weight basis) and the cells are very small. The removal of water from the media to obtain microalgal paste (≈10–15% solid) is the main issue.

The commonly used methods for removal of liquid from microalgal culture to produce concentrated microalgal paste are centrifugation and chemical flocculation.

In centrifugation, the algal culture is pumped into a high-speed large centrifuge with the outer wall fitted with a filter for collecting microalgal cells. The water is forced out and algae remain as paste on the filter. This is a proven technology; however, the high cost of operation is a limiting factor. Use of settling tanks/ponds is highly recommended before centrifugation as this will save much of the energy used for centrifugation.

In chemical flocculation, chemicals causing charge neutralization are used for clumping algal cells together. The chemicals used are lime, alum, or chitosan. This is also expensive, because of the large amounts of chemicals required. Froth flotation procedure is also used for harvesting algae from dilute suspensions (Gilbert *et al*., 1962). Here, harvesting is carried out in a long column containing the feed solution, which is aerated from below. The cell concentration of the harvest is dependent on pH, aeration rate, aerator porosity, feed concentration and height of foam in the harvesting column. This seems to be an economically favourable process for mass harvesting of microalgae. Interrupting the CO_2 supply to an algal system can cause algae to flocculate on its own, which is called autoflocculation. Ultrasound-based methods of algae harvesting are currently under development, and other, additional methods are currently being developed. One of the simple and cost-effective methods now used for harvesting microalgae is microstraining. However, clogging of pores due to algal settling is the only problem associated with it. The use of a sand filtration technique is now also gaining interest, because of its easy and very low-cost technology.

Oil Extraction from Microalgae

In order to be converted into a liquid fuel the oil contained in the algae must be extracted. The simplest method of extracting oil from algae is mechanical crushing. As algal strains vary in their physical characteristics, various press types are used for specific types of algae. Algae is dried and then pressed out to yield oil. Algal oil can be extracted using chemicals, such as benzene, hexane, ether, dichloromethane etc., and through repeated washing, or percolation, under reflux in special glassware such as a Soxhlet apparatus. Often, mechanical crushing in conjunction with chemical extraction gives excellent results. Enzymatic extraction uses enzymes to degrade the cell walls, with water acting as the solvent, making fractionation of the oil much easier. However, the costs of this extraction process are estimated to be much greater than hexane extraction. Carbon dioxide is also used for algal oil extraction; CO_2 is liquefied under pressure and heated to the point where it will have properties of both liquid and gas, and this fluid is then used as the solvent in extracting the oil. Osmotic shock is also used to release microalgal oil. In ultrasonic extraction, ultrasonic waves are used to create cavitation bubbles in a solvent material; when these bubbles collapse near the cell walls, it creates shock waves and liquid jets, causing the algal cell walls to break and release their contents into the solvent.

The extracted oil must be transesterified in order to become biodiesel. It is a simple four-step chemical reaction requiring two chemicals. Here, methanol or ethanol and sodium hydroxide are mixed to prepare sodium methoxide, which is then mixed with algal oil. The mixture is allowed to settle for 8 h and biodiesel is obtained on draining glycerine. In this process, 86% methyl esters or biodiesel and 9% glycerine is obtained (Tickell, 2003).

Economics of Microalgal Biofuel

The Aquatic Species Program (ASP) of the US Department of Energy (DOE) faced three main limiting factors for commercial algal production: (i) difficulty of maintaining desirable species in the culture system as many contaminating species polluted them; (ii) the low yield of algal oil; and (iii) the high cost of harvesting the algal biomass. Due to recent history of fuel price increases, algal biofuel production has gained renewed interest with a final goal of commercial algal biofuel production. There are two routes to achieving this goal: to genetically and metabolically alter algal species and to develop new growth technologies so as to obtain sufficient biomass with high oil content. However, to date, no significant breakthrough has been achieved in either method.

The production cost of algal oil depends on biomass yield, microalgal oil content, scale of production systems and the cost of recovering oil from algal biomass etc. Algal-oil production is still far more expensive than petroleum diesel fuels.

Neenan et al. (1986) reported a cost of algal oil as US$0.67 per pound in 2009 dollar equivalents. Assuming the oil content of the algae to be approximately 30% and algal oil has roughly 80% of the energy content of crude petroleum, algae oil production from a photo-bioreactor with an annual production capacity of 10,000 t year^{-1} will be US$2.80 l^{-1} (US$10.50 per gallon) as estimated by Chisti (2007). This estimation excluded costs of converting algal oil to biodiesel, distribution and marketing costs for biodiesel, and taxes. Chisti (2007) considered algal oil economics as correlated to petroleum oil price and used the following equation to estimate the cost of algal oil:

$$C_{\text{algal oil}} = 6.9 \times 10^{-3} \, C_{\text{petroleum}}$$

where: $C_{\text{algal oil}}$ is the price of microalgal oil in dollars per gallon and $C_{\text{petroleum}}$ is the price of crude oil in dollars per barrel. The recovery process contributes 50% to the cost of the final recovered oil.

Huntley and Redalje (2007) proposed a hybrid system combining raceways and photo-bioreactors, in which the inoculums are generated in the photo-bioreactor. Once initial culture reached a certain concentration, it is then passed to the raceways for further growth. They estimated a cost of US$84/barrel if raceways cost US$74,782 and photo-bioreactor costs were US$197,000 ha^{-1}. This was estimated on the basis that 80% of the culture facility was raceways and 20% photo-bioreactors. Hassannia (2009) reported a cost of production of algal oil as about US$0.82 per pound. Richardson *et al*. (2010), using a Monte Carlo simulation model for a commercial-scale microalgae farm in the US desert Southwest, reported a total cost of algal oil in the range from US$0.85 to US$3.67/pound, with an average of US$1.61 (with by-product credits) for the conventional wisdom input/output coefficients.

However, the cost of microalgal biodiesel can be reduced by using a biorefinery concept and by genetically and metabolically altering algal species to develop stable engineered strains along with suitable photo-bioreactor engineering. Moreover, biochemical and environmental factors triggering oil accumulation might be identified.

Conclusion

Microalgal biofuel can become a reality only if its production cost can be minimized while enhancing algal biology (in terms of biomass yield and oil content) along with an efficient harvesting technology. Currently, algal-biofuel production is still too expensive to be commercialized. In addition, producing value-added products besides algal fuel is an attractive way to lower the cost of algal-biofuel production. Other than oil, microalgae contain large quantities of proteins, carbohydrates and other nutrients (Spolaore *et al*., 2006), making the residue suitable for use as animal feed or in other value-added products.

Photo-bioreactors can be considered as an initial inoculum generator, as these provide a controlled environment for high initial cell density of desired unialgal cultures. The follow up large-scale culture in raceways will reduce the cost, making the production of microalgal biodiesel affordable in near future. There is a need for a large-scale supply of purified microalgae and suitable technology should be developed for *in situ* collection of algal biomass and extraction of fats and lipids. Moreover, the glycerine by-product can be successfully used in the pharmaceutical and cosmetic industry, adding to the cost effectiveness of microalgal biofuel. Other than biodiesel, microalgal systems should be extensively evaluated for the production of substances of industrial and pharmaceutical importance.

References

Ambati, R.R., Ravi, S. and Aswathanarayana, R.G. (2010) Enhancement of carotenoids in green alga *Botryococcus braunii* in various autotrophic media under stress conditions. *International Journal of Biomedical and Pharmaceutical Sciences* 4, 87–92.

Ashokkumar, V. and Rengasamy, R. (2012) Mass culture of *Botryococcus braunii* Kutz. under open raceway pond for biofuel production. *Bioresource Technology* 104, 394–399.

Banerjee, A., Sharma, R., Chisti, Y. and Banerjee, U.C. (2002) *Botryococcus braunii*: a renewable source of hydrocarbons and other chemicals. *Critical Reviews in Biotechnology* 22, 245–279.

Barclay, W.R., Terry, K.L., Nagle, N.J., Weissman, J.C. and Goebel, R.P. (1987) Potential of new strains of marine and inland saline-adapted microalgae for aquaculture. *Journal of World Aquaculture Society* 18, 218–228.

Barnwal, B.K. and Sharma, M.P. (2005) Prospects of biodiesel production from vegetables oils in India. *Renewable and Sustainable Energy Reviews* 9, 363–378.

Carvalho, A.P., Meireles, L.A. and Malcata, F.X. (2006) Microalgal reactors: a review of enclosed system designs and performances. *Biotechnology Progress* 22, 1490–1506.

Chisti, Y. (1980–1981) An unusual hydrocarbon. *Journal of Ramsay Society* 27–28, 24–6.

Chisti, Y. (2007) Biodiesel from microalgae. *Biotechnology Advances* 25, 294–306.

Dayananda, C., Sarada, R., Bhattacharyya, S. and Ravisankar, G.A. (2005) Effect of media and culture conditions on growth and hydrocarbon production by *Botryococcus braunii*. *Process Biochemistry* 40, 3125–3131.

Dayananda, C., Sarada, R., Srinivas, P., Shamala, T.R. and Ravisankar, G.A. (2006) Presence of methyl branched fatty acids and saturated hydrocarbons in botryococcene producing strain of *Botryococcus braunii*. *Acta Physiologiae Plantarum* 28, 251–256.

Dunahay, T.G. (1993) Transformation of *Chlamydomonas reinhardtii* with silicon carbide whiskers. *BioTechniques* 15, 452–460.

Fedorov, A.S., Kosourov, S., Ghirardi, M.L. and Seibert, M. (2005) Continuous H2 photoproduction by *Chlamydomonas reinhardtii* using a novel two-stage, sulfate-limited chemostat system. *Applied Biochemistry Biotechnology* 121–124, 403–412.

Fukuda, H., Kondo, A. and Noda, H. (2001) Biodiesel fuel production by transesterification of oils. *Journal of Bioscience and Bioengineering* 92, 405–416.

Gavrilescu, M. and Chisti, Y. (2005) Biotechnology – a sustainable alternative for chemical industry. *Biotechnol Advances* 23, 471–499.

Gilbert, V.L., John, R.C., Ahron, G. and Frederick, D.B. (1962) Harvesting of Algae by Froth Flotation. *Applied Microbiology* 10, 169–175.

Grobbelaar, J.U. (2004) Algal nutrition. In: Richmond, A. (ed.) *Handbook of Microalgal Culture: biotechnology and applied phycology*. Blackwell, Oxford, UK, pp. 97–115.

Hassannia, J. (2009) Algae biofuels economic viability: A project-based perspective. *Diversified Energy Newsletter* 20, 1-4.

Hill, J., Nelson, E., Tilman, D., Polasky, S. and Tiffany, D. (2006) Environmental, economic, and energetic costs and benefits of biodiesel and ethanol biofuels. *Proceedings of the National Academy of Sciences* 103, 11206–11210.

Huntley, M.E. and Redalje, D.G. (2007) CO_2 mitigation and renewable oil from photosynthetic microbes: A new appraisal. *Mitigation and Adaptation Strategies for Global Change* 12, 573–608.

Jarvis, E.E., Dunahay, T.G. and Brown, L.M. (1992) DNA nucleoside composition and methylation in four classes of microalgae. *Journal of Phycology* 28, 356–362.

Kapdan, I.K. and Kargi, F. (2006) Bio-hydrogen production from waste materials. *Enzyme and Microbial Technology* 38, 569–582.

Kaur, S., Gogoi, H.K., Srivastava, R.B. and Kalita, M.C. (2009) Algal diversity as a renewable feedstock for biodiesel. *Current Science* 96, 182.

Knothe, G., Dunn, R.O. and Bagby, M.O. (1997) Biodiesel: the use of vegetable oils and their derivatives as alternative diesel fuels. *ACS Symposium Series* 666, 172–208.

Kulkarni, M.G. and Dalai, A.K. (2006) Waste cooking oil – an economical source for biodiesel: A review. *Industrial and Engineering Chemistry Research* 45, 2901–2913.

Lorenz, R.T. and Cysewski, G.R. (2003) Commercial potential for *Haematococcus* microalga as a natural source of astaxanthin. *Trends In Biotechnology* 18, 160–167.

Mata, T.M., Martins, A.A. and Caetano, N.S. (2010) Microalgae for biodiesel production and other applications: A review. *Renewable and Sustainable Energy Review* 14, 217–232.

Melis, A. (2002) Green alga hydrogen production: progress, challenges and prospects. *International Journal of Hydrogen Energy* 27, 1217–1228.

Metzger, P. and Largeau, C. (2005) *Botryococcus braunii*: a rich source for hydrocarbons and related ether lipids. *Applied Microbiology Biotechnology* 66, 486–496.

Miao, X. and Wu, Q. (2006) Biodiesel production from heterotrophic microalgal oil. *Bioresource Technology* 97, 841–846.

Molina Grima, E., Acién Fernández, F.G., García Camacho, F. and Chisti, Y. (1999) Photobioreactors: light regime, mass transfer, and scale up. *Journal of Biotechnology* 70, 231–247.

Nagle, N., Chelf, P., Lemke, P. and Barclay, W. (1988) Conversion of lipids through biological pretreatment. *Proceedings of Energy from Biomass and Waste XII*, 15–19 Feb., Institute of Gas Technology, Chicago, Illinois.

Nagle, N. and Lemke, P. (1990) Production of methyl ester fuel from microalgae. *Applied Biochemistry Biotechnology* 24/25, 355–361.

Neenan, B., Feinberg, D., Hill, A., McIntosh, R. and Terry, K. (1986) *Fuels from microalgae: Technology status, potential, and research requirements (SERI/SP-231-2550 DE86010739)*. Solar Energy Research Institute, US DOE, Golden, Colorado.

Veziroglu, T.N. and Sahin, S. (2008) 21st Century's energy: Hydrogen energy system. *Energy Conversion and Management* 49, 1820–1831.

Oil and Gas Journal (2005) Worldwide look at reserves and production. *Oil and Gas Journal* 103, 24–25.

Pulz, O. (2001) Photobioreactors: production systems for phototrophic microorganisms. *Applied Microbiology Biotechnology* 57, 287–293.

Richardson, J.W., Outlaw, J.L. and Allison, M. (2010) The Economics of Microalgae Oil. *AgBioForum* 13, 119–130.

Roessler, P.G. (1990) Purification and Characterization of Acetyl-CoA Carboxylase from the Diatom *Cyclotella cryptica*. *Plant Physiology* 92, 73–78.

Roessler, P.G. and Ohlrogge, J.B. (1993) Cloning and characterization of the gene that encodes acetyl-coenzyme A carboxylase in the alga *Cyclotella cryptica*. *Journal of Biological Chemistry* 268, 19254–19259.

Roessler, P.G., Brown, L.M., Dunahay, T.G., Heacox, D.A., Jarvis, E.E., Schneider, J.C. et al. (1994) Genetic-engineering approaches for enhanced production of biodiesel fuel from microalgae. *ACS Symposium Series* 566, 255–270.

Sánchez Mirón, A., Contreras Gómez, A., García Camacho, F., Molina Grima, E. and Chisti, Y. (1999) Comparative evaluation of compact photobioreactors for large-scale monoculture of microalgae. *Journal of Biotechnology* 70, 249–270.

Sawayama, S., Inoue, S., Dote, Y. and Yokoyama, S.Y. (1995) CO_2 fixation and oil production through microalga. *Energy Conversion Management* 36, 729–731.

Sheehan, J., Dunahay, T., Benemann, J. and Roessler, P. (1998) A look back at the US Department of Energy's Aquatic Species Program – biodiesel from algae. *Report NREL/TP-580–24190*. National Renewable Energy Laboratory, Golden, Colorado.

Singh, S., Kate, B.N. and Banerjee, U.C. (2005) Bioactive compounds from cyanobacteria and microalgae: an overview. *Critical Reviews in Biotechnology* 25, 73–95.

Sivasubramanian, V. (2009) Current status of Research on algal bio-fuels in India. *Journal of Algal Biomass Utilization* 1, 1–8.

Spolaore, P., Joannis-Cassan, C., Duran, E. and Isambert, A. (2006) Commercial applications of microalgae. *Journal of Bioscience and Bioengineering* 101, 87–96.

Terry, K.L. and Raymond, L.P. (1985) System design for the autotrophic production of microalgae. *Enzyme and Microbial Technology* 7, 474–487.

Thomas, W.H., Tornabene, T.G. and Weissman, J. (1984) Screening for lipid yielding microalgae: activities for 1983. *SERI/STR* 231-2207.

Tickell, J. (2003) *From the Fryer to the Fuel Tank the Complete Guide to Using Vegetable Oil as an Alternative Fuel*. Tickell Energy Consultants, Covington, Louisiana.

Tredici, M.R. (1999) Bioreactors. In: Flickinger, M.C. and Drew, S.W. (eds) *Encyclopedia Of Bioprocess Technology: Fermentation, Biocatalysis and Bioseparation*. John Wiley & Sons, New York, pp. 395–419.

Tripathi, U., Sarada, R. and Ravisankar, G.A. (2001) A culture method for micro algal forms using two-tier vessel providing carbon-dioxide environment: studies on growth and carotenoids production. *World Journal of Microbiology and Biotechnology* 17, 325–329.

Walter, T.L., Purton, S., Becker., D.K. and Collet, C. (2005) Microalgae as bioreactor. *Plant Cell Reports* 24, 629–641.

Weissman, J.C. and Tillett, D.M. (1989) Design and operation of an outdoor microalgae test facility. In: Bollmeier, W.S. and Sprague, S., *Aquatic Species Program Annual Report, September 1989*. SERI/SP 2321-3579. Solar Energy Research Institute, Golden, Colorado.

Weissman, J.C. and Tillett, D.M. (1992) Design and operation of an outdoor microalgae test facility: Large-scale system results. In: Brown, L.M. and Sprague, S. *Aquatic Species Project Report, Jan. 1992*. NREL MP-232-4174. National Renewable Energy Laboratory, Golden, Colorado, pp. 32–56.

Xu, H., Miao, X. and Wu, Q. (2006) High quality biodiesel production from a microalga *Chlorella protothecoides* by heterotrophic growth in fermenters. *Journal of Biotechnology* 126, 499–507.

Chapter 14

Biodegradation of Petroleum Hydrocarbons in Contaminated Soil

Aniefiok E. Ite and Kirk T. Semple

Introduction

Petroleum hydrocarbon contamination of soil is a widespread and well recognized global environmental problem associated with oil and gas industries. Release of petroleum hydrocarbons into the environment whether accidentally or due to human activities is a major cause of controlled water and soil pollution (Benka-Coker and Olumagin, 1996; Holliger *et al.*, 1997; Kharaka *et al.*, 2007). Petroleum hydrocarbons can be divided into four classes: saturates (pentane, hexadecane, octacosane, cyclohexane), aromatics (naphthalene, phenanthrene, benzene, pyrene), asphaltenes (phenols, fatty acids, ketones, esters and porphyrins) and resins (pyridines, quinolines, carbazoles, sulfoxides and amides) (Fig. 14.1; Colwell, 1977).

Soil and sediments have become the ultimate sink for most petroleum contaminants, such as benzene, toluene, ethylbenzene and xylenes (BTEX), aliphatic and polycyclic aromatic hydrocarbons (PAHs). PAHs containing from two to five fused aromatic rings are of significant concern because of the mutagenicity and carcinogenicity of several of these compounds and tendency to bioaccumulate in organic tissues due to their lipophilic character and electrochemical stability. PAHs are a widespread class of environmental chemical contaminants which make up about 5% by volume (Block *et al.*, 1991), while aliphatic hydrocarbons are significant contaminants in some parts of the world. For instance, predominant oil pollution (petrol and diesel) in the UK contains high volumes of aliphatic hydrocarbons (Stroud *et al.*, 2007), while petroleum pollution in the tropical region like the Nigeria's Niger Delta contains complex mixtures of both the aliphatic and aromatic hydrocarbons (Olajire *et al.*, 2005; Osuji and Ozioma, 2007).

Hydrocarbon contamination of soil is well studied in terms of its toxic effects and there is evidence of their potential toxic and negative impacts on the environment (Holdway, 2002). However, indigenous soil microbes possess the inherent ability to transform organic contaminants into less toxic or non-toxic end products, thereby mitigating or eliminating contamination from the environment.

Biodegradation is the general term used to describe the biological conversion of organic contaminants to products that are generally lower in free energy (Surampalli and Ong, 2004). In fact, this term is usually used loosely and interpreted in various ways. Biodegradation involves either partial or complete mineralization of environmental

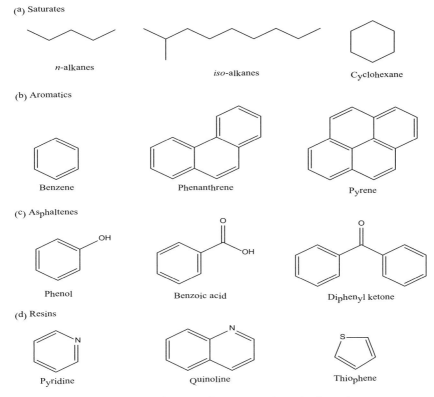

Fig. 14.1. Structural classification of some petroleum hydrocarbon components.

contaminants by complex, genetically regulated physiological reactions catalysed largely by microorganisms (Madsen, 1991, 1998b; Young and Cerniglia, 1995; Alexander, 1999) and plants (Burken and Schnoor, 1998; Bhadra et al., 1999; Bizily et al., 1999; Siciliano et al., 2001). Microbial degradation is often a growth-linked process that brings about mineralization and represents the primary mechanism by which petroleum hydrocarbon contaminants can be removed from the environment (Leahy and Colwell, 1990). Biodegradation of petroleum hydrocarbons in the environment has been well studied for decades and several factors that influence the rate of oil biodegradation have been reviewed (Leahy and Colwell, 1990; Banat, 1995; Okoh, 2006; Brassington et al., 2007; Das and Chandran, 2011). The success of petroleum hydrocarbon biodegradation depends on the ability to establish and maintain conditions that favour enhanced oil biodegradation rates in the contaminated environment. The transformation of organic contaminants in petroleum-contaminated soils to less toxic end products or decontamination often relies on microbial degradation, both naturally and technologically.

In this chapter, we will provide a clear insight into microbial degradation of petroleum hydrocarbons in contaminated soil that have emerged from the growing body of bioremediation research and its applications in practice. Understanding the biodegradability of several classes of petroleum contaminants is of critical importance for developing bioremediation strategies for risk mitigation.

Microbial Degradation of Petroleum Hydrocarbons

Microbial degradation studies have been undertaken to assess the fate of *n*-alkanes (Haines and Alexander, 1974; Sepic *et al.*, 1995), cycloalkanes (Singer and Finnerty, 1984) and aromatics (Bossert and Bartha, 1986; Sepic *et al.*, 1995; Juhasz *et al.*, 1996a,b; Wammer and Peters, 2005; Couling *et al.*, 2010). Hydrocarbons differ in their susceptibility to microbial attack and the susceptibility of hydrocarbons to microbial degradation can be generally ranked as follows: *n*-alkanes > branched-chain alkanes > branched alkenes > low-molecular-weight *n*-alkyl aromatics > monoaromatics > cyclic alkanes > polynuclear aromatics > asphaltenes (Huesemann, 1995).

A number of studies have dealt with biotransformation, biodegradation and bioremediation of petroleum hydrocarbons by microorganisms. For example, Bouwer and Zehnder (1993), Chen *et al.* (1999) and Mishra *et al.* (2001b) have investigated the biotransformation of organic contaminants in the environment to understand microbial ecology, physiology and evolution for potential biodegradation strategies. Biodegradation of petroleum hydrocarbons in the environment is complex, which is often dependent on the behaviour of the organic contaminant in a particular environment, concentration, chemical structure, contaminant bioavailability and bioaccessbility. Biodegradability of petroleum hydrocarbons in both the laboratory and underfield conditions are widely reported (Chaîneau *et al.*, 1995, 2003a; Huesemann, 1995; Salanitro *et al.*, 1997; Gogoi *et al.*, 2003; Huesemann *et al.*, 2003, 2004; Chaillan *et al.*, 2004, 2006). For example, different contaminated environments with varying level of contamination often support different groups of hydrocarbon-degrading microorganisms (Mueller *et al.*, 1989; Grosser *et al.*, 1991; Ramsay *et al.*, 2000; Tam *et al.*, 2001, 2002). The biodegradation potential of strains isolated from hydrocarbon-contaminated environments is often more active than those isolated from non-contaminated systems, as degrading microorganisms have acclimatized to the contaminated environment (Wild and Jones, 1993; Chaîneau *et al.*, 1999).

Petroleum hydrocarbons in the natural environment are biodegraded primarily by a diverse group of bacteria, yeast and fungi (Table 14.1).

The reported efficiency of biodegradation ranged from 6% (Jones *et al.*, 1970) to 82% (Pinholt *et al.*, 1979) for soil fungi, 0.13% (Jones *et al.*, 1970) to 50% (Pinholt *et al.*, 1979) for soil bacteria, and 0.003% (Hollaway *et al.*, 1980) to 100% (Mulkins-Phillips and Stewart, 1974) for marine bacteria. Although many microbes are able to metabolize organic contaminants, mixed populations with overall broad enzymatic capabilities are required to degrade complex mixtures of hydrocarbons, such as crude oil, in soil (Bartha and Bossert, 1984; Mishra *et al.*, 2001a), fresh water (Cooney, 1984) and marine environments (Floodgate, 1984; Atlas, 1985). Biodegradation of hydrocarbons can occur under oxic and anoxic conditions (Zengler *et al.*, 1999; Zhou *et al.*, 2009). Biodegradation of petroleum hydrocarbons under anaerobic and aerobic condition have been discussed (Heider *et al.*, 1998; Spormann and Widdel, 2000; Widdel and Rabus, 2001; Grossi *et al.*, 2008) and its pathways have been reviewed (Atlas, 1981; Bartha, 1986; Grishchenkov *et al.*, 2000; Van Hamme *et al.*, 2003).

Biodegradation Pathways for Aliphatic and Aromatic Hydrocarbons

The *n*-alkanes are the main constituents of petroleum hydrocarbon contaminations. Long-chain *n*-alkanes (C_{10}–C_{24}) are degraded most rapidly by the chemical pathways

Table 14.1. Predominant hydrocarbon-degrading microbes in the petroleum-contaminated soil environment (Watkinson and Morgan, 1990; Cerniglia, 1992; Šašek *et al.*, 1993; Fritsche and Hofrichter, 2005, 2008).

Bacteria			
Gram-negative bacteria	Gram-positive bacteria	Yeast	Fungi
Pseudomonas spp.	*Nocardia* spp.	*Aureobasidium*	*Trichoderma*
Acinetobacter spp.	*Mycobacterium* spp.	*Candida*	*Mortiecerella*
Alcaligenes sp.	*Corynebacterium* spp.	*Rhodotorula* spp.	*Penicillium*
Flavobacterium/	*Arthrobacter* spp.	*Sporobolomyces*	*Aspergillus*
Cytophaga group		*Exophiala*	*Fusarium* spp.
Xanthomonas spp.	*Bacillus* spp.	*Trichosporon* spp.	

shown in Fig. 14.2. Short-chain alkanes (less than C_9) are toxic to many microorganisms, but being volatile they are rapidly lost from petroleum-contaminated sites (Fritsche and Hofrichter, 2005, 2008). The principal mechanism of *n*-alkane degradation involves terminal oxidation to form alcohols, aldehydes, or fatty acids (Sepic *et al.*, 1995; Van Hamme *et al.*, 2003). The most common pathway used to catabolize fatty acids is known as β-oxidation, a pathway that cleaves off consecutive two-carbon fragments (Fig. 14.2). Each two-carbon fragment is removed by coenzyme A as acetyl-CoA. *n*-Alkanes having an odd number of carbon atoms are degraded to propionyl-CoA, which is in turn carboxylated to methylmalonyl-CoA and further converted to succinyl-CoA (Fritsche and Hofrichter, 2005, 2008). Fatty acids having a physiological chain length may be directly incorporated into membrane lipids, but most degradation products are fed into the tricarboxylic acid cycle. Subterminal oxidation occurs with lower (C_3–C_6) and longer alkanes, with formation of a secondary alcohol and subsequently of a ketone. A certain amount of unsaturated 1-alkenes are oxidized at the saturated end of the chains to form the corresponding 1,2-epoxyalkane and a minor pathway proceeds via an epoxide, which is converted to a fatty acid (Fritsche and Hofrichter, 2005, 2008). Although branching reduces the rate of biodegradation, methyl side groups do not noticeably decrease the biodegradability, whereas complex branched chains, e.g. the tertiary butyl group, hinder the action of the degradative enzymes.

Cycloalkanes represent minor components of mineral oil and are relatively resistant to microbial attack. The absence of an exposed terminal methyl group complicates the primary attack. A few species can use cyclohexane as a sole carbon source, but it is more commonly co-metabolized by mixed cultures (Fritsche and Hofrichter, 2005, 2008). The pathway of cyclohexane degradation is shown in Fig. 14.3.

Cycloalkanes are biodegraded by oxidative processes that lead to linearization, usually resulting in a dicarboxylic fatty acid, which is further degraded. In general, the presence of alkyl sidechains on cycloalkanes facilitates their degradation.

Since bacteria initiate PAH degradation by the action of intracellular dioxygenases, the PAHs must be taken up by the cells before degradation can take place. Bacteria most often oxidize PAHs to *cis*-dihydrodiols by incorporation of both atoms of molecular oxygen. The *cis*-dihydrodiols are further oxidized, first to the aromatic dihydroxy compounds (catechols) and then channelled through the *ortho*- or *meta* cleavage pathways (Cerniglia, 1984a; Smith, 1990). Figure 14.4 shows the pathways of oxygenolytic ring cleavage to either intermediates of the central metabolism. At the branchpoint, catechol is oxidized by

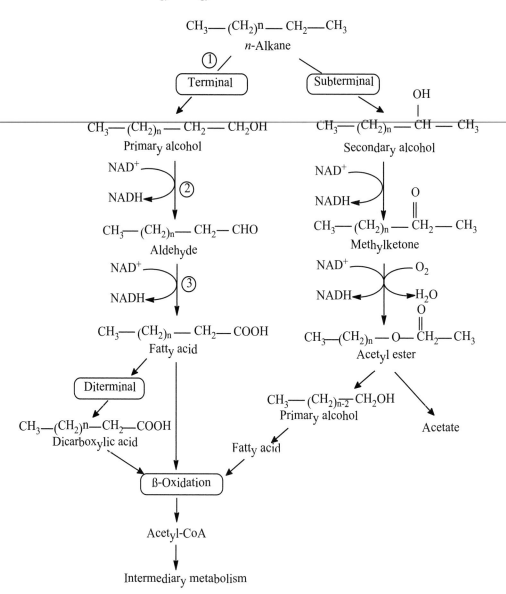

Fig. 14.2. Pathways of alkane degradation. The main pathway is their terminal oxidation to fatty acids catalysed by (1) *n*-alkane monooxygenase, (2) alcohol dehydrogenase and (3) aldehyde dehydrogenase (Fritsche and Hofrichter, 2005, 2008).

intradiol *ortho* cleavage or extradiol *meta* cleavage. Both ring-cleavage reactions are catalysed by specific dioxygenases. The product of the *ortho* cleavage – *cis,cis*-muconate – is transformed into an unstable enol-lactone, which is in turn hydrolysed to oxoadipate. This dicarboxylic acid is activated by transfer to CoA, followed by thiolytic cleavage to acetyl-CoA and succinate. Protocatechuate is metabolized by a homologous set of enzymes. The

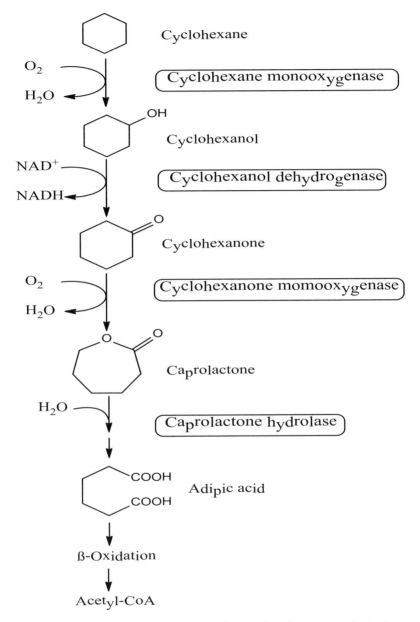

Fig. 14.3. Metabolic pathways for degradation of cycloaliphatic compounds (cycloparaffins) (Fritsche and Hofrichter, 2005, 2008).

additional carboxylic group is decarboxylated, and the double bond is simultaneously shifted to form oxoadipateenol-lactone. The oxygenolytic meta cleavage yields 2-hydroxymuconic semialdehyde, which is metabolized by hydrolytic enzymes to formate, acetaldehyde and pyruvate. These are then utilized in the central metabolism. In general, a wealth of aromatic substrates are degraded by a limited number of reactions: hydroxylation, oxygenolytic ring cleavage, isomerization, and hydrolysis (Fritsche and Hofrichter, 2005, 2008). Generally, the inducible nature of the enzymes and their substrate specificity enable bacteria having high degradation potential to adapt their metabolism to the effective utilization of substrate mixtures in polluted soils and to grow at a high rate.

To date, there have been several studies that have demonstrated the differences in the extents to which aliphatic and aromatic contaminants may be degraded. In a study by Huesemann *et al.* (2004), biodegradation of hydrocarbons was investigated in a soil containing 1.6% hydrocarbon contamination. Less than 20% of the initial concentration of hexadecane remained after 27 days of remediation. However, phenanthrene concentrations decreased to less than 5%, showing the higher residual fractions of an aliphatic hydrocarbon in the contaminated soil.

In a similar study, Chaîneau *et al.* (1995) investigated the biodegradation of drill cuttings, showing that the concentration of saturated hydrocarbons decreased from 1350 mg kg^{-1} TPH to about 400 mg kg^{-1} TPH. The aromatic fraction decreased from an initial 600 mg kg^{-1} TPH to about 200 mg kg^{-1} TPH. Thus, whilst the percentage of biodegradation was similar, double the concentration of saturated hydrocarbons remained in the soil after biodegradation.

A key study illustrating the degradability of hydrocarbons was carried out by Loser *et al.* (1999), in which a microporous, sandy soil was spiked with 0.3% hexadecane or phenanthrene, and biodegradation was measured to study a pilot scale percolator.

The degree of biodegradation of hexadecane was 80%, highly consistent with the degree of biodegradation of the aliphatic fraction found in remediation studies, equating to a residual fraction of 600 mg kg^{-1}. However, the degree of phenanthrene biodegradation was significantly higher (96%) with a residual concentration of only 100 mg kg^{-1}. A range of phenanthrene concentrations was tested, which showed different percentages of biodegradation; none the less, a residual concentration was always detected at about 100 mg kg^{-1} TPH. Thus, the aliphatic hydrocarbon had a significantly higher concentration remaining in the soil after biodegradation as compared with phenanthrene, under the same experimental conditions.

According to Stroud *et al.*(2007), these results indicate that aliphatic hydrocarbons may be constrained by factors affecting bioaccessibility to a greater extent than PAHs. However, biodegradation strategy can sometimes becomes technically complex because the variety of associated chemicals in the petroleum contamination is extremely wide (Li *et al.*, 2004).

Factors Influencing Petroleum Hydrocarbon Degradation

The rates and extents of biodegradation and microbial growth in soil are influenced by a variety of abiotic factors, including the complexity and concentration of the organic contaminant mixtures, contaminant bioavailability/bioaccessibility and contaminant interactions in soil, organic matter, temperature, pH, nutrient availability (particularly nitrogen and phosphorus), soil moisture content, availability of oxygen and redox potential (Dibble and Bartha, 1979; Cerniglia, 1984a, 1992; Volkering *et al.*, 1997; Del'Arco and de França, 2001; Ehlers and Luthy, 2003; Semple *et al.*, 2003, 2004; Chaillan *et al.*, 2006). This section summarizes the different factors that may affect the biodegradation of the petroleum hydrocarbons.

Overall the rate and extent of biodegradation depends not only on microbial activity and the environmental conditions that affect them, but also on the physicochemical constraints that control contaminant bioavailability (i.e. the fraction of contaminant

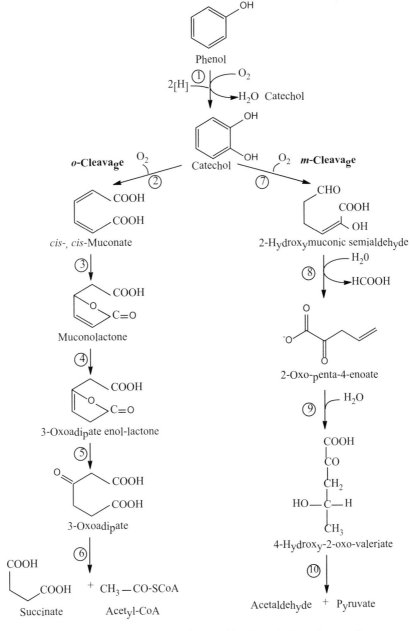

Fig. 14.4. The two alternative pathways for aerobic degradation of aromatic compounds: *ortho* and *meta* cleavage. (1) phenol monooxygenase, (2) catechol 1,2-dioxygenase, (3) muconate-lactonizing enzyme, (4) muconolactoneisomerase, (5) oxoadipateenol-lactone hydrolase, (6) oxoadipatesuccinyl-CoA transferase, (7) catechol 2,3-dioxygenase, (8)

hydroxymuconicsemialdehyde hydrolase, (9) 2-oxo-penta-4-enoic acid hydrolase, (10) 4-hydroxy-2-oxo-valeriate aldolase (Fritsche and Hofrichter, 2005, 2008).

molecules that are dissolved and available to the microorganisms) (Ramaswami and Luthy, 1997).Recent advances in the understanding and prediction of the bioremediation of organic contaminants in different soils explicitly include the concept of bioavailability and bioaccessbility. According to Semple *et al.*(2003), bioavailability is the fraction of a given contaminant of concern or analyte that is in a form which is biodegradable. Recently, emphasis has been on defining and attempting to characterize their bioavailability and bioaccessibility. Bioavailability is defined as the contaminant fraction 'which is freely available to cross an organism's (cellular) membrane from the medium the organism inhabits at a given point in time', whereas bioaccessibility encompasses what is actually bioavailable now plus what is 'potentially bioavailable' (Semple *et al.*, 2004, 2007). Extraction procedures that mimic or parallel bioavailability/bioaccessibility, often referred to as biomimetic techniques, have been sought in order to assess exposure and bioremediation potential (Semple *et al.*, 2007).The factors reducing the hydrocarbon bioavailability involve sorption to the soil organic matter, their presence in the non-aqueous phase liquids and their entrapment with the physical matrix of soil (Fig. 14.5) (Semple *et al.*, 2003). Several hypotheses have been offered to explain the sequestration and resultant decline in bioavailability of organic chemicals in soil. The amount of organic contaminant that is bioavailable depends on: (i) the rate of transfer of the compound from the soil to the living cell (mass transfer); and (ii) the rate of uptake and metabolism (the intrinsic activity of the cell). Mass transfer of hydrocarbons into microbial cells is a significant determinant of biodegradation rates and extents. Reduction in bioavailability of hydrocarbons can limit biodegradation, particularly in aged soils that have been contaminated for many years and during the final stages of a soil bioremediation treatment process (Alexander, 2000). Partitioning of compounds into soil organic matter has been suggested as a mechanism of their sequestration (Freeman and Cheung, 1981; Bouchard *et al.*, 1988); and the bioavailability of contaminants in soil had been discussed, particularly in relation to contact time with the soil (Semple *et al.*, 2003). There are several constraints that can limit the bioavailability of organic compounds in the environment.

Moisture may limit microbial activity in soil as soil microbes normally need 25–28% of water holding capacity to support their metabolic processes (Uhlířová *et al.*, 2005). Generally, excess moisture conditions are unfavourable for microbial growth and metabolism (Orchard and Cook, 1983; Leirós *et al.*, 1999; Belnap *et al.*, 2005). Soil moisture levels in the range of 20–80% of saturation generally allow biodegradation to take place (Bossert and Bartha, 1984), while 100% saturation often inhibits aerobic biodegradation due to low availability of oxygen. The optimum moisture level for a given situation is a function of the soil properties, contaminant characteristics and oxygen requirement (Alexander, 1999). However, the ability for soil microflora to withstand or endure extreme dryness or drought-like conditions varies and the desiccation-resistance of soil microflora are as follows: fungi > actinomycetes > bacteria.

It is widely known that undisturbed soils usually have soil pH within the range of 6–8, and most soil microbes functions best with 5.5–8.8 pH range. Several studies reported that heterotrophic bacteria and fungi thrive well in a neutral pH, with fungi showing more tolerance to acidic conditions (Dragun, 1998; Venosa and Zhu, 2003). The pH of the soil can inhibit microbial activity (Atlas, 1981) and also affect the solubility of important nutrients such as phosphorus (Eweis *et al.*, 1998; Hyman and Dupont, 2001). Several studies showed that degradation of oil increased with increasing pH, with the optimum degradation occurring under slightly alkaline conditions (Dibble and Bartha, 1979; Atlas, 1981; Foght and Westlake, 1987). Giles *et al.* (2001) reported a soil pH of 6.1 during the bioremediation of weathered oil sludge, suggesting that the 'typical' bioremediation pH range is likely to be suitable for weathered petroleum hydrocarbons. In addition, soil pH can change the structure of organic contaminants and strongly influences contaminant's bioavailability. Biodegradation studies are typically conducted at pH values 5.0–9.0 (depending on the microbial species involved), with a pH of 7.0 being preferable.

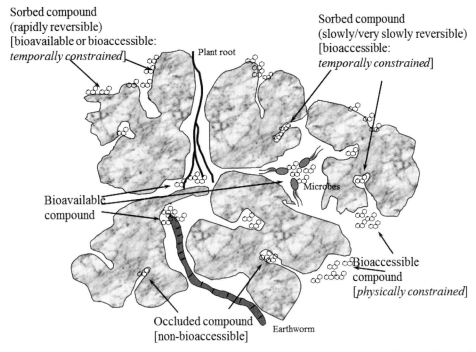

Fig. 14.5. This conceptual diagram illustrates the bioavailable and bioaccessible fractions of a contaminant in soil as defined by physical location (Semple *et al.*, 2004).

Aerobic respiration is the most effective pathway for bioremediation (Trinidade *et al.*, 2002) and the initial steps in the catabolism of aliphatic, cyclic and aromatic hydrocarbons by bacteria and fungi involve the oxidation of the substrate by oxygenases, for which molecular oxygen is required (Cerniglia, 1984b; Leahy and Colwell, 1990; Atlas, 1995). Optimum condition for microbial activity require >0.2 mg l^{-1}DO and 10% air-filled pore space for aerobic degradation. The degradative microorganisms need oxygen at two sites, at the initial attack of the substrate and at the end of the respiratory chain. Aerobic conditions are therefore necessary for this route of microbial oxidation of hydrocarbons in the environment (Leahy and Colwell, 1990; Yeung *et al.*, 1997). Therefore, in order to maintain metabolic activities of microbial cells, the oxygen supply rate must match the overall oxygen consumption rate under equilibrium conditions (10–40%). Nutrient availability plays an important role in adaptation of microbes and their growth on petroleum hydrocarbons. Two major nutrients, nitrogen (N) and phosphorus (P), are considered to be the most important, as they are required for microbial growth (Prince et al., 2002). However, it is important to note that any limiting nutrient will reduce microbial activity in the soil. The release of hydrocarbons into ecosystems containing low concentrations of inorganic nutrients often produces high carbon/nitrogen or carbon/phosphorus ratios, or both, which are unfavourable for microbial growth (Atlas, 1981; Cooney et al., 1985; Leahy and Colwell, 1990). Adjustment of carbon/nitrogen/phosphorus/potassium (C-N-P-K) ratios by the addition of urea, phosphate, N-P-K fertilizers, and ammonium and phosphate salts often accelerate the biodegradation of crude oil or gasoline in soil and groundwater (Leahy and Colwell, 1990; Abu and Ogiji, 1996). The C:N:P ratio of 100:10:1 is widely accepted as optimal for hydrocarbon biodegradation in most contaminated soils. Some studies, however, have observed no increase or an inhibition in biodegradation rates in response to nutrient amendment (Swindoll et al., 1988).

Temperature influences petroleum biodegradation by its effect on the physical nature and chemical composition of the oil, rate of hydrocarbon metabolism by microorganisms, and composition of the microbial community (Venosa and Zhu, 2003). Higher temperature increases the rate of hydrocarbon metabolism to a maximum, typically in the range of 30 to 40°C. Above this temperature, enzyme activity is reduced and membrane toxicity of hydrocarbons is increased (Bossert and Bartha, 1984). The optimum temperature for biodegradation has been reported in the range of 25 to 40°C (Van Hamme et al., 2003), although degradation by psychrophilic microorganisms can thrive under cold condition.

Extremely high total petroleum hydrocarbon (TPH) concentrations have been shown to inhibit microbial activity, thus limiting the biodegradation potential (Admon et al., 2001), while extremely low TPH concentrations can limit biodegradation because the carbon supply may be too low to support microbial growth (Leahy and Colwell, 1990). As such, soils that are considered suitable for bioremediation are typically contaminated with hydrocarbons at levels of 0.2–55% by volume of oil concentration (Huesemann et al., 2002). In addition, soils with a mixture of sand, silt and clay offer a more favourable habitat for microflora because they hold more nutrients and provide for better water and air flow. Soils that are considered suitable for bioremediation are soils that typically contain low clay or silt content. The presence of other inhibitory contaminants is of great importance for biodegradation suitability.

In practice, some of the rate limiting factors may be optimized during engineered bioremediation to enhance the biodegradation of petroleum hydrocarbons in the contaminated environments.

Bioremediation Strategies and Their Application

Bioremediation is a process of elimination, attenuation or transformation of contaminants using biodegradation processes to mitigate risks. Bioremediation functions by exploiting the diverse metabolic capabilities of microorganisms to detoxify or remove organic contaminants. Bioremediation technologies use the catabolic potential of microorganisms and plants to decontaminate or reduce the concentration of environmental hazardous contaminants in field sites to levels of risk that are acceptable to site owners and/or regulatory agencies (Tadesse *et al.*, 1994; Siciliano and Germida, 1997, 1998; Burken and Schnoor, 1998; Madsen, 1998a; Siciliano *et al.*, 2001; Lynch and Moffat, 2005).

Bioremediation strategies can be categorized based on the location of contaminant biodegradation and the aggressiveness of the remediation (Madsen, 1997). For example, *in situ* bioremediation is performed with the contaminated material left in its natural or original position (e.g. bioventing, biosparging, monitored natural attenuation and phyto-/rhizo-remediation), whereas *ex situ* bioremediation involves the removal of contaminated material from its original position and its treatment either on-site or at another location (e.g. biopiling, bioreactors, composting and land farming) (Vidali, 2001; Pavel and Gavrilescu, 2008). Both *in situ* and *ex situ* technologies can be applied to solid-, slurry- and vapour-phase systems. At the passive end of the spectrum, there is intrinsic or natural attenuation, which relies solely on the innate capabilities of naturally occurring microorganisms to degrade the contaminants *in situ*. In more active bioremediation systems, actions are taken to modify the site or contaminated material to promote and enhance the biodegradative activities of microorganisms (Madsen, 1998b; Siciliano and Germida, 1998). These technologies are typically referred to as engineered or enhanced bioremediation approaches. In practice, the selection of a method is often driven by economics and available environmental technologies. However, heavy molecular weight hydrocarbons in soil are often difficult to bioremediate and there are some limitations of the use of bioremediation technology. Where appropriate, bioremediation is effective for remediating soils contaminated with petroleum hydrocarbons (Flathman *et al.*, 1994; Hyman and Dupont, 2001). The choice of bioremediation technique will depend on a number of site-specific factors, including the type, mobility, concentration and volume of a contaminant, the soil structure, surrounding geology, the proximity to structures and potential receptors, and intended end use (Cookson, 1995; Eweis *et al.*, 1998). Several studies have highlighted the effectiveness of different bioremediation techniques in achieving the acceptable levels of biodegradation of most petroleum hydrocarbons in contaminated soil (Brar *et al.*, 2006; Mohan *et al.*, 2006; Peng *et al.*, 2008). The processes and logistics of bioremediation are relatively simple from an engineering perspective (Eweis *et al.*, 1998) and bioremediation strategies for various environmental organic contaminants can be achieved. However, there is no single method for remediation of every situation and combinations of techniques are often implemented at sites with multiple contamination sources. A selection of the more commonly used *in situ* and *ex situ* bioremediation techniques are described below.

In situ bioremediation strategies

Biostimulation

The use of biostimulation (also referred to as environmental modification) to remove petroleum hydrocarbons in the soil has been developed around strategies for delivering aeration, nutrients and moisture in order to optimize/enhance indigenous microbial activity and degradation (MacNaughton *et al.*, 1999; Kaplan and Kitts, 2004; Ayotamuno *et al.*, 2006b; Stroud *et al.*, 2007). A number of studies have documented the positive effects of biostimulation in the attenuation of total petroleum hydrocarbons (Rosenberg *et al.*, 1992; Rhykerd *et al.*, 1999; Sarkar *et al.*, 2005; Kogbara, 2008). Although there appear to be inconsistencies in the published literature about biostimulation, inorganic nutrients (fertilizers) are the most widely used amendment, and success has been reported in several studies (Whyte et al., 2001; Chaîneau et al., 2003b; Trindade et al., 2005; Ayotamuno et al., 2006b, 2009a; Perfumo et al., 2006). For example, the detailed studies on the effect of different types of biodegradation strategy on the coastal region around Prince Sound in Alaska contaminated by the *Exxon Valdez* oil spill revealed that the addition of fertilizers – composed mainly of nitrogen and phosphorus – accelerated removal of oil approximately five times more rapidly (Bragg et al., 1994; Swanell and Head, 1994). In a related study, Ayotamuno et al. (2006b) found that the addition of NPK fertilizer to a polluted agricultural soil significantly enhanced the biodegradation rate of the crude oil, initially present at 84 mg kg^{-1} TPH and reduced by 50–95% in the test cells. Several studies have also investigated the effect of different types of fertilizer and different delivery strategies in a low-energy, sandy beach or in a salt marsh (Lee and Levy, 1989, 1991; Lee and Trembley, 1993; Lee et al., 1995, 1997). These studies demonstrated that biostimulation using periodic addition of inorganic fertilizers (e.g. ammonium nitrate and triple superphosphate) increased the rate of petroleum hydrocarbon degradation on beaches. Further, Lee et al. (1995) compared the performance of inorganic nutrients with organic fish bone-meal fertilizer in a field study. The results showed that the organic fertilizer had the greatest effect on microbial growth and activity, while the inorganic nutrients were much more effective in crude oil degradation; however, the traditional addition of fertilizers or urea has been reported to have no effect (Nyman, 1999; Bento et al., 2005; Sarkar et al., 2005; Fernandez-Alvarez et al., 2006). Traditional application techniques have been found to have a negative impact on bacteria (Sarkar et al., 2005) and fungi (Chaillan et al., 2006) and, thus, may inhibit biodegradation. For example, Sarkar et al.(2005) reported that fertilizer addition caused either NH_3 overdosing and/or fertilizer induced acid toxicity, which significantly affected the microbial community. Further, Chaillan et al. (2006) observed that the addition of urea had a fungicidal effect and caused a toxic concentration of ammonia gas, significantly limiting the degradation extent of soil contaminated with weathered oils and drill cuttings.

Several studies have highlighted the effectiveness of biostimulation of petroleum-contaminated soils (Odokwuma and Dickson, 2003; Adesodun and Mbagwu, 2008; Tanee and Kinako, 2008; Ayotamuno et al., 2009a). For example, Ayotamuno et al.(2009a) investigated biostimulation supplemented with phytoremediation in which biostimulation of indigenous microbes through the addition of NPK fertilizer and tillage were then utilized in the remedial treatment. Comparative studies of biostimulation using local materials and phytoremediation of 5% crude oil contaminated soil in the tropical region have been investigated (Tanee and Kinako, 2008). For this study, the highest levels of crude oil biodegradation and improvement in nutrient content of the soil were observed in NPK 15:15:15 and phytoremediation treatments, indicating that these are good remedial treatment options in the mitigation of crude oil toxicity (Tanee and Kinako, 2008). In a similar study, Adesodun and Mbagwu (2008) evaluated the applicability of some organic wastes from animal droppings as bioremediation alternative for soils spiked with waste-lubricating oil (spent oil). The total hydrocarbon content (THC) with respect to sampling time was significantly reduced with addition of cow dung (CD), poultry manure (PM) and pig wastes (PW). From this study, the general trend in the first year showed that PW stimulated the highest net percentage loss in THC for soils polluted with 5000 mg kg^{-1} (0.5%) and 50,000 mg kg^{-1} (5%) spent oil levels. Poultry manure stimulated the highest reduction in soils polluted with medium oil concentration, i.e. 2.5% SP (25,000 mg kg^{-1}). The overall net loss mediated by each organic waste in the second year showed that PM addition was better irrespective of total oil loading. For instance, at 3 months PM led to 16.1% and 14.6% net reduction in THC for soils treated with 50,000mg kg^{-1} (5%) and 100,000 mg kg^{-1} (10%) total oil loading, respectively; whereas at the same period, the performance of the organic wastes were relatively similar in soils with 10,000 mg kg^{-1} oil loading. According to Adesodun and Mbagwu (2008), PW was better at low oil pollution level, while PM performed better at high oil pollution levels. Overall, the differential performance of these organic amendments followed the sequence PM>PW>CD. Considering the complex nature of nutrient amendment during remediation of contaminated soil, biostimulation is sometimes combined with other strategies in attempts to develop effective bioremediation strategies.

Bioaugmentation

Bioaugmentation relies on the introduction of catabolically active microbes into a contaminated soil when indigenous hydrocarbon degradation activity is low. The use of bioaugmention as a potential bioremediation strategy for treatment of contaminated soils has been reported in a number of studies (Nwachukwu, 2001; Odokwuma and Dickson, 2003; Okoh, 2003; Bento *et al.*, 2005; Ojumu *et al.*, 2005; Oboh *et al.*, 2006; Ojo, 2006; Ayotamuno *et al.*, 2007, 2009b; Odjadjare *et al.*, 2008; Abdulsalam and Omale, 2009; George-Okafor *et al.*, 2009; Okparanma *et al.*, 2009). In a comparative bioremediation study, Bento *et al.* (2005) reported up to a four-fold increase in the microbial activity of bioaugmented diesel-contaminated soils with a corresponding higher extent (75.2 ± 0.2%) of light hydrocarbon fraction (C_{12}–C_{23}) degradation in the bioaugmented treatment compared with the degradation extent of 48.7 ± 0.3% observed in the naturally attenuated treatment. In the heavy hydrocarbon fraction (C_{23}–C_{40}), 45.7 ± 0.4% degradation was found in the biostimulated condition as compared with 72.7 ± 0.4% degradation in the bioaugmented condition using Long Beach soil, which was initially contaminated with 2800 mg kg^{-1} TPH (C_{12}–C_{23}) and 9450 mg kg^{-1} TPH (C_{23}–C_{40}). In another study, Ayotamuno *et al.*(2007) reported that the total hydrocarbon content (THC) reduction in oily sludge (69,372 mgkg^{-1}THC) varied between 40.7% and 53.2% within 2 weeks as well as between 63.7% and 84.5% within 6 weeks of applying the catabolic microbial inoculum. The CFUs of the added biopreparation varied between 1.2×12^{12} and 3.0×10^{12} CFUg^{-1} of sludge and decreased to 7.0×10^{11} CFUg^{-1} of sludge by the end of the sixth week. When compared with the performance of the indigenous microbes in the control sample, the added biopreparation evidently increased the THC reduction rate in the oily sludge (Ayotamuno *et al.*, 2007). In a similar study, Odjadjare *et al.*(2008) reported that specific growth rates of axenic cultures of the bacteria during degradation of Escravos light crude oil ranged between 0.0037 and 0.0505 h^{-1}, while that of the mixed cultures varied from 0.0144 to 0.1301 h^{-1}. In addition, crude oil biodegradation ranged from 28.71 to 99.01% for single cultures and from 12.38 to 91.58% for the mixed cultures.

In a similar study, Oboh *et al.*(2006) investigated the relevance of 15 hydrocarbon-degrading bacterial and fungal isolates; the predominant species isolated primarily belonged to the genera *Pseudomonas* and *Aspergillus*. The results showed maximal increase in optical densities and total viable counts concomitant with a decrease in pH of the culture media and, according to Oboh *et al.*(2006), typical generation times varied between 0.64 and 1.09 days, 0.97 and 3.03 days, 0.88 and 2.97 days for kerosene, diesel and naphthalene, respectively. All the isolates utilized the hydrocarbons as sole carbon and energy sources with no statistical difference ($P>0.05$) in the utilization rates, suggesting close genetic similarities in respect of each isolate's oil degradation capabilities (Oboh et al., 2006). Similarly, Okoh (2003) described the degradation rates of different strains of *Pseudomonas aeruginosa* on crude oil, with evidence of significant reductions of major peak components of the oil. Virtually all of these studies utilized organisms isolated from petroleum-contaminated sites and degradation competencies were tested on various petroleum hydrocarbon sources. Recently, George-Okafor et al.(2009) investigated hydrocarbon degradation potential of indigenous fungal isolates from petroleum-contaminated soils and found that two isolates (*A. versicolor* and *A. niger*) exhibited >98% degradation efficiency for PAHs when grown in a culture medium containing 1% crude oil and 0.1% Tween 80 for 7 days. The degradation potentials of the isolates were not only demonstrated on crude oil and petroleum feedstock, including kerosene and diesel, but naphthalene, which is usually more degradable in comparison to the high molecular weight (HMW) PAHs (Oboh et al., 2006). In general, some researchers have suggested that mixed populations with broad enzymatic capacities are necessary in treating complex hydrocarbon mixtures (Ghazali et al., 2004; Sathishkumar et al., 2008). Effective understanding of the individual roles played by each consortium member is, therefore, essential in influencing the efficacy of microbial consortium and their exploitation in the successful implementation of bioaugmentation strategies to treat contaminated land.

Phytoremediation

Phytoremediation involves the use of plant-based processes and their associated microflora to remove, transfer, stabilize, or destroy contaminants in soil, sediment and groundwater (Alkorta and Garbisu, 2001; Reichenauer and Germida, 2008; Ayotamuno et al., 2009a). Plants can stimulate contaminant loss/removal by accumulation and transformation (Schneider et al., 1996), by extracellular transformation (Garcia et al., 1997; Siciliano et al., 1998) and by stimulating microbial degradative activity in the rhizosphere (Banks et al., 1999; Siciliano and Germida, 1999). Bert et al. (2009) reviewed phytoremediation as a management option for contaminated sediments in tidal marshes, flood control areas and dredged sediment landfill sites. Phytoremediation is an emerging technology that offers efficient tools and environmentally friendly solutions for the cleanup of contaminated sites and water.

In a study, Njoku et al. (2009)investigated plants that could be used to enhance the remediation of crude oil-contaminated soil and observed that the pH, moisture and organic matter contents of soils contaminated with crude oil were significantly affected by the growth of *Glycine max* at different levels of significance ($P<0.001$, $P<0.01$ and $P<0.05$). Degradation of petroleum hydrocarbons was enhanced in soil contaminated with 25 g crude oil in the presence of *G. max*. Although the growth of *G. max* did not significantly affect the crude oil level in the 50g and 75g treatments, the soils became more favourable for plant growth as weeds sprouted from the contaminated soil planted in the presence of *G. max* (Njoku et al., 2009). The findings of this study suggested that growth of certain plants, such as *G. max,* could be effective in cleanup of crude oil-contaminated soil, as well as in the reduction in crude oil toxicity in soil. Phytoremediation is a promising technology for the clean-up of petroleum-contaminated soils, especially in the tropics where climatic conditions favour plant growth and microbial activity and where financial resources can be limited (Merkl et al., 2005). Several studies have demonstrated an enhanced microbial degradation of organic contaminants in planted in comparison to unplanted soil (Wiltse *et al.*, 1998; Hutchinson *et al.*, 2001; Chaîneau *et al.*, 2003b; White *et al.*, 2006; Lu *et al.*, 2010). To date, very few studies have investigated supplementing biostimulation with phytoremediation in the cleanup of oil-polluted soils. Ina few studies, biostimulation strategies were combined with phytoremediation to enhance oil degradation in soil(Chaîneau *et al.*, 2003; Ayotamuno *et al.*, 2006a). However, these were largely cases of phytoremediation where fertilizers were applied to support plant growth, and both techniques were applied concurrently, not one after the other. Ecological rehabilitation with vetiver grass (*Vetiveria zizanioides*) has been found to enhance the phytoremediation of an oil shale-mined land contaminated with heavy metals (Xia, 2004). More recently, Lu *et al.*(2010)found that goose grass (*Eleusine indica*) significantly enhanced phytoremediation of soil contaminated with TPH and PAHs. Compared to many other plants, grasses have characteristics of rapid growth, large biomass, strong resistance to the contaminated environment, and effective stabilization to soils and, thus, result in excellent restoration effects in contaminated lands, particularly in the tropics and subtropics with high temperatures and levels of precipitation (Xia, 2004). The potential of common tropical grasses, such as elephant grass (*Pennisetum purpureum*), to enhance the decontamination of a crude oil-polluted soil has been reported (Ayotamuno *et al.*, 2006a).In petroleum-contaminated sites, phytoremediation can be applied at moderate contamination levels or after the application of other remediation measures as a polishing-step to further degrade residual hydrocarbons and to improve soil quality (Frick *et al.*, 1999). According to Cherian and Oliveira (2005), phytoremediation efficiency of plants and their potential use in environmental clean-up may be substantially improved using genetic engineering technologies.

Ex situ bioremediation strategies

Biopiling

Biopiling is an engineered system, in which air is pumped through the material, via a network of sparger pipes, to facilitate the remediation process for the particular hydrocarbon and soil matrix (Wu and Crapper, 2009). Biopile treatments have been applied to the treatment of non-chlorinated volatile organic compounds (VOCs) and petroleum hydrocarbon-contaminated soil (Schoefs *et al.*, 1998; Iturbe *et al.*, 2004, 2007; Li *et al.*, 2004; Kogbara, 2008; Sanscartier *et al.*, 2009; Coulon *et al.*, 2010). In a study, Iturbe *et al.* (2004) evaluated the treatment of a soil contaminated with about 4600 mg kg^{-1} of TPH, and four PAHs (phenanthrene, naphthalene, fluorene and pyrene) at different concentrations (total PAH concentration of 13.07 mg kg^{-1}). The authors employed a 100 m^3 biopile to treat the soil for 66 days. At the end of the process, a total of 691 mg kg^{-1} of TPH remained as residue in the biopile (i.e. a TPH removal of 85.2%). The degradation rate for the system was about 60 mg kg^{-1}day^{-1}, and a rough calculation of building and operation costs for the 100 m^3 biopile gave a cost of US$92m^{-3} (Iturbe *et al.*, 2004). In a similar study, Iturbe *et al.* (2007) discussed the capability of using a biopile to treat soils contaminated with about 40,000 mgkg^{-1} of TPH. In addition to the design and operation of a 27m^{3}biopile description, a final average TPH concentration of 7300 mg kg^{-1} was achieved in 22 weeks, which represents a removal efficiency of 80 %.

More recently, Kogbara (2008) reported a combination of experimental cells consisting of some agro-technical methods aimed at accelerating the biodegradation of petroleum-contaminated soils. The methods of treatment included variations of tilling, watering and nutrient application, plus biopile and phytoremediation treatments. Analysis of soil parameters, after a 6-week study period, showed an increase in total heterotrophic bacteria (THB) counts across all the treatments, with THB counts increasing with increment in soil nutrient level and initial concentration of the contaminant. Analysis of the hydrocarbon content, based on a performance index introduced in this study, indicated that on the average, variations in nutrient application, tilling and watering facilitated the attenuation of hydrocarbons at the rate of 429.4 mg $kg^{-1} day^{-1}$, 653.2 mg $kg^{-1} day^{-1}$ and 327.5 mg $kg^{-1} day^{-1}$, respectively. While the combined effect of various levels of nutrients, tilling and watering performed at the rate of 558.7 mg $kg^{-1} day^{-1}$, biopile and phytoremediation treatments recorded 427.9 mg $kg^{-1} day^{-1}$ and 489.3 mg $kg^{-1} day^{-1}$, respectively (Kogbara, 2008). In a recent field study, Coulon *et al.*(2010) compared windrow turning and biopile techniques for the remediation of soil contaminated with Bunker C fuel oil. The particular soil used, which was heavy in texture and historically contaminated with the bunker fuel, was more effectively remediated by windrowing, but coarser textures may be more amendable to biopiling (Coulon *et al.*, 2010). From these results, although nutrient application, watering and other factors affect the biodegradation process, frequent tilling for maximum oxygen exposure is the most important factor that affects the biodegradation of petroleum-hydrocarbons in soils.

Bioreactors

A bioreactor is an innovative technology that involves degradation of organic contaminants in water with microorganisms through attached or suspended biological systems. A bioreactor is an engineered slurry phase bioremediation system, in which biochemical transformation of materials is promoted by optimizing the activity of microorganisms, or by '*in vitro*' cellular components of the microbial cells. Slurry phase biological treatment can be applied to a variety of contaminated wastes amenable to microbial degradation, including manufactured gas plant wastes, wood-treating wastes, refinery wastes, petroleum hydrocarbons and select chlorinated compounds (Woodhull and Jerger, 1994).Slurry phase bioreactors for treating petroleum hydrocarbon-contaminated soils have been evaluated (Cassidy, 2001) and this is considered as one of the fastest bioremediation technologies because contaminants can be effectively transported to the microbial cells (Mueller *et al.*, 1991). Some limiting factors affecting the slurry phase bioreactor process during decontamination of oil-contaminated soil can be controlled (Khan *et al.*, 2004). An attractive alternative to the slurry bioreactors for treating oil-contaminated soils is the rotating drum bioreactors, since they can handle soils with high concentrations of petroleum hydrocarbons (Gray *et al.*, 1994; Banerjee *et al.*, 1995). The fluid phase enhances transport of nutrients and 'solubilized' or dispersed TPH contaminants to the degrading bacteria. With a bioreactor, temperature, pH and some other parameters are usually optimized for degradation. The contaminated soil must be excavated, mixed with water and introduced into the reactor. Generally, the rate-limiting factors in any bioreactor system used for crude oil degradation are the degree of TPH solubilisation through bio surfactant production and the level or concentration of active biomass of hydrocarbonoclastic bacteria maintained in the system (Stroo, 1992). Degradation products in bioreactors are often monitored and various inputs regulated. Bioreactors are, however, intrinsically more expensive than *in situ* or land treatment technologies because they are specialized. Studies on degradation of organic compounds have shown that some microorganisms are extremely versatile at catabolizing recalcitrant molecules. Harnessing this catabolic potential could make it possible to bioremediate some chemically contaminated environmental systems.

Composting

Composting relies on mixing/aeration of the primary ingredients of composting with the contaminated soil, and the contaminants are degraded by the active microflora within the mixture, but without an added microbial inoculums (Jørgensen *et al.*, 2000; Semple *et al.*, 2001). Composting has been evaluated to be effective in bioremediation of both explosive compounds and PAHs in soils (Kaplan and Kaplan, 1982; Williams *et al.*, 1992; USEPA, 1996). Composting of petroleum-contaminated soil and petroleum-derived chemical wastes has been demonstrated by few researchers (Stegmann *et al.*, 1991; Milne *et al.*, 1998). However, impacts of composting strategies on the treatment of soils contaminated with a variety of organic pollutants have been reviewed (Semple *et al.*, 2001). The effectiveness of composting as a remediation strategy for petroleum hydrocarbon-contaminated soil has been demonstrated in some studies (Udosen *et al.*, 2001; Ayotamuno *et al.*, 2009b). Udosen *et al.*(2001) investigated the effect of organic amendments on the oil content, heavy metals concentration and pH of petroleum-contaminated sandy loam ultisol obtained from the Rumuekpe oilfield in the Niger Delta. The addition of organic amendments resulted in a significant (at 95% probability level) decrease in oil content by 92% for composting, 81% for soil treated with sawdust and 58% for soil with ash supplementation, over 6 months. According to the authors, amending the soil with wood ash raised the pH from 5.6 to 6.2 and soil treatment with compost generally gave the best remediation results, followed by sawdust and then ash. Udosen *et al.* (2001) suggested that adjusting the pH of petroleum hydrocarbon-contaminated soil to high acidic levels may promote the availability and migration of heavy metals in remediated soils and not necessarily the rate of oil biodegradation.

Ayotamuno *et al.* (2009b) compared the feasibility of using bioaugmentation and composting as bioremediation technologies for the removal of PAHs from oilfield drill cuttings. According to Ayotamuno *et al.*(2009b), there were far more PAH reductions with the bioaugmentation method than with the composting method, particularly at the initial and final stages of the experiment, which clearly indicates that the performance of the bioaugmentation method was better than the composting method. A maximum composting temperature of $54^{\circ}C$ was observed during the study. Considering the impact of the environmental bacterial isolates (i.e. *Bacillus* sp. and *Pseudomonas* sp.) used to degrade the PAH of the drill cuttings, the *Pseudomonas* sp. degraded the 3-ring and 4-ring PAHs more effectively than the *Bacillus* sp. The *Bacillus* sp. degraded the 5-ring PAHs better than the *Pseudomonas* sp. only in the first 2 weeks of treatment. Ayotamuno *et al.*(2009b) also observed that the co-metabolism of the 3- or 4-ring PAHs could not have a synergetic effect on the 5-ring PAHs when the mixed culture was tested. However, this resulted in the limited degradation of the 5-ring PAHs, particularly in the fourth week of the experiment.

Composting has been demonstrated to be effective in biodegrading petroleum hydrocarbons in soils, but full-scale applications are yet to be implemented in the bioremediation industries. However, either the success or failure of a composting remediation strategy depends on a number of factors, the most important of which are pollutant bioavailability and biodegradability.

Landfarming

Landfarming is a bioremediation treatment process that is performed in the upper soil zone or in biotreatment cells. Contaminated soils, sediments, or sludges are incorporated into the soil surface and periodically stimulated aerobically by tilling or ploughing the soil (Kelly *et al.*, 1998; Pearce and Ollermann, 1998). Turning the soil regularly provides oxygen transportation needed for biostimulation and increases the opportunity of contact by mixing microbes with organic contaminants and water. The use of landfarming to remediate hydrocarbon-contaminated soils has been demonstrated in several studies (Prado-Jatar *et al.*, 1993; Genouw *et al.*, 1994; Kuyukina *et al.*, 2003; Ebuehi *et al.*, 2005; Rubinos *et al.*, 2007) and petroleum hydrocarbon-contaminated wastes (Hejazi *et al.*, 2003; Hejazi and Husain, 2004a, b). Prado-Jatar *et al.*(1993)showed that total crude oil biodegradation rates of the sludges were stabilized within 6–9 months, with a mean total hydrocarbon degradation of 80%. Saturated hydrocarbon fraction degradation rates were significantly lower (overall slope *t*-tests $P<0.001$) in treatments where aeration (harrowing), watering and fertilization were applied, contributing the most to microbial activation of site-specific soil and sludge. According to Prado-Jatar *et al.*(1993), the effectiveness of each method was also evaluated by comparison with 'control' crop yields.

Remediation by enhanced natural attenuation (RENA), a landfarming treatment technology for remediation of petroleum hydrocarbon-contaminated soils, has been demonstrated (Ebuehi *et al.*, 2005). Ebuehi *et al.*(2005) investigated remediation of crude-oil contaminated soil using an enhanced natural attenuation technique. A TPH concentration of 1.1004×10^4 mgkg^{-1} of the sandy soil was achieved after spiking and tilling. From this study, there was a reduction in the TPH level from 300 mg kg^{-1} after 8 weeks, to 282 mgkg^{-1} after 10 weeks. However, no significant reduction in the TPH level was observed after the tenth week. The nitrogen and phosphorus levels of the sandy soil were 24.6 and 22.8 mg kg^{-1}, respectively. This suggests that the nitrogen and phosphorus levels could no longer support biodegradation at the residual TPH levels of 282 mg kg^{-1} and 22.8 mg kg^{-1} after spiking and tilling, respectively, which further reduced to 0.12 mg kg^{-1} and 1.7 mg kg^{-1}, respectively, after 10 weeks. The total hydrocarbon-utilizing bacteria (THUB) increased from 3.0×10^4 CFUg^{-1} to 8.55×10^4 CFUg^{-1} and finally reduced to 5.38×10^4 CFU g^{-1}, while the total heterotrophic bacteria (THB) reduced from 1.22×10^8 CFUg^{-1} to 5.98×10^5 CFUg^{-1} (Ebuehi *et al.*, 2005). From this study, the available data indicated that the remediation-enhanced natural attenuation technique could be employed to remediate a farm settlement contaminated by crude oil. In a similar study, Compeau *et al.*(1991) reported that the landfarming strategy had achieved an enhancement in the decontamination of 50 cm topsoil of an area previously polluted with crude oil.

Large-scale landfarming experiments have been performed by Genouw *et al.*(1994) and the case-study revealed that oil sludge can effectively be treated by landfarming, if appropriate technical measures are taken and a sufficient time (minimum 15 years) for bioremediation is provided. On-site 'landfarming' methods have been used successfully (and within a reasonable period of time) to degrade only those PAHs with three or fewer aromatic rings (Wilson and Jones, 1993). Although long treatment time in the landfarming method is often due to the lack of control of parameters affecting microbial activity, microorganisms' role in the bioremediation process should be monitored with an accurate molecular biotechnology. The landfarming method used in the bioremediation of oil-contaminated soil is an effective, economic and promising technology for cleaning up hydrocarbon-contaminated soil (Rittmann and McCarty, 2001). However, a land treatment site must be managed properly to prevent both on-site and off-site problems with groundwater, surface water, air, or food-chain contamination.

Conclusion

Although several widely available bioremediation strategies have been discussed, there are many *ex situ* and *in situ* bioremediation strategies for treatment of hydrocarbon-contaminated soil. These include the use of bioventing (low airflow rates to provide only enough oxygen to sustain microbial activity and sometimes combines an increased oxygen supply with vapour extraction) and biosparging (involves injection of air into an aquifer or the soil to increase the biological activity in soil by increasing the O_2 supply). A detailed analysis of some of these remediation technologies has been discussed by Khan *et al.* (2004). Many microorganisms possess the inherent ability to transform or utilize organic contaminants (e.g. petroleum hydrocarbons) as a carbon and energy source. This has been demonstrated in both laboratory and field-scale studies. Ithas been shown that biodegradation is a potentially important option for dealing with hydrocarbon contamination and biodegradation strategy can be used as a clean-up approach by exploiting the activities of indigenous microflora with the capacity to degrade these organic contaminants. Biodegradation depends on several physical and chemical factors that need to be properly controlled to optimize the environmental conditions for the microflora community and successfully remediate the contaminated soils. Generally, the success of a biodegradation strategy depends on predominant environmental conditions, the chemical structure of the pollutants, contaminant bioavailability, the presence of catabolically active microbes, and the organic pollutant–matrix interaction in contaminated soil (Volkering *et al.*, 1997; Del'Arco and de França, 2001; Semple *et al.*, 2001, 2007). Since each contaminated site will be different, it is necessary to have a deeper understanding of the microbial ecology of contaminated sites so that bioremediation strategies could be improved further. Successful application of bioremediation technologies to hydrocarbon-contaminated systems requires understanding of various organic contaminant's biological, chemical and physical processes in the environment.

References

Abdulsalam, S. and Omale, A.B. (2009) Comparison of biostimulation and bioaugmentation techniques for the remediation of used motor oil contaminated soil. *Brazilian Archives of Biology and Technology* 52, 747–754.

Abu, G.O. and Ogiji, P.A. (1996) Initial test of a bioremediation scheme for the clean up of an oil-polluted waterbody in a rural community in Nigeria. *Bioresource Technology* 58, 7–12.

Adesodun, J.K. and Mbagwu, J.S.C. (2008) Biodegradation of waste-lubricating petroleum oil in a tropical alfisol as mediated by animal droppings. *Bioresource Technology* 99, 5659–5665.

Admon, S., Green, M. and Avnimelech, Y. (2001) Biodegradation kinetics of hydrocarbon in soil during land treatment of oily sludge. *Bioremediation Journal* 5, 193–209.

Alexander, M. (1999)*Biodegradation and Bioremediation*, 2nd edn. Academic Press, New York.

Alexander, M. (2000) Aging, Bioavailability, and Overestimation of Risk from Environmental Pollutants. *Environmental Science & Technology* 34, 4259–4265.

Alkorta, I. and Garbisu, C. (2001) Phytoremediation of organic contaminants in soils. *Bioresource Technology* 79, 273–276.

Atlas, R.M. (1981) Microbial degradation of petroleum hydrocarbons: an environmental perspective. *Microbiology and Molecular Biology Reviews* 45, 180–209.

Atlas, R.M. (1985) Effects of hydrocarbons on micro-organisms and biodegradation in Arctic ecosystems. In: Engelhardt, F.R. (ed.) *Petroleum Effects in the Arctic Environment*. Elsevier, London, UK, pp. 63–99.

Atlas, R.M. (1995) Petroleum biodegradation and oil spill bioremediation. *Marine Pollution Bulletin* 31, 178–182.

Ayotamuno, J.M., Kogbara, R.B. and Egwuenum, P.N. (2006a) Comparison of corn and elephant grass in the phytoremediation of a petroleum-hydrocarbon-contaminated agricultural soil in Port Harcourt, Nigeria. *Journal of Food, Agriculture & Environment* 4.

Ayotamuno, M.J., Kogbara, R.B., Ogaji, S.O.T. and Probert, S.D. (2006b) Bioremediation of a crude-oil polluted agricultural-soil at Port Harcourt, Nigeria. *Applied Energy* 83, 1249–1257.

Ayotamuno, M.J., Okparanma, R.N., Nweneka, E.K., Ogaji, S.O.T. and Probert, S.D. (2007) Bio-remediation of a sludge containing hydrocarbons. *Applied Energy* 84, 936–943.

Ayotamuno, J., Kogbara, R. and Agoro, O. (2009a) Biostimulation supplemented with phytoremediation in the reclamation of a petroleum contaminated soil. *World Journal of Microbiology and Biotechnology* 25, 1567–1572.

Ayotamuno, J.M., Okparanma, R.N. and Araka, P.P. (2009b) Bioaugmentation and composting of oil-field drill-cuttings containing polycyclic aromatic hydrocarbons (PAHs). *Journal of Food Agriculture & Environment* 7, 658–664.

Banat, I.M. (1995) Biosurfactants production and possible uses in microbial enhanced oil recovery and oil pollution remediation: A review. *Bioresource Technology* 51, 1–12.

Banerjee, D., Fedorak, P., Hashimoto, A., Masliyah, J., Pickard, M. and Gray, M. (1995) Monitoring the biological treatment of anthracene-contaminated soil in a rotating-drum bioreactor. *Applied Microbiology and Biotechnology* 43, 521–528.

Banks, M.K., Lee, E. and Schwab, A.P. (1999) Evaluation of Dissipation Mechanisms for Benzo[a]pyrene in the Rhizosphere of Tall Fescue. *Journal of Environmental Quality* 28, 294–298.

Bartha, R. (1986) Biotechnology of Petroleum Pollutant Biodegradation. *Microbial Ecology* 12, 155–172.

Bartha, R. and Bossert, I. (1984) The treatment and disposal of petroleum wastes. In: Atlas, R.M. (ed.) *Petroleum Microbiology*. Macmillan, New York, pp. 553–578.

Belnap, J., Welter, J.R., Grimm, N.B., Barger, N. and Ludwig, J.A. (2005) Linkages between Microbial and Hydrologic Processes in Arid and Semiarid Watersheds. *Ecology* 86, 298–307.

Benka-Coker, M.O. and Olumagin, A. (1996) Effects of waste drilling fluid on bacterial isolates from a mangrove swamp oilfield location in the Niger Delta of Nigeria. *Bioresource Technology* 55, 175–179.

Bento, F.M., Camargo, F.A.O., Okeke, B.C. and Frankenberger, W.T. (2005) Comparative bioremediation of soils contaminated with diesel oil by natural attenuation, biostimulation and bioaugmentation. *Bioresource Technology* 96, 1049–1055.

Bert, V., Seuntjens, P., Dejonghe, W., Lacherez, S., Thuy, H. and Vandecasteele, B. (2009) Phytoremediation as a management option for contaminated sediments in tidal marshes, flood control areas and dredged sediment landfill sites. *Environmental Science and Pollution Research* 16, 745–764.

Bhadra, R., Spanggord, R.J., Wayment, D.G., Hughes, J.B. and Shanks, J.V. (1999) Characterization of oxidation products of TNT metabolism in aquatic phytoremediation systems of *Myriophyllum aquaticum*. *Environmental Science & Technology* 33, 3354–3361.

Bizily, S.P., Rugh, C.L., Summers, A.O. and Meagher, R.B. (1999) Phytoremediation of methylmercury pollution: merB expression in *Arabidopsis thaliana* confers resistance to organomercurials. *Proceedings of the National Academy of Sciences of the United States of America* 96, 6808–6813.

Block, R., Allworth, N. and Bishop, M. (1991) Assessment of diesel contamination in soil. In: Calabrese, E. and Kostecki, P. (eds) *Hydrocarbon Contaminated Soils, Vol. I: Remediation Techniques, Environmental Fate, Risk Assessment, Analytical Methodologies, Regulatory Considerations*. Lewis Publishers, Chelsea, Michigan, pp. 135–148.

Bossert, I.D. and Bartha, R. (1984) The fate of petroleum in soil ecosystems. In: Atlas, R.M. (ed.) *Petroleum Microbiology*. Macmillan Publishing Co., New York, pp. 435–474.

Bossert, I.D. and Bartha, R. (1986) Structure-biodegradability relationships of polycyclic aromatic hydrocarbons in soil. *Bulletin of Environmental Contamination and Toxicology* 37, 490–495.

Bouchard, D.C., Wood, A.L., Campbell, M.L., Nkedi-Kizza, P. and Rao, P.S.C. (1988) Sorption nonequilibrium during solute transport. *Journal of Contaminant Hydrology* 2, 209–223.

Bouwer, E.J. and Zehnder, A.J.B. (1993) Bioremediation of organic compounds--putting microbial metabolism to work. *Trends in Biotechnology* 11, 360–367.

Bragg, J.R., Prince, R.C., Harner, E.J. and Atlas, R.M. (1994) Effectiveness of bioremediation for the Exxon Valdez oil spill. *Nature* 368, 413–418.

Brar, S.K., Verma, M., Surampalli, R.Y., Misra, K., Tyagi, R.D., Meunier, N. and Blais, J.F. (2006) Bioremediation of Hazardous Wastes – A Review. *Practice Periodical of Hazardous, Toxic, and Radioactive Waste Management* 10, 59–72.

Brassington, K.J., Hough, R.L., Paton, G.I., Semple, K.T., Risdon, G.C., Crossley, J., Hay, I., Askari, K. and Pollard, S.J.T. (2007) Weathered hydrocarbon wastes: A risk management primer. *Critical Reviews in Environmental Science and Technology* 37, 199–232.

Burken, J.G. and Schnoor, J.L. (1998) Predictive relationships for uptake of organic contaminants by hybrid poplar trees. *Environmental Science & Technology* 32, 3379–3385.

Cassidy, D.P. (2001) Biological Surfactant Production in a Biological Slurry Reactor Treating Diesel Fuel Contaminated Soil. *Water Environment Research* 73, 87–94.

Cerniglia, C.E. (1984a) Microbial metabolism of polycylic aromatic hydrocarbons. *Advances in Applied Microbiology* 30, 31–71.

Cerniglia, C.E. (1984b) Microbial transformation of aromatic hydrocarbons. In: Atlas, R.M. (ed.) *Petroleum Microbiology*. Macmillan, New York, pp. 99–128.

Cerniglia, C.E. (1992) Biodegradation of polycyclic aromatic hydrocarbons. *Biodegradation* 3, 351–368.

Chaillan, F., Le Fleche, A., Bury, E., Phantavong, Y.H., Grimont, P., Saliot, A. and Oudot, J. (2004) Identification and biodegradation potential of tropical aerobic hydrocarbon-degrading microorganisms. *Research in Microbiology* 155, 587–595.

Chaillan, F., Chaîneau, C.H., Point, V., Saliot, A. and Oudot, J. (2006) Factors inhibiting bioremediation of soil contaminated with weathered oils and drill cuttings. *Environmental Pollution* 144, 255–265.

Chaîneau, C.H., Morel, J.L. and Oudot, J. (1995) Biodegradation of fuel oil hydrocarbons in soil contaminated by oily wastes produced during onshore drilling operations. In: vandenBrink, W.J.B., Bosman, R. and Arendt, F. (eds) Contaminated Soil '95. Proceedings. Academic Publishers, Maastricht, the Netherlands, pp. 1173–1174.

Chaîneau, C.-H., Morel, J., Dupont, J., Bury, E. and Oudot, J. (1999) Comparison of the fuel oil biodegradation potential of hydrocarbon-assimilating microorganisms isolated from a temperate agricultural soil. *Science of the Total Environment* 227, 237–247.

Chaîneau, C.-H., Indonesie, T.E.P., Vidalie, J.-F., DGEP, T., Hamzah, U.S., MIGAS, B., Suripno, G.A., Najib, M. and Indonesie, T.E.P. (2003a)*Bioremediation of Oil-Based Drill Cuttings under Tropical Conditions, SPE 13th Middle east Oil Show & Conference*, 9–12 June 2003, Bahrain. Society of Petroleum Engineers, Bahrain.

Chaîneau, C.H., Yepremian, C., Vidalie, J.F., Ducreux, J. and Ballerini, D. (2003b) Bioremediation of a Crude Oil-Polluted Soil: Biodegradation, Leaching and Toxicity Assessments. *Water, Air, & Soil Pollution* 144, 419–440.

Chen, D.R., Bei, J.Z. and Wang, S.G. (1999) Study on biodegradation behavior of polycaprolactone microparticles. *Acta Polymerica Sinica* 620–622.

Cherian, S. and Oliveira, M.M. (2005) Transgenic Plants in Phytoremediation: Recent Advances and New Possibilities. *Environmental Science & Technology* 39, 9377–9390.

Colwell, R.R. (1977) Ecological aspects of microbial degradation of petroleum in the marine environment. *CRC Critical Reviews in Microbiology* 5, 423–445.

Compeau, G.C., Mahaffey, W.D. and Patras, L. (1991) Full scale bioremediation of contaminated soil and water. In: Sayler, G.S., Fox, R. and Blackbourn, J.W. (eds) *Environmental Biotechnology for Waste Treatment*. Plenum Press, New York, pp. 91–109.

Cookson, J.T. (1995)*Bioremediation Engineering: design and application*. McGraw-Hill, New York.

Cooney, J.J. (1984) The fate of petroleum pollutants in fresh water ecosystems. In: Atlas, R.M. (ed.) *Petroleum Microbiology*. Macmillan, New York, pp. 399–434.

Cooney, J.J., Silver, S.A. and Beck, E.A. (1985) Factors influencing hydrocarbon degradation in three freshwater lakes. *Microbial Ecology* 11, 127–137.

Couling, N.R., Towell, M.G. and Semple, K.T. (2010) Biodegradation of PAHs in soil: Influence of chemical structure, concentration and multiple amendment. *Environmental Pollution* 158, 3411–3420.

Coulon, F., Al Awadi, M., Cowie, W., Mardlin, D., Pollard, S., Cunningham, C., Risdon, G., Arthur, P., Semple, K.T. and Paton, G.I. (2010) When is a soil remediated? Comparison of biopiled and windrowed soils contaminated with bunker-fuel in a full-scale trial. *Environmental Pollution* 158, 3032–3040.

Das, N. and Chandran, P. (2011) Microbial Degradation of Petroleum Hydrocarbon Contaminants: An Overview. *Biotechnology Research International* 2011, 13.

Del'Arco, J.P. and de França, F.P. (2001) Influence of oil contamination levels on hydrocarbon biodegradation in sandy sediment. *Environmental Pollution* 112, 515–519.

Dibble, J.T. and Bartha, R. (1979) Effect of Environmental Parameters on the Biodegradation of Oil Sludge. *Applied and Environmental Microbiology* 37, 729–739.

Dragun, J. (1998) *The Soil Chemistry of Hazardous Materials*, 2nd edn. Amherst Scientific Publishers, Amherst, Massachusetts.

Ebuehi, O.A.T., Abibo, I.B., Shekwolo, P.D., Sigismund, K.I., Adoki, A. and Okoro, I.C. (2005) Remediation of crude-oil contaminated soil by enhanced natural attenuation technique. *Journal of Applied Sciences & Environmental Management* 9, 103–106.

Ehlers, L.J. and Luthy, R.G. (2003) Peer Reviewed: Contaminant Bioavailability in Soil and Sediment. *Environmental Science & Technology* 37, 295A–302A.

Eweis, J.B., Ergas, S.J., Chang, D.P.Y. and Schroeder, E.D. (1998) *Bioremediation Principles*, 2nd edn. McGraw-Hill, Boston.

Fernandez-Alvarez, P., Vila, J., Garrido-Fernandez, J.M., Grifoll, M. and Lema, J.M. (2006) Trials of bioremediation on a beach affected by the heavy oil spill of the Prestige. *Journal of Hazardous Materials* 137, 1523–1531.

Flathman, P.E., Jerger, D.E. and Exner, J.H. (eds) (1994) *Bioremediation Field Experience*. Lewis Publishers, Boca Raton, Florida.

Floodgate, G. (1984) The fate of petroleum in marine ecosystems. In: Atlas, R.M. (ed.) *Petroleum Microbiology*. Macmillion, New York, pp. 355–398.

Foght, J.M. and Westlake, D.S.W. (1987) Biodegradation of hydrocarbon in fresh water. In: Vandermeulen, J.H. and Hrudey, S.E. (eds) *Oil in Freshwater: Chemistry, Biology, Countermeasure Technology*. Proceedings of the Symposium [on] Oil Pollution in Freshwater, Edmonton, Alberta, Canada, 1984. Pergamon Press, pp. 217-230.

Freeman, D.H. and Cheung, L.S. (1981) A Gel Partition Model for Organic Desorption from a Pond Sediment. *Science* 214, 790–792.

Frick, C.M., Germida, J.J. and Farrell, R.E. (1999) *Assessment of Phytoremediation as an In-Situ Technique for Cleaning Oil-contaminated Sites*. Petroleum Technology Alliance of Canada, Calgary.

Fritsche, W. and Hofrichter, M. (2005) Aerobic Degradation of Recalcitrant Organic Compounds by Microorganisms. In: Jördening, H.-J. and Winter, J. (eds) *Environmental Biotechnology: Concepts and Applications*. Wiley-VCH Verlag GmbH & Co. KGaA, Weinheim, Germany, pp. 203–227.

Fritsche, W. and Hofrichter, M. (2008) Aerobic Degradation by Microorganisms. In: Rehm, H.-J. and Reed, G. (eds) *Biotechnology: Environmental Processes*, 2nd edn. Wiley-VCH Verlag GmbH, Weinheim, Germany, pp. 144–167.

Garcia, C., Roldan, A. and Hernandez, T. (1997) Changes in Microbial Activity after Abandonment of Cultivation in a Semiarid Mediterranean Environment. *Journal of Environmental Quality* 26, 285–292.

Genouw, G., Naeyer, F., Meenen, P., Werf, H., Nijs, W. and Verstraete, W. (1994) Degradation of oil sludge by landfarming – a case-study at the Ghent harbour. *Biodegradation* 5, 37–46.

George-Okafor, U., Tasie, F. and Muotoe-Okafor, F. (2009) Hydrocarbon Degradation Potentials of Indigenous Fungal Isolates from Petroleum Contaminated Soils. *Journal of Physical and Natural Sciences* 3(1).

Ghazali, F.M., Rahman, R.N.Z.A., Salleh, A.B. and Basri, M. (2004) Biodegradation of hydrocarbons in soil by microbial consortium. *International Biodeterioration & Biodegradation* 54, 61–67.

Giles, W.R., Jr, Kriel, K.D. and Stewart, J.R. (2001) Characterization and Bioremediation of a Weathered Oil Sludge. *Environmental Geosciences* 8, 110–122.

Gogoi, B.K., Dutta, N.N., Goswami, P. and Krishna Mohan, T.R. (2003) A case study of bioremediation of petroleum-hydrocarbon contaminated soil at a crude oil spill site. *Advances in Environmental Research* 7, 767–782.

Gray, M.R., Banerjee, D.K., Fedorak, P.M., Hashimoto, A., Masliyah, J.H. and Pickard, M.A. (1994) Biological remediation of anthracene-contaminated soil in rotating bioreactors. *Applied Microbiology and Biotechnology* 40, 933–940.

Grishchenkov, V.G., Townsend, R.T., McDonald, T.J., Autenrieth, R.L., Bonner, J.S. and Boronin, A.M. (2000) Degradation of petroleum hydrocarbons by facultative anaerobic bacteria under aerobic and anaerobic conditions. *Process Biochemistry* 35, 889–896.

Grosser, R.J., Warshawsky, D. and Vestal, J.R. (1991) Indigenous and enhanced mineralization of pyrene, benzo[a]pyrene, and carbazole in soils. *Applied and Environmental Microbiology* 57, 3462–3469.

Grossi, V., Cravo-Laureau, C., Guyoneaud, R., Ranchou-Peyruse, A. and Hirschler-Réa, A. (2008) Metabolism of n-alkanes and n-alkenes by anaerobic bacteria: A summary. *Organic Geochemistry* 39, 1197–1203.

Haines, J.R. and Alexander, M. (1974) Microbial degradation of high-molecular-weight alkanes. *Applied Microbiology* 1084–1085.

Heider, J., Spormann, A.M., Beller, H.R. and Widdel, F. (1998) Anaerobic bacterial metabolism of hydrocarbons. *FEMS Microbiology Reviews* 22, 459–473.

Hejazi, R.F. and Husain, T. (2004a) Landfarm performance under arid conditions. 1. Conceptual framework. *Environmental Science & Technology* 38, 2449–2456.

Hejazi, R.F. and Husain, T. (2004b) Landfarm performance under arid conditions. 2. Evaluation of parameters. *Environmental Science & Technology* 38, 2457–2469.

Hejazi, R.F., Husain, T. and Khan, F.I. (2003) Landfarming operation of oily sludge in and region – human health risk assessment. *Journal of Hazardous Materials* 99, 287–302.

Holdway, D.A. (2002) The acute and chronic effects of wastes associated with offshore oil and gas production on temperate and tropical marine ecological processes. *Marine Pollution Bulletin* 44, 185–203.

Hollaway, S.L., Faw, G.M. and Sizemore, R.K. (1980) The bacterial community composition of an active oil field in the northwestern Gulf of Mexico. *Marine Pollution Bulletin* 11, 153–156.

Holliger, C., Gaspard, S., Glod, G., Heijman, C., Schumacher, W., Schwarzenbach, R.P. and Vazquez, F. (1997) Contaminated environments in the subsurface and bioremediation: organic contaminants. *FEMS Microbiology Reviews* 20, 517–523.

Huesemann, M.H. (1995) Predictive Model for Estimating the Extent of Petroleum Hydrocarbon Biodegradation in Contaminated Soils. *Environmental Science & Technology* 29, 7–18.

Huesemann, M.H., Hausmann, T.S. and Fortman, T.J. (2002) Microbial Factors Rather Than Bioavailability Limit the Rate and Extent of PAH Biodegradation in Aged Crude Oil Contaminated Model Soils. *Bioremediation Journal* 6, 321–336.

Huesemann, M.H., Hausmann, T.S. and Fortman, T.J. (2003) Assessment of bioavailability limitations during slurry biodegradation of petroleum hydrocarbons in aged soils. *Environmental Toxicology and Chemistry* 22, 2853–2860.

Huesemann, M.H., Hausmann, T.S. and Fortman, T.J. (2004) Does bioavailability limit biodegradation? A comparison of hydrocarbon biodegradation and desorption rates in aged soils. *Biodegradation* 15, 261–274.

Hutchinson, S.L., Banks, M.K. and Schwab, A.P. (2001) Phytoremediation of Aged Petroleum Sludge: Effect of Inorganic Fertilizer. *Journal of Environmental Quality* 30, 395–403.

Hyman, M. and Dupont, R.R. (2001) *Groundwater and Soil Remediation: Process Design and Cost Estimating of Proven Technologies*. American Society of Civil Engineers Press, Reston, Virginia.

Iturbe, R., Flores, R., Flores, C. and Torres, L. (2004) TPH-contaminated Mexican Refinery Soil: Health Risk Assessment and the First Year of Changes. *Environmental Monitoring and Assessment* 91, 237–255.

Iturbe, R., Flores, C. and Torres, L.G. (2007) Operation of a 27-m^3 biopile for the treatment of petroleum-contaminated soil. *Remediation Journal* 17, 97–108.

Jones, J.G., Knight, M. and Byrom, J.A. (1970) Effect of Gross Pollution by Kerosine Hydrocarbons on the Microflora of a Moorland Soil. *Nature* 227, 1166–1166.

Jørgensen, K.S., Puustinen, J. and Suortti, A.M. (2000) Bioremediation of petroleum hydrocarbon-contaminated soil by composting in biopiles. *Environmental Pollution* 107, 245–254.

Juhasz, A.L., Britz, M.L. and Stanley, G.A. (1996a) Degradation of benzo[a]pyrene, dibenz[a,h]anthracene and coronene by *Burkholderia cepacia*. *International Conference on Environmental Biotechnology*, Palmerston North, New Zealand, pp. 45–51.

Juhasz, A.L., Britz, M.L. and Stanley, G.A. (1996b) Degradation of high molecular weight polycyclic aromatic hydrocarbons by *Pseudomonas cepacia*. *Biotechnology Letters* 18, 577–582.

Kaplan, C.W. and Kitts, C.L. (2004) Bacterial Succession in a Petroleum Land Treatment Unit. *Appl. Environ. Microbiol.* 70, 1777–1786.

Kaplan, D.L. and Kaplan, A.M. (1982) Thermophilic biotransformations of 2,4,6-trinitrotoluene under simulated composting conditions. *Applied and Environmental Microbiology* 44, 757–760.

Kelly, R.L., Liu, B. and Srivastava, V. (1998) Bioremediation: Principles and Practice-Bioremediation Technologies.In: Sikdar, S.K. and Irvine, R.L. (eds) *Landfarming: A practical guide*. Technomic Publishing, Lancaster, Pennsylvania, pp. 223–243.

Khan, F.I., Husain, T. and Hejazi, R. (2004) An overview and analysis of site remediation technologies. *Journal of Environmental Management* 71, 95–122.

Kharaka, Y.K., Hanor, J.S., Heinrich, D.H. and Karl, K.T. (2007)*Deep Fluids in the Continents: I. Sedimentary Basins, Treatise on Geochemistry*. Pergamon, Oxford, pp. 1–48.

Kogbara, R.B. (2008) Ranking agro-technical methods and environmental parameters in the biodegradation of petroleum-contaminated soils in Nigeria. *Electronic Journal of Biotechnology* 11.

Kuyukina, M.S., Ivshina, I.B., Ritchkova, M.I., Philp, J.C., Cunningham, C.J. and Christofi, N. (2003) Bioremediation of crude oil-contaminated soil using slurry-phase biological treatment and land farming techniques. *Soil & Sediment Contamination* 12, 85–99.

Leahy, J.G. and Colwell, R.R. (1990) Microbial degradation of hydrocarbons in the environment. *Microbiological Reviews* 54, 305–315.

Lee, K. and Levy, E.M. (1989) Enhancement of the natural biodegradation of condensate and crude oil on beaches of Atlantic Canada.*Proceedings of the 1989 Oil Spill Conference*. American Petroleum Institute, Washington, DC, pp. 479–486.

Lee, K. and Levy, E.M. (1991) Bioremediation: Waxy crude oils stranded on low-energy shorelines. *Proceedings of the 1991 Oil Spill Conference*. American Petroleum Institute, Washington, DC, pp. 541–554.

Lee, K. and Trembley, G.H. (1993) Bioremediation: application of slow-release fertilizers on low energy shorelines. *Proceedings of the 1993 Oil Spill Conference*. American Petroleum Institute, Washington, DC, pp. 449–454.

Lee, K., Tremblay, G.H. and Cobanli, S.E. (1995) Bioremediation of oiled beach sediments: assessment of inorganic and organic fertilizers.*Proceedings of 1995 Oil Spill Conference*. American Petroleum Institute, Washington, DC, pp. 107–113.

Lee, K., Lunel, T., Wood, P., Swannell, R. and Stoffyn-Egli, P. (1997) Shoreline cleanup by acceleration of clay-oil flocculation processes. *Proceedings of 1997 International Oil Spill Conference*. American Petroleum Institute, Washington, DC, pp. 235–240.

Leirós, M.C., Trasar-Cepeda, C., Seoane, S. and Gil-Sotres, F. (1999) Dependence of mineralization of soil organic matter on temperature and moisture. *Soil Biology and Biochemistry* 31, 327–335.

Li, L., Cunningham, C.J., Pas, V., Philp, J.C., Barry, D.A. and Anderson, P. (2004) Field trial of a new aeration system for enhancing biodegradation in a biopile. *Waste Management* 24, 127–137.

Loser, C., Seidel, H., Hoffmann, P. and Zehnsdorf, A. (1999) Bioavailability of hydrocarbons during microbial remediation of a sandy soil. *Applied Microbiology and Biotechnology* 51, 105–111.

Lu, M., Zhang, Z., Sun, S., Wei, X., Wang, Q. and Su, Y. (2010) The Use of Goosegrass (*Eleusine indica*) to Remediate Soil Contaminated with Petroleum. *Water, Air, & Soil Pollution* 209, 181–189.

Lynch, J.M. and Moffat, A.J. (2005) Bioremediation – prospects for the future application of innovative applied biological research. *Annals of Applied Biology* 146, 217–221.

MacNaughton, S.J., Stephen, J.R., Venosa, A.D., Davis, G.A., Chang, Y.-J. and White, D.C. (1999) Microbial Population Changes during Bioremediation of an Experimental Oil Spill. *Applied and Environmental Microbiology* 65, 3566–3574.

Madsen, E.L. (1991) Determining *In Situ* Biodegradation: Facts and Challenges.*Environmental Science & Technology* 25, 1663–1673.

Madsen, E. (1997) Methods for determining biodegradability. In: Hurst, C.J., Knudsen, G.R., McInerney, M.J., Stetzenbach, L.D. and Walter, M.V. (eds) *Manual of Environmental Microbiology*. ASM Press, Washington, DC, pp. 177–212.

Madsen, E.L. (1998a) Epistemology of environmental microbiology. *Environmental Science & Technology* 32, 429–439.

Madsen, E.L. (1998b) Theoretical and applied aspects of bioremediation: The influence of microbiological processes on organic compounds in field sites.In: Burlage, R., Atlas, R., Stahl, D., Geesey, G. and Sayler, G. (eds) *Techniques in Microbial Ecology*. Oxford University Press, New York, pp. 354–407.

Merkl, N., Schultze-Kraft, R. and Infante, C. (2005) Assessment of tropical grasses and legumes for phytoremediation of petroleum-contaminated soils. *Water, Air, & Soil Pollution* 165, 195–209.

Milne, B.J., Baheri, H.R. and Hill, G.A. (1998) Composting of a heavy oil refinery sludge. *Environmental Progress* 17, 24–27.

Mishra, S., Jyot, J., Kuhad, R.C. and Lal, B. (2001a) Evaluation of inoculum addition to stimulate *in situ* bioremediation of oily-sludge-contaminated soil. *Applied and Environmental Microbiology* 67, 1675–1681.

Mishra, V., Lal, R. and Srinivasan.(2001b) Enzymes and operons mediating xenobiotic degradation in bacteria. *Critical Reviews in Microbiology* 27, 133–166.

Mohan, S., Kisa, T., Ohkuma, T., Kanaly, R. and Shimizu, Y. (2006) Bioremediation technologies for treatment of PAH-contaminated soil and strategies to enhance process efficiency. *Reviews in Environmental Science and Biotechnology* 5, 347–374.

Mueller, J.G., Chapman, P.J. and Pritchard, P.H. (1989)Creosote-contaminated sites – Their potential for bioremediation. *Environmental Science & Technology* 23, 1197–1201.

Mueller, J.G., Lantz, S.E., Blattmann, B.O. and Chapman, P.J. (1991) Bench-scale evaluation of alternative biological treatment processes for the remediation of pentachlorophenol- and creosote-contaminated materials: Slurry-phase bioremediation. *Environmental Science and Technology* 25, 1055–1061.

Mulkins-Phillips, G.J. and Stewart, J.E. (1974) Distribution of hydrocarbon-utilizing bacteria in Northwestern Atlantic waters and coastal sediments. *Canadian Journal of Microbiology* 20, 955–956.

Njoku, K.L., Akinola, M.O. and Oboh, B.O. (2009) Phytoremediation of crude oil contaminated soil: The effect of growth of *Glycine max* on the physico-chemistry and crude oil contents of soil. *Nature and Science* 7, 79–87.

Nwachukwu, S.U. (2001) Bioremediation of Sterile Agricultural Soils Polluted with Crude Petroleum by Application of the Soil Bacterium, *Pseudomonas putida*, with Inorganic Nutrient Supplementations. *Current Microbiology* 42, 231–236.

Nyman, J.A. (1999) Effect of Crude Oil and Chemical Additives on Metabolic Activity of Mixed Microbial Populations in Fresh Marsh Soils. *Microbial Ecology* 37, 152–162.

Oboh, B.O., Ilori, M.O., Akinyemi, J.O. and Adebusoye, S.A. (2006) Hydrocarbon Degrading Potentials of Bacteria Isolated from a Nigerian Bitumen (Tarsand) Deposit. *Nature and Science* 4, 51–57.

Odjadjare, E.E.O., Ajisebutu, S.O., Igbinosa, E.O., Aiyegoro, O.A., Trejo-Hernandez, M.R. and Okoh, A.I. (2008) Escravos light crude oil degrading potentials of axenic and mixed bacterial cultures. *Journal of General and Applied Microbiology* 54, 277–284.

Odokwuma, L.O. and Dickson, A.A. (2003). Bioremediation of a crude-oil polluted tropical mangrove environment. *Journal of Applied Science&Environmental Management* 7, 23–29.

Ojo, O.A. (2006) Petroleum-hydrocarbon utilization by native bacterial population from a wastewater canal Southwest Nigeria. *African Journal of Biotechnology* 5, 333–337.

Ojumu, T.V., Bello, O.O., Sonibare, J.A. and Solomon, B.O. (2005) Evaluation of microbial systems for bioremediation of petroleum refinery effluents in Nigeria. *African Journal of Biotechnology* 4, 31–35.

Okoh, A.I. (2003) Biodegradation of Bonny light crude oil in soil microcosms by some bacterial strains isolated from crude oil flow stations saver pits in Nigeria. *African Journal of Biotechnology* 2, 104–108.

Okoh, A.I. (2006) Biodegradation alternative in the cleanup of petroleum hydrocarbon pollutants. *Biotechnology and Molecular Biology Reviews* 1, 38–50.

Okparanma, R.N., Ayotamuno, M.J. and Araka, P.P. (2009) Bioremediation of hydrocarbon contaminated-oil field drill-cuttings with bacterial isolates. *African Journal of Environmental Science Technology* 3, 131–140.

Olajire, A.A., Altenburger, R., Küster, E. and Brack, W. (2005) Chemical and ecotoxicological assessment of polycyclic aromatic hydrocarbon – contaminated sediments of the Niger Delta, Southern Nigeria. *Science of the Total Environment* 340, 123–136.

Orchard, V.A. and Cook, F.J. (1983) Relationship between soil respiration and soil moisture. *Soil Biology and Biochemistry* 15, 447–453.

Osuji, Leo C. and Ozioma, A. (2007) Environmental Degradation of Polluting Aromatic and Aliphatic Hydrocarbons: A Case Study. *Chemistry & Biodiversity* 4, 424–430.

Pavel, L.V. and Gavrilescu, M. (2008) Overview of *ex situ* decontamination techniques for soil cleanup. *Environmental Engineering and Management Journal* 7, 815–834.

Pearce, K. and Ollermann, R.A. (1998) Status and scope of bioremediation in South Africa.In: Sikdar, S.K. and Irvine, R.L. (eds) *Bioremediation: Principles and Practice-Bioremediation Technologies*. Technomic Publishing, Lancaster, Pennsylvania, pp. 155–182.

Peng, R.-H., Xiong, A.-S., Xue, Y., Fu, X.-Y., Gao, F., Zhao, W., Tian, Y.-S. and Yao, Q.-H. (2008) Microbial biodegradation of polyaromatic hydrocarbons. *FEMS Microbiology Reviews* 32, 927–955.

Perfumo, A., Banat, I.M., Canganella, F. and Marchant, R. (2006) Rhamnolipid production by a novel thermophilic hydrocarbon-degrading *Pseudomonas aeruginosa* AP02-1. *Applied Microbiology and Biotechnology* 72, 132–138.

Pinholt, Y., Struwe, S. and Kjøller, A. (1979) Microbial Changes during Oil Decomposition in Soil. *Holarctic Ecology* 2, 195–200.

Prado-Jatar, M., Correa, M., Rodriguez-Grau, J. and Carneiro, M. (1993) Oil Sludge Landfarming Biodegradation Experiment Conducted At a Tropical Site in Eastern Venezuela. *Waste Management Research* 11, 97–106.

Prince, R.C., Clark, J.R. and Lee, K. (2002) Bioremediation effectiveness: Removing hydrocarbons while minimizing environmental impact. *9th International Petroleum Environmental Conference*, IPEC (Integrated Petroleum Environmental Consortium), Albuquerque, New Mexico.

Ramaswami, A. and Luthy, R.G. (1997) Measuring and modeling physicochemical limitations to bioavailability and biodegradation. In: Hurst, C.J., Knudsen, G.R., Mcinerney, M.J., Stetzenbach, L.D. and Walter, M.V. (eds) *Manual of Environmental Microbiology*. ASM Press, Washington, DC, p. 894.

Ramsay, M.A., Swannell, R.P.J., Shipton, W.A., Duke, N.C. and Hill, R.T. (2000) Effect of bioremediation on the microbial community in oiled mangrove sediments. *Marine Pollution Bulletin* 41, 413–419.

Reichenauer, T.G. and Germida, J.J. (2008) Phytoremediation of Organic Contaminants in Soil and Groundwater. *ChemSusChem* 1, 708–717.

Rhykerd, R., Crews, B., McInnes, K. and Weaver, R.W. (1999) Impact of bulking agents, forced aeration, and tillage on remediation of oil-contaminated soil. *Bioresource Technology* 67, 279–285.

Rittmann, B.E. and McCarty, P.L. (2001) *Environmental Biotechnology: Principles and Applications*. McGraw-Hill, New York.

Rosenberg, E., Legmann, R., Kushmaro, A., Taube, R., Adler, E. and Ron, E.Z. (1992) Petroleum bioremediation – a multiphase problem. *Biodegradation* 3, 337–350.

Rubinos, D., Villasuso, R., Muniategui, S., Barral, M. and Díaz-Fierros, F. (2007) Using the Landfarming Technique to Remediate Soils Contaminated with Hexachlorocyclohexane Isomers. *Water, Air, & Soil Pollution* 181, 385–399.

Salanitro, J.P., Dorn, P.B., Huesemann, M.H., Moore, K.O., Rhodes, I.A., Jackson, L.M.R., Vipond, T.E., Western, M.M. and Wisniewski, H.L. (1997) Crude oil hydrocarbon bioremediation and soil ecotoxicity assessment. *Environmental Science & Technology* 31, 1769–1776.

Sanscartier, D., Zeeb, B., Koch, I. and Reimer, K. (2009) Bioremediation of diesel-contaminated soil by heated and humidified biopile system in cold climates. *Cold Regions Science and Technology* 55, 167–173.

Sarkar, D., Ferguson, M., Datta, R. and Birnbaum, S. (2005) Bioremediation of petroleum hydrocarbons in contaminated soils: Comparison of biosolids addition, carbon supplementation, and monitored natural attenuation. *Environmental Pollution* 136, 187–195.

Šašek, V., Volfová, O., Erbanová, P., Vyas, B.R.M. and Matucha, M. (1993) Degradation of PCBs by white rot fungi, methylotrophic and hydrocarbon utilizing yeasts and bacteria. *Biotechnology Letters* 15, 521–526.

Sathishkumar, M., Binupriya, A.R., Baik, S.-H. and Yun, S.-E. (2008) Biodegradation of Crude Oil by Individual Bacterial Strains and a Mixed Bacterial Consortium Isolated from Hydrocarbon Contaminated Areas. *CLEAN - Soil, Air, Water* 36, 92–96.

Schneider, K., Oltmanns, J., Radenberg, T., Schneider, T. and Pauly-Mundegar, D. (1996) Uptake of nitroaromatic compounds in plants. *Environmental Science and Pollution Research* 3, 135–138.

Schoefs, O., Deschenes, L. and Samson, R. (1998) Efficiency of a new covering system for the environmental control of biopiles used for the treatment of contaminated soils. *Journal of Soil Contamination* 7, 753–771.

Semple, K.T., Reid, B.J. and Fermor, T.R. (2001) Impact of composting strategies on the treatment of soils contaminated with organic pollutants. *Environmental Pollution* 112, 269–283.

Semple, K.T., Morriss, A.W.J. and Paton, G.I. (2003) Bioavailability of hydrophobic organic contaminants in soils: fundamental concepts and techniques for analysis. *European Journal of Soil Science* 54, 809–818.

Semple, K.T., Doick, K.J., Jones, K.C., Burauel, P., Craven, A. and Harms, H. (2004) Defining bioavailability and bioaccessibility of contaminated soil and sediment is complicated. *Environmental Science & Technology* 38, 228a–231a.

Semple, K.T., Doick, K.J., Wick, L.Y. and Harms, H. (2007) Microbial interactions with organic contaminants in soil: Definitions, processes and measurement. *Environmental Pollution* 150, 166–176.

Sepic, E., Leskovsek, H. and Trier, C. (1995) Aerobic bacterial degradation of selected polyaromatic compounds and n-alkanes found in petroleum. *Journal of Chromatography* A 697, 515–523.

Siciliano, S.D. and Germida, J.J. (1997) Bacterial inoculants of forage grasses that enhance degradation of 2-chlorobenzoic acid in soil. *Environmental Toxicology and Chemistry* 16, 1098–1104.

Siciliano, S.D. and Germida, J.J. (1998) Mechanisms of phytoremediation: biochemical and ecological interactions between plants and bacteria. *Environmental Reviews* 6, 65–79.

Siciliano, S.D. and Germida, J.J. (1999) Enhanced phytoremediation of chlorobenzoates in rhizosphere soil. *Soil Biology and Biochemistry* 31, 299–305.

Siciliano, S.D., Goldie, H. and Germida, J.J. (1998) Enzymatic activity in root exudates of dahurian wild rye (*Elymus dauricus*) that degrades 2-chlorobenzoic acid. *Journal of Agricultural and Food Chemistry* 46, 5–7.

Siciliano, S.D., Fortin, N., Mihoc, A., Wisse, G., Labelle, S., Beaumier, D., Ouellette, D., Roy, R., Whyte, L.G., Banks, M.K., Schwab, P., Lee, K. and Greer, C.W. (2001) Selection of specific endophytic bacterial genotypes by plants in response to soil contamination. *Applied and Environmental Microbiology* 67, 2469–2475.

Singer, M. and Finnerty, W. (1984) Microbial metabolism of straight-chain and branched alkanes. In: Atlas, R.M. (ed.) *Petroleum Microbiology*. Macmillan, New York, pp. 1–59.

Smith, M.R. (1990) The biodegradation of aromatic hydrocarbons by bacteria. *Biodegradation* 1, 191–206.

Spormann, A.M. and Widdel, F. (2000) Metabolism of alkylbenzenes, alkanes, and other hydrocarbons in anaerobic bacteria. *Biodegradation* 11, 85–105.

Stegmann, R., Lotter, S. and Heerenklage, J. (1991) Biological treatment of oil contaminated soils in bioreactors. In: Hinchee, R.E. and Offenbuttel, R.E. (eds) *On-Site Bioremediation: Processes for Xenobiotic and Hydrocarbon Treatment*. Butterworth-Heinemann, Oxford, UK, pp. 188–208.

Stroo, H.F. (1992) Biotechnology and Hazardous Waste Treatment. *Journal of Environmental Quality* 21, 167–175.

Stroud, J.L., Paton, G.I. and Semple, K.T. (2007) Microbe-aliphatic hydrocarbon interactions in soil: implications for biodegradation and bioremediation. *Journal of Applied Microbiology* 102, 1239–1253.

Surampalli, R.Y. and Ong, S. (2004) *Natural Attenuation of Hazardous Wastes*. American Society of Civil Engineers, Reston, Virginia.

Swanell, R.P.J. and Head, I.M. (1994) Bioremediation come of age. *Nature* 396–397.

Swindoll, C.M., Aelion, C.M. and Pfaender, F.K. (1988) Influence of inorganic and organic nutrients on aerobic biodegradation and on the adaptation response of subsurface microbial communities. *Applied and Environmental Microbiology* 54, 212–315.

Tadesse, B., Donaldson, J.D. and Grimes, S.M. (1994) Contaminated and polluted land: A general review of decontamination management and control. *Journal of Chemical Technology & Biotechnology* 60, 227–240.

Tam, N.F.Y., Guo, C.L., Yau, W.Y. and Wong, Y.S. (2001) *Preliminary study on biodegradation of phenanthrene by bacteria isolated from mangrove sediments in Hong Kong*. Pergamon-Elsevier Science Ltd, Hong Kong, pp. 316–324.

Tam, N.F.Y., Guo, C.L., Yau, W.Y. and Wong, Y.S. (2002) Preliminary study on biodegradation of phenanthrene by bacteria isolated from mangrove sediments in Hong Kong. *Marine Pollution Bulletin* 45, 316–324.

Tanee, F.B.G. and Kinako, P.D.S. (2008) Comparative studies of biostimulation and phytoremediation in the mitigation of crude oil toxicity in tropical soil. *Journal of Applied Science Environmental Management* 12, 143–147.

Trinidade, P., Sobral, L.G., Rizzo, A.C., Leite, S.G.F., Lemos, J.L.S., Milloili, V.S. and Soriano, A.U. (2002) Evaluation of the biostimulation and bioaugmentation techniques in the bioremediation process of petroleum hydrocarbon contaminated soils. *9th International Petroleum Environmental Conference*, IPEC (Integrated Petroleum Environmental Consortium), Albuquerque, New Mexico.

Trindade, P.V.O., Sobral, L.G., Rizzo, A.C.L., Leite, S.G.F. and Soriano, A.U. (2005) Bioremediation of a weathered and a recently oil-contaminated soils from Brazil: a comparison study. *Chemosphere* 58, 515–522.

Udosen, E.D., Essien, J.P. and Ubom, R.M. (2001) Bioamendment of petroleum contaminated ultisol: effect on oil content, heavy metals and pH of tropical soil. *Journal of Environmental Sciences-China* 13, 92–98.

Uhlířová, E., Elhottová, D., Tříska, J. and Šantrůčková, H. (2005) Physiology and microbial community structure in soil at extreme water content. *Folia Microbiologica* 50, 161–166.

USEPA (1996) Engineering Bulletin: Composting (EPA/540/S-96/502).

Van Hamme, J.D., Singh, A. and Ward, O.P. (2003) Recent advances in petroleum microbiology. *Microbiology and Molecular Biology Reviews* 67, 503–549.

Venosa, A.D. and Zhu, X.Q. (2003) Biodegradation of crude oil contaminating marine shorelines and freshwater wetlands. *Spill Science & Technology Bulletin* 8, 163–178.

Vidali, M. (2001) Bioremediation. An overview. *Pure Applied Chemistry* 73, 1163–1172.

Volkering, F., Breure, A.M. and Rulkens, W.H. (1997) Microbiological aspects of surfactant use for biological soil remediation. *Biodegradation* 8, 401–417.

Wammer, K.H. and Peters, C.A. (2005) Polycyclic aromatic hydrocarbon biodegradation rates: A structure-based study. *Environmental Science & Technology* 39, 2571–2578.

Watkinson, R.J. and Morgan, P. (1990) Physiology of aliphatic hydrocarbon-degrading microorganisms. *Biodegradation* 1, 79–92.

White, P., Wolf, D., Thoma, G. and Reynolds, C. (2006) Phytoremediation of alkylated polycyclic aromatic hydrocarbons in a crude oil-contaminated soil. *Water, Air, & Soil Pollution* 169, 207–220.

Whyte, L.G., Goalen, B., Hawari, J., Labbé, D., Greer, C.W. and Nahir, M. (2001) Bioremediation treatability assessment of hydrocarbon-contaminated soils from Eureka, Nunavut. *Cold Regions Science and Technology* 32, 121–132.

Widdel, F. and Rabus, R. (2001) Anaerobic biodegradation of saturated and aromatic hydrocarbons. *Current Opinion in Biotechnology* 12, 259–276.

Wild, S.R. and Jones, K.C. (1993) Biological and abiotic losses of polynuclear aromatic hydrocarbons(PAHs) from soils freshly amended with sewage sludge. *Environmental Toxicology and Chemistry* 12, 5–12.

Williams, R.T., Ziegenfuss, P.S. and Sisk, W.E. (1992) Composting of explosives and propellant contaminated soils under thermophilic and mesophilic conditions. *Journal of Industrial Microbiology and Biotechnology* 9, 137–144.

Wilson, S.C. and Jones, K.C. (1993) Bioremediation of soil contaminated with polynuclear aromatic hydrocarbons (PAHs): a review. *Environmental Pollution* 81, 229–249.

Wiltse, C.C., Rooney, W.L., Chen, Z., Schwab, A.P. and Banks, M.K. (1998) Greenhouse Evaluation of Agronomic and Crude Oil-Phytoremediation Potential among Alfalfa Genotypes. *Journal of Environmental Quality* 27, 169–173.

Woodhull, P.M. and Jerger, D.E. (1994) Bioremediation using a commercial slurry-phase biological treatment system: Site-specific applications and costs. *Remediation Journal* 4, 353–362.

Wu, T. and Crapper, M. (2009) Simulation of biopile processes using a hydraulics approach. *Journal of Hazardous Materials* 171, 1103–1111.

Xia, H.P. (2004) Ecological rehabilitation and phytoremediation with four grasses in oil shale mined land. *Chemosphere* 54, 345–353.

Yeung, P.Y., Johnson, R.L. and Xu, J.G. (1997) Biodegradation of petroleum hydrocarbons in soil as affected by heating and forced aeration. *Journal of Environmental Quality* 26, 1511–1516.

Young, L.Y. and Cerniglia, C.E. (1995) *Microbial Transformation and Degradation of Toxic Organic Chemicals.* Wiley Liss, New York.

Zengler, K., Richnow, H.H., Rossello-Mora, R., Michaelis, W. and Widdel, F. (1999) Methane formation from long-chain alkanes by anaerobic microorganisms. *Nature* 266–269.

Zhou, Q.L., McCraven, S., Garcia, J., Gasca, M., Johnson, T.A. and Motzer, W.E. (2009) Field evidence of biodegradation of N-Nitrosodimethylamine (NDMA) in groundwater with incidental and active recycled water recharge. *Water Research* 43, 793–805.

Chapter 15

Bioremediation of Polycyclic Aromatic Hydrocarbons

Carl G. Johnston and Gloria P. Johnston

General Description of Polycyclic Aromatic Hydrocarbons

Polycyclic aromatic hydrocarbons (PAHs) are among the most common organic pollutants in soil, water and sediments worldwide. They are a large, diverse class of persistent hydrophobic organic compounds. PAHs are considered potential carcinogens, mutagens and teratogens by the US Environmental Protection Agency (USEPA, 2011), the US Agency for Toxic Substances and Disease Registry (ATSDR, 2010), and are listed by the National Waste Minimization Program of the US EPA (USEPA, 2009) and the European Union as priority pollutants.

PAHs are composed of carbon, hydrogen, nitrogen, sulfur and oxygen atoms. PAHs can be colourless or pale yellow with low solubility in water, high melting points and low vapour pressure (Haritash and Kaushik, 2009). PAHs are classified as semivolatile organic compounds (SVOC) because of their high boiling points (greater than 200°C; USEPA, 2011). The simplest PAH is naphthalene ($C_{10}H_8$), consisting of two fused benzene rings. The most complex PAH is coronene ($C_{24}H_{12}$) with seven fused benzene rings. In between, there are at least 100 identified species of PAHs, all of which differ in physical and chemical properties and biodegradability. The recalcitrance of PAHs increases with mass. Low molecular weight (LMW) PAHs usually have higher solubility, lower water-octanol partition coefficients (K_{ow}) and are more likely to be biodegraded. High molecular weight (HMW) PAHs, conversely have low solubility, higher K_{ow}, tend to sorb into organic matter and soils and persist longer in the environment.

Sources of PAHs

PAHs can be categorized as pyrolytic or petrogenic. Pyrolytic PAHs result from incomplete fuel combustion, wood and coal burning, car emissions, smoking tobacco and grilling meat. This type of PAH is also found in coal tar, creosote, roofing tar and parking lot sealcoats (ATSDR, 2010). Petrogenic PAHs are mainly derived from crude oil, unburned fuel and refinery products.

Most terrestrial natural sources of PAHs come from wildfires. Emissions from volcanic eruptions and thermal geological reactions are considered negligible (European Commission, 2011). PAHs are also naturally produced components of surface waxes of leaves, plant oils and cuticles of insects (Jaward *et al.*, 2004). Naphthalene is produced by magnolia flowers (Azuma *et al.*, 2010), Annonaceae flowers (Jürgens *et al.*, 2000), *Muscodor vitigenus*, an endophytic fungus (Daisy *et al.*, 2002) and the subtropical North American termite *Coptotermes formosanus* (Chen *et al.*, 1998).

Both anthropogenic and natural PAHs that are released into the atmosphere enter aquatic and terrestrial ecosystems via atmospheric deposition. In both urban and rural areas, PAHs can be transported as gases or aerosols over great distances. PAH deposition in sediments and waters is the result of point sources in industrialized and urbanized zones, such as harbours, railways, steel mills and shipping ports. This type of contamination is usually characterized by high PAH concentrations, affect coastal and riverine systems directly and present a more hazardous condition for humans, fish and wildlife.

Exposure to PAHs

According to the US EPA (2011) the majority of PAHs are harmful toxins, and some are considered potential carcinogens. In laboratory studies, PAHs cause tumours in animals exposed through contaminated food, breathing contaminated air, and by skin contact with contaminated soils or sediments. PAHs in air, soil, water and groundwater are usually present in mixtures rather than as single compounds (Gan *et al.*, 2009).

Exposure through air is mainly due to PAH vapours that are carried along with dust and other particles. Carcinogenic PAHs in air are produced by activities such as cigarette smoking, car exhaust, asphalt roads, agricultural and household wood burning, cooking meat or other foods at high temperatures and waste incineration.

Soil exposure occurs via contact with previously contaminated soils (e.g. after wood, coal or gasoline was burned at a particular site). Soils in and near coal tar production sites, petroleum facilities, asphalt and aluminium production plants, coal-gasification and wood-preserving facilities usually have elevated levels of PAHs (Wehrer and Totsche, 2009; Van Metre and Mahler, 2010).

While contact bans can be issued to minimize human contact with contaminated sediments, wildlife can get exposed regardless. Bottom-dwelling fish can develop deformations, erosions, lesions and tumours (DELTs) in PAH-contaminated sediments as a result of indirect exposure (i.e. via resuspension). In 2011, the United States Fish and Wildlife Service in a monitoring study of the South River, a tributary of the Chesapeake Bay, found that 53% of brown bull heads, *Ameirus nebulosus* had skin lesions and/or squamous carcinomas and 20% had liver tumours. They concluded that the tumours were likely a result of exposure to PAHs in the sediments. Similarly, a survey conducted from 1997 to 2000 in industrially contaminated Lake Erie tributaries (the Detroit, Ottawa, Ashtabula and Niagara Rivers) revealed that concentrations of PAH metabolites in bile of *A. nebulosus* were positively associated with concentrations of PAHs in sediments (Yang and Baumann, 2005). Also, reptiles, such as turtles, can be exposed to contaminated sediments with PAHs. A study from 2000 through 2003 showed a high incidence of deformities in embryos of snapping turtles (*Chelydra serpentina*) and painted turtles (*Chrysemys picta*) in the John Heinz National Wildlife Refuge, Philadelphia, Pennsylvania (Bell *et al.*, 2006). They found that from 13 to 19% of snapping turtle embryos and from 45 to 71% of painted turtle embryos showed lethal deformities. Snapping turtle adults and embryos had high levels of PAHs in their fat, indicating bioaccumulation of PAHs. The study concluded that the high deformity rates were primarily caused by sediment pollution.

Benthic macro-invertebrates that inhabit sediments can also develop morphological deformities due to direct exposure to heavy metals and organic compounds in sediments (Boonyatumanond et al., 2006).

In groundwater, PAHs usually occur in combination with other mono-aromatic hydrocarbons (benzene, toluene, ethyl benzene and xylenes) most commonly known as BTEX. BTEX and PAHs are associated with gasoline and other petroleum-derived fuels in underground storage tanks. It has been estimated that 65% of these tanks in the USA leak (Kane et al., 2001), and thus they are the main source of groundwater contamination. This situation is no different in other countries. For instance, groundwater monitoring wells of gasoline stations in the city of Rio de Janeiro, Brazil exceeded PAHs and BTEX maximum concentrations limits for drinking water (Rego and Pereira, 2007).

PAHs in Aquatic Environments

PAHs are ubiquitous toxins in different ecosystems. However, sediments are the major sink of PAHs in both freshwater and marine aquatic systems (Stark et al., 2003). Because of their hydrophobic nature, PAHs accumulate in bottom sediments and tend to sorb on to organic material. For example, in intertidal estuarine wetlands, oil spills with PAHs accumulated in mangrove sediments (Ke et al., 2002). In these ecosystems, concentrations of PAHs are usually elevated (>10,000 ng g^{-1} dry weight), and consequently benthic communities are impaired (Tian et al., 2008). High concentrations of petroleum compounds and PAHs are commonly found in harbours with intensive shipping activities (Hayes et al., 1999). For instance, in Lake Erie, USA, the highest levels of PAHs were found in urbanized harbours in the cities of Detroit, Cleveland and Buffalo (DeBruyn et al., 2009).

PAH contamination in lake sediments is widespread. The US Geological Survey National Water-Quality Assessment surveyed 38 lakes from 1970 to 2001 (Van Metre and Mahler, 2005). The study revealed that 42% of the lakes surveyed had increased PAH concentrations in sediments. These lakes were located in urbanized watersheds. The study concluded that rapid urbanization in the USA contributed to this trend. In support of this conclusion, analysis of the sediments in Lake Taihu, one of the five largest freshwater lakes in China, contained high concentrations of total PAHs (1207 to 4754 ng g^{-1} dry weight) from high temperature pyrolytic origins (Qiao et al., 2006). Indeed, heavy industrialization and increased populations in the last decade contributed to the sediment contamination and threatened water supply for nearby areas.

Recreational activities can also cause PAH contamination of aquatic systems. For instance, power boating, water skiing and jet skiing on Brown Lake, Australia resulted in PAH contamination (Mosisch and Arthington, 2001). Ten PAHs, including benzo(a)pyrene, flouranthene and chrysene (known indicators of fossil fuel combustion processes) were found in sediments. In addition, a high concentration of benzo(a)pyrene (1070 µg kg^{-1} dry weight) exceeded sediment quality guidelines, presenting a risk for aquatic organisms. The study concluded that the level of contamination was a consequence of four decades of unregulated motorized recreational activities.

Contaminated river sediments, as illustrated in the examples below, are abundant worldwide. In the USA, some of these ecosystems are highly contaminated and therefore designated 'superfund sites', the highest pollution category of the US EPA. For example, Fields Brook, a tributary stream of the Ashtabula River in north-east Ohio and a superfund site, contains elevated sediment concentrations of uranium, PAHs, polychlorinated biphenyls (PCBs) and heavy metals (Li *et al*., 2001). One of the most industrialized regions in China, the Pearl River Delta, has been heavily impacted with anthropogenic PAHs (138 to 6973 ng g^{-1} dry weight total PAHs) in the last three decades as a result of urbanization and development (Luo *et al*., 2008). Similarly, sediments of the Yellow River, the second largest river in China, are highly contaminated with PAHs (464 to 2621 ng g^{-1} dry weight total PAHs) because the river is used as an industrial sewer receiving wastes from oil refineries, paper mills and pharmaceutical companies (Xu *et al*., 2007). In Europe, sediments of the Oder River, situated in a heavily industrialized district in Poland (with numerous power stations, metallurgical and chemical industries) are heavily contaminated (633 to 146,400 ng g^{-1} dry weight total PAHs) with pyrogenic and petrogenic PAHs (Witt and Gründler, 2005).

Bioremediation of PAHs of Soils and Sediments

Bioremediation is a process that relies on biodegradation, which is the partial or total transformation, detoxification or removal of pollutants by microorganisms, plants and/or enzymes. According to the US EPA (2011), biodegradation as well as sorption and other chemical reactions that reduce concentrations, toxicity and mobility of contaminants in the environment are considered natural attenuation technologies. Application of bioremediation strategies in treatment of contaminated soils and sediments with PAHs can be simple, cost-effective, reduce environmental risks, eliminate further contamination and, more importantly, can be applied on site and *in situ* (Chauhan *et al*., 2008; Pazos *et al*., 2010). However, limiting factors, length of treatment, existence of microbial populations with degrading capabilities, and lack of nutrients for optimal growth are not necessarily well understood and can obscure bioremediation potential.

PAHs are recalcitrant compounds with low water solubility and high affinity for organic matter and soil particles. These characteristics often diminish bioavailability of PAHs to microorganisms and therefore their biodegradability. In addition, geochemical characteristics of the sediment matrix (i.e. concentration and presence of metals in sediments, age of the contaminants, organic matter and sorption) and environmental parameters (e.g. redox, pH, temperature) also determine the extent of biodegradation (Cerniglia *et al*., 1992; Nguyen *et al*., 2005; Xia *et al*., 2006). To overcome these limitations, two processes, biostimulation (adding nutrients or bulking agents) and bioaugmentation (addition of degrader microorganisms), are typically included in most bioremediation efforts (Bamforth and Singleton, 2005). For instance, bioremediation of groundwater contaminated with PAHs and/or BTEX often include both biostimulation, bioaugmentation and supplying oxygen into the aquifer to stimulate indigenous microbes (Alexander, 1994).

In this chapter, bioremediation approaches will be divided in two main categories: aerobic and anaerobic degradation. There are several approaches to bioremediate contaminated soils and sediments under aerobic conditions including composting, landfarming, microbial degradation (algal, fungal and bacterial) and phytoremediation. The success of each strategy depends on microbial degradation, both aerobic and anaerobic, and is discussed in detail in a subsequent section.

Aerobic Bioremediation

Aerobic bioremediation of PAHs is based on oxidation of benzene rings followed by breakdown into less complex metabolites and, sometimes, by complete mineralization into water and carbon dioxide (CO_2) (Bamforth and Singleton, 2005; Haritash and Kaushik, 2009). To achieve biodegradation, contaminated soils and sediments can be treated aerobically *in situ* or *ex situ*. Aerobic strategies have been developed to enhance microbial degradation and consequently PAH removal. Some of these strategies have been performed on bench studies, field scales and technological scales.

Composting

Composting is a biological process done under conditions that allow thermophilic heterotrophic microbes to decompose organic matter (by products and wastes) into a stable, useful final product (Iranzo *et al*., 2004). Compost piles are monitored for moisture and temperature and are periodically aerated. For bioremediation purposes, contaminated soils or sediments can be mixed with supplemental materials (organic waste and/or nutrients) and aerobic microorganisms with degradative capabilities. A combination of poultry manure and wood chips (Atagana, 2004) and mushroom compost with wheat straw, chicken manure and gypsum (Sasek *et al*., 2004) yielded high removal (~80%) of 3- and 4-ring PAHs from contaminated soils. The length of treatments can vary (100 days to 19 months) depending upon the carbon source, initial concentration, composition of PAHs and temperature.

Landfarming

Landfarming is the incorporation of soil or contaminated sediments into surfaces of non-contaminated soil (Sylvia *et al*., 1999). Clean soils are mixed with contaminated soils or sediments, then tilled to improve aeration and promote degradation by indigenous microbes. When the native microbial population lacks degradation potential, biostimulation and/or bioaugmentation is included to increase efficiencies of the process. As in composting, waste products are usually utilized as bulking agents (e.g. rice hulls) and as nutrient sources (e.g. dried blood as a slow-release nitrogen source). Known PAH degraders such as *Pseudomonas aeruginosa* strain 64 (Straube *et al*., 2003), *P. citronelloslis* 222A and *P. aeruginosa* isolate 312A (Rodrigo *et al*., 2005) have been used in landfarming to treat contaminated highly contaminated sediments such as found at superfund sites. High removal efficiency can be achieved by landfarming in moderate times. For instance, Mphekgo and Cloete (2004) found that after 11 months of treatment of soil, total PAHs decreased by 87%. Landfarming can be a relatively simple, low cost and low maintenance approach. However, unfavourable conditions such as limited aeration, mobility of PAHs in soils, toxicity, and transformation of PAHs into more complex by-products can restrict or prolong the remediation process.

Microbial degradation

Algae, fungi and bacteria can degrade PAHs by mineralization or cometabolism.

Algal degradation of PAHs

Some algae, including cyanobacteria, green algae and diatoms, are able to degrade PAHs (Abed and Koster, 2005; Hong et al., 2008). Mechanisms of removal of PAHs by algae follow similar steps to that of other pollutants and heavy metals (fast removal initially, followed by slow absorption, accumulation and degradation; Hong et al., 2008). Algal degradation is species-specific because size, cell wall composition and cell morphology influence degradation (Lee et al., 1999). Cell surface area to cell volume ratios and initial cell density of freshwater algae correlates with removal rates. Chan et al. (2006) showed that cells of *Selenastrum capricornutum* could degrade a mixture of phenanthrene, fluoranthene and pyrene at higher rates when cell density was greater than 5×10^4. A number of algal species from the division Chlorophyta have been used in experiments of PAH removal. Pure cultures of *Chlorella vulgaris*, *Scenedesmus platydiscus*, *Scenedesmus quadricauda* and *Selenastrum capricornutum* were able to remove fluoranthene (47%, 48%, 56% and 81%, respectively) and pyrene (41%, 49%, 35% and 74%, respectively) after 7 days of incubation (Lei et al., 2007). *Nitzschia* sp. and *Skeletonema costatum* were able to degrade mixtures of fluoranthene and phenanthrene more efficiently than when provided with these same molecules as single compounds (Hong et al., 2008). Degradation by algae, unlike bacterial degradation, often requires consortia. Borde et al. (2003) reported 85% removal of phenanthrene when algae, *Chlorella sorokini-ana*, and two bacteria, *Pseudomonas migulae* and *Sphingomonas yanoikuyae*, were combined in a microcosm experiment.

Bacterial degradation of PAHs

Aerobic bacterial degradation of PAHs has been demonstrated in a variety of environments, including marshes, marine sediments, brackish sediments and contaminated soils (Langenhoff et al., 1996; Lovley, 2000; Amellal et al., 2001; Verrhiest et al., 2002; Meckenstock et al., 2004; Abbondanzi et al., 2005; Ambrosoli et al., 2005; Chang et al., 2005, 2008). Several strains of *Pseudomonas*, *Agrobacterium*, *Bacillus*, *Burkholderia* and *Sphingomonas* have been isolated from PAH-contaminated soils and can use PAHs as a sole source of carbon and energy. Some bacteria isolated from oil-polluted desert soils are of special importance for bioremediation because at high temperatures (~50°C) solubility and mass transfer rates of PAHs increase (Zeinali et al., 2007a). One example is the hydrocarbon-degrading thermophilic bacterium *Nocardia otitidiscaviarum* strain TSH1, which can grow on straight chain aliphatic hydrocarbons and on PAHs (pyrene, phenanthrene, anthracene and naphthalene) as sole sources of carbon and energy (Zeinali et al., 2007b). Other extremophiles such as *Halomonas aromativorans*, a moderate halophile (Garcia et al., 2004), and the thermophiles *Thermus brockii* (Feitkenhauer et al., 2003) and *Bacillus thermooleovorans* (Annweiler et al., 2000) can also grow on pyrene and naphthalene as the sole sources of carbon and energy.

Bacterial biodegradation of PAHs is mainly achieved by two mechanisms: mineralization and cometabolism. Bacterial taxonomic groups involved in mineralization of PAHs include Nocardoforms, Sphingomonads, *Burkholderia*, *Pseudomonas* and *Mycobacterium* (Johnsen et al., 2005). The latter (well known PAH degrader) can mineralize sorbed phenanthrene and other PAHs with low solubility (Bastiaens et al., 2000). Research has also shown that some species of the ß- and γ-Proteobacteria and *Flavobacteria* can mineralize rapidly (~70%) naphthalene and phenanthrene within 40 days in aerobic microcosms (Rogers et al., 2007).

Some PAHs can be degraded partially or totally when other PAHs are used as primary energy or carbon source. This process is called co-metabolism and has been reported as one of the main processes involved in PAH degradation. *Sphingomonas* LB126 can degrade phenanthrene, fluoranthene and anthracene completely (without accumulation of metabolites) co-metabolically through multiple pathways when growing on glucose and pyruvate (van Herwijnen *et al*., 2003). Fluoranthene degraders (i.e. *Sphingomonas* spp.) can co-metabolize pyrene, fluorine, anthracene and benzo[a]pyrene after exposure to phenanthrene (Ho *et al*., 2000). *Sphingomonas paucimobilis* can co-metabolically degrade high molecular weight PAHs while *Rhodococcus* species can degrade flouranthene in the presence of anthracene (Dean-Ross *et al*., 2002).

Most bacterial degradation is accomplished by oxidation of benzene rings by dioxygenase enzymes, although some bacteria can use mono-oxygenases when degrading pyrene (Sanghvi, 2005). Mono-oxygenases reduce dioxygen and incorporate one oxygen into a PAH and a second one into water by oxidating NADPH (Husain, 2008). Dioxygenases (aromatic ring hydroxylating dioxygenases and ring-cleaving dioxygenases) incorporate both oxygen atoms into PAHs. Generally, after the initial reaction of dioxygenases, *cis*-dihydrodiol compounds are formed and dehydrogenated to produce dihydroxylated intermediates or catechols. Catechols then can be further metabolized into CO_2 and water (Bamforth and Singleton, 2005). Formation of different intermediates and enzymatic reactions depend on which degradation pathway is followed. Most PAH-degrading bacteria follow either the ortho or meta pathways. Other pathways such the upper and lower pathway, the Evans and Kiyohara pathway, gentisate, *o*-phthalate and the beta-ketoadipate pathway (Husain, 2008) have been described for few bacterial species. In addition, cytochrome P450s are used by bacteria and fungi in the oxidative process of PAHs.

Genes that code for enzymes involved in aerobic bacterial degradation of PAHs belong to Gram-negative and some Gram-positive bacteria (Habe and Omori, 2003). The majority of these genes are on circular plasmids (Habe *et al*., 2003; Stingley *et al*., 2004; Kim *et al*., 2008) but there are exceptions. For instance, the gene *pcaD* in *Terrabacter* sp. strain DBF63 involved in fluorene degradation is a linear plasmid (Habe *et al*., 2003). The most well-studied gene is *nahAc*, which codes for the α subunit ring-hydroxylating-dioxygenase (RHDα) in *Pseudomonas putida* NCIB 9816-4. RHDs genes (composed of small (α) and large (β) subunits) are widespread in bacteria species from the α-Proteobacteria, β-Proteobacteria, and γ-Proteobacteria (Cébron *et al*., 2008). Other genes include the *nidBA* genes present in *Mycobacterium* sp. strain PYR-1 during pyrene degradation.

Aerobic bacterial degradation of PAHs has been extensively accomplished with slurry systems and bioreactors. In a slurry reactor soils or sediments are mixed with a liquid that could also contain inoculum, nutrients or both. The success of this approach depends on aeration, mixing, temperature, dissolved oxygen, pH, nutrients and controlled addition of single or multiple microorganisms. Bioreactors, vessels that carry out chemical and biological processes, have been used mainly in the treatment of sewage and waste water but have been modified and adapted to treat contaminated soils (Gan *et al*., 2009). Bioreactors can be continuous flow tanks, fixed bioreactor columns, biofilters and bioscrubbers (Alexander, 1994). Bioreactors need to be monitored and controlled to efficiently remove PAHs. For instance, production of CO_2, dilution rate, temperature, hydraulic retention time, generation of toxic gases and by-products need to be carefully evaluated.

Fungal degradation of PAHs

Most bioremediation research is with bacteria, however, the filamentous nature of fungi and their ability to produce extracellular degradative enzymes allows them to penetrate solid substrates and to translocate nutrients in heterogeneous environments, traits which can confer an advantage in many heterogeneous solid systems such as soils. Some fungi are better able to degrade higher molecular weight PAHs than are bacteria. Fungi have been shown to degrade 'naphthalene, phenanthrene, anthracene, pyrene, benzo[a]pyrene, fluorene, dibenzothiophene, catechol, benzo[a]anthracene, benzo[-ghi]-erylene, chrysene, benzo[b]fluoranthene and benzo[k]-fluoranthene' (Cerniglia, 1997; Zheng and Obbard, 2002, 2003; Peng *et al.*, 2008). Fungi able to mineralize PAHs can be grouped as either ligninolytic or nonligninolytic (Cerninglia, 1997), where ligninolytic, fungi, also known as white rot fungi (WRF), are those that can degrade lignin. The dark-coloured lignin in wood is broken down by WRF, exposing cellulose and leaving a bleached appearance, i.e. a white rot, which gives these fungi their name. Lignin degradation is an aerobic process that allows fungal access to plant cellulose, the major carbon and energy source for the fungi. WRF are particularly suitable for bioremediation of a multitude of organic pollutants (including PAHs) under the right conditions. WRF can oxidize very insoluble, high molecular weight PAHs with five or more aromatic rings (Bumpus *et al.*, 1985; Wolter *et al.*, 1997; Baldrian, 2000). Many species of WRF are associated with bioremediation of PAHs. Among the most studied WRFs, *Phanearochaete chrysosporium, Pleurotus ostreatus, Irpex lacteus, Trametes versicolor* and *Bjerkandera* spp. are all are capable of metabolizing multiple PAHs to differing degrees (Novotny *et al.*, 1999; Baldrian, 2000). Of these, *P. ostreatus* is often chosen to study bioremediation of PAHs because of its ability to grow aggressively, outcompete other soil microbes, and produce ligninolytic enzymes in soil (Lang *et al.*, 1997; Martens and Zadrazil, 1998). In addition, WRF that cause white rot of humic material, i.e. litter-degrading fungi, show promise for soil bioremediation because they produce lignin-degrading enzymes and are better adapted to soil conditions than many wood-degrading WRF (Steffen *et al.*, 2003). A mixture of benzo[a]anthracene, chrysene, benzo[b]fluoranthene, benzo[k]fluoranthene, benzo[a]pyrene, dibenzo[a,h]anthracene and benzo[ghi]perylene, all high molecular weight PAHs, were significantly degraded in nonsterile soil after 15 weeks of incubation with *P. ostreatus* (Baldrian *et al.*, 2000). In this same study, cadmium and mercury affected ligninolytic enzyme production but this did not appear to decrease PAH degradation. Creosote-contaminated soils treated with *P. ostreatus* or *Irpex lacteus* showed higher degradation of 4–6 ring PAHs than soils treated with indigenous microbes (Byss *et al.*, 2008). Soils spiked with radiolabelled PAHs treated with *P. ostreatus* mineralized 5-ring PAH to a greater extent than in soils treated with indigenous microbes (In Der Wiesche *et al.*, 2003). The authors concluded that mineralization of lower molecular weight PAHs was reduced because *P. ostreatus* was antagonistic towards indigenous microbes.

The lignin-degrading system is the main process that WRF use to degrade organic contaminants (Barr and Aust, 1994). This system consists of extracellular peroxidases, lignin peroxidases (LiPs) and manganese-requiring peroxidases (MnPs), as well as laccases. WRF produce extracellular hydrogen peroxide, which oxidizes LiPs and MnPs. WRF also produce veratryl alcohol (VA), which is oxidized to a highly reactive cation radical by LiPs and MnPs. Thus LiPs and MnPs, supplied by hydrogen peroxide: (i) generate oxygen free radicals; (ii) oxidize mediator molecules such as VA; or (iii) oxidize MnII to MnIII (only in the case of MnPs), all of which lead to the non-specific one-electron oxidation of aromatic rings within lignin molecules or in organic pollutants (Barr and Aust, 1994). An aromatic ring oxidized by this process forms an unstable aryl cation radical that can break down to open the ring. Oxidation of benzo(a)pyrene into quinones using an enzyme preparation of LiP and hydrogen peroxide was the first demonstration of *in vitro* degradation of a PAH by a lignin-degrading enzyme (Haemmerli *et al.*, 1986). Enzyme preparations of MnPs are also able to degrade PAHs (Hofrichter *et al.*, 1998). Laccases, similar to LiPs and MnPs, catalyse one electron oxidation reactions, however in contrast to the peroxidases, the electron acceptor is molecular oxygen rather than hydrogen peroxide (Muñoz *et al.*, 1997). The electron donor for laccases can be from a wide spectrum of phenolic compounds.

While the ability of WRF to degrade many pollutants is still most often attributed to lignin-degrading enzymes, initial steps of PAH degradation by *P. ostreatus* was shown to involve cytochrome P450s (Bezalel *et al.*, 1996, 1997). Most recently, preparations of cloned P450 enzymes of *P. chrysosporium* were found to mediate the first rate limiting step in PAH degradation: i.e. they hydroxylate HMW-PAHs such as phenanthrene, pyrene and benzo(a)pyrene (Syed *et al.*, 2010). These studies show that WRF have multiple enzyme systems (besides LiPs, MnPs, and laccases) capable of metabolizing high molecular weight PAHs, but many are not yet identified. Cytochrome P-450 mono-oxygenases are hemethiolate proteins that catalyse multiple reactions such as reduction, dealkylation, epoxidation and carbon hydroxylation (Husain, 2008; Syed *et al.*, 2010). P450 systems are associated with cell membranes and consist of a mono-oxygenase and an oxidoreductase. Cytochrome P-450 mono-oxygenase incorporates one atom of oxygen into a xenobiotic molecule (i.e. the PAH) forming an epoxide; the other oxygen atom is reduced to water. These xenobiotic epoxides are unstable and rearrange into hydroxyl derivatives or become hydrated into *trans*-dihydrodiols (Cerniglia *et al.*, 1992). This is in contrast to bacterial cytochrome P-450s, which produce *cis*-dihydrodiols. These metabolites can undergo further enzyme reactions such as addition of glutathione, sulfate or sugar residues, generally making them more water soluble and less toxic.

Although the majority of fungal degradation of PAHs is performed by wood-degrading (ligninolytic) fungi, PAH degradation by non-ligninolytic fungi has also been observed. Non-ligninolytic fungal degradation is often accomplished by oxidation of benzene rings by the cytochrome P450 mono-oxygenase enzyme (described above). The majority of non-lygnolytic fungi cannot mineralize PAHs but generate more soluble and less toxic metabolites (Bamforth and Singleton, 2005). Non-ligninolytic fungi such as *Cunninghamella elegans* (Pothuluri *et al.*, 1995), *Penicillium janthinellum* (Launen *et al.*, 1999) and *Aspergillus niger* (Bamforth and Singleton, 2005) all have PAH-degrading capabilities. Two fungi isolated from a PAH contaminated gas work plant, *Coniothyrium* sp. and *Fusarium* sp., both showed improved PAH degradation (particularly for PAHs with more than three rings) when re-inoculated into contaminated soil (Potin *et al.*, 2004). This indicates that fungi indigenous to contaminated sites have great potential for PAH bioremediation, but have not yet been fully investigated.

Phytoremediation

Phytoremediation, the use of plants for degrading pollutants in soils, sediments, wetlands and groundwater is a cost-effective and environmentally friendly technology that has been successfully applied at sites containing a variety of organic compounds (Fiorenza *et al.*,

2000; Spriggs et al., 2005; Lin and Mendelssohn, 2009). Efficacy varies among plant species and depends on physicochemical properties of PAHs and environmental conditions. Because of their higher water solubility and volatility, low molecular weight PAHs can be more effectively removed by phytoremediation. Limitations of phytoremediation techniques include accurate measurement of degradation rates, prediction of treatment times, and effective monitoring programmes (Gan et al., 2009), all of which are made more complex due to the inherent heterogeneity of both contaminant distribution and plant growth.

Phytoremediation of contaminated soils occurs within two compartments: (i) the plant (within roots and vegetative parts of the plants); and (ii) the rhizosphere, the soil associated with the plant roots (Shimp et al., 1993; Siciliano and Germida, 1998). Mechanisms in the plant compartment include absorption to the root biomass, active translocation from roots to shoots, phytoaccumulation in roots and vegetative parts, transformation of PAHs into less toxic metabolites (detoxification) within plant cells, and phytoevaporation from plant leaves (Zeeb et al., 2006; Whitfield-Aslund et al., 2007). Plant-mediated abiotic removal of PAHs from soil include leachate, volatilization, photo-degradation and irreversible sorption (Pan et al., 2008). In the rhizosphere, PAHs are directly removed and/or transformed by peroxidases or dehydrogenases produced by the root system (Chu et al., 2006), indirectly removed by increased leaching due to root exudates, and indirectly mineralized or transformed by enhancement of indigenous microbes (Sun et al., 2010). Mineralization or transformation of PAHs by rhizosphere microbes is stimulated by root exudates (Chekol et al., 2004; Leigh et al., 2006). For example, Yoshitomi and Shann (2001) showed higher rates of pyrene removal in soils incubated with maize exudates than in soils without addition of exudates.

Ideal plants for phytoremediation have a large root surface area, are adapted to the soil conditions and require low maintenance (little fertilizer, no trimming). Grasses, such as species of the family Graminaceae, fulfil many of these considerations and have been widely studied. A greenhouse study of historically PAH-contaminated soils demonstrated that phytoremediation using tall fescue or yellow sweet clover reduced the labile PAH concentrations measured by PAH association into a solid organic polymer, however substantial amounts of the PAHs in the soil were unavailable to the plants for degradation (Parrish et al., 2005). A similar greenhouse study showed that phytoremediation of contaminated soils by tall fescue and switchgrass decreased by 40–90% the amount of labile PAHs (Cofield et al., 2008). These decreased PAH concentrations correlated with several toxicity assays.

One of the problems with phytoremediation is the heterogeneity of degradation (e.g. plant roots are not uniformly distributed in the soil, different depths have better root penetration and growth). For instance, Lee et al. (2008), in a phytoremediation study of PAH spiked soils, showed that two Korean grasses grew well and were associated with high phenanthrene (~99%) and pyrene (~90%) degradation, though unplanted soil controls also showed relatively high phenanthrene (99%) and pyrene (69%) degradation. Similarly, Rezek et al. (2008) showed that ryegrass was associated with PAH degradation (~50%) after 18 months. The lowest final PAH concentrations were found in bottom soils (18 cm) and were associated with greater root density.

Marine sediments, often contaminated with PAHs from multiple sources, can also be ameliorated by phytoremediation. Marine sediments contaminated with PAHs and PCBs were treated with eelgrass during 60 weeks (Huesemann et al., 2009). Much higher removal of PAHs was observed (73%) compared to unplanted control sediments (25%). Eelgrass probably stimulated microbial PAH degradation by increasing oxygen, providing root exudates and releasing plant enzymes. In a microcosm study of pyrene spiked sediments, *Kandelia candel* and *Bruguiera gymnorrhiza*, two mangrove plant species native to the South China Sea, removed 96% and 93%, respectively, of the pyrene without significant plant uptake within a 6 month treatment (Ke *et al.*, 2003).

Arbuscular mycorrhizal fungi (AMF) have been shown to promote plant establishment and survival in PAH-contaminated soils (Liao *et al.*, 2003). Supplementing plants with AMF can greatly enhance phytoremediation. Gao *et al.* (2011) showed high removal of phenanthrene (~99%) and pyrene (88%) when contaminated soils were treated with a host plant lucerne (*Medicago sativa*) and the AMF *Glomus mosseae* and *G. etunicatum*. Mycorrhizal colonization promoted accumulation of PAHs in plant roots while plant uptake was negligible. Similarly, residual PAH concentrations were significantly lower in soils inoculated with various AMF (*G. versiforme*, *G. intraradices*, *G. etunicatum*, *G. mosseae* and *G. constrictum*) than those without AMF (Jankong and Viosoottiviseth, 2008).

Anaerobic Bioremediation

Although much PAH degradation takes place in the presence of oxygen (as discussed above), research has clearly showed degradation of PAHs under anaerobic conditions. In anaerobic environments, with high biological oxygen demand of organic material and low solubility of oxygen in water, alternative electron acceptors are used by obligate or facultative anaerobic bacteria for respiration (Quantin *et al.*, 2005). When oxygen is not available, electron acceptors are depleted in the following sequence: nitrate, manganese, iron, sulfate and labile organic compounds, based on available free energy potential and thermodynamics (Stumm and Morgan, 1981).

Anaerobic biodegradation of PAHs in sediments and soils has been demonstrated under denitrifying, sulfate-reducing and methanogenic conditions (Lovley *et al.*, 1995; Coates *et al.*, 1996, 1997; Hayes *et al.*, 1999; Rothermich *et al.*, 2002). PAHs such as phenanthrene and fluorene (Ambrosoli *et al.*, 2005), and pyrene and naphthalene (Hutchins *et al.*, 1991) can be anaerobically degraded when nitrate is used as terminal electron acceptor by native bacterial consortia. Denitrifying bacteria of the genera *Azoarcus* and *Thauera* have been isolated during anaerobic degradation of toluene (Leuthner *et al.*, 1998) while *Clostridium pascui* strain MSA3 can use phenanthrene and pyrene as sole carbon sources for growth anaerobically (Chang *et al.*, 2008). Mineralization of PAHs can also be achieved by anaerobic bacteria, although it is limited to some species. Pure bacterial cultures (NAP-3-2 and NAP-4) can mineralize naphthalene anaerobically without producing nitrogen gas. Analysis of 16S ribosomal DNA sequences revealed that NAP-4 was closely related to *Vibrio pelagius* (Rockne *et al.*, 2000). In the same manner, enriched cultures isolated from creosote-contaminated marine sediments were able to mineralize (~96%) radiolabelled phenanthrene. This process was coupled with dissimilatory nitrate reduction in a fluidized bed reactor experiment (Rockne and Strand, 2001).

Anaerobic degradation of PAHs is also feasible under sulfate-reducing conditions. Naphthalene can be anaerobically degraded by microbes related to the phylotype NaphS2, a sulfate reducer from the δ-Proteobacteria isolated from marine sediments (Hayes and Lovley, 2002). Mineralization of fluorene, phenanthrene and fluoranthene has been achieved by indigenous bacteria from PAH-contaminated marine and river sediments when sulfate was used as terminal electron acceptor (Lei *et al.*, 2005). Respiratory reduction of iron and manganese oxides has been also linked to oxidation of PAHs. Specifically, anaerobic degradation of pyrene from creosote-contaminated soils was demonstrated when iron oxyhydroxide and manganese oxide were used as electron acceptors (Nieman *et al.*, 2001). Lovley *et al.* (1994) showed mineralization of ^{14}C-labelled toluene and benzene by iron-reducing bacteria.

Conclusion

PAHs are common persistent toxic organic pollutants found worldwide in petroleum products and from incomplete combustion of carbon. PAHs have low solubility in water and sorb to organic matter in water, soils and sediments. They are often found in high concentrations near industrial and urban point sources of pollution. Environmental exposure to PAHs is linked with deformities and cancer in many animals including humans, fish and benthic organisms.

Bioremediation by aerobic bacteria, the most commonly studied approach, has been applied successfully in soils, sediments, fresh- and saltwater systems. Algal biodegradation of PAHs occurs in aqueous systems and works best with high cell density, with single PAHs, or in consortia with bacteria. Bioremediation by fungi is perhaps best applied in soils and involves extracellular lignin-degrading enzymes. Phytoremediation of PAHs is largely accomplished by plant uptake, volatilization and enhancement of degradation by microbes associated with roots.

Anaerobic biodegradation of PAHs can occur mainly with nitrate and sulfate as terminal electron acceptors. However, reduction of manganese and iron oxides has been linked to oxidation of PAHs. Bioremediation of contaminated soils and sediments with PAHs can be simple, cost-effective and *ex situ* and *in situ*. Although biostimulation and bioaugmentation are typically included in most bioremediation projects, more research is needed to fully understand mechanisms of bacterial metabolism in the degradation of PAHs.

Acknowledgements

The authors gratefully acknowledge Laura Leff, PhD, Kent State University, Kent, Ohio 44242, USA for providing editorial comments.

References

Abbondanzi, F., Campisi, T., Focanti, M., Guerra, R. and Iacondini, A. (2005) Assessing degradation capability of aerobic indigenous microflora in PAH- contaminated brackish sediments. *Marine Environmental Research* 59, 419–434.

Abed, R. and Koster, J. (2005) The direct role of aerobic heterotrophic bacteria associated with cyanobacteria in the degradation of oils compounds. *International Biodeterioration and Biodegradation* 55, 29–37.

Agency for Toxic Substances and Disease Registry (2010) Polycyclic Aromatic Hydrocarbons (PAHs). Agency for Toxic Substances and Disease Registry, Atlanta, Georgia, USA. Available at: http://www.atsdr.cdc.gov/toxfaqs/index.asp

Alexander, M. (1994) *Biodegradation and Bioremediation*. Academic Press, San Diego, California.

Ambrosoli, R., Petruzzelli, L., Minati, J. and Marsan, F. (2005) Anaerobic PAH degradation in soil by a mixed bacterial consortium under denitrifying conditions. *Chemosphere* 60, 1231–1236.

Amellal, N., Portal, J., Vogel, T. and Berthelin, J. (2001) Distribution and location of polycyclic aromatic hydrocarbons (PAHs) and PAH-degrading bacteria within polluted soil aggregates. *Biodegradation* 12, 49–57.

Annweiler, E., Richnow, H., Antranikian, S., Hedenbrock, C., Garms, C., Franke, S., Francke, W. and Michaelis, W. (2000) Naphthalene degradation and incorporation of naphthalene-derived carbon into biomas by the thermophile *Bacillus thermoleovorans*. *Applied Environmental Microbiology* 66, 518–523.

Atagana, H. (2004) Co-composting of PAH-contaminated soil with poultry manure. *Letters in Applied Microbiology* 39, 163–168.

Azuma, H., Toyota, M., Asakawa, Y. and Kawano, S. (2010) ChemInform Abstract: Naphthalene – A constituent of Magnolia flowers. *Wiley Online Library ChemInform* 27, 40.

Baldrian, P., Der Wiesch, C., Garbiel, J., Nerud, F. and Zadrazil, F. (2000) Influence of cadmium and mercury on activities of ligninolytic enzymes and degradation of polycyclic aromatic hydrocarbons by *Pleurotus ostreatus* in soil. *Applied Environmental Microbiology* 66, 2471–2478.

Bamforth, S. and Singleton, I. (2005) Bioremediation of polycyclic aromatic hydrocarbons: current knowledge and future directions. *Journal of Chemical Technology and Biotechnology* 80, 723–736.

Barr, D.P. and Aust, S.D. (1994) Mechanisms white-rot fungi used to degrade pollutants. *Environmental Science and Technology* 28, 78–87.

Bastiaens, L., Springael, D., Wattiau, P., Harms, H., deWatcher, R., Verachert, H. and Diels, L. (2000) Isolation of adherent polycyclic aromatic hdrocarbon (PAH)-degrading bacteria using PAH-sorbing carriers. *Applied Environmental Microbiology* 66, 1834–1843.

Bell, B., Spotila, J.R. and Congdon, J. (2006) High incidence of deformity in aquatic turtles in the John Heinz National Wildlife Refuge. *Environmental Pollution* 142, 457–465.

Bezalel, L., Hadar, Y. and Cerniglia., C.E. (1996) Mineralization of polycyclic aromatic hydrocarbons by the white rot fungus *Pleurotus ostreatus*. *Applied Environmental Microbiology* 62, 292–295.

Bezalel, L., Hadar, Y. and Cerniglia, C.E. (1997) Enzymatic mechanisms involved in phenanthrene degradation by the white rot fungus *Pleurotus ostreatus*. *Applied Environmental Microbiology* 63, 2495–2501.

Boonyatumanond, R., Gullaya, W., Ayako, T. and Hideshige, T. (2006) Distribution and origins of polycyclic aromatic hydrocarbons (PAHs) in riverine, estuarine, and marine sediments in Thailand. *Marine Pollution Bulletin* 52, 942–956.

Borde, X., Guieysse, B., Delgado, O., Munoz, R., Hatti-Kaul, R., Nugier-Chauvin, C., Patin, H. and Mattiasson, B. (2003) Synergistic relationships in algal-bacterial microcosms for the treatment of aromatic pollutants. *Bioresource Technology* 86, 293–300.

Bumpus, J.A., Tien, M., Wright, D. and Aust, S.D. (1985) Oxidation of persistent environmental pollutants by a white rot fungi. *Science* 228, 1434–1436.

Byss, M., Elhottova, D. and Baldrian, P. (2008) Fungal bioremediation of the creosote-contaminated soil: Influence of *Pleurotus ostreatus* and *Irpex lacteus* on polycyclic aromatic hydrocarbons removal and soil microbial community composition in the laboratory-scale study. *Chemosphere* 73, 1518–1523

Cébron, A., Norini, M., Beguiristain, T. and Leyval, C. (2008) Real-time PCR quantification of PAH-ring hydroxyting dioxygenase (PAH-RHDα) genes from Gram positive and Gram negative bacteria in soil and sediment samples. *Journal of Microbiological Methods* 73, 148–159.

Cerniglia, C.E. (1997) Fungal metabolism of polycyclic aromatic hydrocarbons: past, present and future applications in bioremediation. *Journal of Industrial Microbiology and Biotechnology* 19, 324–333.

Cerniglia, C.E., Sutherland, J.B. and Crow, S.A. (1992) Fungal metabolism of aromatic hydrocarbons. In: Winkelmann, G. (ed.) *Microbial Degradation of Natural Products*. VCH Press, Weinheim, Germany, pp. 193–217.

Chan, S.M.N., Luan, T.G. and Wong, M.H. (2006) Removal and biodegradation of polycyclic aromatic hydrocarbons by *Selenastrum capricornutum*. *Environmental Toxicology and Chemistry* 25, 1772–1779.

Chang, B., Chang, I. and Yuan, S. (2008) Anaerobic degradation of phenanthrene and pyrene in mangrove sediment. *Bulletin of Environmental Contamination and Toxicology* 80, 145–149.

Chang, W., Um, Y. and Holoman, T. (2005) Molecular characterization of anaerobic microbial communities from benzene-degrading sediments under methanogenic conditions. *Biotechnology Progress* 21, 1789–1794.

Chauhan, A., Fazlurrahman., Oakeshott, J. and Jain, R. (2008) Bacterial metabolism of polycyclic aromatic hydrocarbons: strategies for bioremediation. *Indian Journal of Microbiology* 48, 95–113.

Chekol, T., Vough, L.R. and Chaney, R.L. (2004) Phytoremediation of polychlorinated biphenyl-contaminated soils. *Environment International* 30, 799–804.

Chen, J., Henderson, G., Grimm, C.C., Lloyd, S.W. and Laine, R.A. (1998) Termites fumigate their nests with naphthalene. *Nature* 392, 558–559.

Chu, W.K., Wong, M.H. and Zhang, J. (2006) Accumulation, distribution and transformation of DDT and PCBs by *Phragmites australis* and *Oryza sativa* L. II. enzyme study. *Environmental Geochemistry and Health* 28, 169–181.

Coates, J., Anderson, R. and Lovley, D. (1996) Oxidation of polycyclic aromatic hydrocarbons under sulfate-reducing conditions. *Applied and Environmental Microbiology* 62, 1099–1101.

Coates, J., Woodward, J., Allen, P. and Lovley, D. (1997) Anaerobic degradation of polycyclic aromatic hydrocarbons and alkanes in petroleum-contaminated marine harbor sediments. *Applied Environmental Microbiology* 63, 3589–3593.

Cofield, N., Banks, M.K. and Schwab, A.P. (2008) Lability of polycyclic aromatic hydrocarbons in the rhizosphere. *Chemosphere* 70, 1644–1652.

Daisy, B.H., Strobel, G.A., Castillo, U., Ezra, D., Sears, J., Weaver, D. and Runyon, J. (2002) Naphthalene, an insect repellent, is produced by *Muscodor vitigenus*, a novel endophytic fungus. *Microbiology* 148, 3737–3741.

Dean-Ross, D., Moody, J. and Cerniglia, C. (2002) Utilization of mixtures of polycyclic aromatic hydrocarbons by bacteria isolated from contaminated sediment. *FEMS Microbiology Ecology* 41, 1–7.

DeBruyn, J., Mead, T., Wilhelm, S. and Sayler, G. (2009) PAH Biodegradative genotypes in Lake Erie sediments: evidence for broad geographical distribution of pyrene-degrading *Mycobacteria*. *Environmental Science and Technology* 43, 3467–3473.

European Commission (2011) European Commission Environment, European Commission. Available at: http://ec.europa.eu/environment/air/pdf/pp_pah.pdf

Feitkenhauer, H., Müller, R.M. and Märkl, H. (2003) Degradation of polycyclic aromatic hydrocarbons and long chain alkanes at 60 70 (sic)°C by *Thermus* and *Bacillus* spp. *Biodegradation* 14, 367–372.

Fiorenza, S., Oubre, C.L. and Ward, C.H. (2000) *Phytoremediation of Hydrocarbon-Contaminated Soil*. Lewis Publishers, Boca Raton, Florida.

Gan, S., Lau, E.V. and Ng, H.K. (2009) Remediation of soils contaminated with polycyclic aromatic hydrocarbons (PAHs). *Journal of Hazardous Materials* 172, 532–549.

Gao, Y., Li, Q., Ling, W. and Zhu, X. (2011) Arbuscular mycorrhizal phytoremediation of soils contaminated with phenanthrene and pyrene. *Journal of Hazardous Materials* 185, 703–709.

Garcia, M., Mellado, E., Ostos, J. and Ventosa, A. (2004) *Halomonas organivorans* sp. nov., a moderate halophile able to degrade aromatic compounds. *International Journal of Systematic and Evolutionary Microbiology* 54, 1723 1728.

Habe, H. and Omori, T. (2003) Genetics of polycyclic aromatic hydrocarbon metabolism in diverse aerobic bacteria. *Bioscience, Biotechnology and Biochemistry* 67, 225–243.

Habe, H., Miyakoshi, M., Chung, K., Kasuga, K., Yoshida, T., Nojiri, H. and Omori, T. (2003) Phthalate catabolic gene cluster is linked to the angular dioxygenase gene in *Terrabacter* sp strain DFB63. *Applied Microbiology and Biotechnology* 61, 44–54.

Haemmerli, S.D., Leisola, M.S.A., Sanglard, D. and Fiechter, A. (1986) Oxidation of benzo(a)pyrene by extracellular ligninases of *Phanerochaete chrysosporium*. *Journal of Biological Chemistry* 261, 6900–6903.

Haritash, A. and Kaushik, C. (2009) Biodegradation aspects of polycyclic aromatic hydrocarbons: a review. *Journal of Hazardous Materials* 169, 1–15.

Hayes, L. and Lovley, D. (2002) Specific 16S rDNA sequences associated with naphthalene degradation under sulfate-reducing condition in harbor sediments. *Microbial Ecology* 43, 134–145.

Hayes, L., Nevin, K. and Lovley, D. (1999) Role of prior exposure on anaerobic degradation of naphthalene and phenanthrene in marine harbor sediments. *Organic Geochemistry* 30, 937–945.

Ho, Y., Jackson, M., Yang, Y., Mueller, J. and Pritchard, P. (2000) Characterization of fluoranthene and pyrene degrading bacteria isolated from PAH-contaminated soils and sediments. *Journal of Industrial Microbiology and Biotechnology* 24, 100–112.

Hong, Y., Dong-Xing, Y., Lin, Q. and Yang, T. (2008) Accumulation and biodegradation of phenanthrene and fluoranthene by the algae enriched from a mangrove aquatic ecosystem. *Marine Pollution Bulletin* 56, 1400–1405.

Hofrichter, M., Schneibner, K., Schneegab, I. and Fritzche, W. (1998) Enzymatic combustion of aromatic and aliphatic compounds by manganese peroxidase from *Nematoloma frowardii*. *Applied Environmental Microbiology* 64, 399–404.

Huesemann, M.H., Hausmann, T.S., Fortman, T.J., Thom, R.M. and Cullinan, V. (2009) *In situ* phytoremediation of PAH- and PCB-contaminated marine sediments with eelgrass (*Zostera marina*). *Ecological Engineering* 35, 1395–1404.

Husain, S. (2008) Literature overview: Microbial metabolism of high molecular weight polycyclic aromatic hydrocarbons. *Remediation Journal* 18, 131–161.

Hutchins, S., Sewell, G., Kovacs, D. and Smith, G. (1991) Biodegradation of aromatic hydrocarbons by aquifer microorganisms under denitrifying conditions. *Environment Science and Technology* 25, 68–76.

In Der Wiesche, C., Martens, R. and Zadrazil, F. (2003) The effect of interaction between white-rot fungi and indigenous microorganisms on degradation of polycyclic aromatic hydrocarbons in soil. *Water, Air, and Soil Pollution: Focus* 3, 73–79.

Iranzo, M., Cañizares, J., Roca-Perez, L., Sainz-Pardo, I., Mormeneo, S. and Boluda, R. (2004) Characteristics of rice straw and sewage sludge as composting materials in Valencia (Spain). *Bioresource Technology* 95, 107–112.

Jankong, P. and Visoottiviseth, P. (2008) Effects of arbuscular mycorrhizal inoculation on plants growing on arsenic contaminated soil. *Chemosphere* 72, 1092–1097.

Jaward, F., Barber, J.L., Booij, K. and Jones, K.C. (2004) Spatial distribution of atmospheric PAHs and PCNs along a north-south Atlantic transect. *Environmental Pollution* 132, 173–181.

Johnsen, A., Wick, L. and Harms, H. (2005) Principles of microbial PAH-degradation in soil. *Environmental Pollution* 133, 71–84.

Jürgens, A., Webber, A.C. and Gottsberger, G. (2000) Floral scent compounds of Amazonian *Annonaceae* species pollinated by small beetles and thrips. *Phytochemistry* 55, 551–558.

Kane, S.R., Beller, H.R., Legler, T.C., Koester, C.J., Pinkart, H.C., Halden, R.U. and Happel, A.M. (2001) Aerobic biodegradation of methyl tert-butyl ether by aquifer bacteria from leaking underground storage tank sites. *Applied and Environmental Microbiology* 67, 5824–5829.

Ke, L., Wong, T.W., Wong, Y.S. and Tam, N.F. (2003) Fate of polycyclic aromatic hydrocarbon (PAH) contamination in a mangrove swamp in Hong Kong following an oil spill. *Marine Pollution Bulletin* 45, 339–347.

Kim, S., Kweon, O., Jones, R., Edmonson, R. and Cerniglia, C. (2008) Genomic analysis of polycyclic aromatic hydrocarbon degradation in *Mycobacterium vanbaalenii* PYR-1. *Biodegradation* 19, 859–881.

Lang, E., Eller, G. and Zadrazil, F. (1997) Lignocellulose decomposition and production of ligninolytic enzymes during interaction of white rot fungi with soil microorganisms. *Microbial Ecology* 34, 1–10.

Langenhoff, A., Zehnder, A. and Schraa, G. (1996) Behavior of toluene, benzene and naphthalene under anaerobic conditions in sediment columns. *Biodegradation* 7, 267–274.

Launen, L.A., Pinto, L.J. and Moore, M.M. (1999) Optimization of pyrene oxidation by *Penicillium janthinellum* using response-surface methodology. *Microbial Biotechnology* 51, 510–515.

Lee, P.-H., Ong, S.K., Golchin, J. and Nelson, G.L. (1999) Extraction method for analysis of PAHs in coal-tar-contaminated soils. ASCE Pract. Period. *Hazardous Toxic Radioactive Waste Management* 3, 155–162.

Lee, S.H., Lee, W.S., Lee, C.H. and Kim, J.G. (2008) Degradation of phenanthrene and pyrene in rhizosphere of grasses and legumes. *Journal of Hazardous Materials* 153, 892–898.

Lei, A., Hu, Z., Wong, Y. and Tam, N. (2007) Removal of fluoranthene and pyrene by different microalgal species. *Bioresource Technology* 98, 273–280.

Lei, L., Khodadoust, A., Suidan, M. and Tabak, H. (2005) Biodegradation of sediment-bound PAHs in field-contaminated sediment. *Water Research* 39, 349–361.

Leigh, M.B., Prouzova, P., Mackova, M., Macek, T., Nagle, D.P. and Fletcher, J.S. (2006) Polychlorinated biphenyl (PCB) degrading bacteria associated with trees in a PCB-contaminated site. *Applied Environmental Microbiology* 72, 2331–2343.

Leuthner, B., Leutwein, C., Schultz, H., Horth, P., Haehnel, W., Schiltz, E., Schagger, H. and Heider, J. (1998) Biochemical and genetic characterization of benzylsuccinate synthase from *Thauera aromatica*: a new glycyl radical enzyme catalyzing the first step in anaerobic toluene metabolism. *Molecular Microbiology* 28, 652–660.

Li, K., Christensen, E.R., Camp, R.P. and Imamoglu, I. (2001) PAHs in dated sediments of Ashtabula River, Ohio, USA. *Environmental Science and Technology* 35, 2896–2902.

Liao, J.P., Lin, X.G., Cao, Z.H., Shi, Y.Q. and Wong, M.H. (2003) Interactions between arbuscular mycorrhizae and heavy metals under sand culture experiment. *Chemosphere* 50, 847–853.

Lin, Q.X. and Mendelssohn, I.A. (2009) Potential of restoration and phytoremediation with *Juncus roemerianus* for diesel-contaminated coastal wetlands. *Ecological Engineering* 35, 85–91.

Lovley, D. (2000) Anaerobic benzene degradation. *Biodegradation* 11, 107–116.

Lovley, D., Woodward, J. and Chapelle, F. (1994) Stimulated anoxic biodegradation of aromatic hydrocarbons using Fe (III) ligands. *Nature* 370, 128–131.

Lovley, D., Coates, J., Woodward, J. and Phillips, E. (1995) Benzene oxidation coupled to sulfate reduction. *Applied and Environmental Microbiology* 61, 953–958.

Luo, S., Chen, S., Mai, B., Sheng, G., Fu, J. and Zeng, E. (2008) Distribution, source apportionment, and transport of PAHs in sediments from the Pearl River Delta and the Northern South China Sea. *Archives of Environmental Contamination and Toxicology* 55, 11–20.

Martens, R. and Zadrazil, F. (1998) Screening of white-rot fungi for their ability to mineralize polycyclic aromatic hydrocarbons in soil. *Folia Microbiology* 43, 97–103.

Meckenstock, R., Safinowski, M. and Griebler, C. (2004) Anaerobic degradation of polycyclic aromatic hydrocarbons. *FEMS Microbiology Ecology* 49, 27–36.

Mosisch, T.D. and Arthington, A.H. (2001) Polycyclic aromatic hydrocarbon residues in the sediments of a dune lake as a result of power boating. *Lakes & Reservoirs: Research and Management* 6, 21–32.

Mphekgo, M. and Cloete, T. (2004) Bioremediation of petroleum hydrocarbons through landfarming: Are simplicity and cost-effectiveness the only advantages? *Reviews in Environmental Science and Bio/Technology* 3, 349–360.

Muñoz, C., Guillen, F., Martinez, A.T. and Martinez, M.J. (1997) Laccase isoenzymes of *Pleurotus eryngii*: characterization, catalytic properties, and participation in activation of molecular oxygen and Mn^{2+} oxidation. *Applied Environmental Microbiology* 63, 2166–2174.

Nguyen, T.H., Goss, K.U. and Ball, W.P. (2005) Polyparameter linear free energy relationships for estimating the equilibrium partition of organic compounds between water and the natural organic matter in soils and sediments. *Environmental Science and Technology* 39, 913–924.

Nieman, J., Sims, R., McLean, J., Sims, J. and Sorensen, D. (2001) Fate of pyrene in contaminated soil amended with alternate electron acceptors. *Chemosphere* 44, 1265–1271.

Novotny, C., Erbanova, P., Sasek, V., Kubatova, A., Cajthaml, T., Lang, E., Krahl, J. and Zadrazil, F. (1999) Extracellular oxidative enzyme production and PAH removal in soil by exploratory mycelium of white rot fungi. *Biodegradation* 10, 159–168.

Pan, S.W., Wei, S.Q., Yuan, X. and Cao, S.X. (2008) The removal and remediation of phenanthrene and pyrene in soil by mixed cropping of alfalfa and rape. *Agricultural Sciences in China* 7, 101–110.

Parrish, Z.D., Banks, M.K. and Schwab, A.P. (2005) Assessment of contaminant lability during phytoremediation of polycyclic aromatic hydrocarbon impacted soil. *Environmental Pollution* 137, 187–197.

Pazos, M., Rosales, E., Gomez, J. and Sanroman, M. (2010) Decontamination of soils containing PAHs by electroremediation: A Review. *Journal of Hazardous Materials* 177, 1–11.

Peng, R.H., Xiong, A.S., Xue, Y., Fu, X.Y., Gao, F., Zhao, W., Tian, Y.S. and Yao, Q.H. (2008) Microbial biodegradation of polyaromatic hydrocarbons. *FEMS Microbiology Reviews* 32, 927–955.

Pothuluri, J.V., Selby, A., Evans, F.E., Freeman, J.P. and Cerniglia, C.E. (1995) Transformation of chrysene and other polycyclic aromatic hydrocarbon mixtures by the fungus *Cunninghamella elegans*. *Canadian Journal of Botany* 73, 1025–1033.

Potin, O., Rafin, C. and Veignie, E. (2004) Bioremediation of an aged polycyclic aromatic hydrocarbons (PAHs)-contaminated soil by filamentous fungi isolated from the soil. *International Biodeterioration and Biodegradation* 54, 45–52

Qiao, M., Wang, C., Huang, S., Wang, D. and Wang, Z. (2006) Composition, sources and potential toxicological significance of PAHs in the surface sediments of the Meiliang Bay, Taihu Lake, China. *Environment International* 32, 28–33.

Quantin, C., Joner, E., Portal, J. and Berthelin, J. (2005) PAH dissipation in a contaminated river sediment under oxic and anoxic conditions. *Environmental Pollution* 134, 315–322.

Rego, E.C. and Pereira, A.D. (2007) PAHs and BTEX in groundwater of gasoline stations from Rio de Janeiro City, Brazil. *Bulletin of Environmental Contamination and Toxicology* 79, 660–664.

Rezek J., in der Wiesche, C., Mackova, M., Zadrazil, F. and Macek, T. (2008) The effect of ryegrass (*Lolium perenne*) on decrease of PAH content in long term contaminated soil. *Chemosphere* 70, 1603–1608.

Rockne, K., Chee-Sanford, J., Sanford, R., Hedlund, B., Staley, J. and Strand, S. (2000) Anaerobic naphthalene degradation by microbial pure cultures under nitrate-reducing conditions. *Applied and Environmental Microbiology* 66, 1595–1601.

Rockne, K. and Strand, S. (2001) Anaerobic biodegradation of naphthalene, phenanthrene, and biphenyl by a denitrifying enrichment culture. *Water Research* 35, 291–299.

Rodrigo, J., Santos, E., Bento, F., Peralba, M., Selbach, P., Sa, E. and Camargo, F. (2005) Anthracene biodegradation by *Pseudomonas* sp. isolated from a petrochemical sludge landfarming site. *International Biodeterioration and Biodegradation* 56, 143–150.

Rogers, S., Ong, S. and Moorman, T. (2007) Mineralization of PAHs in coal-tar impacted aquifer sediments and associated microbial community structure investigated with FISH. *Chemosphere* 69, 1563–1573.

Rothermich, M., Hayes, L. and Lovley, D. (2002) Anaerobic, sulfate-dependent degradation of polycyclic aromatic hydrocarbons in petroleum-contaminated harbor sediment. *Environment Science & Technology* 36, 4811–4817.

Sanghvi, S. (2005) Bioremediation of polycyclic aromatic hydrocarbon contamination using *Mycobacterium vanbaalenii*. *Basic Biotechnology eJournal* MMG445. Online source available at: www.msu.edu

Sasek, V., Bhatt, M., Cajthaml, T., Malachova, K. and Lednicka, D. (2004) Compost-mediated removal of polycyclic aromatic hydrocarbons from contaminated soil. *Archives of Environmental Contamination and Toxicology* 44, 0336–0342.

Shimp, J.F., Tracy, J.C., Davis, L.C., Lee, E., Huang, W., Erickson, L.E. and Schnoor, J.L. (1993) Beneficial effects of plants in the remediation of soil and groundwater contaminated with organic materials. *Critical Reviews in Environmental Science and Technology* 23, 41–77.

Siciliano, S.D. and Germida, J.J. (1998) Mechanisms of phytoremediation: biochemical and ecological interactions between plants and bacteria. *Environmental Reviews* 6, 65–79.

Spriggs, T., Banks, M.K. and Schwab, P. (2005) Phytoremediation of polycyclic aromatic hydrocarbons in manufactured gas plant-impacted soil. *Journal of Environmental Quality* 34, 1755–1762.

Stark, A., Abrajano, T., Hellou, J. and Metcalf-Smith, J. (2003) Molecular and isotopic characterization of polycyclic aromatic hydrocarbon distribution and sources at the international segment of the St. Lawrence River. *Organic Goechemistry* 34, 225–237.

Steffen, K.T., Hatakka, A. and Hofrichter, M. (2003) Degradation of benzo[a]pyrene by the litter-decomposing basidiomycete *Stropharia coronilla*: role of manganese peroxidase. *Applied Environmental Microbiology* 69, 3957–3964.

Stingley, R., Kahn, A. and Cerniglia, C. (2004) Molecular characterization of a phenanthrene degradation pathway in *Mycobacterium vanbaalenii* PYR-1. *Biochemical and Biophysical Research Communications* 322, 133–146.

Straube, W.L., Nestler, C., Hansen, L., Ringleberg, D., Pritchard, P. and Jones-Meehan, J. (2003) Remediation of polycyclic hydrocarbons (PAHs) through landfarming with biostimulation and bioaugmentation. *Acta Biotechnologica* 23, 179–196.

Stumm, W. and Morgan, J. (1981) *Aquatic Chemistry*, 2nd edn. John Wiley & Sons, New York.

Sun, T.R., Cang, L., Wang, G.Y., Zhou, D.M., Cheng, J.M. and Xu, H. (2010) Roles of abiotic losses, microbes, plant roots, and root exudates on phytoremediation of PAHs in a barren soil. *Journal of Hazardous Materials* 176, 919–925.

Syed, K., Doddapaneni, H., Subramanian, V., Lam, Y.W. and Yadav, J.S. (2010) Genome-to-function characterization of novel fungal P450 monooxygenases oxidizing polycyclic aromatic hydrocarbons (PAHs). *Biochemical and Biophysical Research Communications* 399, 492–497

Sylvia, D.M., Fuhrmann, J.J., Hartel, P.G. and Zuberer, D.A. (1999) *Principles and Applications of Soil Microbiology*. Prentice Hall, Upper Saddle River, New Jersey, USA.

Tian, Y., Liu, H., Zheng, T., Kwon, K., Kim, S. and Yan, C. (2008) PAHs contamination and bacterial communities in mangrove surface sediments of the Jiulong River Estuary, China. *Marine Pollution Bulletin* 57, 707–715.

United States Environmental Protection Agency (2009) National Priority List, United States Environmental Protection Agency. Available at: http://www.epa.gov/osw/hazard/wastemin/priority.htm

United States Environmental Protection Agency (2011) Integrated Risk Information System, Environmental Protection Agency. Available at: http://www.epa.gov/IRIS

Van Herwijnen, R., van de Sande, B., van der Wielen, F., Springael, D., Govers, H. and Parsons J. (2003) Influence of phenanthrene and fluoranthene on the degradation of fluorene and glucose by *Sphingomonas* sp. strain LB126 in chemostat cultures. *FEMS Microbiology Ecology* 46, 105–111.

Van Metre, P. and Mahler, B. (2005) Trends in hydrophobic organic contaminants in urban and reference lake sediments across the United States, 1970–2001. *Environment Science and Technology* 39, 5567–5574.

Van Metre, P.C. and Mahler, B.J. (2010) Contribution of PAHs from coal-tar pavement sealcoat and other sources to 40 US lakes. *Science of the Total Environment* 409, 334–344.

Verrhiest, G., Clément, B., Volat, B., Montuelle, B. and Perroding, Y. (2002) Interactions between a polycyclic aromatic hydrocarbon mixture and the microbial communities in a natural freshwater sediment. *Chemosphere* 46, 187–196.

Wehrer, M. and Totsche, K. (2009) Difference in PAH release processes from tar-oil contaminated soil materials with similar contamination history. Chemie der Erde-Geochemistry-Interdisciplinary. *Journal for Chemical Problems of the Geosciences and Geoecology* 69, 109–124.

Whitfield-Aslund, M.L., Zeeb, B.A., Rutter, A. and Reimer, K.J. (2007) In situ phytoextraction of polychlorinated biphenyl (PCB) contaminated soil. *Science of the Total Environment* 374, 1–12.

Witt, G. and Gründler, P. (2005) The consequences of the Oder flood in 1997 on the distribution of polycyclic aromatic hydrocarbons in the Oder River. *Acta Hydrochimica et Hydrobiologica* 33, 301–314.

Wolter, M., Zadrazil, F., Martens, R. and Bahadir, M. (1997) Degradation of eight highly condensed polycyclic aromatic hydrocarbons by *Pleurotus* sp. Florida in solid wheat straw substrate. *Applied Microbiology and Biotechnology* 48, 398–404.

Xia, X., Yu, H., Yang, Z. and Huang, G. (2006) Biodegradation of polycyclic aromatic hydrocarbons in the natural Waters of the Yellow River: effects of high sediment content on biodegradation. *Chemosphere* 65, 457–466.

Xu, J., Yu, Y., Wang, P., Guo, W., Dai, S. and Sun, H. (2007) Polycyclic aromatic hydrocarbons in the surface sediments from Yellow River, China. *Chemosphere* 67, 1408–1414.

Yang, X. and Baumann, P. (2005) Biliary PAH metabolites and the hepatosomatic index of brown bullheads from Lake Erie tributaries. *Ecological Indicators* 6, 567–574.

Yoshitomi, K.J. and Shann, J.R. (2001) Corn (*Zea mays* L) root exudates and their impact on ^{14}C-Pyrene mineralization. *Soil Biology and Biochemistry* 33, 1769–1776.

Zeeb, B.A., Amphlett, J.S., Rutter, A. and Reimer, K.J. (2006) Potential for phytoremediation of polychlorinated biphenyl-(PCB)-contaminated soil. *International Journal of Phytoremediation* 8, 199–221.

Zeinali, M., Vossoughi, M. and Ardestani, S. (2007a) Characterization of a moderate thermophilic *Nocardia* species able to grow on polycyclic aromatic hydrocarbons. *Letters in Applied Microbiology* 45, 622–628.

Zeinali, M., Vossoughi, M., Ardestani, S., Babanezhad, E. and Masoumian, M. (2007b) Hydrocarbon degradation by thermophilic *Nocardia otitdiscaviarum* strain TSH1: physiological aspects. *Journal of Basic Microbiology* 47, 534–539.

Zheng, Z. and Obbard, J.P. (2002) Polycyclic aromatic hydrocarbon removal from soil by surfactant solubilization and *Phanerochaete chrysosporium* oxidation. *Journal of Environmental Quality* 31, 1842–1847.

Zheng, Z. and Obbard J.P. (2003) Oxidation of polycyclic aromatic hydrocarbons by fungal isolates from an oil contaminated refinery soil. *Environmental Science and Pollution Research* 10, 173–176.

Chapter 16

The Role of Biological Control in the Creation of Bioremediation Technologies

Yana Topalova

Introduction

The creation of efficient bioremediation technologies for the purification of waste water, containing toxic compounds, requires knowledge and management of the biological system that carries out the biodegradation (Alexander, 1985). In these cases the applied biological design is significant, using suitable indicators such as time and spatial critical control points (CCP). The purposeful biological control is the basic factor for the prognostication, modelling and the proposal for the modules and parameters of the future bioremediation technologies.

Prognostication of the Biodegradation of Varied Xenobiotics by means of Simulation Modelling and Determination of the Biodegradation Index

The prognostication of the biodegradation of xenobiotics with different chemical structure in the course of the wastewater purification processes depends on several different parameters: (i) accident load of the xenobiotics; (ii) the mechanisms of their biodegradation in aerobic or anaerobic conditions; and (iii) the adaptation of the biological system, in this case the activated sludge (AS). For the creation of detoxication technologies processes of simulation modelling are necessary, in the course of which the processes of accumulation and biodegradation are prognosticated in the particular conditions of the water purification process. So on the basis of the limits of plasticity of the AS as a universally used biological system in the detoxication technologies, an effective biodegradation process can be prognosticated and later managed. In the experimental programme, initially in bioreactors with complete mixing and within a batch process, an accident overload of xenobiotic with a different chemical structure was simulated. The processes of xenobiotic accumulation and biodegradation were studied in order to calculate the biodegradation index, which further can be used in the biodesign of the detoxication technology.

The biodegradation index as an indicator in our analysis was created in analogy to the highly important indicator from the biological water purification–biochemical index (Buitron *et al.*, 1998; Byrns, 2001). The biochemical index measures how easily degradable are the pollutants in the wastewater, and it is a correlation between BOD_5 and COD. In this particular case the indicator 'Biodegradation index' was formulated as a correlation between the degraded quantity of xenobiotic and the total eliminated quantity of xenobiotic, including adsorption and absorption. The biodegradation index data are presented in Fig. 16.1.

The analysis of the data of the biodegradation index of the studied xenobiotics confirms that oNP (ortho-nitrophenol), TCP (trichlorophenol) and DCP (dichlorophenol) degrade comparatively easier both in the early (24–48 h) and later phase (144 h) of the studied biodetoxication process. The values of the biodegradation index are higher than 0.6 (vary from 0.93 to 1) (Topalova and Dimkov, 2003; Topalova, 2009). The biodegradation index for PCP (pentachlorophenol) and the combination of PCP with oNP is extremely low and varies from 0.17 to 0.33. This index also confirms that in trivial detoxication technologies PCP will be accumulated in the biological and the water-purification system, but the biodegradation will be minimal. In a subsequent de-accumulation, the water-receivers will be polluted with toxic pollutants.

This necessitates a special and extended programme for studying the adaptation potential of the AS to PCP. The development of special biological systems is especially important in this case, as well as of algorithms and technologies, aiming to increase the real biodegradation of PCP.

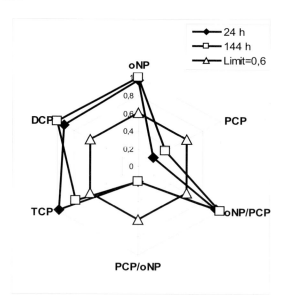

Fig. 16.1. Dynamics of biodegradation index of the different studied xenobiotics (the biodegradation index is calculated as a ratio between a degraded quantity of xenobiotic and total eliminated quantity of xenobiotic).

Creating an Algorithm for Adaptation of the Activated Sludge to a Biodegradation of PCP in Anaerobic Conditions and Starting a Bioreactor with a Sequence Batch Reactor for Detoxication of PCP

In this part of the experimental programme the emphasis was put on the development of a specialized adaptation algorithm, in order to ensure and create the best conditions for the stage development of the biodegradation potential of the activated sludge (AS) (Buitron *et al.*, 1998; Hinchee *et al.*, 1995). We were guided by several supporting points:

1. We worked on the mechanisms of PCP degradation and the known fact that the initial dehalogenation runs at a higher speed under anaerobic conditions, and the following cleavage of the benzene ring of the low-halogenated derivatives is more effective under aerobic conditions (Otte *et al.*, 1995; Topalova *et al.*, 1998). This directed us to the idea of a two-step technology – anaerobic and aerobic in correspondence to the sequence and the speed of transformation of PCP.

2. In the experimental setting, the results from the study of the adaptation mechanisms of the AS were used, according to which the low-halogenated products of PCP – 2,4,5-trichlorophenol (TCP) and 3,4-dichlorophenol (DCP) – have a high biodegradation index and are easily degraded, which can be achieved by increasing the stay in the aerobic facility.

3. According to literary data the concentration of the toxicants in the watery phase and in the biological system play a significant role in the stage development of the biodegradation potential. The high accident load induces resistance more quickly in correspondence with the life-saving defence mechanisms, but limit the possibilities of evolution and manifestation of the biodegrading qualities of the system.

4. All this confirmed the necessity of making specialized rules – for purposeful development of the adaptive properties, while considering its critical concentration and the remaining quantities of xenobiotic (Douglas, 1995; Trably *et al.*, 2003). An anaerobic module for intensive biodegradation of PCP was created. The initiation and confirmation of this module is ensured also by the respective adaptation algorithm.

In automated 4 l anaerobic bioreactors with full mixing, an adaptation algorithm was developed with the aim to: (i) find out the smallest time interval during which PCP is degraded in concentration (2 mg l^{-1}) in batch regime; and (ii) find out the smallest time interval for degradation of PCP in increasing concentration from 2 mg l^{-1} to 10 mg l^{-1}. In the bioreactor strict control was exercised on every portion of xenobiotic to be degraded completely and not to register residue quantity of PCP in the medium and, therefore, the concentration of PCP was increased to 10 mg l^{-1}.

Control of the Dynamics of the Detoxication Process

The transformation of PCP in the first phase of the adaptation algorithm is shown in Fig. 16.2. The first portion of 2 mg l^{-1} PCP was fully degraded for 5 days. A second portion of 2 mg l^{-1} PCP was entered and was degraded for 2 days, the third portion for 38 h, the fourth for 24 h and so the degradation interval of the given portion (2 mg l^{-1}) of PCP quickly shortened with the degree induction of the biodegradation potential. We achieved an interval of degradation of PCP in concentration 2 mg l^{-1} in 4 h. This time is commensurable with the stay in a real facility.

The second phase of the adaptation algorithm was started, where the PCP concentration was increased to 4 mg l^{-1}. The procedure was repeated many times with the aim of developing and confirming the higher portion of xenobiotic. In this way two parameters for management of the adaptation process were used simultaneously and coordinately – the degree and careful increase of the PCP concentration and the shortening of the interval, for which the given portion of xenobiotic can be fully degraded.

In this way the biodegradation potential of the AS was increased in stages and purposefully. An increase of the residue concentration of xenobiotic in the watery phase was not allowed, as well as inhibition of part of the biodegradation properties of the system. A decisive moment in the adaptation algorithm was the degree increase of the critical concentration, where the biological system can work (Ribarova *et al.*, 2001; Andreozzi *et al.*, 2006).

~~The described results confirm two basic points:~~

1. The increase of the PCP concentration should be in harmony with the speed of development of the biodegradation properties of the system, subjected to adaptation. In these cases both factors of adaptation should be used very skillfully – time and concentration according to the Holdein equation and the peculiarities of the biological system (Topalova and Dimkov, 2003).

2. In the course of the adaptation process the residue concentration of the toxicant in the watery phase and the biological system should be followed strictly.

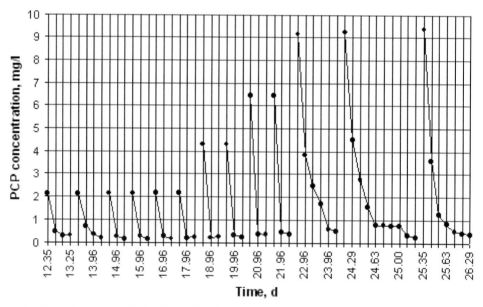

Fig. 16.2. Summarized algorithm of activated sludge (AS) to biodegradation of PCP in Bioreactor 1: the adaptation procedure is carried out in degrees and under control on increasing the PCP concentration (from 2 to 10 mg l^{-1}) and shortening the time of degradation in a semi-continuous regime.

The analysis of the data shows that the described adaptation algorithm for creating a specialized biological system has skillfully followed the course of the Holdein dependence in the linear part (Ribarova *et al.*, 2001; Topalova and Dimkov, 2003). The critical concentration for these experimental conditions is about 10 mg l^{-1} PCP. Exceeding this concentration is not recommended, since it will result in the inhibition and destruction of the biodegradation system.

Microbiological Control

Particularly interesting were the changes in the structure and functions of the biological system in the course of the biodegradation. The deciphering of the biological essence of the occurred changes contributes to the explanation, the better control, regulation and application of the adaptation algorithm. According to the management theory a microbiological and enzymological control was carried out in the course of adaptation at six critical control points (CCP):

1. CCP-1: AS in 0 moment without interaction with PCP (control).
2. CCP-2: AS adapted to biodegradation of PCP (2 mg l^{-1}) in a batch regime.
3. CCP-3: AS adapted to biodegradation of PCP (2 mg l^{-1}) in a semi-continuous regime.
4. CCP-4: AS adapted to biodegradation of PCP (4.35 mg l^{-1}) in a semi-continuous regime.
5. CCP-5: AS adapted to biodegradation of PCP (9.2 mg l^{-1}) in a semi-continuous regime.
6. CCP-6: AS adapted to biodegradation of PCP (10 mg l^{-1}) in a semi-continuous regime.

The anaerobic heterotrophs and the aerobic heterotrophs increased insignificantly in the course of adaptation. A more significant increase was shown by the quantity of the bacteria from the genera *Pseudomonas* and *Acinetobacter* – the increase is between 5 and 20% (Fig. 16.3).

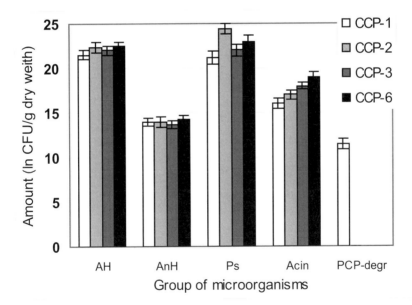

Fig. 16.3. Quantity of the key groups of organisms for the purification and detoxication process at four critical control points: CCP1, activated sludge (AS) in 0 moment without PCP; CCP-2, AS adapted to 2 mg l^{-1} PCP in batch regime; CCP-3, AS adapted to 2 mg l^{-1} PCP in a semi-continuous regime; CCP-6, AS adapted to 10 mg l^{-1} PCP in a semi-continuous regime.

With the increase of the speed of biodegradation, the bacteria from the genus *Acinetobacter* increase more synchronically. This confirms our assumption that the bacteria from this genus play a more important role in the adaptation to PCP biodegradation in anaerobic processes (Topalova, 2009). In the microbiological monitoring a paradox was found. With the advance of adaptation to biodegradation and with the increase of the concentration of the degraded PCP, the PCP-degrading bacteria did not increase in quantity; in CCP-2, CCP-3 and CCP-6 they were not registered at all. This seemingly surprising fact found its logical explanation in the microscopic and transmission electronic microscopic analyses of the AS in the course of its adaptation to PCP-biodegradation.

The data from the microscopic analysis of the AS are shown in Fig. 16.4. The presented images prove that in the course of the adaptation process the AS not only acquires new, valuable properties, but also undergoes significant changes in its structure. In the initial phases very tight zones are found, resembling pellet formation. With the progress of the adaptive changes this process intensifies and at CCP-4, CCP-5 and CCP-6 AS was found with a completely different structure in comparison to the initial state. The biological system is two-phase: it consists of well-formed pellets (Fig. 16.4.d and f) and a background of homogeneous cells. The homogeneous cells most probably consist of the bacterial genera, which increase their quantity in the adaptation process. In our study these homogeneous cells did not possess PCP-degrading properties. They probably perform minimal functions in degrading the benzene ring of the low halogenated and non-halogenated phenol derivatives. It is still unclear why at the highest speed of degradation of PCP the PCP-biodegradation microbe complex is not found. The answer to this question was given by the pictures from a section of the pellets (Fig. 16.4). It can be clearly seen that inside the pellets structures of microbe consortiums were formed, which are highly specialized and resemble tissues in higher organisms. Most probably in the course of adaptation in the PCP-degrading complex, specific, highly productive in relation to their dehalogenase complex intrabacterial relations were formed – from co-metabolic cooperation to syntrophy. These relations ensure the high speed of dehalogenation of PCP. After the destruction of the relations between the PCP-degrading bacteria, an attempt to isolate and cultivate them as pure cultures of a PCP-containing medium was unsuccessful.

So the acquisition of new biodegradation functions in the course of the specialized adaptation algorithm is accompanied by the creation of an entirely different two-phase AS. It consists of pellets, containing syntrophic PCP-degrading consortium and of homogeneous cells, belonging to the genera *Acinetobacter* and *Pseudomonas*. In the two phases of the AS different functions are localized. The dehalogenation is carried out mainly in an anaerobic mechanism deep in the pellets, and the following minimal degradation of the benzene nucleus of the low-halogenated and non-halogenated phenols is carried out in the homogeneous phase of the activated sludge, i.e. we come across newly formed specialized microniches in the AS.

Enzymological Control

The data from the microbiological studies were completed and further deciphered by means of the enzymological analysis of the biological system in the different CCP. The dynamics of the dehalogenase activity in the course of the adaptation process are shown in Fig. 16.5.

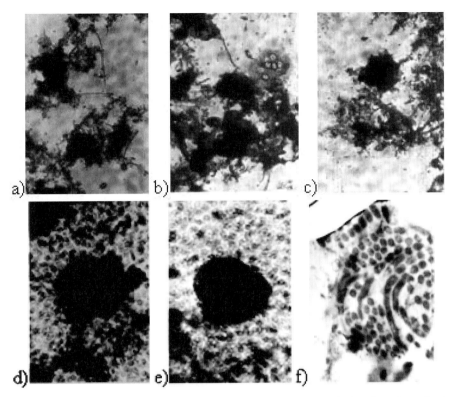

Fig. 16.4. Transitional phases in the structure of the activated sludge from floccular to pellet type in the course of its adaptation to biodegradation of PCP in high concentration (10 mg l^{-1}) for the shortest period (4 h): (a) floccular structure with forming pellets (LM 660×); (b) pellets in process of forming and algae from *Volvocales* (LM 660×); (c) transition phase of forming of pellets (LM 660×); (d) and (e) pellets with free-swimming cells (two-phase structure) (LM 800×); (f) internal synergetic structure of the pellets (TEM 12,000×) (Photos are our originals).

It is clear that the development of the biodegradation properties of the activated sludge is accompanied by a degree increase of the dehalogenase activity. The highest dehalogenase activity is registered at CCP-4 and CCP-5. These are the points where the adaptation procedure is close to the critical for the biological system PCP concentration, but it has not yet reached it. The course of the dehalogenase activity is an experimental confirmation of the already presented suggestion that the critical concentration of PCP for this biological system is 9.2 mg l^{-1}. The PCP concentration (10 mg l^{-1}) slightly exceeds the critical concentration and this leads to a decrease of the dehalogenase activity. Most probably the increase of the concentration above 10 mg l^{-1} will lead to an accumulation of PCP because of its incomplete degradation and to a decrease of the biodegrading efficiency of the AS. The recommended concentration for the work of this system is 9.2 mg l^{-1} PCP.

The course of the oxygenase activity deserves a more extensive commentary (Fig. 16.6). Most of all it is necessary to point out that in this part of the analyses we come across a second paradox. Under anaerobic conditions oxygenase activity is registered and developed, which strongly depends on the presence of diluted oxygen in the environment

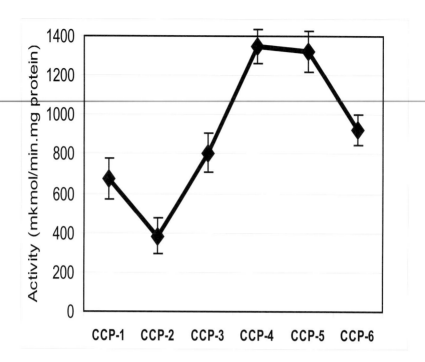

Fig. 16.5. Dynamics of the dehalogenase activity in different CCP: CCP-1, AS in 0 moment without PCP; CCP-2, AS adapted to 2 mg l^{-1} PCP in a batch regime; CCP-3, AS adapted to 2 mg l^{-1} PCP in a semi-continuous regime; CCP-4, adapted to 4.3 mg l^{-1} PCP in a semi-continuous regime; CCP-6, AS adapted to 10 mg l^{-1} PCP in a semi-continuous regime.

(Gibson and Parales, 1999, 2000). An explanation of this paradox is given again by the microscopic structure of the newly formed biological system. In the AS along with the pellets we found the presence of algae: *Spirulina*, *Volvox* and *Phacus*. At a low redox potential in the facility they create microniches with minimal quantities of oxygen as a result of their metabolic activity. The minimal quantities of oxygen enable the homogeneous bacteria nearby to develop and reveal their oxygenase apparatus. This proves the presence of processes of opening the benzene ring while still in the anaerobic reactor.

The dynamics of the oxygenase activity is interesting at the initial CCP-2 and CCP-3. At a lower PCP concentration, catechol 1,2-dioxygenase (C12DO) and catechol 2,3-dioxygenase (C23DO) activities are higher. In these CCP the AS still has a transitional floccular structure. With the advance of the process of pellet-formation the oxygenase function decreases and is limited only in the homogeneous bacterial part of the AS and mainly in those microniches where there are minimal quantities of oxygen. Though not high and localized in specific microniches, the oxygenase activity plays its role in the cleavage of the benzene ring and the decrease of the concentration of the chlorine-containing xenobiotics still in the anaerobic stage of the technology.

The biological control of the AS shows that when using a selective algorithm in the AS adaptive changes go off multilaterally and on different levels, which are directed towards maximum development of the new functions.

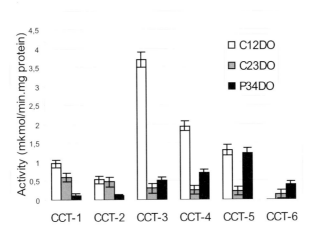
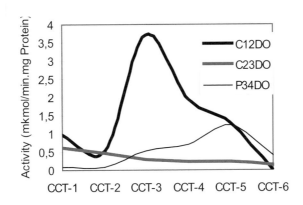

Fig.16.6. Oxygenase activity in the six CCP: (a) in absolute values; (b) general course of the oxygenase activites. CCP-1, AS in 0 moment without PCP; CCP 2, AS adapted to 2 mg l^{-1} PCP in a batch regime; CCP-3, AS adapted to 2 mg l^{-1} PCP in a semi-continuous regime; CCP-6, AS adapted to 10 mg l^{-1} PCP in a semi-continuous regime. C12DO, catechol 1,2-dioxygenase; C23DO, catechol 2,3-dioxygenase; P34DO, protocatechate 3,4-dioxygenase.

Aerobic Degradation of the Dichlorophenols and Trichlorophenols in a Complete Mixing Bioreactor

We also modelled a process of biodegradation of the low halogenated metabolic products of PCP in an aerobic bioreactor with complete mixing. The results show that these metabolites can be degraded by adapted AS with an increased residence time (Amor *et al.*, 2005; Chong and Lin, 2007) (Fig. 16.7).

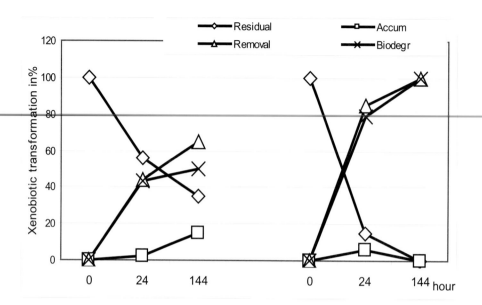

Fig. 16.7. Dynamics of the transformation processes (percentage in relation to the initial concentration of the xenobiotic) in the course of adaptation of the activated sludge with different xenobiotic pollutants: (a) for TCP (20mg l)$^{-1}$; (b) for DCP (20 mg l^{-1}).

Microbiological and Enzymological Control

The microbiological control shows an initial destruction of the AS, but a following considerable increase in the quantities of the bacteria from the genera *Pseudomonas* and *Acinetobacter*. All this is accompanied by an increase of the biodegradation activity of the AS in relation to the low halogenated metabolites of PCP and an accompanying increase of the oxygenase activity of the AS (Figs 16.8 and 16.9).

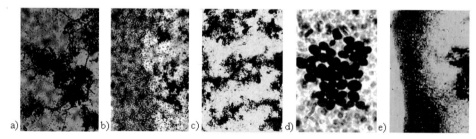

Fig. 16.8. Changes in the microscopic structure of the activated sludge during the adaptation process after an accident overload of the activated sludge with xenobiotics: (a) floccular activated sludge before the xenobiotic activity (LM 660×); (b) wide aerophilic zone of homogeneous cells from bacteria from the genera *Pseudomonas*, *Acinetobacter* and from the xenobiotic-degrading complex and abnormal 'pint-pont' flocculi (LM 660×); (c) disintegrated flocculi (LM 660×); (d) image of an active colony from the genus *Acinetobacter* (TEM 5000×); (e) aerophilic zone in AS (EM 660×).

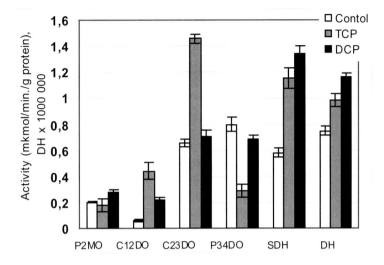

Fig. 16.9. Enzyme activity of key enzymes for the adaptation of the activated sludge on the 144th hour under accident overload with xenobiotics without a preliminary adaptation. C, control AS without xenobiotic on the 144th hour; TCP, AS with 20 mg l^{-1} TCP on the 144th hour; DCP, AS with 20 mg l^{-1} DCP on the 144th hour. P2MO, phenol 2-monoxygenase; C12DO, catechol 1,2-dioxygenase; C23DO, catechol 2,3-dioxygenase; P34DO, protocatechuate 3,4-dioxygenase; SDH, succinate dehydrogenase; DH, dehalogenase.

An Overall Design of a Two-step Detoxication Technology

By simulating model processes, based on the mechanisms of biodegradation and biological control, and by following the adaptation mechanisms of the complex systems (the AS), a specialized anaerobic module was created for a successful and complete 100% biodegradation of PCP at 9.2 mg l^{-1} in the influent. An adaptive algorithm was also developed to this module with a description and explanation of the separate steps of the adaptation of the AS, which realizes the biodegradation.

The module functions as a sequencing batch reactor (SBR) with two phases, while the entire technology is tri-phase.

1. Phase of mixing in the bioreactor and intensive biodegradation of PCP in concentration 9.2 mg l^{-1} for 4 h.
2. Stopping the mixer and settling down of the AS. The filtered water is led away in an aerobic module (conventional biobasin or bioreactor for total degradation of the low-halogenated and non-halogenated phenols). The anaerobic module is turned on for a second cycle of detoxication of a new portion of PCP for 4 h.
3. Aerobic bioreactor (or biobasin) – finalizing phase of biodegradation of the low-halogenated or non-halogenated phenol derivatives and final degradation of the trivial pollutants.

The two-step technology is shown in Fig. 16.10.

308   *Microbial Biotechnology: Energy and Environment*

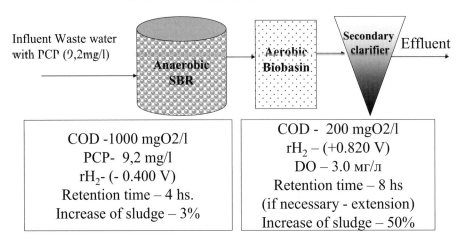

Fig. 16.10. Scheme of a two-step technology of detoxication of PCP.

Conclusion

The role of biological control carried out purposefully and adequately is extremely important (Swindoll *et al.*, 1995; Topalova, 2005). It is not limited only to evaluation of the processes of biodegradation and biodetoxication. The biological control is a key mechanism in the design and deciphering of the detoxication technologies. In this aspect the biological control, carried out professionally, is a significant factor in the creation and management of the bioremediation technologies and above all in the management of the most dynamic element from the ecobiotechnologies – the biological system.

Acknowledgements
This study was funded by the National Scientific Fund to the Ministry of Education and Science, Project No. DVU 01/0013, 'Ecological Modeling of Water Purification Processes in Critical Control Points of Upper Subcatchment of Iskar River'.

References

Alexander, M. (1985) Biodegradation of organic chemicals. *Environmental Science and Technology* 18, 106–111.
Amor, L., Eiroa, M., Kennes, C. and Veiga, M.C. (2005) Phenol biodegradation and its effect on the nitrification process. *Water Research* 39(13), 915–920.
Andreozzi, R., Cesar, R. and Pirozzi, F. (2006) Evaluation of biodegradation kinetic of aromatic compounds by means of aerobic experiments. *Chemosphere* 62(9), 1436–1487.
Buitron, G., Gonzales, A. and Lopes-Martin, L.M. (1998) Biodegradation of phenolic compounds by acclimated activated sludge and isolated bacteria. *Water Science and Technology* 37(4–5), 371–378.
Byrns, G. (2001) The fate of xenobiotic organic compounds in wastewater treatment plants. *Water Research* 35(10), 2523–2533.
Chong, N.M. and Lin, T.Y. (2007) Measurement of the degradation capacity of activated sludge for a xenobiotic organic. *Bioresource Technology* 98(5), 1124–1127.

Douglas, G. (1995) In: Hinchee, R.E., Douglas, G.S. and ONG, S.K. (eds) *Monitoring and Verification of Bioremediation.* Battelle Press, Columbus, Ohio, 286 pp.

Gibson, D.T. and Parales, R.E. (1999) Oxygenases – environmental role. *Current Opinion in Biotechnology* 10, 145–158.

Gibson, D.T. and Parales, R.E. (2000) Aromatic hydrocarbondioxygenases in Environmental Biotechnology. *Current Opinion in Biotechnology* 11, 236–243.

Hinchee, R., Vogel, C. and Brockman, F. (eds) (1995) *Microbial Processes for Bioremediation.* Battelle Press, Columbus, Ohio, 361 pp.

Otte, M.P., Gagnon, J., Comeau, Y., Matte, N., Greer, C.W. and Samson, R. (1995) Activation of indigenous microbial consortium for bioaumentation of pentachlorphenol/creosote contaminated soils. *Applied Microbiology and Biotechnology* 40, 926–932.

Ribarova, I., Topalova, J., Ivanov, I., Kozuharov, D., Dimkov, R. and Cheng, C. (2001) Anaerobic sequencing batch reactor as an initiating stage in complete PCP biodegradation. *Water Science and Technology* 46 (1/2), 565–569.

Swindoll, C.M., Perkins, R.E., Gannon, J.T., Holmes, M. and Fisher, G.A. (1995) Assessment of Bioremediation of a Contaminated Wethland. In: Hinchee, R.E., Wilson, J.T. and Dawney, D.C. (eds) *Intrinsic Bioremediation.* Battelle Press, Columbus, Ohio, pp. 163–169.

Topalova, Y. (2005) Challenges in wastewater treatment technologies in 21st century. *Environmental Engineering and Environment Protection* 3–4, 32–41.

Topalova, Y. (2009) *Biological Control and Management of Water Treatment.* Ed.PublishSiteSet-Eco. ISBN 978-954-749-042-0, 352 pp.

Topalova, Y. and Dimkov, R. (2003) *Biodegradation of Xenobiotics.* Sofia, Bulgaria, 120 pp.

Topalova, Y., Van Keer, C., Ivanov, I. and Dimkov, R. (1998) Bioremediation potential of aerobic sludge towards combination oNP/PCP. Med. Fac. Landbouww.Univ. Gent, 63/4b, *12th Forum on Applied Biotechnology*, Sept. 98, Ghent, 63/4/, pp. 1901–1905.

Trably, E., Patureau, D. and Delgenes, J.P. (2003) Enhancement of polycyclic aromatic hydrocarbons removal during anaerobic treatment of urban sludge. *Water Science and Technology* 48(4), 53–60.

Chapter 17

Bioremediation of Uranium, Transuranic Waste and Fission Products

Evans M.N. Chirwa

Introduction

Since the industrial revolution, the world has seen an unprecedented increase in energy demand due to population growth, urbanization and rapid industrialization. The increase in industrial activities to support the booming population has exerted pressure on the planet's ecological and energy resources. In order to ensure future sustainability of the human race on Earth, it has become increasingly clear that we will need to find less-polluting energy resources to reduce our current reliance on carbon-based fuels. Renewable and cleaner energy sources such as solar, wind, nuclear and geothermal energy could help halt the current trends of carbon inputs into the atmosphere. Geothermal energy is considered the cleanest among the list of cleaner alternatives and could be the most sustainable form of energy in the long-term. However, its actual implementation remains a theoretical curiosity due to non-existence of viable technologies to harness it. In order to abandon carbon-based fuels as the primary energy source, humankind will require an alternative concentrated energy source that could reliably replace the fossil fuels to power existing energy grids. Currently, only nuclear energy appears to satisfy the criteria of low carbon footprint, stability and concentration (Mourogov *et al.*, 2002).

Nuclear energy has so far been implemented in 31 countries around the world, providing about 14% of the world's primary energy supply (IAEA, 2009). The world's nuclear generating capacity currently stands at about 372 GWe, with the USA and France as the leading producers at 27% and 17% of the world nuclear capacity, respectively. The worldwide application of this technology is mostly hindered by fears of nuclear accidents, proliferation of nuclear weapons, and radioactive waste pollution.

Potential radioactive pollution to the environment does not only concern nuclear power plants. Other activities such as radioisotope manufacturing and biomedical research also release large amounts of potentially harmful radioisotopes (Macaskie and Lloyd, 2002; Nazina *et al.*, 2004; Tikilili and Chirwa, 2011). Most of the radioactive pollutants from the latter activities are organic in nature and are amenable to biological degradation (Cerniglia *et al.*, 1984; Bouwer and Zehnder, 1993). However, due to the toxic nature of the waste stream, an additional effort is required to isolate specialized bacteria that are resistant to the toxic effects of the released compounds and that are capable of breaking the complex structures of the organic compounds (Tikilili and Chirwa, 2011).

This chapter provides a concise review of the waste compounds originating from nuclear power generation and other radioisotope-releasing activities and how these could be treated for beneficial use. Recent developments in the bioremediation of radionuclides such as uranium(VI) and technicium(VII) and related biochemical pathways in bacteria are evaluated. Biochemical processes that have yielded positive results are presented as part of the review and their impact on the future of power generation is evaluated.

Characterization of Nuclear Waste

Several categories of radioactive waste are produced in the nuclear industry ranging from highly radioactive waste to low radiation-level waste. A detailed categorization of the radioactive waste is provided by the United States Nuclear Regulatory Commission (http://www.nrc.gov/waste). The main categories are summarized below.

Low-level waste (LLW)

Low-level waste (LLW) is generated from hospitals and industry, as well as the nuclear fuel cycle. It comprises paper, rags, tools, clothing and filters, etc., which contain small amounts of mostly short-lived radioactivity. It does not require shielding during handling and transport and is suitable for shallow land burial. To reduce its volume, it is often compacted or incinerated before disposal. The bulk of graphitic waste from Generation IV fast reactors could fall into this category. In this case, the limitation is space requirements rather than shielding and treatability.

Intermediate level waste (ILW)

Intermediate level waste (ILW) contains higher amounts of radioactivity and some require shielding. It typically comprises resins, chemical sludges and metal fuel cladding, as well as contaminated materials from reactor decommissioning. It may be solidified in concrete or bitumen for disposal. Generally, short-lived waste (mainly from reactors) is buried in a shallow repository, whereas long-lived waste, for example, waste from fuel reprocessing, could be buried deep underground.

High level and transuranic waste

High-level waste (HLW) arises from the use of uranium fuel in nuclear reactors and nuclear weapons processing. It contains the fission products and transuranic elements generated in the reactor core. It is highly radioactive and 'hot'. In the parlance of the nuclear industry, it is regarded as the 'ash' from 'burning' uranium. HLW accounts for over 95% of the total radioactivity produced in the process of nuclear electricity generation.

Transuranic (TRU) waste is contaminated with alpha-emitting transuranium radionuclides with half-lives greater than 20 years, and concentrations greater than $100 nCig^{-1}$ but not including HLW. In the USA, HLW and TRU arises mainly from weapons production, and consists of clothing, tools, rags, residues, debris and other such items contaminated with small amounts of radioactive elements, mostly plutonium. These elements have an atomic number greater than uranium thus are transuranic (beyond uranium). Because of the long half-lives of these elements, this waste is not disposed of as either LLW or ILW.

A typical example of waste volumes produced in the power generation industry is shown for the Low Enriched Uranium Once Through (LEU-OT) and Mixed-Oxide Once Through (MOX-OT) fuel processing cycles (Tables 17.1 and 17.2). In the tables, it is shown that the majority of waste in the nuclear power generation industry originates from the uranium mining and milling operations.

Waste from high temperature fast reactors

High temperature gas-cooled reactors (HTGR), also known as fast reactors, mostly utilize graphite as the fission reaction moderator. Graphite in the fast reactors is used either as part of the structural material for the reactor core vessel or as fuel containment elements in the form of pebbles (spheres). The graphite used from natural sources contains non-carbon impurities within the carbon matrix. Among these impurities are oxygen and nitrogen from entrapped air, cobalt, chromium, calcium, iron and sulfur (Khripunov et al., 2006). Upon exposure to high neutron flux, most of the impregnated impurities are expected to transmute to unstable radioactive forms. Examples of transmutation products are shown in Table 17.3. Other impurities such as transitional metals Cr^{6+} and Co^{2+} are also commonly found in the radioactive forms.

Radioactive fission products created in fuel elements migrate through grain boundaries and then through microscopic cracks in the graphic matrix (Fig.17.1). Most of the fission products are entrained in the matrix – a small proportion escapes through the outer layers into the gas phase. The challenge of reprocessing involves the separation of the metallic radionuclides from the graphite matrix and reduction of the radioisotope C-14 content of the containment graphite. In the case of pebble-bed reactors, the pebble is deemed 'expired' when the C-14 and impurities surpass a predetermined operational level that is accompanied by loss of efficiency of the graphite as a neutron flux regulator.

Impurities trapped in the graphite matrix consist mainly of transitional and transuranic products of the nuclear chain reaction including: (i) metallic fission products (such as Mo, Tc, Ru, Rh and Pd), which occur in the grain boundaries as immiscible micron to nanometre-sized metallic precipitates (ε-particles); (ii) fission products that occur as oxide precipitates of Rb, Cs, Ba and Zr; and (iii) fission products that form solid solutions with the UO_2 fuel matrix, such as Sr, Zr, Nb and the rare earth elements. This results in complexity in the characteristics of the resulting spent-fuel elements depending on thermal history, period spent in the reactor and physical structure of the original fuel element (Buck et al., 2004; Bruno and Ewing, 2006).

Treatment Options

Treatment is performed on nuclear waste to achieve one or all of the four targets for handling of waste, i.e.: (i) waste minimization; (ii) toxicity reduction; (iii) volume reduction; and/or (iv) security (deterrence of proliferation). Items (i) to (iii) can be achieved through a combination of physical-chemical and biological processes whereas item (iv) is achieved through both operational and policy mechanisms.

Table 17.1. Characteristics of waste generated from a Low Enriched Uranium Once Through (LEU-OT) processing cycle ($m^3 GWe^{-1} year^{-1}$).

Steps	SF[b]	ILW	LLW	Tailings	Comments
Mining and milling	–	–	–	65,000	Currently rated as the most hazardous step in the nuclear fuel chain, disproportionately impacting indigenous communities.
Conversion	–	–	32–112	–	Hazards include chemicals such as hydrofluoric acid, nitric acid and fluorine gas.
Enrichment	–	–	3–40	–	Typically buried at dump sites with a high risk of leaching radionuclides into the groundwater. Typically contains PCBs, chlorine, ammonia, nitrates, zinc and arsenic.
Fuel fabrication	–	–	3–9	–	Does not involve the production of liquid waste, thus its effects are mainly restricted to workers and are on the same order as for workers in the reprocessing sector.
Reprocessing and vitrification	not applicable	not applicable	not applicable	not applicable	Waste may contain chlorinated aromatic compounds due to the use of decontamination reagents such as CCl_4 together with phenolic tar (Gad Allah, 2008).
Reactor operations	–	22–33	86–130	–	Boiling water reactors have considerable emissions of radioactive noble gases.
Spent fuel storage and encapsulation	–	2	0.2	–	Considerable quantities of 'low-level' waste are created due to fission products leaking into the spent fuel pools from cracks in the fuel cladding (Choi *et al.*, 1997).
Spent fuel final disposal	26	–	–	–	Insufficient treatment can cause continued exposure to environment and local population.
Decommissioning	–	9	333	–	Most of the radioactivity from reactor decommissioning waste is in a relatively small volume of intensely radioactive material.
Totals	26	33–44	457–624	65,000	

[a] SF, spent fuel

Table 17.2. Characteristics of waste generated from a Mixed-Oxide Once Through (MOX-OT) processing cycle ($m^3 GWe^{-1} year^{-1}$).

Steps	SF[b]	HLW	ILW	LLW	Tailings	Comments
Mining and milling	–	–	–	–	50,060	Mill tailings account for over 95% of the total volume of the radioactive waste from MOX-OT processing cycle. This does not include mine wastes. Many tailings sites all over the world remain unremediated.
Conversion	–	–	–	25–86	–	Hazards include chemicals such as hydrofluoric acid, nitric acid and fluorine gas.
Enrichment	–	–	–	3–25	–	Buried at dump sites. Typically contains PCBs, chlorine, ammonia, nitrates, zinc and arsenic.
Fuel fabrication	–	–	13	7.4–12.5	–	Workers are at the highest risk of exposure. Very low liquid waste generated.
Reprocessing and vitrification	–	2–4[c]	17–39	8,016–8,037[d]	–	As in LUE-OT system, wastes from reprocessing discharged together with spent fuel produce the highest risk. The waste is high inorganic content.
Reactor operations	–	–	22–33	86–130	–	Boiling water reactors have considerable emissions of radioactive noble gases.
Spent fuel storage and encapsulation	–	–	0.3	0.03	–	As in the LUE-OT system, large quantities of 'low-level' waste are created due to fission products leaking into the spent fuel pools from cracks in the fuel cladding.
Spent fuel final disposal	26	–	–	–	–	Insufficient treatment can cause continued exposure to environment and local population.
Decommissioning	–	–	10.1	315	–	Most of the radioactivity from reactor decommissioning waste is in a relatively small volume of intensely radioactive material.
TOTALS	26	2–4	62–95	8,452–8,615	50,060	

[a] SF, spent fuel

Table 17.3. Carbon-14 production mechanisms and cross-sections.

Target isotope	Mechanism	Thermal cross-section (burns)	Isotopic abundance (%)*
^{14}N	$^{14}N(n, p)^{14}C$	1.81	99.6349
^{12}C	$^{12}C(n, \gamma)^{14}C$	n/k	n/k
^{13}C	$^{13}C(n, \gamma)^{14}C$	0.0009	1.103
^{17}O	$^{17}O(n, \alpha)^{14}C$	0.235	0.0383

Source: Molokwane and Chirwa (2007).
n/k: Not known.

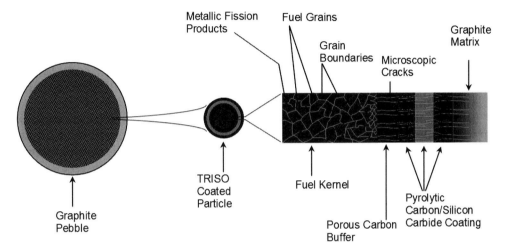

Fig. 17.1. Propagation of fission products and impurities in a graphite-regulated high temperature gas-cooled reactor fuel element (Adapted from Chirwa, 2011).

Physical and chemical treatment technologies

Physical chemical processes utilize chemicals or chemical properties of surfaces to chemically reduce or physically extract the radionuclides based on chemical charge or size. Processes that have been tried include the use of specialized ion exchange resins, chemical oxidation and adsorption columns. Mechanisms and materials employed include the following.

Adsorption

Adsorption is primarily used for the removal of soluble organics using common adsorbents such as activated carbon. Removal of organics on the adsorbent surfaces is mostly predicted by the Freundlich isotherm:

$$y = kc^{1/n} \quad (17.1)$$

where y = adsorbent capacity, mass pollutant/mass adsorbent, c = concentration of the pollutant in waste (mass/volume), and k and n = empirical constants. The other non-linear models: Langmuir and Langmuir-Hinshelwood have been used for single-layer adsorption as opposed as the double-layer model of the Freundlich isotherm.

Ion exchange

Several ion exchange resins have been tested for the removal of metallic elements and radiocarbon-14 from waste streams. Such a process was attempted for recovery of C-14 from nuclear process water in the so-called Swedish Process (Van Lam et al., 2000). Although sometimes proven successful, the regeneration of the resin for reuse is often a problem. For this reason, there have been limited efforts to develop this technology further.

Stripping

Air stripping is applied for the removal of volatile substances from water. Henry's law is used in the design of stripping systems. The gas: liquid ratio required to remove a semi-volatile pollutant from liquid is estimated by:

$$\frac{Q_g}{Q_L} = \frac{C_{in} - C_{out}}{(H) C_{in}} \quad (17.2)$$

where Q_g = the flow rate of the stripping gas (volume/time), Q_L = the flow rate of the liquid carrying the waste, C_{in} and C_{out} = the influent and effluent concentration across the control volume, and H = Henry's constant for the pollutant.

Membrane processes

The emerging technology of nanopore membranes holds promise for the future as one of the processes that could be used to separate radioisotopes from water or gas streams. Membrane processes are normally easy to operate and offer the additional advantages of immediate capture of the pollutant for cleaning and recycling. An additional advantage is that membrane reactors tend to be portable and can be built to occupy the smallest land footprint possible.

Biological treatment

Biological systems can be used in remediation of natural environments includingthe aquatic, soils, and sediments and treatment of effluent streams (Boopathy, 2006). Toxic organic pollutants are generally treated by mineralization(oxidation to form CO_2 and water). On the other hand, toxic metals cannot be degraded, but they can be transformed into less mobile valency states (Gadd, 2004). Microorganisms use one of the following four processes to immobilize metals and radionuclides in the environment: biosorption, bioaccumulation, bioprecipitation and bioreduction (Nancharaiah et al., 2006).

Biosorption

This is a technique in which the uranium-bearing water is brought into contact with living or resting cells of bacteria, fungi, yeast, algae, or other forms of biological material that possess abundant functional groups on their surfaces. Uranyl species react with these sites through passive, physical-chemical mechanisms. Microbial enzymatic activity is not directly involved although polymers secreted by many metabolizing microbes also immobilize metals (Francis *et al.*, 1994). *Pseudomonas aeruginosa* strain CSU, a non-genetically engineered bacterial strain known to bind dissolved hexavalent uranium (UO_2^{2+}), was characterized with respect to its sorptive activity (equilibrium and dynamics) (Hu *et al.*, 1996). In the study by Hu and co-workers (1996), living, heat-killed, permeable and unreconstituted lyophilized cells were all capable of binding uranium. U(VI) removal by biomass was comparable to removal by commercial cation-exchange resins, particularly in the presence of dissolved transition metals as it adsorbed a large amount of uranium. The binding of metals by the biomass in this case was pH dependent.

Desorption and recovery of the biosorbed radionuclides is easy. Radionuclide-binding to cell surfaces and polymers is a promising technology for remediating contaminated waters. However, the effectiveness of biosorbants, for example, fungal or bacterial biomass, is affected by poor selectivity against competing ions and saturation at high radionuclide concentrations (Ashley and Roach, 1990). Other limitations of the biosorption methods include the complexation of the metal with carbonates, resulting in slower adsorption rates. Although it is possible to regenerate some of the biomass by extracting the uranium with salt solutions or acids, this extraction is expensive and can result in a large volume of corrosive uranium-containing waste. And lastly, biosorption poorly extracts uranium when it is present at low concentrations (Lloyd *et al.*, 2002).

Bioaccumulation

This is an active process whereby metals are taken up into living cells and sequestered intracellularly by complexation with specific metal-binding components or by precipitation. All classes of microorganisms have the capability to accumulate metals intracellularly by an energy-dependent transport system. Localizing the metal within the cells permits its accumulation from bulk solution, although the metals cannot be easily desorbed and recovered (Macaskie *et al.*, 1992). A major drawback associated with the use of active uptake systems is the requirement of metabolically active cells. This may prohibit their use in the treatment of highly toxic waste. Regardless, this approach is promising as a means of remediating fission products from dilute waste streams. The process involves uptake of U(VI) via the K^+-transport system (Tsuruta, 2006).

Biocrystallization

Biocrystallization, also known as bioprecipitation or biomineralization, is the generation of metal precipitates and minerals by bacterial metabolism. Somespecies of bacteria have been shown to immobilize and concentrate metal ions with the possibility of purifying minerals. Microorganisms growing in biofilms for example, have the capability to bindlarge quantities of metallic ions.Biofim may function as a template for the precipitation of insoluble mineral phases. The biochemistry of the interactions of metal ions with bacterial cell walls, extracellular biopolymers and microfossil formations in immobilizing toxic metals has been extensively studied (Appukuttan *et al.*, 2006). In 2000, Macaskie *et al*. also investigated *Citrobacter* sp. accumulation of uranyl ion (UO_2^{2+}) via precipitation with phosphate ligand liberated by phosphatase activity. This yielded a novel approach for monitoring the cell-surface-associated changes using a transmission electron microscope.

Using this method, Macaskie et al. (2000) illucidated that metal deposition occurs via an initial nucleation with phosphate groups localized within the lipopolysaccharide (LPS). Accumulation of metal phosphate within the LPS was suggested to prevent fouling of the cell surface by the accumulated precipitate and localization of phosphatase exocellularly (Macaskie et al., 2000).

Although this process involves an enzymatic reaction, it does not involve the metal directly. One shortcoming of the biosorptive process is that it may be hindered by the presence of carbonate. Second, the amount of uranium that can be sorbed onto the cell surface is limited (Gadd, 1992). In addition, the biosorptive process is non-selective and therefore may not be effective where single species recovery is desired.

Reduction

Reduction involves the transformation of an element from a higher valency state to a lower valency state. Reduction of an element may facilitate precipitation or volatilization (Lovley et al., 1991). For uranium(VI), reduction of hexavalent state has been achieved in pure and mixed cultures of iron-reducing and sulfate-reducing bacteria grown under anaerobic and sulfate-reducing conditions. U(VI) reduction in microbial species is microbiologically and metabolically diverse (Table 17.4).

Microorganisms are known to have evolved biochemical pathways for degrading or transforming toxic compounds from their immediate environment, either simply for survival or to derive energy by using the toxic compounds as electron donors or electron sinks (Rothschild and Mancinelli, 2001; Bush, 2003). The biotransformation pathways commonly take advantage of the advanced and well conserved membrane electron transport respiratory apparatus within the organisms (Dickerson, 1980; Li and Graur, 1991). For example, the redox reactions involving some of the metallic pollutants are coupled to the electron transport through electron carriers in the cytoplasmic membrane and the flux of protons through the ATP-synthase. The proton flux and production of ATP through ATP-synthase generates the required energy equivalents for use in cellular metabolism (Alberts et al., 1994).

Researchers have demonstrated that the reduction of toxic forms of Fe, Cr, Mn, U, Tc and other toxic metals could be catalysed by enzymes associated with the membrane electron transport respiratory pathway (Wade and DiChristina, 2000; Lloyd, 2003). The catalytic redox processes for metallic species have been put to beneficial use in engineering applications such as bioleaching and recovery of precious metals from waste streams and tailing dams (Demergasso et al., 2010; Bakhtiari et al., 2011; Dong et al., 2011; Qiu et al., 2011). These processes have been used in place of chemical processes, which are considered costly and environmentally intrusive.

Biological U(VI) Reduction

Radioactive heavy metal species in the actinide family such as U(VI) and Tc(VII) may be treated biologically in microbial processes involving the membrane electron transport system. These compounds readily serve as electron sinks in biochemical processes commonly found in sulfate-reducing bacteria (Wade and DiChristina, 2000; Macaskie and Lloyd, 2002). U(VI) is reduced to the less toxic and less mobile tetravalent state (U(IV)) by a variety of sulphate-reducing organisms growing under strictly anaerobic conditions.

Researchers such and Lloyd, Macaskie, and Lovley and co-workers have over theyears demonstrated that several species of sulfate-reducing bacteria utilize the transmembrane sulfate shuttle system to transport the oxyionic forms of U(VI) (UO_2^{2+}, $U_2O_8^{2+}$, etc.) into the cells where they are reduced to U(IV) (Merroun and Selenska-Pobell, 2008) or truncated membrane respiratory pathway through NADH-dehydrogenase (Horitsu *et al.*, 1987; Ishibashi *et al.*, 1990; Shen and Wang, 1993). Since U(VI) is predominantly reduced internally in these organisms, subsequent recovery of the reduced uranium for reclamation and further reprocessing could be challenging.

Recently, the role of cytoplasmic elements such as thioredoxin and NADH was elucidated in *Desulfovibriodesulfuricans* mutant strain G20 (Li and Krumholz, 2009). In these species, the reduction of U(VI) was linked to the presence of thioredoxin, thioredoxin reductase and metal oxidoreductase. Electron microscopic images of wild-type *D. desulfuricans* showed reduced uranium species predominantly in the cytoplasm and periplasm of non-disrupted cells (Sani *et al.*, 2006; Li *et al.*, 2009).

Uranium(VI)-Reducing Mechanisms

Enzymatic mechanisms of radionuclide reduction by *Geobacter* and *Shewanella*

The mechanisms by which Fe(III)-reducing bacteria transfer electrons to insoluble Fe(III) oxides during anaerobic growth have been extensively studied in *Geobacter* and *Shewanella* species (Lloyd *et al.*, 2002). In both organisms, an electron transfer chain containing *c*-type cytochromes is thought to pass through the periplasm and terminate at the outer membrane, facilitating electron transfer to the extracellular solid phase substrate (Gaspard *et al.*, 1998; Lloyd *et al.*, 2002). Given that U(VI) is reported to precipitate outside the cell in *Geobacter* and *Desulfovibrio* species, it was hypothesized that a similar electron transfer pathway terminating at the cell surface may be important in U(VI) reduction by *Geobacter sulfurreducens*.

*Shewanella*reductase(s)

To date, only four strains have been reported to gain sufficient energy from U(VI) respiration to support growth: *Shewanella putrefaciens*, *Geobacter metallireducens*, *Desulfotomaculum reducens*, and *Thermoterrabacterium ferrireducens* (Lovley and Phillips, 1992; Shelobolina *et al.*, 2004; Marshall *et al.*, 2005). The above species were shown to be capable of reducing U(VI) to U(IV) as part of the energy conversion mechanisms under certain conditions. Early work with *S. putrefaciens* showed that cells limited for Fe were unable to use Fe(III) as a terminal electron acceptor (Beliaev and Saffarini, 1998). These cells also lost their orange colour and this indicated a major decrease in *c*-type cytochrome content (Obuekwe and Westlake, 1982).

The interpretation of these observations was that cytochromes were involved in the transfer of electrons to the terminal electron acceptor or were the terminal reductases. Subsequently, various cytochromes of *Shewanella* were shown to localize in the periplasm and with either the cytoplasmic or the outer membrane (Myers and Myers, 1992). Figure 17.2 shows the suggested U(VI) reduction pathways based on electron transport and deposition of reduced forms in the periplasm and outer membrane systems in *Shewanella oneidensis* MR-1.

Mutation of the *S. putrefaciens* 200 enzyme, tetraheme c-type cytochrome (SO3980),implicated the nitrite reductase in U(VI) reduction because of the simultaneous loss of U(VI) and NO_2^- reduction in the absence of this reductase (Wade and DiChristina, 2000). Transposon mutagenesis of *S. putrefaciens* identified a decaheme outer membrane c-type cytochrome, MtrA, as necessary for Fe(III) and Mn(IV) reduction (Beliaev and

Table 17.4. U(VI) reducing bacteria, their source and preferred environmental conditions.

Bacterium	Source of culture	Growth conditions|energy source
Anaeromyxobacter dehalogenans str. 2CP-C	Stream sediment, Lansing, Michigan	Anaerobic|2-chlorophenol
Cellulomonas flaigena ATCC 482	Sugarcane field	Aerobic/anaerobic |glucose and others
Clostridium sphenoides ATCC 19403	Mine pit water	Anaerobic|citric acid and glucose
Deinococcus radiodurans R1	Irradiated ground pork and beef	Aerobic|non-fermentable carbon sources
Desulfomicrobium norvegicum DSM 765	Sediment core	Anaerobic|acetate and others
Desulfotomaculum reducens	Saltwater, California, USA	Anaerobic|butyrate and lactate
Desulfovibrio baarsii DSM 2075	Ditch mud, Germany	Anaerobic|butyrate, ethanol and others
Desulfovibrio desulfuricans ATCC 29577	Tar-sand mixture, UK	Anaerobic|acetate and lactate
Desulfovibrio desulfuricans strain G20	Soured oil reservoir, Alaska	Anaerobic|lactate, acetate, glucose and others
Desulfovibrio sp. UFZ B 490	Uranium dump, Saxony, Germany	Anaerobic|TCA metabolites and ethanol
Desulfovibrio vulgaris Hildenborough	Wealden clay, England	Anaerobic|lactate
Geobacter metallireducens GS-15	Sediment, Potomac River, USA	Anaerobic|phenol, acetate, formate
Geobacter sulfurreducens	Surface sediments, Norman, Oklahoma	Anaerobic|fumarate and acetate
Pseudomonas putida	Uranium mill tailing sites	Anaerobic|pyruvate, glucose
Pseudomonas sp. CRB5	Chromate-containing sewage	Anaerobic|lactate
Pyrobaculum islandicum	Icelandic geothermal power plant	Anaerobic|iron, thiosulfate and elemental sulfur
Salmonella subterranea sp. nov. strain FRC1	Uranium-contaminated sediment	Aerobic|citrate, acetate and others
Shewanella alga BrY	Estuary Sediment, New Hampshire	Facultative anaerobic|insoluble mineral oxides
Shewanella oneidensis MR-1	Sediment, Oneida Lake, New York	Anaerobic|lactate
Shewanella putrefaciens strain 200	Oil pipeline, Alberta, Canada	Anaerobic|formate, lactate
Thermoanaerobacter sp.	Geothermal spring	Anaerobic|glucose, pyruvate, peptone
Thermus scotoductus	Hot tap water, Selfoss, Iceland	Aerobic|acetate
Thermoterrabacterium ferrireducens	Hot springs in Yellowstone National Park, USA	Anaerobic| glycerol and citrate

* Adapted from Chabalala (2011).

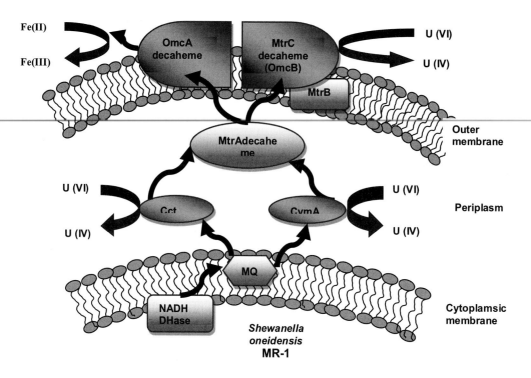

Fig. 17.2. A concept model showing electron pathways for U(VI) reduction and possible reaction sites in the periplasm and outer membrane. MQ, menaquinone; CymA, tetraheme membrane-bound cytochrome; Cct, tetraheme periplasmic cytochrome; OmcA, decaheme outer membrane cytochrome; MtrA, decaheme periplasmic cytochrome; MtrB, outer membrane structural protein; MtrC (OmcB), decaheme outer membrane cytochrome.

Saffarini, 1998). The function of these electron carriers was only recently evaluated as a part of the analysis of global transcriptional responses to U(VI) (Bencheikh-Latmani et al., 2005). Genome sequencing of *S. oneidensis* MR-1 revealed the presence of 42 putative *c*-type cytochromes (Heidelberg et al., 2002; Beliaev et al., 2005). Global transcript analysis of these cytochrome genes during growth on different metal and non-metal electron acceptors (but not uranium or chromium) showed only one cytochrome, SO3300, to be significantly increased in expression during metal reduction (Beliaev et al., 2005). In contrast, when these cells were incubated under non-growing conditions with 0.1 mM U(VI) or Cr(VI) present, of the 32 genes that increased (three-fold) in both cultures, 12 were cytochromes, but SO3300 was not among them (Bencheikh-Latmani et al., 2005).

Several proteins, including one involved in menaquinone biosynthesis (MenC), an outer membrane protein (MtrB), a periplasmic decaheme cytochrome (MtrA), an outer membrane decaheme cytochrome (MtrC, also named OmcB), and a tetraheme cytochrome (CymA) anchored in the cytoplasmic membrane have been shown to be required for optimal U(VI) reduction in *S. oneidensis* MR-1 (Wall and Krumholz, 2006).

Also of vital importance was the observation that the mutants lacking one or more of these electron transfer components were all still capable of U(VI) reduction with lactate as electron donor. Thus multiple pathways for electron delivery to U(VI) are available in *Shewanella*. Comparison of UO_2(s) deposition by omcA or mtrC mutants lacking outer membrane decaheme *c*-type cytochromes showed accumulation predominantly in the periplasm versus the deposition of uraninite external to wild-type cells (Marshall *et al.*, 2005). This result is consistent with the observation that U(VI) reduction is not eliminated by any of the single mutants analysed and supports the hypothesis that uranium reductases are likely non-specific, low potential electron donors present in both the periplasm and the outer membrane. It remains to be determined whether the mutants altered for U(VI) reduction are similarly affected in their ability to use U(VI) as terminal electron acceptors for growth (Wall and Krumholz, 2006).

Cellular location of UO_2 precipitates

Because of the insoluble nature of U(IV) oxide, the site of deposition should give an indication of the location of the reductase. Many researchers have confirmed the accumulation of uraninite in the periplasm of dissimilatory metal-reducing bacteria (DMRB) using transmission electron microscopy (TEM) imaging. These observations have been used to validate the dependence of U(VI) reduction on transmembrane proteins.

Remarkably, for the Gram-positive bacterium *Desulfosporosinus*, uraninite was found in a similar location, concentrated in the region between the cytoplasmic membrane and the cell wall (Suzuki *et al.*, 2004). These results would point to a uranium reductase on the periplasmic (outer) face of the cytoplasmic membrane or in the periplasm itself. Uraninite deposits within the cytoplasm of a pseudomonad and *D. desulfuricans* strain G20 have also been observed (McLean and Beveridge, 2001; Sani *et al.*, 2002). The pseudomonad isolated from a site formerly used for treating wood for preservation removed U(VI) from solution under aerobic or anaerobic conditions. When TEM thin sections of those cells were examined, U(IV) was found inside as well as concentrated at the envelope. As uranium has no biological function and is toxic, the observation of its precipitation in the cytoplasm was surprising. McLean and Beveridge (2001) suggested that the polyphosphate granules present in the pseudomonad might protect the cell by forming strong complexes with uranium, thus impounding it in the cytoplasm.

The internal deposition of uraninite observed in *D. desulfuricans* G20 occurred in cells that were grown in a medium intended to limit heavy metal precipitation and maximize toxicity (Sani *et al.*, 2002). In order to prevent the formation of strong complexes, the medium had no specifically added carbonate or phosphate. Amendments such as these could also alter the physiology of the bacterium, stimulating uptake systems that might allow access of the toxic metal to the cytoplasm.Cytoplasmic deposition of U(IV) has not been reported from other studies with *Desulfovibrio* (Lovley and Phillips, 1992; Barton *et al.*, 1996), and future studies on the effects of nutritional stresses on U(VI) reduction may prove interesting.

With the exception of these unusual reports of cytoplasmic uraninite, the localized precipitation of insoluble U(IV) in the periplasm and outside of both Gram-negative and Gram-positive cells suggests that U(VI) complexes do not generally have access to intracellular enzymes. Therefore, the best candidates for reductases would be electron-carrier proteins or enzymes exposed to the outside of the cytoplasmic membrane, within the periplasm, and/or in the outer membrane (Wall and Krumholz, 2006).

Characterization of Facultative U(VI) Reducers

Initial evaluation of the bacteria isolated from soil samples from an abandoned uranium mine showed that cells in anaerobic cultures were predominantly Gram-negative (Fig. 17.3a). However, cultures grown under aerobic conditions were predominantly Gram-positive (Fig. 17.3b). These results were almost predictable since most bacilli thrive better under aerobic conditions. Close inspection revealed a myriad of Gram-positive cocci and Gram-negative rod-shaped cells in both systems.

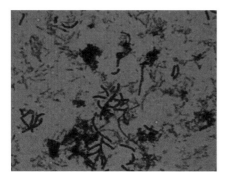 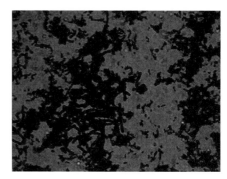

Fig.17.3. Optical micrographs of (A) anaerobically grown mine soil consortium and (B) aerobically grown mine soil consortium taken under the Zeiss Axioskop II microscope (Carl-Zeiss, Oberkochen, Germany) (Chabalala, 2011).

The cultures were later characterized phylogenetically and cell types belonging to the genera *Pseudomonas*, Bacilli, Enterobacteriaceae and Pantoeae were shown to reduce U(VI) under anaerobic conditions (Figs17.4–17.6).

U(VI) reduction capacity was confirmed in pure cultures of *Pantoea agglomerans* and three other Gram-negative isolates. *Pantoea agglomerans* was confirmed as a member of
the family Enterobacteriaceae within the gamma subdivision of the Proteobacteria, which was demonstrated earlier by other researchers to catalyse the reduction of several metals, including Fe(III), Mn(IV) and Cr(VI) under anaerobic conditions (Tebo *et al.*, 2000).

The species *Pseudomonas stutzeri* also reduces U(VI) under anaerobic conditions, but this species required citric acid as an electron donor to reduce uranium. The biotic processes in the uranium(VI) reduction can sometimes be masked by the complexation processes. Many organic compounds form stable complexes with actinides, likewise, microbial metabolites and the products or intermediates from waste degradation, may be an important source of agents affecting the bioavailability of the radionuclides.

Microbial Uranium (VI) Reduction Kinetics

Uranium (VI) reduction experiments

Bacteria grown overnight in nutrient broth was harvested by centrifugation at 6000 rpm (2860×g) for 10 min. The pellet was washed three times with 0.85% NaCl solution and was re-suspended in 100 ml BMM solution in 250 ml Erlenmeyer flasks for aerobic batch

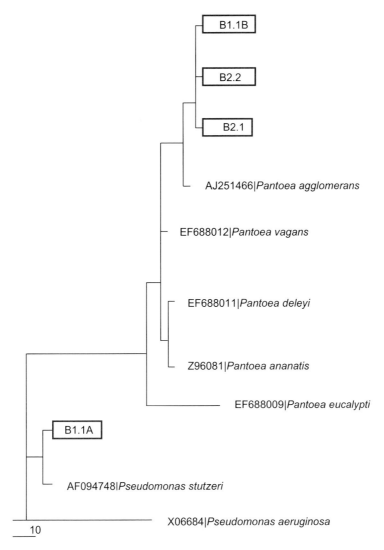

Fig. 17.4. Phylogenetic analysis of the dominant cultivable bacteria (Gram-negative) species present in the mine soil sample. Three identities closely associated with *P. stutzeri* and *P. agglomerans* were U(VI) tolerant at 75 mg l^{-1} and showed the capability to reduce U(VI) while incubated at 30°C in basal mineral medium (BMM) with D-glucose as the sole added carbon source.

experiments or in 100 ml serum bottles for the anaerobic experiments. The average biomass concentration of 9.3 mgml^{-1} was determined initially.

A preliminary experiment using consortium cultures from soil showed U(IV) oxidation activities in some of the species, thus experiments were conducted with purified cultures to isolate U(VI)-reducing species. Initial (added) U(VI) concentrations ranging from 30 to 400 mgl^{-1} were tested with pure cultures of bacteria. The experiments were conducted at 30°C

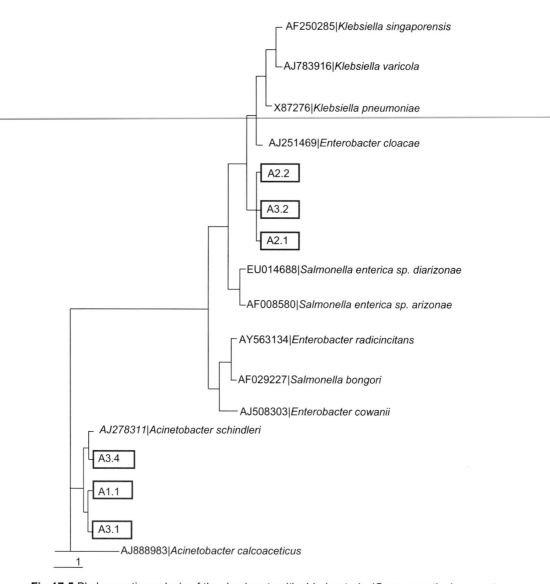

Fig.17.5. Phylogenetic analysis of the dominant cultivable bacteria (Gram-negative) present in the soil sample. The identities closely associated with *E. cloacae* were also U(VI) tolerant at 75 mgl^{-1} and showed the capability to reduce U(VI). Identities associated with *A. schindleri* were U(VI) tolerant at 75 mgl^{-1} but were not able to reduce U(VI) under anaerobic conditions.

at determined time interval (24–48 h) at 120 rpm on the orbital shaker (Labotec, Gauteng, South Africa). These were then purged with 99.9% pure nitrogen for 5 min each. After reduction, the solution was centrifuged at 10,000 rpm for 10 min. A syringe was used to draw samples at intervals up to 48 h, followed by a uranium analysis in solution. An Arsenazo III spectrophotometric method developed by Chabalala and Chirwa (2010) was used to analyse U(VI) and U(IV) concentration in the samples.

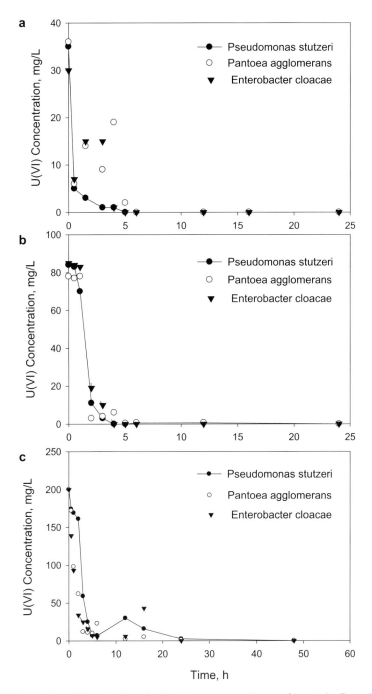

Fig. 17.6. Uranium(VI) reduction for the three pure cultures of bacteria *Pseudomonas stutzeri*, *Pantoea agglomerans* and *Enterobacter cloacae* under an initial concentration of (a) 30 mgl^{-1}, (b) 85 mgl^{-1} and (c) 200 mgl^{-1}.

Results showed complete removal of U(VI) in batch cultures at initial U(VI) concentration up to 200 mg l^{-1}. Up to 96% of this activity was observed within the first 5 h of incubation (Fig. 17.6). U(VI) reduction proceeded to approximately 80% in cultures started with an initial concentration of 400 mgl^{-1}, showing the inhibitory effect of U(VI) on the cultures at high concentrations (Fig. 17.7).

Effect of U(VI) Inhibition

Dissimilatory respiratory U(VI) reduction by microorganisms has been studied in sulphate-reducing organisms such as *Desulfovibrio desulfuricans* and *Desulfovibrio vulgaris* (Lovely and Phillips, 1992). During dissimilatory U(VI) reduction chemical energy is conserved from electron flow during the reduction process. In the study by Chabalala and Chirwa (2010), it was assumed that U(VI) reduction is carried out by viable cells and that viable cells concentration of U(VI)-reducing organisms could produce U(VI) reductase proportional to the amount of cells present in the system. If the amount of enzyme produced E is only associated with viable cells X in the system, then uranium (VI) reduction rate can be represented by enzyme kinetics:

$$-\frac{dU}{dt} = \frac{k_u \cdot U}{K_u + U} \cdot X \quad (17.3)$$

where U is the concentration of U(VI) (ML^{-3}) at time t (T); X is viable cell concentration (ML^{-3}) at time t; k_u is the maximum specific U(VI) reduction rate coefficient (($M_U M_X^{-1} T^{-1}$); and K_u is the half velocity (saturation) constant (ML^{-3}).

Due to the toxic effect of U(VI) on the viable cells, a certain amount of cells added will kill a proportional number of cells during the process. For this reason, due to different levels of tolerance in different species, each species will have its own U(VI) reduction capacity (T_c g U(VI) reducedg^{-1} cell inactivated).

The cell inactivation phenomenon was observed earlier by Shen and Wang (1994) while studying biological Cr(VI) reduction in resting cells of *Escherichia coli* ATCC 33456. The cell inactivation term is thus represented by:

$$X = X_0 - \frac{U_0 - U}{T_c} \quad (17.4)$$

where U_0 is the initial concentration of U(VI) (ML^{-3}); X_0 is the initial cells density of U(VI)-reducing strains (ML^{-3}); and T_c is the maximum U(VI) reduction capacity of cells ($M_U M_X^{-1}$). Substituting Eqn 17.4 into Eqn 17.3 yields the following equation:

$$-\frac{dU}{dt} = \frac{k_u \cdot U}{K_u + U} \left(X_0 - \frac{U_0 - U}{T_c} \right) \quad (17.5)$$

The model fitted the experimental data well, as shown in the example from *Pantoea agglomerans* (Fig. 17.8). The goodness of fit of the model for the rest of the range of data is shown with associated statistical errors in Table 17.5.

Bioremediation of Fission Products

Radioisotope fission products are routinely or accidentally discharged into the environment with wastewaters from various activities, such as mining and milling of nuclear fuel, fallout from nuclear weapon testing or leakage from storage facilities at

Table 17.5. Kinetic parameters for U(VI) reduction in *Pantoea agglomerans*.

Initial U^{6+} mgl^{-1}	K_u mgl^{-1}	k_u mg U^{6+}mg^{-1} cellh^{-1}	T_c mg U^{6+}mg^{-1} cell	X_o mg celll^{-1}	χ^2
30	99.7	0.39	0.4	210	491
75	99.6	0.31	0.49	340	2,037.8
100	99.3	0.3	0.4	2,400	246.5
200	98.6	0.3	0.49	663.5	1,360
400	98	0.39	0.4	919.3	14,078

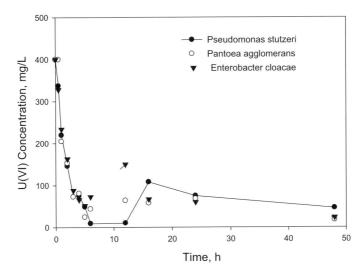

Fig. 17.7. Uranium(VI) reduction for the three pure cultures of bacteria *Pseudomonas stutzeri*, *Pantoea agglomerans* and *Enterobacter cloacae* under an initial concentration of 400 mgl^{-1}.

nuclear installations as well as from industrial and medical facilities (Singh *et al.*, 2008).

Fission products mainly consist of elements lighter than uranium. They exist in nature in a range of ionic species and oxidation states. Some of the cationic species are highly mobile and have significantly long half-lives of radioactive decay. Conventional adsorbents and ion-exchange resins are used to extract and recover cationic fission products (Chaalal and Islam, 2001). In recent research, microbial cells have been investigated as potential biosorbents. Functional groups on the microbial cell walls, such as –OH, –NH$_2$ and –COOH, have been shown to selectively adsorb different ionic species making it possible to purify and recover targeted metals (Chubar *et al.*, 2008; Ngwenya and Chirwa, 2011).

Ngwenya and Chirwa (2011) investigated the biosorption of the common fission products – Cs-137, Sr-90 and Co-60 – using a consortium of sulphate-reducing organisms. These radionuclides, especially the divalent forms, Sr-90 and Co-60, are of environmental concern due to relatively long half-lives and similarity with the ubiquitous cation, Ca^{2+}.

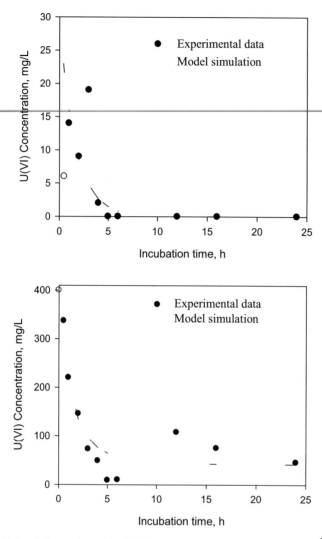

Fig.17.8. Model simulation using optimized parameters: $k_u = 0.34\pm0.05$ mg U^{6+} mg^{-1} cellsh^{-1}, $K_u = 99.04\pm0.72$ mg U^{6+}l^{-1}, and $T_c = 0.44\pm0.05$ mg U^{6+}mg^{-1} cells in *Pantoea agglomerans* at 30 and 400 mg U^{6+}l^{-1}, respectively, and initial cell concentration $X_0 = 906\pm88$ mgl^{-1}.

The biosorption process was designed to remove metallic species, especially those originating from the nuclear reaction process in power plants, for further processing and recovery. The majority of these are the products of nuclear fission of uranium to form lighter elements. During the first several hundred years after the fuel is removed from a reactor, fission products are considered the most hazardous elements to living organisms in the environment.

In the detailed biosorption study using sulphate-reducing bacteria, Ngwenya and Chirwa (2011) evaluated the cell surface properties that determine the biosorptive efficiency of a microbial cell. Sulfate-reducing bacteria (SRB) were used as an example due to their demonstrated ability in adsorbing a range of metals including palladium(II) and calcium(II).

Biosortive processes in SRB

Biosorption of Sr^{2+} was observed under equilibrium conditions in 2 l bench-scale anaerobic bioreactors. The initial Sr^{2+} concentration in the experiments was varied between 75 and 1000 mg l^{-1}, while the SRB cell density was kept constant at 1 mgl^{-1}. The suspensions were agitated for 3 h and then samples were withdrawn for residual Sr^{2+} concentrations analysis.

Sr^{2+} concentration was measured in the medium and a desorption process was conducted to determine the amount sorbed on cells. Based on the amounts in the medium and the amount recovered from cells, it was observed that adsorption on the cells followed the Langmuir model. This suggests that Sr^{2+} removal occurred until equilibrium was reached, as opposed to precipitation reactions where the data cannot be fitted with a simple Langmuir model.

Straightforward precipitation follows a multi-layer model including surface precipitates. The SRB cells were then demonstrated conclusively to adsorb Sr^{2+} sorbent. The sorption capacity of the cells (q_{max}) was relatively high (444 mgg^{-1}), a value much higher than values obtained from other cell types (Shaukat et al., 2005; Dabbagh et al., 2007; Chegrouche et al., 2009) and purified cultures of sulphate-reducing bacteria (Vijayaraghavan and Yun, 2008).

The solution to the classical Langmuir model is shown in Eqn 17.6 and the Freundlich isotherm in Eqn 17.7 with the optimum values obtained for the sulphate-reducing consortium shown in Table 17.6.

$$\frac{1}{q} = \frac{1}{q_{max}} + \frac{1}{C_{eq}} \cdot \frac{1}{bq_{max}} \quad (17.6)$$

$$log(q) = log(k) + \frac{1}{n} log(C_{eq}) \quad (17.7)$$

where q = sorption uptake (MM^{-1}), q_{max} = maximum sorbate uptake (MM^{-1}), b = coefficient related to the affinity between the sorbent and sorbate, C_{eq} = equilibrium concentration of the sorbate remaining in the solution (ML^{-3}), k = constant corresponding to the binding capacity and n = coefficient related to the affinity between the sorbent and sorbate.

Table 17.6. Langmuir and Freundlich model parameters for the equilibrium sorption of Sr^{2+} by a SRB biomass (1gl^{-1}).

Langmuir model			Freundlich model		
q_{max} (mgg^{-1})	b	R^2	k	n	R^2
444	0.011	0.993	17.2	1.95	0.986

Cell surface characterization

Acid/base properties of SRB cells with regard to H$^+$ and OH$^-$ ions were studied by potentiometric titrations. The titrations were performed according to published procedures. Titrations were performed using an automated titration system comprising of a burette system, glass electrode and a pH meter (Metrohm 718 STAT-Titrino model, Metrohm, UK). Prior to titration, 0.3 g (wet weight) SRB cells were suspended in 25 ml of the electrolyte solution, which had been purged with 99.9% pure N_2 for 60 min to eliminate CO_2. The suspension was immediately placed into a sealed titration vessel with continuous stirring at 140 rpm under pure nitrogen.

Results from the titration analysis depicted a four-site reaction behaviour. Equilibrium constants obtained from auto-titration data for the SRB cells suggested a four-site non-electrostatic model with site 1 corresponding to carboxylic acid functional groups (pK_a = 4–5), the near-neutral site 2 corresponding to phosphates (pK_a =6–7), and sites 3 and 4 (pK_a = 8–12) corresponding to basic sites containing either hydroxyl groups at around pH 10.5–12 or amine groups at pH 8.5–9.0.

Biosorptive recovery of divalent fission products

The study using SRBs showed that SRB cells could achieve up 68% removal of strontium from the medium (Ngwenya and Chirwa, 2010). Most of the solid phase Sr^{2+} species were easily desorbed from biomass using $MgCl_2$ (Fig. 17.9). Metal species in the desorbed fraction gave an indication of the amount of Sr^{2+} that is bound on the biomass surface by relatively weak electrostatic interactions, which are easily released by the ion-exchange process (Dahl et al., 2008). The elevated concentrations of Sr^{2+} in the desorbed fraction may be due to the release from the complexing agents on the microbial cell surface.

These experiments showed that the SRB cells were excellent cation exchangers. The significance of these findings was that Sr^{2+} and other divalent cationic fission products could be extracted from water using bacteria under natural biological conditions. Separation of bacteria from water is relatively easy and cost effective. The bacteria could then be treated with an eluant to reverse the process thereby recovering the metals. Afterwards, the bacteria could be returned to the biosorption reactors.

Treatment of Mixed Waste

Radioactive waste streams are not homogenous in nature: they contain a mixture of metallic elements, fission products and a large amount of tough-to-degrade halogenated compounds such as polychlorinated biphenyls (PCBs), chlorophenols, chlorotoluenes, chloropropanes, phosphotyrated organics and polynuclear aromatic hydrocarbons (PAHs) (Castillo et al., 1997). In order to treat metallic radionuclides and fission products using the processes presented in this chapter, the organic component must be considered. Sometimes the organic component can be utilized as the energy source for biological processes taking place in the system to achieve reduction-precipitation removal of toxic metals (Chirwa and Wang, 2000). Unfortunately, the majority of the compounds are too toxic to organisms commonly used in wastewater treatment plants. The biodegradation process of these compounds can be inhibited, not only by the chemical toxicity, but also by the radiotoxicity from radioactive components within the wastewater (Cacace et al., 1960; Xu and Obbard, 2004).

In order to treat nuclear wastewater effectively, organisms capable of degrading specific compounds can be isolated and applied in mineralizing the toxic organic component. Such a process was demonstrated by Tikilili and Chirwa (2011) in which polynuclear aromatic hydrocarbon (PAH)-degrading organisms isolated from a landfill site

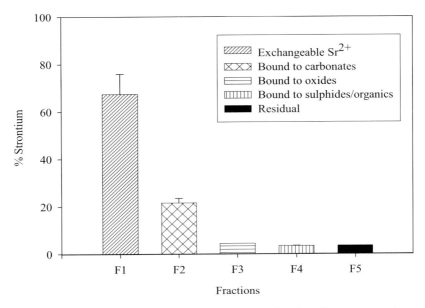

Fig.17.9. Partitioning of strontium species in the solid fraction after exposure to an SRB consortium.

completely degraded up to 60 mg l^{-1} of double-ring PAHs and removed others at different extents depending on loading, molecular complexity and solubility (Fig. 17.10).

The biodegradation observed in Fig. 17.10 was achieved under a background radiation as high as 1.067 Mega-Bq l^{-1}, enough to inhibit cell growth in aerobic mesophilic bacteria commonly found at wastewater treatment plants. The radiation source was simulated by adding particles of expired graphite from nuclear power plant. With these results, the researchers demonstrated that the organic component of a radioactive wastewater can either be degraded in a pretreatment process before removing the metals or could be used as an electron donor or carbon source during the biological reduction of the toxic radioactive metals.

Performance could be improved by applying an acclimation stage to build resistance in the cultures especially when treating wastes with a high concentration of higher molecular weight compounds.

Emerging Treatment Technologies

Biofilm systems

Microorganisms in nature and in reactor systems rarely grow as separate cells. The microorganisms form complex communities either in the form of agglomerations called flocs or as biofilm on the surfaces of inanimate objects and other organisms (Chirwa, 2011). The performance of a microbial culture is not only a function of its capability to degrade or transform a pollutant but also the configuration of the community in which it resides. There are complex interrelationships that occur within the microstructure that affect the availability of substrates, symbiotic existence through toxicity shielding of more sustainable species, and transfer of metabolites to organisms that could otherwise not grow on the only primary substrate in the bulk liquid. The biofilm itself is often a complex structure constructed by the bacteria. The formation of the biofilm especially in the initial stages is believed to be an active process coupled to the cell's central metabolism (Kjelleberg and Hermanson, 1984; Paul, 1984). Within the biofilm, complex processes take place, such as nutrient cycling, mass transport resistance, cell and substrate diffusion, and biofilm loss at the surface, that make prediction of the performance of the culture in the biofilm mathematically challenging (Nkhalambayausi-Chirwa and Wang, 2001).

In pure culture biofilms, 80% of the biofilm weight is occupied by exopolysaccharides (EPS) (Nelson et al., 1996). In continuous-flow systems, shear forces at the liquid/biofilm surface cause cell detachment and loss of EPS. Viable cells in the biofilm matrix must continuously replace the displaced biofilm materials, thereby drawing on the cell's energy resources.

Additionally, cell attachment plays an important role in the cell division cycle inside the biofilm. For example, Meadows (1971) observed that *Pseudomonas fluorescens* and *Aeromonas liquifaciens* cells undergo cell division only during their most stable attachment phase (when lying longitudinal to the solid surface). In more dramatic cases such as the prosthecate bacterium *Caulobacter crescentus*, cell division occurs only during the attachment phase of the cells life cycle (Neidhardt et al., 1990). In *Caulobacter*, the prosthecate (stalked) form undergoes cell division giving rise to a swarm cell equipped witha flagellum while the other daughter cell remains attached to the surface with a singleprostheca (stalk). The life cycle is completed by differentiation of swarm cells to prosthecates, followed by attachment and cell division.

Scientists and mathematicians have found biofilms too complex to analyse and model, especially when working with mixed-culture communities. Biofilm systems in laboratory studies are often oversimplified by using systems with defined chemical and microbial species composition.

The complexity in biofilms presented above, sometimes presents an advantage when complex metabolic processes and cooperation between different species in the community of organisms is required to remove a particular compound. For example, Nkhalambayausi-Chirwa and Wang (2001) achieved simultaneous removal of phenol and Cr(VI) in a biofilm environment with phenol serving as the sole electron donor in the system. To achieve optimum removal of the two pollutants, the slightly facultative Cr(VI)-reducer, *E. coli*, grew in the inner layers of the biofilm, whereas the obligately aerobic phenol degrader, *Pseudomonas putida*, colonized the outer layers. The effect of the biofilm structure on the distribution of species was best demonstrated in a laboratory study conducted using pure cultures of *Bacillus* sp. (Nkhalambayausi-Chirwa and Wang, 2004).

The study by Nkhalambayausi-Chirwa and Wang (2001) showed that the metal removal in the biofilm layer was mass transport-limited under low mass loading conditions and reaction rate-limited under inhibiting conditions. The mass transport of dissolved species through a thin liquid layer (L_w) and the biofilm matrix (L_f) was represented by Fick's second law of molecular diffusion (Eqn 17.8) (Fig.17.11):

$$\frac{\partial \mathbf{u}_i}{\partial t} = D_{wi} \frac{\partial^2 \mathbf{u}_i}{\partial z^2} \quad (17.8)$$

where $\mathbf{u}_i$ = concentration of dissolved species (ML^{-3}), D_{wi} = diffusivity of dissolved species in water ($L^2 T^{-1}$), t = time (T) and z is the spatial coordinate (L).

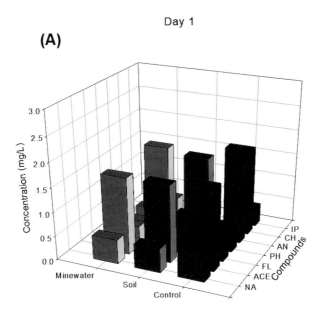

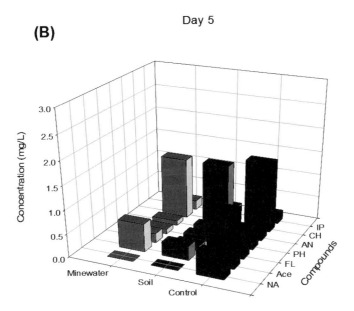

Fig. 17.10. PAH degradation during incubation of a mixed compound sample with bacteria (A) after 1 day of inoculation and (B) after 5 days of inoculation.

Viable and inert cell concentrations in the bulk liquid zone are represented by the vector $\mathbf{x}_B(t) = \{X_{aB}(t), X_{iB}(t)\}$ whereas the respective biomass densities are represented by the vector $\mathbf{x}_f(t) = \{X_{af}(t), X_{if}(t)\}$ in the biofilm zones, where X_{aB} and X_{iB} = the viable and inert cell concentration in the bulk liquid (ML^{-3}), and X_{af} and X_{if} = viable and inert cell density in the biofilm (ML^{-3}). The mass balance across the bulk liquid (submerged-bed) is represented by an ordinary differential equation as shown below:

$$\frac{\partial(u_B V_B)}{\partial t} = Q(u_{in} - u_B) - r_u \cdot V_B - j_u \cdot A_f \quad (17.9)$$

$$\frac{\partial(x_B V_B)}{\partial t} = -Q x_B + \lambda(u) \cdot x_f L_f A_f - b_x x_B V_B, \quad x_B = \begin{vmatrix} X_{aB} \to 0 \\ X_{iB} \gg 0 \end{vmatrix} \quad (17.10)$$

where j_u = flux of dissolved species into the biofilm ($ML^{-2}T^{-1}$), r_u = rate of removal of dissolved species by suspended cells in the bulk liquid ($ML^{-3}T^{-1}$), λ = cell detachment rate coefficient (T^{-1}) – a function of the interfacial velocity u (LT^{-1}), b_x = cell death rate coefficient (T^{-1}), A_f = biofilm surface area (L^2), and V_B = the bulk liquid volume (L^3). Due to the dependence of cell attachment on metabolism, dead cells are expected to detach more easily from the biofilm than live cells. Thus, the viable cell concentration in the bulk liquid is expected to be much lower than the inert cell concentration, such that $x_B \approx X_{iB}$.

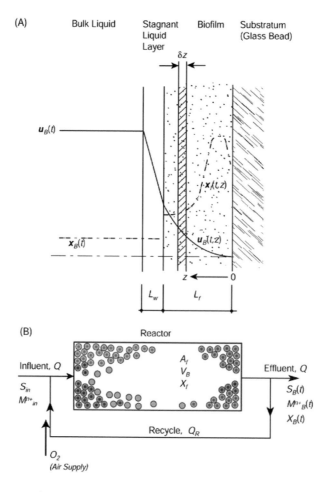

Fig. 17.11. A conceptual mass balance model representing (A) concentration profiles across the biofilm zone and (B) removal or accumulation of species across the entire reactor.

The boundary conditions at the liquid/biofilm interface depend on the mass transport rates across the liquid film:

$$j_s = k_{LS} \cdot \left(S_B(t) - D_{fs}(t, L_f)\right) \quad (17.11)$$

$$j_{Xa} = \lambda_a \cdot X_{af} \cdot L_f \quad (17.12)$$

$$j_{Xi} = \lambda_i \cdot X_{if} \cdot L_f \quad (17.13)$$

$$j_M = k_{LM} \cdot \left(M_B^{n+}(t) - M_{fs}^{n+}(t, L_f)\right) \quad (17.14)$$

where $S\ (ML^{-3})$ = the concentration of the energy source in the reactor system; M^{n+} = concentration of the cationic species of nth-state targeted for removal; $k_{Ls} = D_{ws}/L_w$, the mass transport rate coefficient of substrate (LT^{-1}); $k_{LM} = D_{wM}/L_w$ is the mass transport rate coefficient of metal to be removed (LT^{-1}); S_{fs} = glucose concentration at the liquid/biofilm interface; and M^{n+}_{fs} = metal concentration at the liquid/biofilm interface.

For impermeable support media (glass, rock, sand) the boundary conditions of the system will apply as follows:

$$D_{ws} = \frac{\partial S(t,0)}{\partial z} = 0, \quad D_{Xa}\frac{\partial X_{af}(t,0)}{\partial z} = 0, \quad D_{Xi}\frac{\partial X_{if}(t,0)}{\partial z} = 0, \quad D_{wc}\frac{\partial M^{n+}(t,0)}{\partial z} = 0 \quad (17.15)$$

assuming no flux of dissolved species and particulate matter across the biofilm/glass bead interface.

The system was solved numerically for removal of a model pollutant (Cr(VI)) and substrate (glucose). The removal of a target metal pollutant (Cr(VI)) and a substrate (glucose), and the accumulation of attached biomass (*Bacillus* sp.) in a fixed-film bioreactor system was predicted using optimized parameters as shown in Figs 17.12 and 17.13.

In some species of bacteria, Cr(VI) reduction and U(VI) reduction use the same or similar pathways involving reduction and oxidation of the cytochrome b_1 and c_3 enzymes in the inner cell membrane (Lovley and Phillips, 1994). For this reason, we project that a similar process can be applied to engineer reduction and removal of U(VI) in a biofilm environment. The above study thus demonstrated the feasibility of biological removal of toxic heavy metals by selected species of bacteria in a self-generating culture system that does not require frequent reinoculation and restarting of operation (Shen and Wang, 1993, 1994; Wang and Shen, 1997).

Permeable reactive barriers

General principles

Several types of treatment walls have been studied to attenuate the movement of metals in groundwater at contaminated sites. Trench materials that have been investigated include zeolite, hydroxyapatite, elemental iron and limestone (Vidic and Pohland, 1996). Elemental iron has been tested for Cr(VI) reduction and other inorganic contaminants (Powell *et al.*, 1995) and limestone for lead precipitation and adsorption (Evanko and Dzombak, 1997).

Permeable reactive barriers are an emerging alternative to traditional pump-and-treat systems for groundwater remediation. Such barriers are typically constructed from highly impermeable emplacements of materials such as grouts, slurries, or sheet pilings to form a

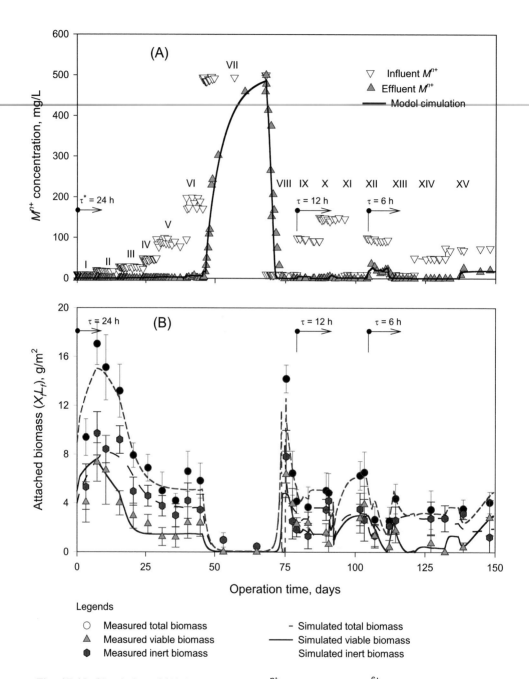

Fig. 17.12. Simulation of (A) the toxic metal M^{n+} concentration (Cr^{6+}) and (B) viable, inert and total biomass in the biofilm zone in a pure culture biofilm reactor under a range of hydraulic loading conditions: 24 h HRT (Phase I–VIII); 12 h HRT (Phase IX–XI); and 6 h HRT (Phase XII–XV).

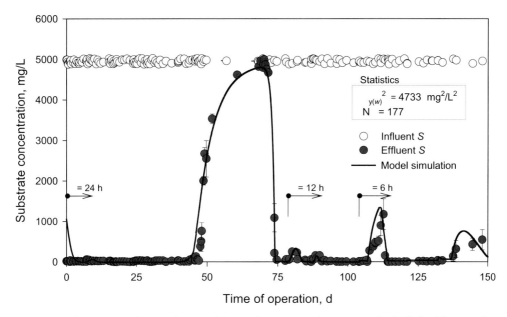

Fig. 17.13. Simulation of the substrate (glucose) concentration across the bulk liquid zone of a pure culture biofilm reactor under a range of hydraulic loading conditions: 24 h HRT (Phase I–VIII); 12 h HRT (Phase IX–XI); and 6 h HRT (Phase XII–XV).

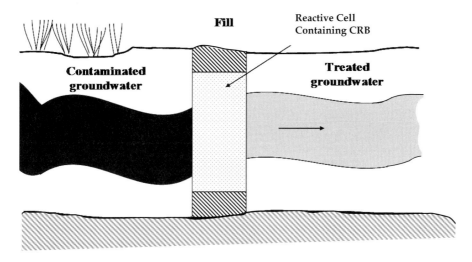

Fig. 17.14. General principle of the operation of a permeable reactive barrier.

subsurface 'wall'. Permeable reactive barriers are created by intercepting a plume of contaminated groundwater with a permeable reactive material (Fig. 17.14).

The properties of the reactive material are selected to promote the attenuation of the contaminant through degradation, precipitation, adsorption or reduction into a sparingly soluble phase. Reactive mixtures for the attenuation of inorganic species are designed to maintain their permeability as secondary precipitates accumulate. The barrier should also be designed in such a way that the contaminant remains immobilized within the aquifer or can be retrieved with the reactive material following treatment.

A wide range of reaction mechanisms can be employed to remove both negatively charged and positively charged contaminants from flowing groundwater. These include adsorption of inorganic anions and cations (Morrison and Spangler, 1993), simple precipitation (McMurty and Elton, 1985), adsorptive precipitation (Baker *et al.*, 1997), reductive precipitation (Blowes and Ptacek, 1992) and biologically mediated transformations (Robertson and Cherry, 1995;Benner *et al.*,1997; Molokwane and Nkhalambayausi-Chirwa, 2009).

So far, permeable reactive barriers have been evaluated for the treatment of inorganic contaminants in groundwater, including As, Cd, Cr, Cu, Hg, Fe, Mn, Mo, Ni, Pb, Se, Te, U, V, NO_3, PO_4 and SO_4. Only chemical systems have been applied *in situ* for metals. Biological permeable reactive barriers (BPRBs) have been used to treat organic compounds in groundwater. Applications of BPRBs in treating metal pollution are still experimental to date.

Biological permeable reactive barriers

Biological permeable reactive barriers (BPRBs) have been successful employed in the treatment of organic pollutants in groundwater. A typical design comprises a double-layer with an aeration zone followed by the bioremediation zone. One such system was evaluated against the removal of methyl-tert-butyl-ether (MTBE)-contaminated groundwater (Liu *et al*, 2006). The aeration in this case was achieved chemically by the oxidation of calcium peroxide (CaO_2) to release oxygen into the medium. Other growth nutrients were added to encourage the growth of MTBE-degrading organisms in the second layer.

Notably, inorganic salts such as potassium dihydrogen phosphate (KH_2PO_4) and ammonium sulfate ($(NH_4)_2SO_4$) can act as buffers against the increase in pH during the oxidation of CaO_2 into carbonates (CO_3^{2-}). Thus, nutrients added in the second layer must include the phosphate buffer for the proper functioning of the barrier.

Applications of *in situ* barriers to treat uranium have only been attempted at the laboratory level. In one suggested application, isolates of sulfate-reducing *Geobacter* sp. were targeted for possible application at a field site in Rifle, Colorado, USA. In this application, acetate could be added to stimulate the growth of *Geobacter* species already present in the soil, thus facilitating *in situ* bioaugmentation of U(VI) at the site. However, further optimization was required in order to promote long-term growth and activity of *Geobacter* species because the sulfate-reducing microorganisms that became predominant with continued acetate injection appeared to be less effective at U(VI) reduction.

Limitations of permeable reactive barrier systems

The long-term stability of biologically reduced uranium is influenced by the complex interplay between the soil/sediment mineralogy, the geochemistry of the environment, and microbial activity in the environment (Ginder-Vogel *et al.*, 2006). Many of these factors have been studied under laboratory conditions. However, the impact of these factors on uranium cycling in natural, subsurface environments around nuclear and radioactive waste dump sitesis still poorly understood.Among the items of concerns are: (i) the difficulty to predict whether the selected organisms will achieve reduction of theradionuclide in a passive operation setting; (ii) the difficulty to determine the correct rates at which carbon sources and nutrients must be replenished to support organisms involved in bioremediation in barrier environments;(iii) the difficulty in sustaining contact between microbes and the target pollutants; and (iv) during long-term operation, loss of permeability of the barrier zone may be encountered due to the accumulation of precipitates of the reduced metal within the barrier zone. This may require remobilization of the reduced metal and flushing with high-rate pumping.

Remobilization of the reduced metal may be counterproductive where the reduced form is significantly toxic or radioactive as this may in turn contaminate the aquifer downstream of the barrier system.To avoid these problems,the operationally simpler pump-and-treat methods have been frequently used.

Future Applications and Improvements

New methods for treating radioactive elements in radioactive wastewater have been proposed, some of which can be applied *in situ* (Chirwa, 2011). Among these futuristic innovations are*in situ* bioaugmentation, molecular bioaugmentation, and biofraction for bioseparation of radioactive elements. These methods are defined below.

*In situ*bioaugmentationentails identifying indigenous species of bacteria within the vicinity of the contaminated site and determining the critical carbon sources and nutrients that could be supplied to encourage the growth of the target species. When the selected nutrients are introduced into the environment, either by injection into boreholes or by spreading on the ground, the target species will out-compete other species and will be able to degrade the contaminants. Such a system is currently under investigation in the laboratories at the University of Pretoria, South Africa (Molokwane, 2010). The potential problem with this form of bioaugmentation is that the nutrients may be viewed as pollutants in their own right, especially at the beginning of the bioaugmentation process when microbial loading is very low. The nutrients such as NO_3^- and SO_4^{2-} have undesirable pollution effects on receiving water bodies such as eutrophication of streams receiving the base flow from the remediated areas.

Molecular bioaugmentationutilizes genetic carriers such as transposons and plasmids to shuttle genetic information for toxic metal remediation into native species in the environment. Several species of bacteria are capable of picking up and retaining circular fragments of DNA called broad-host-range plasmids, which may be engineered to carry specific genes for the degradation of xenobiotic compounds and transformation of toxic metals (Weightman *et al*., 1984; Vincze and Bowra, 2006). The same process can be applied using genetically engineered linear DNA called transposons. Although studies have been conducted using these techniques in laboratory microcosms, the application in actual environments has not been attempted (Hill *et al*., 1994). In the future, it is foreseeable that these methods will find wide application for the new pollutant varieties that may be untreatable by conventional methods.

Biofractionation isa very little understood application,which involves using microorganisms to selectively accumulateradioisotopes by atomic mass. This process has been tested in the fractionation of deutilium and hydrogen (Whiticar, 1999), selenium 82/74 (Herbel *et al*., 2000), and acetic acid in sediments (Habicht and Canfield, 1997). In a more recent study, Molokwane and Chirwa (2007) observed partialfractionation ofC-14 from C-

12 carbon matrix. However, in this preliminary study, the amount of C-14 remaining in solution was not quantified, which could be required to draw a mass balance on C-14 in the system. These preliminary results on C-12/C-14 bioseparation hold promise for development of the decontamination and reuse process for graphite in model HTGR nuclear reactors that produce large amounts of radioactive nuclear graphite from expired fuel containment (pebbles) and for possible reuse of structural graphite from decommissioned plants.

Conclusion

The chapter presents a concise review of recent trends in uranium(VI) bioremediation and biological treatment of radioactive waste containing uranium and its fission products. The biochemistry of U(VI) reduction is presented based on information obtained from studies on U(VI) reduction in sulfate-reducing bacteria (SRB) and *Shewanellaoneidensis*MR-1 by Macaskie, Llyod and other researchers. An improvement of the process – U(VI) reduction by facultatively grown oraganisms – was achieved in the recent studies by Chabalala and Chirwa (2010). The information presented demonstrates the feasibility of replacing current physical-chemical technologies with more environmentally freindly biological processes that can be operated under natural pH and temperature conditions. The efficiency of the proposed processes can be improved by using heterogeneous biofilm reactors and/or *in situ*permeable reactive barriers– the latter still under laboratory investigation. The demand for the above suggested technologies is expected to increase with the increasing demand for cleaner (carbon free) energy sources around the world.

Acknowledgements

The authors thank theNational Research Foundation (NRF) of South Africa for funding the research on the reduction metals and treatment of radioactive wastes through the grant no. FA2007030400002 awarded to Prof. E.M.N. Chirwa and the students' grants through the South African Nuclear Human Asset Research Programme (SANHARP). Thanks are also due to others corporations, such as the Pebble Bed Modular Reactor Company (PBMR), and governmental institutions that provided laboratory equipment and logistic support at different stages of the project.

Nomenclatures and Symbols

A_f	biofilm surface area (L^2)
b	coefficient related to the affinity between the sorbent and sorbate
b_x	cell death rate coefficient (T^{-1})
c	concentration of the pollutant in waste (mass/volume)
C_{eq}	equilibrium concentration of the sorbate remaining in the solution (mass/volume)
C_{in}	effluent concentration across the control volume (mass/volume)
C_{out}	effluent concentration across the controlvolume (mass/volume)
D_{wi}	diffusivity of dissolved species in water (L^2T^{-1})
H	Henry's constant for the pollutant
$\mathbf{j}_u$	flux of dissolved species into the biofilm ($ML^{-2}T^{-1}$)
k	empirical constants

k	constant corresponding to the binding capacity
$k_{LM} = D_{wM}/L_w$	is the mass transport rate coefficient of metal to be removed (LT^{-1})
$k_{Ls} = D_{ws}/L_w$	the mass transport rate coefficient of substrate (LT^{-1})
k_u	maximum specific U(VI) reduction rate coefficient (mass/mass cells/time)
K_u	half velocity (saturation) constant (mass/volume)
λ	cell detachment rate coefficient (T^{-1})
L_w	thickness of stagnant liquid layer (L)
L_f	thickness of biofilm matrix (L)
M^{n+}	concentration of the cationic species of nth-state (ML^{-3})
$M^{n+}{}_{fs}$	metal concentration at the liquid/biofilm interface (ML^{-3})
n	coefficient related to the affinity between the sorbent and sorbate
n	empirical constants
q	sorption uptake (mass pollutant/mass adsorbent), Langmuir
q	sorption uptake (mass pollutant/mass adsorbent), Freundlich
q_{max}	maximum sorbate uptake (mass pollutant/mass adsorbent)
Q_g	flow rate of the stripping gas (volume/time)
Q_L	flow rate of the liquid carrying the waste (volume/time)
$\mathbf{r}_u$	rate of removal of dissolved species by suspended cells in the bulk liquid $(ML^{-3}T^{-1})$
S	concentration of the energy source (limiting substrate) (ML^{-3})
S_{fs}	glucose concentration at the liquid/biofilm interface (ML^{-3})
t	time (T)
T_c	maximum U(VI) reduction capacity of cells (mass U reduced/mass cells)
$\mathbf{u}_i$	concentration of dissolved species (ML^{-3}),
u	interfacial flow velocity (LT^{-1})
U	concentration of U(VI) at time (mass/volume)
U_0	initial concentration of U(VI) (mass/volume)
V_B	the bulk liquid volume (L^3)
X	viable cell concentration at time (mass/volume)
X_{aB}	viable cell concentration in the bulk liquid (ML^{-3})
X_{af}	viable cell density in the biofilm (ML^{-3})
X_{iB}	inert cell concentration in the bulk liquid (ML^{-3})
X_{if}	inert cell density in the biofilm (ML^{-3})
X_0	initial cells density of U(VI)-reducing strains (mass/volume)
y	adsorbent capacity (mass pollutant/mass adsorbent)

References

Alberts, B., Bray, D., Lewis, J., Raff, M., Roberts, K. and Watson, J.D. (1994) *Molecular Biology of the Cell*. Garland Publishing, New York.

Appukuttan, D., Rao, A.S. and Apte, S.K. (2006) Engineering of *Deinococcus radiodurans* R1 for bioprecipitation of uranium from dilute nuclear waste. *Applied and Environmental Microbiology* 72, 7873–7878.

Ashley, N.V. and Roach, D.J. (1990) Review of biotechnology applications to nuclear waste treatment. *Radiology* 53, 261–265.

Baker, M.J., Blowes, D.W. and Ptacek, C.J. (1997) Phosphorous adsorption and precipitation in a permeable reactive wall: applications for wastewater disposal systems. *Land Containment Reclamation* 5, 189–193.

Bakhtiari, F., Atashi, H., Zivdar, M., Seyedbagheri, S. and Fazaelipoor, M.H. (2011) Bioleaching kinetics of copper from copper smelters dust. *Journal of Industrial Engineering and Chemistry* 17, 29–35.

Barton, L.L., Choudhury, K., Thomsom, B.M., Steenhoudt, K. and Groffman, A.R. (1996) Bacterial reduction of soluble uranium: the first step of *in situ* immobilization of uranium.*Radioactive Waste Management and Environmental Restoration* 20, 141–151.

Beliaev, A.S. and Saffarini, D.A. (1998) *Shewanella putrefaciens* mtrB encodes an outer membrane protein required for Fe(III) and Mn(IV) reduction. *Journal of Bacteriology* 180, 6292–6297.

Beliaev, A.S., Klingeman, D.M., Klappenbach, J.A., Wu, L. and Romine, M.F. (2005) Global transcriptome analysis of *Shewanellaoneidensis* MR-1 exposed to different terminal electron acceptors. *Journal of Bacteriology* 187, 7138–7145.

Bencheikh-Latmani, R., Williams, S.M., Haucke, L., Criddle, C.S. and Wu, L. (2005) Global transcriptional profiling of *Shewanella oneidensis* MR-1 during Cr(VI) and U(VI) reduction. *Applied and Environmental Microbiology* 71, 7453–7460.

Benner, S.G., Blowes, D.W. and Ptacek, C.J. (1997) A full-scale porous reactive wall for prevention of acid mine drainage.*Groundwater Monitoring Retention* 17, 99–107.

Blowes, D.W. and Ptacek, C.J. (1992) Geochemical remediation of groundwater by permeable reactive walls: Removal of chromate by reaction with iron-bearing solids. In: *Proc. Subsurface Restoration Conference (3rd International Conference on Groundwater Quality Research)*, Dallas, Texas (June 1992), pp. 214–216.

Boopathy, R. (2006) Factors limiting bioremediation technologies. *Bioresource Technology* 74, 63–67.

Bouwer, E.J. and Zehnder, A.J.B. (1993) Bioremediation of organic compounds-putting microbial metabolism to work. *Trends Biotechnology* 11, 360–367.

Buck, E.C., Hanson, B.D. and McNamara, B.K. (2004) The geochemical behaviour of Tc, Np and Pu in spent nuclear fuel in an oxidizing environment. In: Gieré, R. and Stille, P. (eds) *Energy, Waste, and the Environment: a Geochemical Perspective*, Vol. 236. The Geological Society of London Special Publication, London, pp. 65–88.

Bush, M.B. (2003) *Ecology of a Changing Planet*, 3rd edn. Prentice Hall, New Jersey.

Bruno, J. and Ewing, R.C. (2006).Spent nuclear fuel.*Elements* 2, 343–349.

Cacace, F., Guarino, A., Montefinale, G. and Possagno, E. (1960) Radioactive atoms distribution in aromatic compounds labelled by exposure to tritium gas. *International Journal of Applied Radiation and Isotopes* 8, 82–89.

Castillo, M., Alpendurada, M.F. and Barcelo, D. (1997) Characterization of organic pollutants in industrial effluents using liquid chromatography–atmospheric pressure chemical ionization–mass spectrometry. *Journal of Mass Spectrometry* 32, 1100–1110.

Cerniglia, C.E., Lambert, K.J., Miller, D.W. and Freeman, J.P.(1984) Transformation of 1- and 2-methylnaphthalene by *Cunninghamella elegans*. *Applied and Environmental Microbiology* 47, 111–118.

Chaalal, O. and Islam, M.R. (2001) Integrated management of radioactive strontium contamination in aqueous stream systems.*Journal of Environmental Management* 61, 51–59.

Chabalala, S. (2011) Reduction of uranium-(VI) under microaerobic conditions using an indigenous mine consortium. Master's thesis, University of Pretoria, Pretoria, RSA.Available at: http://upetd.up.ac.za/thesis/available/etd-09222011-102801(accessed on 7/10/2011).

Chabalala, S. and Chirwa, E.M.N. (2010) Uranium(VI) reduction and removal by high performing purified anaerobic cultures from mine soil. *Chemosphere* 78, 52–55.

Chegrouche, S., Mellah, A. and Barkat, M. (2009) Removal of strontium from aqueous solutions by adsorption onto activated carbon: kinetic and thermodynamic studies.*Desalination* 235, 306–318.

Chirwa, E.M.N. (2011) Development of biological treatment processes for the separation and recovery of radioactive wastes. In: Nash, K.L. and Lumetta, G.J. (eds) *Advanced Separation Techniques for Nuclear Fuel Reprocessing and Radioactive Waste Treatment*. Woodhead Publishing, Cambridge, UK, pp. 436–472.

Chirwa, E.M. and Wang, Y.T. (2000) Simultaneous Cr(VI) reduction and phenol degradation in an anaerobic consortium of bacteria. *Water Research* 34, 2376–2384.

Chubar, N., Behrends, T. and Van Cappellen, P. (2008) Biosorption of metals (Cu^{2+}, Zn^{2+}) and anions (F^-, $H_2PO_4^-$) by viable and autoclaved cells of the gram-negative bacterium *Shewanella putrefaciens*. *Colloids and Surfaces B: Biointerfaces* 65, 126–133.

Dabbagh, R., Ghafourian, H., Baghvand, A., Nabi, G.R., Riahi, H. and Ahmadi Faghih, M.A. (2007) Bioaccumulation and biosorption of stable strontium and Sr-90 by *Oscillatoria homogenea* cyanobacterium. *Journal of Radioanalytical and Nuclear Chemistry* 221, 53–59.

Dahl, O., Nurmesniemi, H. and Poykio, R. (2008) Sequential extraction partitioning of metals, sulfur, and phosphorus in bottom ash from a coal-fired power plant.*International Journal of Environmental Analytical Chemistry* 88, 61–73.

Demergasso, C., Galleguillos, F., Soto, P., Serón, M. and Iturriaga, V. (2010) Microbial succession during a heap bioleaching cycle of low grade copper sulfides: does this knowledge mean a real input for industrial process design and control?*Hydrometallurgy* 104, 382–390.

Dickerson, R.E. (1980) Cytochrome *c* and the evolution of energy metabolism.*Scientific America*242, 136–153.

Dong, Y., Lin, H., Wang, H., Mo, X., Fu, K. and Wen, H. (2011) Effects of ultraviolet irradiation on bacteria mutation and bioleaching of low-grade copper tailings.*Minerals Engineering*24, 870–875.

Evanko, C.R. and Dzombak, D.A. (1997) Remediation of Metals-Contaminated Soils and Groundwater. Technology Evaluation Report.Ground-Water Remediation Technologies Analysis Center, Pittsburgh, Pennsylvania, USA.Available at:http://www.clu-in.org/download/toolkit/metals.pdf/.

Francis, A.J., Dodge, C.J., Lu, F., Halada, G.P. and Clayton, C.R. (1994) XPS and XANES studies of uranium reduction by *Clostridium* sp. *Environmental Science and Technology* 28, 636–639.

Gadd, G.M. (1992) Biosorption. *Journal of Chemical Technology and Biotechnology* 55, 302–304.

Gadd, G.M. (2004) Microbial influence on metal mobility and application for bioremediation. *Geoderma* 122, 109–119.

Gaspard, S., Vazquez, F. and Holliger, C. (1998) Localizationand solubilization of the iron(III) reductase of *Geobactersulfurreducens*. *Applied and Environmental Microbiology* 64, 3188–3194.

Ginder-Vogel, M., Wu, W.M., Carley, J., Jardine, P., Fendorf, S. and Criddle, C. (2006) *In situ* biological uranium remediation within a highly contaminated aquifer.SSRL Science Highlight, October 2006, Dunn, L. (ed.), Stanford University, California.Available at:http://ssrl.slac.stanford.edu/research/highlights_archive/u_remed.html(accessed on11/11/2011).

Habicht, K.S. and Canfield, D.E. (1997) Sulfur isotope fractionation during bacterial sulphate reduction in organic-rich sediments.*Geochimica et Cosmochimica Acta* 61, 5351–5361.

Heidelberg, J.F., Paulsen, I.T., Nelson, K.E., Gaidos, E.J. and Nelson, W.C. (2002) Genome sequence of the dissimilatory metal ion-reducing bacterium *Shewanella oneidensis*.*Nature Biotechnology* 20, 1118–1123.

Herbel, M.J., Johnson, T.M., Oremland, R.S. and Bullen, T.D. (2000) Fractionation of selenium isotopes during bacterial respiratory reduction of selenium oxyanions. *Geochimica et Cosmochimica Acta* 64, 3701–3709.

Hill, K.E., Fry, J.C. and Weightman, A.J. (1994) Gene transfer in the aquatic environment: persistence and mobilization of the catabolic recombinant plasmid pDlO in the epilithon.*Microbiology*140, 1555–1563.

Horitsu, H., Futo, S., Miyazawa, Y., Ogai, S. and Kawai, K. (1987) Enzymatic reduction of hexavalent chromium by hexavalent tolerant *Pseudomonas ambigua* G-1.*Agricultural and Biological Chemistry* 51, 2417–2420.

Hu, M.Z., Norman, J.M., Faison, B.D. and Reeves, M.E. (1996) Biosorption of uranium by *Pseudomonas aeruginosa* strain CSU – characterization and comparison studies. *Biotechnology and Bioengineering* 51, 237–247.

IAEA (2009) *Nuclear Technology Review*. International Atomic Energy Agency Scientific and Technical Publication Series, Vienna, Austria.

Ishibashi, Y., Cervantes, C. and Silver, S. (1990) Chromium reduction in *Pseudomonas putida*. *Applied and Environmental Microbiology* 56, 2268–2270.

Khripunov, V.I., Kurbatov, D.K. and Subbotin, M.L. (2006) C-14 Production in CTR materials and blankets.*Proceedings of the 21st Fusion Energy Conference (FEC2006)*, 16–21 October 2006, Chengdu, China, SE/pp. 2–3.

Kjelleberg, S. and Hermanson, M. (1984) Starvation-induced effects on bacterial surface characteristics.*Applied and Environmental Microbiology* 48, 497–503.

Li, W.-H.andGraur, D. (1991) *Fundamentals of Molecular Evolution*. Sinauer Associates, Sunderland, Massachusetts.

Li, X. and Krumholz, L.R. (2009) Thioredoxin is involved in U(VI) and Cr(VI) reduction in *Desulfovibrio desulfuricans* G20. *Journal of Bacteriology* 191, 4924–4933.

Li, X., Luo, Q., Wofford, N.Q., Keller, K.L., McInerney, M.M., Wall, J.D. and Krumholz, L.R. (2009) A molybdopterin oxidoreductase is involved in H_2 oxidation in *Desulfovibrio desulfuricans* G20. *Journal of Bacteriology* 109, 2675–2682.

Liu, S.-J., Jiang, B., Huang, G.-Q.and Li, X.-G. (2006) Laboratory column study for remediation of MTBE-contaminated groundwater using a biological two-layer permeable barrier.*Water Research* 40, 3401–3408.

Lloyd, J.R. (2003) Microbial Reduction of Metals and Radionuclides.*FEMS Microbiology Reviews* 27, 411–425.

Lloyd, J.R., Chesnes, J., Glasauer, S., Bunker, D.J., Livens, F.R. and Lovley, D.R. (2002) Reduction of actinides and fission products by Fe(III)-Reducing bacteria. *Geomicrobiology Journal* 19, 103–120.

Lovley, D.R. and Phillips, E.J.P. (1992) Reduction of uranium by *Desulfovibrio desulfuricans*. *Applied and Environmental Microbiology* 58, 850–856.

Lovley, D.R. and Phillips, E.J.P. (1994) Reduction of chromate by *Disulfovibrio vulgaris* and its c_3 cytochrome. *Applied and Environmental Microbiology* 60, 726–728.

Lovley, D.R., Phillips, E.J.P., Gorby, Y.A. and Landa, E.R. (1991) Microbial reduction of uranium. *Nature* 350, 413–416.

Macaskie, L.E. and Lloyd, J.R. (2002) Microbial interactions with radioactive wastes and potential applications. In: Keith-Roach, M.J. and Livens, F.R. (eds) *Radioactivity in the Environment*, Vol. 2: *Interactions of Microorganisms with Radionuclides*. Elsevier, pp. 343–381.

Macaskie, L.E., Empson, R.M., Cheetham, A.K., Grey, C.P. and Skarnulis, A.J. (1992) Uranium accumulation by a *Citrobacter* sp. as a result of enzyme mediated growth of polysrystalline HUO2PO4. *Science* 257, 782–784.

Macaskie, L.E., Bonthrone, K.M., Yong, P. and Goddard, D.T. (2000) Enzymically mediated bioprecipitation of uranium by a *Citrobacter* sp.: a concerted role for exocellular lipopolysaccharide and associated phosphatase in biomineral formation. *Microbiology* 146, 1855–1867.

Marshall M.J., Kennedy, D.W., Dohnalkova, A., Saffarini, D., Culley, D.E., Romine, M.F., Reed, S.B., Beliaev, A.S., Zachara, J.M. and Fredrickson, J.K.(2005)The role of *Shewanella oneidensis* MR-1 *c*-type cytochromes and Type II secretion system in uranium reduction and localization of nanoparticles. *Proceedings of the 8th Annual US Department of Energy Natural and Accelerated Bioremediation Research Principal Investigator Workshop*, Warrenton, Virginia, 18 April, 2005.

McLean, J. and Beveridge, T.J. (2001) Chromate reduction by a pseudomonad isolated from a site contaminated with chromated copper arsenate. *Applied and Environmental Microbiology* 67, 1076–1084.

McMurty, D. and Elton, R.O. (1985) New approach to *in situ* treatment of contaminated groundwaters. *Environmental Progress* 4, 168–170.

Meadows, P.S. (1971) The attachment of bacteria to solid surfaces. *Archiv fur Mikrobiologie* 75, 374–381.

Merroun, L.M. and Selenska-Pobell, S. (2008) Localisation of cytochromes to the membrane of anaerobically grown *Shewanellaputrefaciens* MR-1. *Journal of Bacteriology* 174, 3429–3438.

Molokwane, P.E. (2010) Simulation of In Situ Bioremediation of Cr(VI) in Groundwater. PhD dissertation, University of Pretoria, Pretoria, South Africa. Available at:http://upetd.up.ac.za/thesis/available/etd-09252010-154146/unrestricted/00front.pdf(accessed on 5/11/2011).

Molokwane, P.E. and Chirwa, E.M. (2007) Development of a carbon-14 bioseperation technique for cleanup of nuclear graphite. In: *Proceedings of 11th International Conference on Environmental Remediation and Radioactive Waste Management (ICEM'07)*, 2–6 September 2007, Bruges, Belgium.

Molokwane, P.E. and Nkhalambayausi-Chirwa, E.M.(2009) Microbial culture dynamics and chromium (VI) removal in packed-column microcosm reactors. *Water Science and Technology* 60, 381–388.

Morrison, S.J. and Spangler, R.R. (1993) Chemical barriers for controlling groundwater contamination. *Environmental Progress* 12, 175–181.

Mourogov, V., Fukuda, K. and Kagramanian, V. (2002) The need for innovative nuclear reactor and fuel cycle systems: Strategy for development and future prospects. *Progress in Nuclear Energy* 40, 285–299.

Myers, C.R. and Myers, J.M. (1992) Localization of cytochromes to the outer membrane of anaerobically grown *Shewanellaputrefaciens* MR-1. *Journal of Bacteriology* 174, 3429–3438.

Nancharaiah, Y.V., Joshi, H.M., Mohan, T.V.K., Venugopalan, V.P. and Narasimhan, S.V. (2006) Aerobic granular biomass: a novel biomaterial for efficient uranium removal. *Current Science* 91, 503–509.

Nazina, T.N., Kosareva, I.M., Petrunyaka, V.V., Savushkina, M.K., Kudriavtsev, E.G., Lebedev, V.A., Ahunov, V.D., Revenko, Y.A., Khafizov, R.R., Osipove, G.A., Belyaev, S.S. and Ivanov, M.V. (2004) Microbiology of formation waters from the deep repository of liquid radioactive wastesSevernyi. *FEMS Microbiology Ecology* 49, 97–107.

Neidhardt, F.C., Ingraham, J.L. and Schaechter, M. (1990) In:*Physiology of the Bacterial Cell: A molecular approach*. Sinauer Associates, Sunderland, Massachusetts, USA, pp. 442–462..

Nelson, Y.M., Lion, L.W., Shuler, M.L. and Ghiorse, W.C. (1996) Modeling oligotrophic biofilm formation and lead adsorption to biofilm components. *Environmental Science and Technology* 30, 2027–2035.

Ngwenya, N. and Chirwa, E.M.N. (2010) Single and binary component sorption of the fission products Sr^{2+}, Cs^+ and Co^{2+} from aqueous solutions onto sulphate reducing bacteria. *Minerals Engineering* 23, 463–470.

Ngwenya, N. and Chirwa, E.M.N. (2011) Biological removal of cationic fission products from nuclear wastewater. *Water Science and Technology* 63, 124–128.

Nkhalambayausi-Chirwa, E.M. and Wang, Y.T. (2001) Simultaneous Cr(VI) reduction and phenol degradation in a fixed-film coculture bioreactor: reactor performance. *Water Research* 35, 1921–1932.

Nkhalambayausi-Chirwa, E.M. and Wang, Y.T. (2004) Modeling Cr(VI) reduction in a *Bacillus* sp. pure culture biofilm reactor. *Biotechnology and Bioengineering* 87, 874–883.

Obuekwe, C. and Westlake, D.W. (1982) Effects of medium composition on cell pigmentation, cytochrome content and ferric iron reduction in a *Pseudomonas* sp. isolated from crude oil. *Canadian Journal of Microbiology* 28, 989–992.

Paul, J.H. (1984) Effects of antimetabolites on the adhesion of an estuarine *Vibrio* sp. to polystyrene. *Applied and Environmental Microbiology* 48, 924–929.

Powell, R.M., Puls, R.W., Hightower, S.K. and Sabatini, D.A. (1995) Coupled iron corrosion and chromate reduction: mechanisms for subsurface remediation. *Environmental Science and Technology* 29, 1913–1922.

Qiu, G., Li, Q., Yu, R., Sun, Z., Liu, Y., Chen, M., Yin, H., Zhang, Y., Liang, Y., Xu, L., Sun, L. and Liu, X. (2011) Column bioleaching of uranium embedded in granite porphyry by a mesophilic acidophilic consortium. *Bioresource Technology* 102, 4697–4702.

Robertson, R.D. and Cherry, J.A. (1995) In situ denitrification of septic system nitrate using reactive porous medium barriers: field trials. *Groundwater* 33, 99–111.

Rothschild, L.J. and Mancinelli, R.L. (2001) Life in extreme environments. *Nature* 409, 1092–1101.

Sani, R.K., Peyton, B.M., Smith, W.A., Apel, W.A. and Petersen, J.N.(2002) Dissimilatory reduction of Cr(VI), Fe(III), and U(VI) by *Cellulomonas* isolates. *Applied Microbiology and Biotechnology* 60, 192–199.

Sani, R.K., Peyton, B.M. and Dohnalkova, A. (2006) Toxic effects of uranium on *Desulfovibrio desulfuricans* G20. *Environmental Toxicology and Chemistry* 25, 1231–1238.

Shaukat, M.S., Sarwar, M.I. and Qadee, R. (2005) Adsorption of strontium ions from aqueous solution on Pakistani coal. *Journal of Radioanalytical and Nuclear Chemistry* 265, 73–79.

Shelobolina, E.S., Sullivan, S.A., O'Neill, K.R., Nevin, K.P. and Lovley, D.R. (2004) Isolation, characterization, and U(VI)-reducing potential of a facultatively anaerobic, acid-resistant bacterium from low-pH, nitrate- and U(VI)-contaminated subsurface sediment and description of *Salmonella subterranea* sp. nov. *Applied and Environmental Microbiology* 70, 2959–2965.

Shen H. and Wang, Y.T. (1993) Characterization of enzymatic reduction of hexavalent chromium by *Escherichia coli* ATCC 33456. Applied Environmental Microbiology 59, 3771–3777.

Shen, H. and Wang Y.T. (1994) Modeling hexavalent chromium reduction in *Escherichia coli* ATCC 33456. *Biotechnology and Bioengineering* 43, 293–300.

Singh, S., Eapen, S., Thorat, V., Kaushik, C.P., Raj, K. and D'Souza, S.F. (2008) Phytoremediation of cesium-137 and strontium-90 from solutions and low-level nuclear waste by *Vetiveria zizanoides*. *Ecotoxicology and Environmental Safety* 69, 306–311.

Suzuki, Y., Kelly, S.D., Kemner, K.M. and Banfield, J.F. (2004) Enzymatic U(VI) reduction by *Desulfosporosinus* species. *Radiochimica Acta* 92, 11–16.

Tebo, B.M., Francis, C.A. and Obraztsova, A.Y. (2000) Dissimilatory metal reduction by the facultative anaerobe *Pantoeaagglomerans* SP1. *Applied and Environmental Microbiology* 66, 543–548.

Tikilili, P.V. and Chirwa, E.M.N. (2011) Characterization and Biodegradation of Polycyclic Aromatic Hydrocarbons in Radioactive Wastewater. *Journal of Hazardous Materials* 192, 1589–1596.

Tsuruta, T. (2006) Removal and recovery of uranium using microorganisms isolated from Japanese uranium deposits. *Journal of Nuclear Science and Technology* 43, 896–902.

Van Lam, P., Van Ngoc, O. and Thinang, N. (2000) Safe operation of existing radioactive waste management facilities at Dalat Nuclear Research Institut. In:*Proceedings of the International Conference on the Safety of Radioactive Waste Management*, 13–17 March 2000, Córdoba, Spain, pp. 152–154.

Vidic, R.D. and Pohland, F.G. (1996) Treatment Walls.*Technology Evaluation Report TE-96-01*, Ground-Water Remediation Technologies Analysis Center, Pittsburgh, Pennsylvania.

Vijayaraghavan, K. and Yun, Y.-S.(2008) Bacterial biosorbents and biosorption. *Biotechnology Advances* 26, 266–291.

Vincze, E. and Bowra, S. (2006) Transformation of *Rhizobia* with broad-host-range plasmids by using a Freeze-Thaw method. *Applied and Environmental Microbiology* 72, 2290–2293.

Wade, R. Jr and DiChristina, T.J. (2000) Isolation of U(VI) reduction-deficient mutants of *Shewanella putrefaciens*. FEMS Microbiology Letters 184, 143–248.

Wall, J. D. and Krumholz, L.R. (2006) Uranium reduction. *Annual Reviews of Microbiology* 60, 149–166.

Wang, Y.T. and Shen, H. (1997) Modelling Cr(VI) reduction by pure bacterial cultures. *Water Research* 41, 727–732.

Weightman, A.J., Don, R.H., Lehrbach, P.R. and Timmis, K.N. (1984) The identification and cloning of genes encoding haloaromatic catabolic enzymes and the construction of hybrid pathways for substrate mineralization. In:Omenn, G.S. and Hollaender, A. (eds) *Genetic Control of Environmental Pollutants*.Plenum Press, New York, pp. 47–80.

Whiticar, M.J. (1999) Carbon and hydrogen isotope systematics of bacterial formation and oxidation of methane.*Chemical Geology* 161, 291–314.

Xu, R. and Obbard, J.P. (2004) Biodegradation of polycyclic aromatic hydrocarbons in oil-contaminated beach sediments treated with nutrient amendments.*Journal of Environmental Quality* 33, 861–867.

Chapter 18

Uranium Bioremediation: Nanotechnology and Biotechnology Advances

Mrunalini V. Pattarkine

Introduction

The use of nuclear fuels in power plants to produce electricity and nuclear weapons invariably leads to high levels of radionuclide wastes (Kumar *et al.*, 2007; Seyrig, 2010), uranium being the most common waste product. Additionally, uranium contamination can occur in low levels in various natural resources (rocks, soil, water bodies), coal burnings and phosphate fertilizer units (Markich, 2002). High-level contamination also occurs through ores and mines from which uranium is extracted for power and weapon production (Seyrig *et al.*, 2010).

Uranium has no known biological function. It is not a rare element; however, it is a heavy metal and is hazardous because it is radioactive. US Department of Energy (DOE) has ongoing projects at 120 sites in 36 states to decontaminate approximately 7200 km^2 of land (NABIR, 2003).

Remediation is the treatment and/or removal of toxic and anthropogenic waste from the environment. Treating contaminated sites with conventional methods using physical, chemical and thermal processes is costly, often involving billions of dollars (Gavrilescu *et al.*, 2009). On the other hand, bioremediation, i.e. any process involving living systems or the enzymes for elimination or attenuation or transformation of polluting or toxic substances from the environment (Shukla *et al.*, 2010), is cost-effective as well as environmentally sustainable (Lovley and Phillips, 1992b; Senko *et al.*, 2002). In these processes, biological systems such as bacteria, yeast and fungi are applied to achieve remediation (Strong and Burgess, 2008).

Many microorganisms have an inherent ability for metal tolerance and detoxification. The chemical and physical properties of metals can be dramatically altered by microbial interaction, while the metals can impact the growth of microbes (Gadd and Raven, 2010). Though uranium is difficult to bioremediate because of its toxicity, its distribution and speciation in the environment can be significantly altered by some microorganisms (Suzuki and Banfield, 1999).

This chapter covers geomicrobiology of uranium, various methods currently applied for uranium bioremediation, microbial and biochemical mechanisms underlining them, and biotechnology- and nanotechnology-based advancements for microbial remediation of uranium.

Uranium Bioremediation – Microbiology

As an adaptation, most microorganisms have developed metal-resistance/tolerance. These microbes effectively remediate the toxic metals by using them as a part of their electron transfer chain, complexing them with extracellular polymers secreted by the cell, internalizing them in the cell cytoplasm, or converting them into a non-toxic biomineral form (Gadd, 2010; Gomathi and Sabrinathan, 2010). Metals have significant impact on microbial growth and metabolism. The metal toxicity decides the course and extent of many metal–microbe interactions and, under toxic conditions, almost all of the interactions can be affected (Gadd, 2010). This metal toxicity in turn is affected by the physico-chemical environment and metal speciation (Gadd and Griffiths, 1978). As a response to the metal toxicity, bacteria have developed mechanisms for metal tolerance and detoxification, such as efflux, enzymatic detoxification (Osman and Cavet, 2008; Rosen, 2002). In some cases, it is seen that the microbes evolved metal-resistant plasmids as reported for various metals such as Cd^+, Hg^{2+}, $CrO_4 2^-$ and many other toxic metals, although none has been reported for uranium. (Silver and Phung, 1996; Rosen, 2002). Additionally, microbes have developed metal binding proteins and peptides for uranium tolerance (Suzuki and Banfield, 1999; Gadd and Raven, 2010). Microbial energy generation, cell adhesion and biofilm production can be influenced by metal–microbe interactions. The microbial metabolism for energy generation primarily involves oxidation of a carbon source (organic matter) to carbon dioxide (CO_2). A terminal electron acceptor is essential for the enzymatic oxidation of the carbon source, a role typically played by metals or metal-containing enzymes (Sen and Chakrabarti, 2009), as is the case with uranium. Various microbial species have been known to effectively reduce the soluble U(VI) to insoluble U(IV) using uranium reductases. Figure 18.1 lists all the known microbial species capable of uranium reduction.

Physiology of Uranium Bioremediation

There are three major mechanisms to bioremediate U(VI): (i) enzymatic conversion of soluble U(VI) into an insoluble U(IV) uraninite by the dissimilatory metal-reducing bacteria (DMRB); (ii) chemical reduction due to by-products of microbial metabolites; and (iii) biosorption on cell surface, biopolymers and dead cell mass.

Bacterial reduction of uranium

Bacterial uranium reduction was first reported as an assay based on consumption of hydrogen, for presence of U(VI) in crude cell extracts (Woolfolk and Whiteley, 1962). After almost 30 years, Lovley and co-workers contributed pioneering findings for bacterial uranium reduction (Lovley *et al.*, 1991, 1993; Gorby and Lovley, 1992; Lovley and Phillips, 1992a). Since then, a number of spectroscopic studies established the need for living cellular system and an electron donor for the bacterial reduction of uranium (Lovley *et al.*, 1993). Three major metal reductases have been studied in detail for uranium reduction, those from *Desulfovibrio vulgaris*, from *Shewanella* and from *Geobacter* (Lovley and Phillips, 1992a; Lovley *et al.*, 1993; Methe *et al.*, 2003). Tetraheme cytochromes along with a periplasmic hydrogenase are believed to be responsible for the uranium reduction by DMRB, though no reductase specific for uranium has been identified

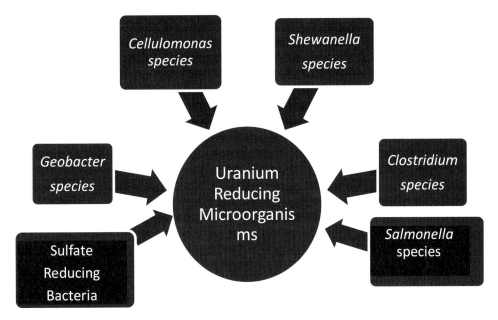

Fig. 18.1. Major species of uranium-reducing microorganisms (inspired by Wall and Krumholz, 2006).

(Wall and Krumholz, 2006). Whether this reduction involves one electron or two is uncertain, though a single electron reduction of U(VI) to U(V) under anaerobic conditions followed by disproportionation is suggested to be the most plausible route (Renshaw *et al.*, 2005) for bacteria.

Uranium re-oxidation

In spite of being one of the most economic modes to bioremediate uranium, bacterial reduction suffers a few drawbacks such as efficiency and stability (Wall and Krumholz, 2006). The availability of electron donors at the site of reduction becomes rate limiting, making this approach inefficient. This is apparent from higher rates of reduction in crude cell extracts or pure protein solutions than in whole cells (Woolfolk and Whiteley, 1962). The second major issue is re-oxidation of the reduced uranium (Senko *et al.*, 2002, 2005a, b; Beller, 2005; Gu *et al.*, 2005a), which can occur abiotically by oxygen or biotically by iron-reducing bacteria such as *Geobacter metallireducens* (Senko *et al.*, 2005a). Two other biotic pathways for re-oxidation are siderophores and microbial by-products such as bicarbonates (Finneran, K.T. *et al.*, 2002; Senko *et al.*, 2005b).

Biosorption

Uranium immobilization is influenced by its interaction with inactive or dead biomass under high actinide concentrations. Dead cells of marine algal species *Cystoseria indica*

(Wan *et al.*, 2005), *Citrobactor freudii* and cellular extracts from *Firmicutes* (Suzuki and Suko, 2006; Shelobolina *et al.*, 2007, 2008) have been shown to effectively adsorb uranium. The thick cell walls rich in peptidoglycans remove the re-oxidized uranium by firmly adsorbing it to the extracellular polymers. This helps in two ways: it effectively removes the re-oxidized uranium from groundwater long after the treatment is over, thus eliminating mobilization of U(VI) back into the groundwater, and the *Firmicutes* bacteria grow on the biomass remaining after the treatment, without any special efforts/stimulation (Markich, 2002). Even though biosorption is significantly more efficient than biotic reduction of uranium, this is not the most efficient way to remediate uranium-enriched sites as uranium toxicity inhibits growth of adequate biomass at these locations. Similar to bacterial reduction, the biosorption can be stimulated by enriching the contaminated site with a high density of active or inactive biomass. One such report by Khani *et al.* (2006) states improved biosorption by pretreatment of the marine alga *Cystoseria* with calcium.

Sometimes biosorption can happen indirectly as a desired result from undesirable intermediate steps. As the desired microbial population grows, either as a native biomass or as a stimulated one, their growth and accumulation can alter water flow. This can change the delivery of nutrients/stimulants; the secondary metabolites can precipitate in the water (N'Guessan *et al.*, 2008) and can negatively impact the concentration of active biomass. The only benefit from this is that inactive/dead biomass can absorb pollutants, thus indirectly remediating or enhancing the overall remediation efforts.

Uranium Bioremediation: Basic Concepts and Technologies

Basic concepts

Microbial energy generation, cell adhesion and biofilm production can be influenced by metal–microbe interactions. The microbial metabolism for energy generation primarily involves oxidation of a carbon source (organic matter) to CO_2. A terminal electron acceptor is essential for the enzymatic oxidation of the carbon source; that role is typically taken up by metals or metal-containing enzymes (Li *et al.*, 2010), as is the case with uranium.

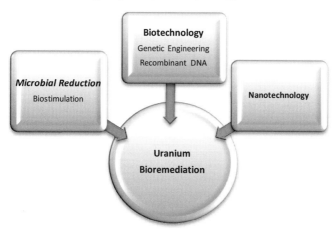

Fig. 18.2. Technologies for uranium bioremediation.

The uranium bioremediation systems (Fig. 18.2) are mainly comprised of microbes and the environment; these are influenced by parameters such as availability of nutrients and oxygen, moisture content, pH and temperature at the site (Anderson *et al.*, 2003).

In situ uranium bioremediation

In this process, the contaminant is treated and remediated at the contamination site itself. The system typically comprises extraction wells, injection wells and filtration systems (to remove biomass that can interfere with the injection system). The extraction wells and their placement is a critical factor in ensuring hydraulic control over the plume to eliminate chances of contaminants spreading into non-contaminated areas (Li *et al.*, 2010). Some tracer elements such as bromide are used to follow the extent of remediation. There are three approaches for *in situ* bioremediation: bioattenuation, bioaugmentation and biostimulation. In the case of uranium, *in situ* bioremediation has been the most common technique so far. At times, the process is made more efficient by combining it with biostimulation and with the use of designer bacterial strains. *In situ* uranium bioremediation is based on immobilization of reduced uranium, first tested in Norman, Oklahoma, USA (Senko *et al.*, 2002). Typically, deep wells are drilled at the site, contaminant solutions are injected into these wells, allowed to incubate for a certain duration, and samples withdrawn periodically to determine the extent of reduction. This system used the microbial populations naturally existing at the site and was the first demonstration of the *in situ* technology. Nevertheless, since this contamination was fabricated and was not pre-existing, the system cannot be considered at a true model for *in situ* bioremediation of uranium.

Following this, studies were undertaken at Rifle, Colorado, USA (Anderson *et al.*, 2003; Ortiz *et al.*, 2004; Senko *et al.*, 2005a). This system comprised two rows of injection wells used to introduce electron donors (acetate) and were placed at right angles to the groundwater flow direction. Additional monitoring wells were installed at various distances from the injection wells. The system showed enrichment of Fe(III)-reducing *Geobacter* species capable of coupling acetate oxidation with U(VI) reduction. After initial reduction in the U(VI) levels, there was subsequent increase in the U(VI) concentrations. It was attributed to either desorption from the sediments or re-oxidation by oxygen or biotic factors. The study also helped understand the relationship and impact of nitrate concentrations for effective and stable reduction of U(VI). The microbial population from the site was subjected to 16S rRNA library screening and phospholipids profiling, which indicated enrichment of sulfate-reducing bacteria (Senko *et al.*, 2005b). In the second acetate injection study at this site, radiolabelled C^{13} acetate was used for stable isotope labelling in addition to 16S rRNA analysis, to identify the bacteria responsible for internalization of acetate and also those involved in uranium reduction (Vrionis *et al.*, 2005). This work showed C^{13}-label uptake by many Proteobacteria, many related to *Geobacter* and *Desufuromonas*, along with *Pseudomonas putida* and *Dechromonas agitates* (Radajewaski *et al.*, 2003). Several recent studies (Chang *et al.*, 2005; Senko *et al.*, 2005a) at the Bear Creek Valley site have explored the role of nitrates in uranium remediation. These established the fact that nitrate reduction must precede reduction of uranium. When groundwater was supplied with adequate levels of nitrates, uranium reduction was observed. 16S rRNA analysis of the microbial samples from the site indicated the presence of *Geobacter* as well as nitrate-reducing bacteria (Istok *et al.*, 2004;

Chang et al., 2005). A more recent study using an ethanol-oxidizing fluidized bed bioreactor demonstrated presence of α- and β-Proteobacteria in the bioreactor biomass (Peacock et al., 2004). The biomass was able to reduce U(VI) when used to inoculate a sediment column (Luo et al., 2005).

Biostimulation

The *in situ* approach for bioremediation is based on the microbial reduction of the radionuclide; therefore it becomes a direct function of availability of biomass in adequate numbers. The *in situ* method can use a bioattenuation approach, i.e. use of native microbial community for remediation of the pollutants, but this may be inefficient and unreliable if the growth is poor (Gu et al., 2005b). The other method called bioaugmentation, where desired microbes are introduced to the polluted site for bioremediation, creates concerns due to introduction of unknown/non-native microbial population into the environment. The biostimulation approach uses already present native microbes to bioremediate the actinide and stimulates their growth by externally providing electron acceptors/donors. Table 18.1 lists common stimulants used for uranium bioremediation.

Table 18.1. Common electron donors and acceptors used in biostimulation (inspired by Farhadian et al., 2008).

Biostimulant	Characteristics
Electron donors	
Acetate	Carbon source
Sulfate	Ability to exchange a larger number of electrons
Nitrate	Higher concentrations can be added to groundwater
	Low electron affinity (less energy required to remove electron)
Ethanol	Carbon source
Electrodes	Results in accumulation of heavy metal on electrode
Electron acceptors	
Oxygen	Most energetically favourable
Sulfate	Higher electron affinity, able to accept more electrons
Nitrate	Higher concentrations can be added to groundwater
Electrodes	Does not involve the addition of a substance to the groundwater

Most of the reports about biostimulation for uranium bioremediation are from studies carried out at the facilities in Rifle (Colorado, USA), Oak Ridge (Tennessee, USA) and at Bear Creek Valley. They have used this approach successfully over the last several years (Anderson et al., 2003; Ortiz et al., 2004; Miller, 2010). Preliminary bench-scale studies (Luo et al., 2005) were successful under laboratory conditions, but field studies showed environmental as well as microbial heterogeneity that could not be simulated in the laboratory settings (Vrionis et al., 2005). The effectiveness of biostimulation for uranium remediation rests on the level of actinide contamination. Uranium concentrations at sites also impact the rates of reduction, with higher rates (days or weeks) at high uranium concentrations (several mM) and slowing down as the soluble uranium levels go down (several months or even years) to less than 5 µM (Cardenas et al., 2008). At these low concentrations, addition of glucose or ethanol stimulated the biomass (sulphate-reducing bacteria, iron-reducing bacteria and denitrifying microbes) to achieve uranium and nitrate remediation. At higher concentrations of uranium, however, the metal toxicity limited the extent of biostimulation for remediation. The biomass density at such sites becomes very low and direct biostimulation is unsuccessful. The solution to this problem came from the work of Wu and co-workers (Wu *et al.*, 2006a, b), where a combination of pre-conditioning and biostimulation was applied. Figure 18.3 depicts the overall scheme for this *in situ* system using biostimulation (Cardenas *et al.*, 2008; Bhowmik *et al.*, 2009).

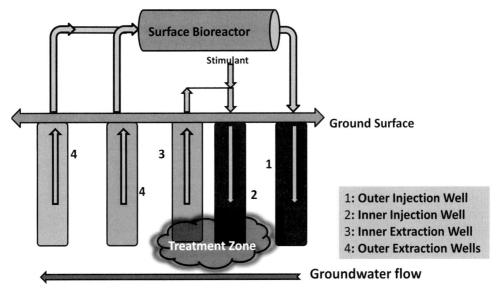

Fig. 18.3. Biostimulation: a schematic flow chart. (Adapted and reprinted with permission from Wu *et al.*, *Environ. Sci. Technol.*, 2006, 40, 3986–3995 copyright 2006, American Chemical Society.)

In this system, a pre-conditioning of the treatment zone was carried out using a double-layered hydraulic system as shown in Fig. 18.3. The outer layer provided shielding from hydrogeological impacts and the inner layer delivered biostimulant (ethanol). This setup was combined with an *ex situ* system to keep the hydraulic recirculation system from clogging because of precipitation (calcium and aluminium) and biomass.

Another adaptation of biostimulation is one applied to a fixed bed reactor system (Gregory and Lovley, 2005; N'Guessan *et al.*, 2008). When sulphate-reducing bacteria (SRB) biofilms were stimulated with high bicarbonate levels, almost 90% of uranium was removed. Although this was not an *in situ* system, it demonstrated an additional option for uranium removal. The success was probably due to high cell mass concentrations in biofilms compared to those in groundwater or sediments.

In another novel approach, instead of usual organic electron donors, *Geobacter* species were stimulated to reduce uranium using electrodes poised at −600 mV (Gregory and Lovley, 2005). This worked in two ways: effective stimulation of biotic uranium reduction and removal of the reduced and precipitated uranium by adsorption on the carbon fibre electrodes.

Challenges for biostimulation

Biostimulation has proved to be a very promising *in situ* remediation strategy for uranium so far and has contributed significantly to the understanding of the process. Nevertheless, there are some challenges faced by scientists with this approach. Since environmental factors such as pH, moisture, dissolved oxygen, etc. impact the bioremediation in general, the same factors can influence efficiency of a stimulant and stability of the reduced uranium as well. The injection of electron donors is aimed at biostimulation of the desired electron acceptors for bioremediation of uranium (Gavrilescu *et al.*, 2009), but these may end up reducing non-target electron acceptors and not uranium. The second challenge for biostimulation is that it is hard to estimate efficiency at contaminated sites under site parameters that are different from normal conditions, such as high salinity due to use of highly acidic solutions used in uranium processing (Cerda *et al.*, 1993; Dou *et al.*, 2008) as reported for a site in Shiprock, New Mexico (Finneran, K. *et al.*, 2002). In this case, uranium reduction was observed upon acetate stimulation and the microbial population of remediating biomass was closely related to *Pseudomonas* and *Desulfosporosinus* (Schippers *et al.*, 1995) instead of the usual Geobateraceae species commonly associated with uranium reduction with acetate biostimulation (Nevin *et al.*, 2003). In yet another study done at the Rifle Colorado site after two acetate injections, significant microbial and geochemical heterogeneity was observed at the site as a function of method of delivery of acetate. Using 16S rRNA analysis and genetic library screening at locations closest to the injection wells, enrichment of sulphate-reducing bacteria in the order of *Desulfobacterales* species was seen in the sediments as well as the groundwater samples (Vrionis *et al.*, 2005). This study also determined heterogeneity in the microbial population in sediments and groundwater samples (Holmes *et al.*, 2005; Vrionis *et al.*, 2005; Cardenas *et al.*, 2008). The heterogeneity found at the site in Rifle Colorado may have resulted from differential exposure to the stimulants of the microbial species from sediments and groundwater. This heterogeneity in the solid phase affected the effective concentrations of acetate, sulfate, soluble U(VI) and insoluble Fe(III), minerals etc. available to the biomass, and probably were the factors that influenced activity and diversity of the microbial populations (Vrionis *et al.*, 2005).

Biotic stabilization of reduced uranium

Given the knowledge of the composition and metabolism of microbes typically found at the uranium contaminated sites, it was realized that the reduced uranium could be stabilized if populations of denitrifying bacteria were enriched at these sites. These remove nitrates and nitrites from the environment and thus stabilize/immobilize the reduced radionuclide by preventing biotic re-oxidation (Vrionis *et al.*, 2005). Another type of microbial population that can help stabilize uranium reduction is SRBs. These produce sulfides as a metabolic product capable of reducing the U(VI) abiotically, thus preventing re-oxidation of the biotically reduced uranium.

Biotechnological Advances

The advancements in biotechnology have contributed novel molecular as well as technical approaches for better mechanisms applicable to uranium bioremediation. This section covers recent advances such as modified PCRs, designer strains, microarrays, immunobiosensors and proteogenomics.

PCR and microarray studies

In 2002, a PCR and kinetics analysis was carried out to study sulfate alone and an accompanied uranium reduction in a mixed culture. The study indicated presence of *Clostridium* and *Desulfovibrio* in the mixed culture (Spear *et al.*, 2000). In the same year, Holmes *et al.* (2002) performed most probable number PCR (MPN-PCR) and Taqman analyses in the San Juan River floodplain in Shiprock New Mexico. In these sediments, U(VI) reduction was accompanied by simultaneous Fe(III) reduction and tremendous enrichment of the members of Geobateriaceae associated with stimulation of dissimilatory sulfate reduction in uranium-contaminated aquifer sediments. In 2005, Holmes *et al.* (2005) performed reverse transcription PCR (RT-PCR) and microarray screening from sediment samples of *Geobacter* to follow gene expression. Once reduced, it is assumed that the U(IV) is stable under constant reducing conditions. Interestingly, the re-oxidation of reduced uranium was documented by Wan *et al.* (2005). In a follow-up study a microarray experiment was performed to determine the contribution due to changes in the microbial community. A high density phylogenetic microarray screening containing 500 probes was carried out (Brodie *et al.*, 2006). This study demonstrated application of the microarray technique for routine detection and monitoring of a large microbial population in an accurate and reproducible manner.

Genetic engineering

Bacteria have inherent abilities to act on waste materials and remediate them. Nevertheless, the complexity of interaction between pollutants and microbes may sometimes lead to inefficient clean-up. Genetic engineering approaches are useful in such cases and designer or over-efficient strains can be generated through these approaches for specific remediation applications (Holmes *et al.*, 2005). *Geobacter sulfurreducens* was engineered for a modified ATP-synthase under IPTG induction to stimulate respiration (Shukla *et al.*, 2010). During this manipulation, the enzymes for the TCA cycle and electron transfer chain were upregulated but the cell numbers/biomass yield was reduced. This was helpful in avoiding clogging of the *in situ* system and, with the higher rates of respiration, helped increase in biotic (Mahadevan *et al.*, 2008; Shukla *et al.*, 2010) reduction of uranium. Another bacterial species as a target for genetic engineering is the c_3 type cytochromes from *Desulfovibrio vulgaris*, which is involved in reduction of U(VI). Many studies trying to establish structure–function relationship in *Desulfovibrio desulfuricans* using various bio-analytical and proteomics techniques have been published (Gavrilescu *et al.*, 2009). Fast kinetics studies (Pattarkine *et al.*, in prep.) have tried to identify amino acid residues from cytochrome c_3 involved in the reduction mechanism. Based on X-ray crystal structure of cytochromes $c3$ in *D. desulfuricans* strain G20, certain amino acid residues were identified as targets for protein engineering (Pattarkine *et al.*, 2006). Using advanced rDNA techniques, research needs to be focused at development of either cell-free immobilized enzyme systems or engineered organisms with improved abilities for U(VI) reduction.

Sensing devices/immunobiosensors

Field-based detection devices is yet another technique that significantly helps in easy monitoring of sites being remediated. Typically, the X-ray absorption near-edge structural analysis (XANES) has been the technique of choice in the past. The procedure is expensive and time-consuming (Komlos *et al.*, 2008). Recently, two immunobiosensors (Melton *et al.*, 2009) were applied for detection of uranium at the Rifle Colorado site. One of the biosensors is a field portable sensor (FPS) and the other an inline sensor giving results on groundwater samples in less than 10 min with sensitivity of 0.33 nM. Using monoclonal antibodies against U(VI)-dicarboxyphenathroline complex (DCP), both sensors were able to detect uranium as low as 79 ppt for FPS and 29 ppt for the inline sensor (regulated drinking water limit is 30 ppb) (Melton *et al.*, 2009). The chelator-pretreated groundwater samples were exposed to DCP, treated with DCP antibodies, and passed through an affinity column that would retain the DCP-antibody complexes.

Nanotechnology advances

With the need for improved technologies to achieve more efficient remediation of uranium, scientists have tried some applied nanotechnology for uranium bioremediation. It was shown in the case of *Shewanella oneidensis* that during the *c*-type cytochrome-dependent reduction of uranium, it actually formed U(IV) nanoparticles on the outer surface of the cells, in association with the extracellular polymeric substance. Some other biominerals deposited by microbes may have catalytic activities and may serve as a starting point for synthesis of novel nanomaterials (Llyod *et al.*, 2008; Theng and Yuan, 2008; Petkov *et al.*, 2009).

The uranium nanowires were discovered in *G. sulfurreducens*. Pili produced by these bacteria were found to be necessary for interaction with iron and manganese oxides. Conductive probe atomic force microscopic analysis of the pili indicated the highly conductive nature of these pili, also termed as nanowires (Reguera *et al.*, 2005; Melton *et al.*, 2009). Similar nanowires were also found in *D. desulfuricans* G20 (Marsili *et al.*, 2005).

Recently, there have been reports about application of zero-valent iron nanoparticles for uranium bioremediation (Dickinson and Scott, 2010). The nanoparticles were allowed to react with the effluent for 28 days with periodic sampling. The solution analysis indicated removal of U to <1.5% of the initial concentration within 1 h of introduction of the nanoparticles and remained at similar concentrations for about 48 h.

A more recent report from the labs of Klimkova *et al.* (2011) has indicated successful application of the zero-valent iron nanoparticles for removal of most heavy metal contaminants from acid mine water.

Conclusion

Uranium contamination of groundwater poses serious problems to the environment. The conventional method of using ion-exchange resins for soluble uranium removal from water has high-cost and technology limitations. Microbial remediation offers cost-effective and ecofriendly technologies that can be optimized. *In situ* bioremediation of uranium allows treatment and remediation of uranium at the contamination site. Biostimulation effectively uses microbial metabolism and biomass for efficient removal of soluble uranium during *in situ* treatments. The microbiologically reduced uranium can be further stabilized by use of denitrifying bacteria in the remediating biomass. This treatment has been further enhanced significantly with the use of recombinant DNA technology involving use of polymerase chain reaction (PCR) to perform gene profiling and genetic engineering to produce designer strains of bacteria for uranium removal. Innovations for field-based detection or monitoring devices such as immunobiosensors have helped tremendously for monitoring the remediation sites. Lastly, the advancements of nanotechnology have led to successful demonstration of nanomaterials such as zero valent iron in removal of uranium from waste effluent. The combination of biotechnology and nanobiotechnology holds the key to new and more efficient technologies for uranium remediation in future.

References

Anderson, R.T., Vrionis, H.A., Ortiz-Berad, I., Resch, C.T. and Long, P.E. (2003) Stimulating the *in situ* activity of *Geobacter* species to remove uranium from the ground water of uranium-contaminated aquifer. *Environmental Science and Technology* 69, 5884–5891.

Beller, H.R. (2005) Anaerobic, nitrate-dependent oxidation of U(IV) oxide minerals by chemilithoautotrophic bacterium *Thiobacillus denitrificans*. *Applied and Environmental Microbiology* 71, 2170–2174.

Bhowmik, A., Aashino, S., Shikari, T., Nakamura, T. and Takamizawa, K. (2009) *In situ* study of tetrachloroethylene bioremediation with different microbial community shifting. *Environmental Technology* 30, 1607–1614.

Brodie, E.L., DeSantis, T.Z., Joyner, D., Baek, S.M., Larsen, J.T., Anderson, G.L., Hazen, T.C., Richardson, P.M., Herman, D.J., Tokunaga, T.K., Wan, J. and Firestone, M.K. (2006) Application of a high density oligonucleotide microarray approach to study bacterial population dynamics during uranium reduction and reoxidation. *Applied and Environmental Microbiology* 72(9), 6288–6298.

Cardenas, E., Wu, W., Leigh, M., Caley, J., Carroll, S., Gentry, T., Luo, J., Watson, D., Gu, B. and Ginder-Vogel, M. (2008) Microbial communities in contaminated sediments associated with bioremediation of uranium in submicromolar levels. *Applied and Environmental Microbiology* 74, 3718–3729.

Cerda, J.S., Gonzalez, J.R. and Quintana, T. (1993) Uranium concentrates bio-production in Spain: a case study. *FEMS Microbiology Review* 11, 253–260.

Chang, Y.J., Long, P.E., Geyer, R., Peacock, A.D. and Resch, C.T. (2005) Microbial incorporation of 13C-labeled acetate at the field scale: detection of microbes responsible for reduction of U(VI). *Applied and Environmental Microbiology* 39, 9039–9048.

Dickinson, S. and Scott, T.B. (2010) The application of zerovalent iron nanoparticles for remediation of uranium contaminated waste effluent. *Journal of Hazardous Materials* 178(1–3), 171–179.

Dou, L., Liu, X., Hu and Deng, D. (2008) Anaerobic BTEX biodegradation linked to nitrate and sulfate reduction. *Journal of Hazardous Materials* 152, 720–729.

Farhadian, M., Vachelard, C., Duchez, D. and Larroche, C. (2008) *In situ* bioremediation of monoaromatic pollutants in groundwater: a review. *Bioresource Technology* 99, 5296–5308.

Finneran, K.T., Housewright, M.E. and Lovley, D.R. (2002) Multiple influences of nitrate on uranium solubility during bioremediation of uranium-contaminated subsurface sediments. *Environmental Microbiology* 4, 510–516.

Finneran, K., Anderson, R., Navin, K. and Lovley, D. (2002) Potential for bioremediation of uranium contaminated aquifers with microbial U (VI) reduction. *Soil Sediment Contamination* 11, 119–157.

Gadd, G.M. (2010) Metals, minerals and microbes: geomicrobiology and bioremediation. *Microbiology* 156, 609–643.

Gadd, G.M. and Griffiths, A.J. (1978) Microorganisms and heavy metal toxicity. *Microbial Ecology* 4, 303–317.

Gadd, G.M. and Raven, J.A. (2010) Geomicrobiology of eukaryotic microorganisms. *Geomicrobiology Journal* 27, 491–499.

Gavrilescu, M., Pavel, L. and Cretescu, I. (2009) Characterization and remediation of soils contaminated with uranium. *Journal of Hazardous Materials* 163, 475–510.

Gomathi, M. and Sabrinathan, K.G. (2010) Microbial Mechanisms of heavy metal tolerance – a review. *Agricultural Review* 31(2), 133–138.

Gorby, Y.A. and Lovley, D.R. (1992) Enzymatic uranium precipitation. *Environmental Science and Technology* 26, 205–207.

Gregory, K.B. and Lovley, D.R. (2005) Remediation and recovery of uranium from contaminated subsurface environments with electrodes. *Environmental Science and Technology* 39, 84943–84947.

Gu, B., Wu, W., Ginder-Vogel, M.S., Yan, H. and Fields, M.W. (2005a) Bioreduction of uranium in a contaminated soil column. *Environmental Science and Technology* 39, 4841–4847.

Gu, B.H., Yan, H., Zhou, P. and Watson, D.B. (2005b) Natural Humics Impact uranium Bioreduction and oxidation. *Environmental Science and Technology* 39, 5268–5275.

Holmes, D., Finnerman, T.K., O'Neil, R.A. and Lovley, D.R. (2002) Enrichment of members of the family Geobactereaceae associated with stimulation of dissimilatory metal reduction in uranium contaminated aquifer sediments. *Applied and Environmental Microbiology* 68(5), 2300–2306.

Holmes, D.E., Nevin, K., O'Neil, R., Ward, J., Adams, L., Woodward, T., Vrionis, H. and Lovley, D.R. (2005) Potential for qualifying expression of the Geobateriaceae citrate synthase gene to assess the activity of Geobateriaceae in the subsurface and on the current harvesting electrodes. *Applied and Environmental Microbiology* 71, 6870–6877.

Istok, J.D., Senko, J.M., Krumholz, L.R., Watson, D. and Bogle, M.A. (2004) *In situ* bioremediation of rechnetium and uranium in a nitrate-contaminated aquifer. *Environmental Science and Technology* 38, 468–475.

Khani, M., Keshtkar, A., Meysami, B. and Jalati, R. (2006) Biosorption of uranium from aqueous solutions by nonliving biomass of marine algae *Cytoseria indica*. *Journal of Biotechnology* 9, 100–106.

Klimkova, S., Cernik, M., Lacinova, L., Filip, J., Jancik, D. and Zboril, R. (2011) Zero valent iron nanoparticles in treatment of acid mine water from *in situ* uranium leaching. *Chemosphere* 82(8), 1178–1184.

Komlos, J., Mishra, B., Lanzirotti, A., Myneni, S. and Jaffe, P. (2008) Real time speciation of uranium during active bioremediation and U(IV) reoxidation. *Journal of Environmental Engineering* 134, 78–86.

Kumar, R., Singh, S. and Singh, O. (2007) Bioremediation of radionuclides: Emerging Technologies. *OMICS* 11, 295 304.

Li, L.S., Kowalsky, C.I., Englert, A. and Hubbard, S.S. (2010) Effects of physical and geochemical heterogeneities on mineral transformation and biomass accumulation during biostimulation experiments at Rifle Colorado. *Journal of Contamination and Hydrology* 112, 45–63.

Llyod, J.R., Pearce, C.I., Coker, V.S., Pattrick, R.A., vander Laan, G., Cutting, R., Vaughan, D.V., Paterson-Beedle, M. and Milkheeno, I.P. (2008) Biomineralization: linking the fossil record to the production of high value functional materials. *Geobiology* 6, 285–297.

Lovley, D.R. and Phillips, E.J.P. (1992a) Reduction of uranium by *Desulfovibrio desulfuricans*. *Applied and Environmental Microbiology* 58, 850–856.

Lovley, D.R. and Phillips, E.J.P. (1992b) Bioremediation of uranium contamination with enzymatic uranium reduction. *Environmental Science and Technology* 26, 2228–2234.

Lovley, D.R., Phillips, E.J.P., Gorby, Y.A. and Landa, E.R. (1991) Microbial Reduction of Uranium. *Nature* 350, 413–416.

Lovley, D.R., Widman, P.K., Woodward, J.C. and Phillips, E.J.P. (1993) Reduction of uranium by cytochromes c3 of *Desulfovibrio vulgaris*. *Applied and Environmental Microbiology* 59, 3572–3576.

Luo, J., Cirpka, O.A., Wu, W., Fiened, M.N. and Jardine, P.M. (2005) Mass-transfer limitations for nitrate removal in a uranium-contaminated aquifer. *Environmental Science and Technology* 39, 8453–8459.

Mahadevan, I.M., Burgard, R., Postier, A., DiDonato, B., Sun, R., Schilling, C.H. and Lovley, D.R. (2008) *Geobacter sulfreducens* strain engineered for increased rates of respiration. *Metabolic Engineering* 10, 267–275.

Markich, S. (2002) Uranium speciation and bioavailability in aquatic systems: an overview. *Science World Journal* 2, 707–729.

Marsili, E., Beyenal, H., Di Palma, L., Merli, C. and Dohnalkova, A. (2005) Uranium removal by sulfate reducing biofilms in presence of carbonates. *Water Science Technology* 52, 49–55.

Melton, S., Yu, H., Williams, K., Morris, S., Long, P. and Blake, D. (2009) Field-based detection and monitoring of uranium in contaminated groundwater using two Immunosensors. *Environmental Science and Technology* 43, 6703–6709.

Methe, B.A., Nelson, K.E., Eisen, J.A., Paulsen, T.E. and Nelson, W. (2003) Genome of *Geobacter sulfreducens*: metal reductions in subsurface environments. *Science* 302, 1967–1969.

Miller, H. (2010) Biostimulation as a form of bioremediation of soil pollutants. *MMG Basic Biotechnology* 6, 1–8.

N'Guessan, A.L., Vrionis, H.A., Resch, C.T., Long, P.E. and Lovley, D.R. (2008) Sustained removal of uranium from contaminated groundwater following a stimulation of dissimilatory metal reduction. *Environmental Science and Technology* 42, 2999–3004.

NABIR (2003) Bioremediation of metals and radionuclides. What is it and how it works. *Rep LBNL–42595.*

Nevin, K., Finnerman, K.T. and Lovley, D.R. (2003) Microorganisms associated with uranium bioremediation in a high-salinity subsurface sediment. *Applied and Environmental Microbiology* 69, 3672–3675.

Ortiz, I., Anderson, R.T., Vrionis, H.A. and Lovley, D.R. (2004) Resistance to solid-phase U(VI) to microbial reduction during in-situ bioremediation of uranium-contaminated groundwater. *Applied and Environmental Microbiology* 70, 7558–7560.

Osman, D. and Cavet, J.S. (2008) Copper homeostasis in bacteria. *Advances in Applied Microbiology* 65, 217–247.

Pattarkine, M.V., Tanner, J., Bottoms, C., Lee, Y.H. and Wall, J. (2006) *Desulfovibrio desulfuricans* G20 Tetraheme Cytochrome Structure at 1.5 Å and Cytochrome Interaction with Metal Complexes. *Journal of Molecular Biology* 358(5), 1314–1317.

Pattarkine, M.V., Tipton, P. and Wall, J. (Manuscript under preparation). Fast Kinetics studies of Uranium interaction with Reduced Cytochrome from *Dedulfovibrio desulfuricans* G20.

Peacock, A., Chang, Y.J., Istok, J.D., Krumholz, L.R. and Geyer, R. (2004) Utilization of microbial biofilms as monitors of bioremediation. *Microbial Ecology* 7, 284–292.

Petkov, V., Ren, Y., Saratovsky, I., Pasten, P., Gurr, S.J., Hayward, M.A., Poeppelmeier, K.R. and Gaillard, J.F. (2009) Atomic scale structure of biogenic material by total X-ray diffraction: a study of bacterial and fungal MnOx. *ACS Nano* 3, 442–445.

Radajewaski, S., McDonald, I.R., and Murrell, J.C. (2003) Stable isotope probing of nucleic acids: a window to the function of uncultured microorganisms. *Current Opinions in Biotechnology* 14, 296–302.

Reguera, G., McCarthy, K.D., Mehta, T., Nicoll, J.S., Tuominen, M.T. and Lovley, D. R. (2005) Extracellular electron transfer via microbial nanowires. *Nature* 435, 1098–10101.

Renshaw, J.C., Butchins, C.J., Livens, E.R., May, I., Charnok, J.M. and Llyod, J.R. (2005) Bioreduction of uranium: environmental implications of a pentavalent intermediate. *Environmental Science and Technology* 39, 5657–5660.

Rosen, B.P. (2002) Transport and detoxification systems for transition metals, heavy metals and metalloids in eukaryotic and prokaryotic microbes. *Computational Biochemical Physiology* 133, 689–693.

Schippers, A.R., Hallmann, S.W. and Sand, W. (1995) Microbial diversity in uranium mine waste heaps. *Applied and Environmental Microbiology* 61, 32930–32935.

Sen, R. and Chakrabarti, S. (2009) Biotechnology: applications to environmental remediation in resource exploitation. *Current Science* 97(6), 768–775.

Senko, J.M., Istok, J.D., Suflita, J.M. and Krumholz, L.R. (2002) *In situ* evidence for remediation of uranium immobilization and remobilization. *Environmental Science and Technology* 36, 1491–1496.

Senko, J.M., Mohamed, Y., Dewers, T.A. and Krumholz, L.R. (2005a) Role for Fe(III) minerals in nitrate-dependent microbial U(IV) oxidation. *Environmental Science and Technology* 39, 2529–2536.

Senko, J.M., Suflita, J.M., Ortiz, I. and Krumholz, L.R. (2005b) Geochemial controls on microbial nitrate-dependent uranium U(IV) oxidation. *Geomicrobiology Journal* 22, 371–378.

Seyrig, G. (2010) Uranium Bioremediation: current knowledge and trends. *MMG Basic Biotechnology* 6, 19–24.

Shelobolina, E.S., Koppi, M., Korenevsky, A., DiDonato, L., Sillivan, S., Konishi, H., Xu, H., Leang, C., Butler, J.E., Kim, B. and Lovley, D.R. (2007) Importance of c-type cytochromes for U(VI) reduction by *Geobactoer sulfrreducens*. *BMC Microbiology* 7, 16–19.

Shelobolina, E.S., Vrionis, F., Findley, R. and Lovley, D.R. (2008) *Geobacter uraniireducens* sp nov, isolated from subsurface sediment undergoing uranium bioremediation. *International Journal of System Evolutionary Microbiology* 58, 1075–1078.

Shukla, K.P., Singh, N.K. and Sharma, S. (2010) Bioremediation: Developments, current practices and perspectives. *Genetic Engineering and Biotechnology Journal* 1–20.

Silver, S. and Phung, L.T. (1996) Bacterial Heavy metal resistance: new surprises. *Annual Review of Microbiology* 50, 753–789.

Spear, J.R., Figuero, L.A. and Honeyman, B.D. (2000) Modeling reduction of uranium by U (VI) under variable sulfate concentrations by sulphate-reducing bacteria. *Applied and Environmental Microbiology* 66(9), 3711–3721.

Strong, P.J. and Burgess, J.E. (2008) Treatment methods for wine-related and distillery wastewaters: a review. *Bioremediation Journal* 12, 70–87.

Suzuki, Y. and Banfield, J.E. (1999) Geomicrobiology of uranium. In: Burns, P.C. and Finch, R. (eds) *Uranium: Mineralogy, geochemistry and the environment. Reviews in Mineralogy* 38, 393–432.

Suzuki, Y. and Suko, T. (2006) Geomicrobial factors that control uranium mobility in the environment: update on recent advances in bioremediation of uranium-contaminated soils. *Journal of Mineral Petrol Science* 101, 299–307.

Theng, B.K. and Yuan, G. (2008) Nanoparticles in the soil environment. *Elements* 4, 395–399.

Vrionis, H.A., Anderson, R.T., Ortiz, I., O'Neill, K.R. and Resch, C.T. (2005) Microbiological and geochemical heterogeneity in an *in situ* uranium bioremediation field site. *Applied and Environmental Microbiology* 71, 6308–6318.

Wall, J.D., and Krumholz, L.R. (2006) Uranium Reduction. *Annual Review of Microbiology* 60, 149–166.

Wan, J., Tokunaga, T.K., Brodie, E., Wang, Z. and Zeng, Z. (2005) Reoxidation of bioreduced uranium under reducing conditions. *Environmental Science and Technology* 39, 6162–6169.

Woolfolk, C.A. and Whiteley, H.R. (1962) Reduction of inorganic compounds with molecular hydrogen by *Micrococcus lactilyticus*. I. Stoichiometry with compounds of arsenic, selenium, tellurium, transition and other elements. *Journal of Bacteriology* 84, 647–658.

Wu, W., Carley, J., Fienen, M., Mehlhorn, T., Lowe, K., Nyman, J., Luo, J., Gentile, M., Rajan, R. and Wagner, D. (2006a) Pilot-scale *in situ* bioremediation of uranium in a highly contaminated aquifer. 1. Conditioning of a treatment zone. *Environmental Science and Technology* 40, 3978–3985.

Wu, W., Carley, J., Gentry, T., Ginder-Vogel, M., Fienen, M., Mehlhorn, T., Yan, H., Caroll, S., Pace, M. and Nyman, J. (2006b) Pilot-scale *in situ* bioremedation of uranium in a highly contaminated aquifer. 2. Reduction of U(VI) and geochemical control of U(VI) bioavailability. *Environmental Science and Technology* 40, 3986–3995.

Chapter 19

Going Extreme for Small Solutions to Big Environmental Challenges

Chris Bagwell

Introduction

Prokaryotes are the most abundant and diverse life form on planet Earth, with estimates of prokaryotic numbers and productivity approaching 10^{30} year^{-1} (Whitman *et al.*, 1998). Prokaryotes are responsible for catalysing important biogeochemical reactions and transformations that sustain the biosphere. Prokaryotes have been evolving for 3–4 billion years (Ernst, 1983) and during this time, have found ways to occupy every conceivable environment on the planet, including the most inhospitable habitats both nature and man have created.

> Microbial ecology is the study of microbial physiology under the worst possible conditions.
>
> T.D. Brock, 1966

Most environmental microbiologists would agree that this popular quote by Thomas Brock straightforwardly applies to the maintenance of microbial diversity and ecosystem functions amid a complex backdrop of ever changing biological, physical and chemical conditions in terrestrial and aquatic biomes. But does it stop there, what about the de facto 'worst possible condition'?

Many of the environmental challenges outlined in this chapter are not exclusive to the USA but rather affect much of the industrialized world (Pedersen, 1999) because of past military activities and continued expansion of the military–industry complex. While not comprehensive, this chapter is devoted to the scale, scope and specific issues confronting the cleanup and long-term disposal of the US nuclear legacy generated during the second world war and the Cold War era. Furthermore, microbial interactions and metabolism in, around and affecting existing and planned geological nuclear waste repositories are serious concerns for safe disposal, future planning and reliable risk assessment for the environment and human health.

Nuclear Legacy Waste

Extensive volumes and complex mixtures of nuclear waste are a lasting legacy of the Cold War era. The USA began building the first atomic bomb in 1942. These efforts generated more than 36 million m^3 of long-lived radioactive and toxic waste by the end of the Cold War in 1989. Irradiated fuels and past nuclear processing streams still await treatment and safe disposal in aged storage configurations. Mixtures of metals, radionuclides, hydrocarbons and ions contaminate soils, sediments and groundwater across the nuclear–industrial complex. The US Department of Energy (DOE) is responsible for management, disposal and long-term stewardship of this lasting legacy. Because of the prohibitive cost and inefficiencies of existing chemical and physical remedial strategies to address large volumes of contaminant mixtures, newly advanced technologies are greatly needed for reduction and treatment of nuclear legacy waste. Bioremediation is a potentially powerful and innovative technology for converting toxic pollutants to benign end-products with a significant cost saving over conventional approaches. The reality, however, is that bioremediation is complicated by the unpredictability of natural ecosystems, complex interactions between co-contaminants and the environmental or containment matrix, and the uncertain behaviour of bacterial populations to periodic shifts, and quite often arduous environmental conditions. Most industrial and DOE contaminated environments contain complex contaminant mixtures of varying concentrations. Successful application, management and ultimate acceptance of bioremediation as a legitimate treatment strategy for nuclear legacy waste demands that a bioremedial candidate be robust: capable of maintaining acceptable rates of metabolism and growth on target pollutants in the presence of co-contaminant toxins, salts and radioactivity. Technological advances are greatly needed to generate novel solutions for legacy waste reduction, elimination and stabilization.

High-level Radioactive Waste (HLW)

During the Cold War, Pu^{239} production for national defence began by irradiating uranium or other target elements in a nuclear reactor. During reprocessing of spent reactor fuel, approximately 99% of U^{235} and Pu^{239} were reclaimed. All remaining radionuclides, fission products, fuel components and nonradioactive chemicals used during reclamation made up the high-level waste stream. High-level waste currently resides at over 100 different sites across the contiguous USA, but the majority of Cold War legacy HLW is located at Hanford (<65 million gallons), near Richland, Washington and the Savannah River Site (SRS) (roughly 35 million gallons), near Aiken, South Carolina USA (Wicks and Bickford, 1989). The current national inventory of HLW has more than 1 billion curies of radioactivity. The Savannah River Site tank waste and much of the Hanford tank waste contains Fe, Al, Si, Ca, F, K, alkali cations, organic solvents, radionuclides, fission products and other governmentally regulated metals. Underground HLW tanks have provided more than half a century of storage, though many of these tanks are well beyond their projected life span. The caustic and corrosive chemistry of HLW has caused some HLW tanks to leak. Many of the oldest single-shell tanks at Hanford have confirmed leaks. An estimated 1 million gallons of HLW has been released into the vadose zone at the Hanford site (Fredrickson *et al*., 2004).

Treatment and safe permanent storage of HLW is an ongoing priority not only for legacy materials, but for all countries that operate nuclear power stations. Approximately 37% of legacy tank waste by volume will enter the Defence Waste Processing Facility (DWPF, in operation at SRS and in construction at Hanford) for conversion to a safe and stable glass form. The balance of the waste volume will be addressed by additional low-level radioactive facilities. Characteristic organic constituents within the aqueous phase are especially problematic for separation and processing of HLW. Organic constituents exist in HLW and in mixed waste in the form of complexants used during separation processes, radiolysis products from degradation of complexants and solvents, and from waste-tank decontamination reagents. One of the preferred decontamination reagents was oxalic acid, which can create problems for storage and final disposal processes due to its unique solubility properties as a sodium salt. Additional reagents in use at SRS and Hanford include glycolic acid, citric acid and formic acid. The potential for *in situ* removal (degradation) of organic constituents within HLW tanks and other storage configurations could greatly improve processing efficiency of HLW and other nuclear waste streams by established chemical methods.

Low-level Contaminated Soil and Groundwater

Another major legacy of the Cold War is enormous volumes of contaminated soil, sediment and groundwater across the DOE–industrial complex. Since long-term waste disposal issues were not fully recognized during the production era, nuclear processing streams were often routed to unlined storage configurations (i.e. landfills, trenches, basins) or simply disposed to the subsurface environment. Over time, contaminants have leached from reservoirs into the subsurface environment and groundwater transport has greatly exacerbated the extent of contamination. Many of the same waste elements and pollutant mixtures found in HLW are also present in the environment at many government, military and industrial installations around the world, though in much lower concentrations. Many of these pollutants are extremely toxic and long-lived in the environment. Environmental and public exposure is a major risk driver; containment and remediation are an immediate priority. The scale of this issue is currently estimated at 79 million m^3 of contaminated soil, nearly 2 billion m^3 of contaminated groundwater, and several million cubic metres of buried waste distributed over more than 100 sites at approximately 20 DOE installations across the USA (Riley and Zachara, 1992; Fioravanti and Makhijani, 1997).

Bioremediation holds great promise for some of our worst problems. There is no compound, man-made or natural, that microorganisms cannot degrade.
Terry Hazen, in response to the Deepwater Horizon oil spill in the Gulf of Mexico, 2010.

Bioremediation of Nuclear Legacy Waste

In situ bioremediation describes the use of biological processes to convert toxic pollutants to safe end-products within the contaminated environmental matrix. There are a number of established methods for doing this. Indigenous microorganisms can either be stimulated to boost rates of pollutant transformation (biostimulation), or if indigenous microbes cannot sustain desired activity, new beneficial populations can be introduced directly into the environment in order to achieve the desired outcome (bioaugmentation). Biological treatments can effectively prevent expansion of subsurface contaminant plumes (Major et al., 2002; Padmanabhan et al., 2003) and, in some cases, provide direct treatment of source zones (Adamson et al., 2003). Tremendous effort has been aimed towards in situ degradation of organic pollutants; however, bioremedial applications have recently been expanded to include metals and radionuclides (Lloyd and Macaskie, 2000). This approach utilizes bacteria that can support energetic metabolism by respiring redox sensitive metals or radionuclides, thereby changing their oxidation state from a water-soluble form to an immobile, mineral precipitate. However, the majority of DOE facilities are co-contaminated by complex mixtures of organic and inorganic compounds, and existing biological technologies proven successful for single classes of pollutants are often rendered ineffective under mixed waste conditions (Ruggiero et al., 2005). As such, biological solutions have so far not been widely applicable for hazardous or mixed waste. High- and low-level nuclear waste is strictly processed by engineered chemical strategies. In certain instances biologically catalysed remedies may be useful as a stand-alone treatment process, but for more complex waste streams their utility may be restricted to pre- or post-treatments or as a polishing step. The utility and acceptance of bioremediation as a valid treatment strategy for nuclear waste necessitates that the biological candidate exerts specific action against target compounds while withstanding the toxic effects of complex contaminant mixtures. New discoveries of naturally occurring 'extreme' bacteria could generate robust, cost-effective biological-based treatment strategies applicable to different nuclear waste categories, ranging from in-tank pretreatment of HLW to the control and size reduction of contaminated soil and groundwater plumes.

The use of the term 'extreme bacteria' in this context is not particularly limited to conventional strains that establish a living at extremes in temperature, pressure, salt saturation, etc., but more precisely microbes that inhabit relevant ecosystems and display meaningful metabolic and/or physiological interactions with nuclear legacy components. Appropriate examples of microorganisms having bioremedial applicability in this context are provided in Table 19.1. The balance of this chapter will be split into two succinct sections for the presentation of original research aimed at teasing apart and going beyond to infer complex microbiological interactions with legacy waste materials generated by past nuclear production activities in the USA. The intended purpose of this research is to identify cost-effective solutions to the specific problems (stability) and environmental challenges (fate, transport, exposure) in managing and detoxifying persistent contaminant species.

High Level Waste Microbiology

Kineococcus radiotolerans was isolated within a shielded cell work area containing highly radioactive nuclear waste at the US DOE Savannah River Site (SRS) (Aiken, South Carolina, USA) (Phillips et al., 2002). *Kineococcus* is an orange pigmented,

Table 19.1. Selected examples of microbial biotechnologies and applications for energy production and environmental restoration. 'Extreme' bacteria are defined here as microbes capable of withstanding excessively toxic environments and having remedial applicability for reducing volume and the inherent toxicity of legacy waste stockpiles and affected lands.

Extreme condition/ phenotype	Representative organism(s)	Biotechnological utility or concern	Reference(s)
Acid (pH <3) Acidophiles	Thiobacilli, Nitrifying bacteria, *Arthrobacter, Bacilllus,* acid mine drainage communities	Metal mobilization, corrosion, bioleaching, bioremediation	Diercks *et al.*, 1991; Martinez *et al.*, 2007; Jonkers, 2008
Alkalinity (pH >9) Alkaliphiles	*Alkaliphilus metalliredigens*	Metal/actinide mineralization	Roh *et al.*, 2007
Salt (> 0.2 M NaCl) Halophiles	*Halomonas* sp. WIPP1A	Radionuclide mobilization	Francis *et al.*, 2000
Radiation-resistant bacteria	*Deinococcus radiodurans*	Bioremediation of radioactive sites	Brim *et al.*, 2000; Fredrickson *et al.*, 2000
Acidophilic, dessication resistance	*Bacillus* spp., *Pseudomonas* spp., WIPP strains	Waste mobilization from solid formations, corrosion	Francis *et al.*, 1980a, b, 1998; Diercks *et al.*, 1991; Horn and Meik, 1995
Organic solvent resistance	*Arthrobacter* spp., *Clostridium thermohydrosulfuricum*	Soil and groundwater remediation, Energy production	Lovitt *et al.*, 1984; Sardessai and Bhosle, 2002
Metal respiring (reducing) bacteria	*Geobacter* spp.	Tc(VII), U(VI), etc. precipitation, energy production	Liu *et al.*, 2002; Anderson *et al.*, 2003; Bond and Lovley, 2003; Lovley and Nevin, 2011
Metal respiring hyper-thermophile	*Geoglobus ahangari*	Bioremediation	Kashefi *et al.*, 2002
Extreme metal resistance	*Arthrobacter* spp.	Bioremediation	Margesin and Schinner, 1996
Psychrotrophic organic solvent	*A. chlorophenolicus* A6	Bioremediation	Backman and Jansson, 2004
Obligate dehalorespiration	*Dehalococcoides ethanogenes*	Soil and groundwater remediation	Lendvay *et al.*, 2003

aerobic, nonsporulating actinomycete belonging to the Kineosporiaceae family. Consistent with its origins in a high-level nuclear waste environment, *K. radiotolerans* has proved to be exceptionally robust and to possess tremendous potential for bioremediation of hazardous and mixed waste environments where toxicity precludes efficient metabolism and survival for other bioremediation candidates (examples included in Table 19.1). While it is certain that *K. radiotolerans* survived the extremely harsh environment of high-level nuclear waste, the specific mechanisms that assured survival are unknown. The possibility that this bacterium may have catabolized organic components from nuclear waste for biomass conversion or maintenance energy is especially intriguing and may lend themselves for development of bioremediation technologies of organo-pollutants from various nuclear legacy waste classifications.

The primary challenges to any applied biotechnology for broad classifications of nuclear legacy waste for reduction and detoxification purposes include the ability to withstand organic and inorganic toxicity, and exposure to γ-radiation. Gamma radiation is one of the most energetic forms of electromagnetic radiation. Gamma rays penetrate tissues and cells, causing direct damage to DNA (namely double strand breaks, DSB), proteins and membranes. The majority (80%) of the resultant damage caused by exposure, however, is indirect and caused by secondary reactions stemming from the ionization of water and formation of free radical species, primarily •OH. DNA lesions block gene transcription and genome replication, and if not correctly repaired, could introduce detrimental mutations or cell death. Relatively few DNA double strand breaks are lethal for most bacteria. *Escherichia coli* succumbs to around 10 DSB and *Shewanella oneidensis* MR-1 dies after 1 DSB (based on calculations of 0.0114 DSBs Gy^{-1} genome^{-1}; Daly *et al.*, 2004). Remarkably, *K. radiotolerans* can accumulate more than 200 DSB (20 kGy γ-radiation) and within 3–4 days all DNA and cellular damage is repaired and cell division resumes (Bagwell *et al.*, 2008a). The cellular and biomolecular phenomena underlying the extreme radioresistance phenotoype in *K. radiotolerans* are unknown, and have been the subject of recent research efforts. Preliminary data indicate important differences compared to the current *Deinococcus radiodurans* model. These differences could suggest that independent evolutionary events are responsible for the radioresistance phenotype in these distinct bacterial lineages.

Reactive oxygen species (ROS) and oxygen free radicals are produced endogenously during aerobic metabolism as O_2 is reduced to H_2O, but are also formed during radiolysis of water. The latter is particularly problematic in radioactive environments given that a bacterial cell can be nearly 90% water. Pathways for cellular injury by exposure to ROS, chiefly •O, H_2O_2 or •OH, involve a number of subcellular cyclic reactions that generate additional reactive oxygen radicals. These highly reactive species oxidize unsaturated fatty acids in cell membranes, RNA, damage proteins and generate DNA lesions (Proctor and Reynolds, 1984; Marnett, 2000). ROS and oxygen free radicals also react with certain transition metals (e.g. Fe^{2+}, Cu^{2+}) in the cell by Fenton or Haber Weiss reactions to produce damaging hydroxide radicals (Imlay, 2003). Thus, radiation resistance and oxidative defence pathways are intricately linked. Genetic experimentation and genome comparisons have quantified the involvement of conventional repair and recombination pathways as well as numerous functionally uncharacterized genes in extreme radioresistance; though the phenotype is not strictly encoded within the genome, a suite of reactive and stabilizing metabolites as well as unusual cellular biochemistry also play critical roles.

A long standing paradigm in radiation biology was that radiation-induced biological effects resulted from direct damage to DNA. This logic stemmed from the acknowledgement that radiation-induced DNA lesions are toxic to living cells, particularly double strand breaks, and that DNA damage (and thus survivability) is radiation dose dependent. For decades, *D. radiodurans* has been studied for heritable trait(s) conferring extreme resistance and high fidelity DNA repair (Battista, 1997; Makarova *et al.*, 2001) and while genes and gene products are clearly one important aspect to extreme radioresistance, our views are expanding to include additional aspects of cell and molecular biology. However, three DNA-centric models have been proposed to explain the extreme resistance phenotype: (i) conventional enzymatic defences operating at extraordinary efficiency; (ii) the involvement of novel repair functions; and (iii) a highly condensed, multigenomic nucleoid (Battista, 1997; Levin-Zaidman *et al.*, 2003; Cox and Battista, 2005; Zimmerman and Battista, 2005). No single hypothesis explains in full the underlying genetic complexity of the extreme resistance phenotype (i.e. Udupa *et al.*, 1994); however, preferential utilization of manganese is thus far the only biochemical strategy broadly conserved among a diverse collection of extreme-resistant bacteria (Daly *et al.*, 2004, 2007). The antioxidative capacity of Mn has been known for decades though with renewed 'discovery' comes an important paradigm shift in radiation biology because manganese, unlike iron, does not catalyse hydroxyl radical formation through Fenton/Haber-Weiss chemistry. Manganese may also mitigate protein oxidation by scavenging oxygen radicals (Daly *et al.*, 2007). Elemental ratios of Mn:Fe have been proposed as a potentially reliable indicator of a cell's susceptibility to oxidative stress (Daly *et al.*, 2004). While Mn-accumulating bacteria accrue comparable levels of DNA damage as Fe-accumulating bacteria for a given dose of γ-radiation (Daly *et al.*, 2004; Granger *et al.*, 2011), manganese appears to quench secondary chemical reactions that produce ROS; thus, cellular damage is minimized, critical enzymatic repair processes are protected and remain active to extend cell survivorship (Daly, 2009).

Kineococcus radiotolerans was evaluated for preferential utilization of Mn, or an analogous molecular role for alternative redox active metals that may similarly function as a cellular antioxidant or to minimize protein damage following environmental assaults (Bagwell *et al.*, 2008b). In a reciprocal experimental design, colony formation during chronic irradiation (4 days at 60 Gy h^{-1}) was measured in response to the addition of individual redox metals at a single concentration (100 μM each of Fe^{2+}, Mn^{2+}, Zn^{2+}, Co^{2+}, Cu^{2+} and Mo^{2+}). Overall, the metal-only treatments had very little effect on colony formation, and the irradiated, metal(s) minus control cultures consistently yielded between 50 and 100 colony forming units (CFUs). Interestingly, colony formation during chronic irradiation in the presence of either Fe^{2+} or Mn^{2+} was negligible compared to the controls; however, the addition of Cu^{2+} combined with chronic radiation resulted in a lawn of bacterial growth, as was observed for the no metal, non-irradiated control cultures. The hypothesized physiological role for Mn^{2+} accumulation for radioresistance of *Deinococcus* and other radiation-resistant bacteria cannot explain this result, as Cu^{2+} does participate in Fenton or Haber-Weiss chemistry for the formation of ROS (Letelier *et al.*, 2005). This study marked the first documented case whereby bacterial growth was legitimately enhanced during chronic irradiation. Growth conditions that were expected to prompt copper-catalysed production of oxygen radicals actually promoted the growth of *K. radiotolerans*, and this response could not be duplicated by chronic irradiation or copper supplementation alone.

Copper is an essential cofactor for a variety of enzymes involved in aerobic respiration and energy production; however, only trace quantities are required and so, intracellular levels are tightly regulated by the cell (Rae et al., 1999). Yet, *K. radiotolerans* actively accumulates Cu^{2+} intracellularly, SEM/EDS spectra for copper were only detected from the cytoplasm of thin section preparations and the extent of accumulation is correlated with aqueous phase concentration. The consequences of copper accumulation are evident, however; Cu^{2+} loaded cultures display increased sensitivity to peroxides and methyl viologen, and post-irradiation recovery is delayed (Bagwell et al., 2008b). These conditions, however, were intended to push the physiological limits for these stressors; copper accumulation did not interfere with cell growth during chronic irradiation and growth rate and biomass yields were unaffected by high levels of copper accumulation (Bagwell et al., 2010). These results imply that *K. radiotolerans* cultures are not burdened by copper accumulation, implying clear capacity for copper coordination and sequestration, and that oxidative defences are responsive to this growth condition. A conventional Cop-type copper homoeostasis pathway has not been deduced from the *K. radiotolerans* genome sequence, though we have partially characterized participatory metal sequestration systems involved in intracellular copper accumulation (e.g. glutathione) and coordinated expression of antioxidants (e.g. amino acids, carbohydrates, organic osmolytes) and enzymatic defences that manage and maintain a proper intracellular environment (Bagwell et al., 2010). A systematic dose–response study has recently revealed that two specific concentrations of cupric sulphate (500 nM and 100 μM) significantly increase the rate of cell division and metabolic respiration in *K. radiotolerans* cultures. Note that 100 μM copper was the concentration used during chronic irradiation that stimulated cell growth and colony formation. We can surmise that perhaps the same phenomenon (i.e. boost in metabolic rate and/or efficiency) may also occur during chronic irradiation though the exact cellular mechanism and molecular function(s) of Cu^{2+} have not been elucidated. As an aside, we have demonstrated conservation of copper-stimulated growth among diverse actinobacteria; for example, 40% of isolated soil actinobacteria from a single shallow subsurface site displayed increased growth rate by copper supplementation. Oxygenic respiration increased markedly in *Lechevalieria xinjiangensis*, a novel actinomycete isolated from radiation polluted soil (Wang et al., 2007) at the following Cu^{2+} concentrations: 1 and 3 μM, 100, 250, 500 nM. The same phenomenon has also been documented at 35 μM and 250 nM Cu^{2+} for *Kineococcus auranticus* (Yokota et al., 1993), the closest known relative to *K. radiotolerans* which does not exhibit the extreme resistance phenotype. The relevance of copper metabolism to *K. radiotolerans* survival in HLW is unclear; none the less, useful applications for this trait can be envisioned. Preferential uptake and intracellular accumulation of copper could be useful as a flow-through biofilter for copper capture from radionuclide-containing waste waters or perhaps some of these bacteria may be useful resources for bioleaching of copper or other precious metals from their ores.

Through a combination of genomics-guided physiological experiments, we have sought to derive an answer to the key question of whether *K. radiotolerans* can actively metabolize organic components of radioactive HLW, or whether it is simply able to survive extremes in environmental conditions. Inferring the potential for bacterial interactions with or metabolism of inorganic pollutants is more complicated as direct and indirect pathways for electron transfer are not strictly conserved within the genome.

Kineococcus radiotolerans is an obligate aerobe but has not been thoroughly examined for specific interactions with or the ability to affect the solubility of metals or radionuclides. The genome of *K. radiotolerans* lacks strongly annotated orthologues for known degradation genes and pathways for pervasive environmental pollutants, aromatic hydrocarbons, petroleum derivatives and volatile organic compounds found in HLW. In addition to many of these components, SRS HLW also contains low molecular weight organic complexants and decontamination reagents (i.e. oxalate, glycolate, citrate and formate), which are noteworthy because they interfere with existing downstream processing of legacy HLW materials. None of these low molecular weight organic compounds are suitable growth substrates for *K. radiotolerans*; however, formate and oxalate each sustained cell viability during periods of prolonged starvation (Bagwell *et al*., 2008a). The genome of *K. radiotolerans* encodes for a single formate dehydrogenase, whose functionality has not be deduced, though in this capacity it may function to generate reducing equivalents for maintenance purposes by oxidizing formate to CO_2. A putative pathway for oxalate mineralization is unknown. It is conceivable that *K. radiotolerans*, and possibly other radioresistant microbes known to inhabit highly radioactive or mixed waste environments (e.g. Francis, 1990; Wolfram *et al*., 1996; Fredrickson *et al*., 2004), may be useful as a pretreatment for scrubbing metabolizable organic constituents from HLW, or other radioactive waste streams, to improve process efficiency and cost effectiveness of existing technologies. Expanded investigation of HLW microbial communities and explicit experimentation of promising microbial species with relevant nuclear waste streams is in large part precluded by cost; however, this research area has the potential for a major return on that investment.

Bacteria Inhabiting Plutonium Laden Soils in the Unsaturated Subsurface

Plutonium is a rare naturally occurring metal on Earth, though its manufacture for military and civil applications has produced vast quantities of fission products whose disposition and ultimate disposal remain major challenges at nuclear production, testing and waste disposal sites (Harley, 1980; Runde, 2000). More than 500 t of plutonium has accumulated from spent nuclear fuel, along with lesser quantities of other actinides. Environmental contamination is widespread due to atmospheric nuclear detonations, however levels are relatively low (<0.4 pg g^{-1}). Few sites around the globe maintain high localized concentrations of actinides, including former production facilities and current repositories (US Department of Energy complex, Waste Isolation Pilot Plant (WIPP), Carlsbad New Mexico, USA), mines, underground testing sites (e.g. Nevada Test Site and Amchitka, USA) as well as sites of accidental releases (e.g. Chernobyl reactor, Ukraine) and natural disasters (Fukushima Daiichi nuclear plant, Japan). Plutonium is a long-lived radioisotope (λ = 24,000 years); thus major environmental issues concern the long-term fate and transport in soil and groundwater (or from storage structures), as well as the rate and extent of accumulation up through the food web (e.g. Au, 1974; Whicker *et al*., 1999; Demirkanli *et al*., 2009; Thompson *et al*., 2009; Kaplan *et al*., 2010).

In surficial sedimentary environments plutonium is highly immobile, Pu(III, IV) are the predominant oxidation states. Pu(IV) adsorbs strongly on to reactive mineral and cellular surfaces, though amorphous complexation (oxides, carbonates, hydroxides), colloidal formations and oxidation can produce mobile forms of Pu. In fact, there are numerous examples of Pu being transported over large distances (Kersting *et al.*, 1999; Dai *et al.*, 2002; Novikov *et al.*, 2006; Xu *et al.*, 2008; Ketterer *et al.*, 2010). Microbial metabolism can also directly (altering oxidation state, bioaccumulation, bioprecipitation, passive mobilization on bacterial surfaces) and indirectly (chemical complexation with metabolites or cell debris, pH, Eh) affect the solubility and/or mobility of plutonium. While none of these biological interactions have been evaluated under realistic or quasi-field conditions, we speculate that a bacterial role could be quantitatively significant to the fate and transport of Pu in the environment (Francis, 2001; Neu *et al.*, 2005).

The work described herein was part of a larger effort to define and measure the biogeochemical controls on plutonium fate and transport in the unsaturated vadose zone. Briefly, an open-top lysimeter gallery was operated at the US DOE Savannah River Site (located in Aiken, South Carolina, USA) for 11 years (1981–1991) to study the effects of natural environmental conditions and soil biogeochemical processes on Pu mobility and speciation (Kaplan *et al.*, 2004, 2006). Lysimeters were backfilled with native vadose soil (kaolinite: Fe-oxide composition, 0.01% OM, 0.1 ppm TOC, low ionic strength, pH 5.5) and characterized sources of weapons grade $^{238, 239, 240}$Pu(III, IV, VI); leachate samples were collected regularly for Pu speciation. Consistent with expectations, Pu(IV) was the predominant oxidation state measured in the lysimeters, though a minor fraction of Pu(III) was also detected. Effectively 99% of the Pu mass moved only 1.25 cm from the source; however, the remaining 1% was transported 21.6 cm to the soil surface and 12 cm down through the lysimeter. The current conceptual model for downward transport suggests that Pu is most often reduced and largely immobile except for short-lived but very important bursts of active transport when Pu(IV) oxidizes to Pu(V) (Demirkanli *et al.*, 2007). Measured upward movement cannot be explained by the model and thus, is an important concern for risk assessment determinations.

Bacteria were grown directly from the Pu(IV) lysimeter soils by overlaying with molten dilute or groundwater plating medium. The majority of pure culture isolates were high percentage G+C Actinobacteria, though strains belonging to the Gammaproteobacteria, *Firmicutes and Deinococcus–Thermus* lineages were also cultivated. Overlapping 16S rRNA genotypes (Proteobacteria, Actinobacteria, *Firmicutes,* Acidobacteria) have been inventoried from SRS vadose soils, as well as from mineralogically and chemically contrasting surface soils from Los Alamos National Laboratory, a former Pu production site in Los Alamos, New Mexico, USA (carbonate-smectite-biotite-vermiculite soil, 0.3% OM, high ionic strength, pH 8.5; courtesy Cheryl Kuske). Recovered strains from many of these lineages was not entirely unexpected as they are each well adapted to low nutrient availability, frequent and prolonged wet–dry cycling, or display insensitivity to contaminant toxicity because of remarkable resistance, spore formation or inactive life stages. Consistent with Zhang *et al.* (1997) who projected that diversity and biological activity in shallow subsurface environments will be restricted most prominently by resource bioavailability, QPCR estimations yielded 1.6 $\pm$ 0.2 $\times$ 10^4 rrn gene copies g^{-1} of SRS vadose soil. This baseline description of microbial diversity in relevant soils from two former Pu production sites helps to focus our attention to those bacterial lineages most likely to encounter and thus affect Pu, or other actinides, at these locations and quite possibly, long-term storage facilities and disposal sites of related geography or storage composition (e.g. WIPP, SRS, Hanford).

We postulate that some of the features that afford survival in the vadose zone may also provide protection against radiation exposure and metal toxicity; in turn, these biological responses have the potential to affect the mobility of actinides in shallow subsurface environments. Representative bacterial strains obtained from the Pu(IV) lysimeter soils were characterized for resistance traits and pathways known to be involved in the complexation and speciation of Pu (Francis, 2001; Neu *et al.*, 2005); results are summarized in Table 19.2. Here, we will pay particular attention to stress resistance and the production of bioactive metabolites.

Radiation and desiccation resistance

The vadose zone is typified by fluctuating moisture regimes, but the mineralogy of the SRS vadose soils (sandy-clay, low organic content) exacerbates this condition; soils warm quickly and maintain low water retention capacity. Dehydration can produce DNA modification(s) and oxidative cellular damage analogous to radiation exposure (UV, IR); as such, a common suite of resistance pathways may provide cross-protection to both stressors. Causal linkages have been proposed for the observed covariance in resistance phenotypes (Mattimore and Battista, 1996; Shukla *et al.*, 2007), however these hypotheses have not been broadly tested. Three of the most UV-C resistant strains (R2A5, R2A7, R2A9) maintained viability following 8 weeks of dehydration, though two of these strains were *Bacillus* spp. and survivability may rely on spore formation. Three isolates, from genera *Arthrobacter* and *Bacillus*, were unaffected by λ radiation or prolonged periods of desiccation exceeding 9 weeks. In general, resistance to λ radiation, but not UV-C, co-varied with desiccation resistance, with the exception of two *Streptomyces*-related bacteria (GW1, GW2), which survived an 8 kGy dose of ionizing irradiation yet showed great sensitivity to desiccation. Isolates closely related by 16S rRNA gene sequence to *Streptomyces* and *Deinococcus* had no tolerance for dehydration longer than 1 week but showed resilience to high levels of λ radiation. Genera related to *Arthrobacter* and *Corynebacterium* showed limited survivability when irradiated yet were resilient to 5 weeks of desiccation.

Radiation resistance, IR and UV-C

Upon exposure to 4 kGy γ-radiation from a ^{60}Co source, only six pure culture strains exhibited 100% survival; four strains survived 8 kGy γ-radiation (GW3, GW7, R2A5, R2A7; data not shown). All strains within our collection proved more resistant to germicidal UV-C than γ-radiation. Four strains (*Actinobacteria*, *Bacillus*) withstood the highest doses of UV-C and IR. It is plausible that spore formation among the *Bacillus* spp. may have exaggerated resistance estimations in spite of our best efforts to conduct experiments with freshly prepared vegetative cultures.

Heavy metal resistance

Plutonium is both chemically and radiologically toxic; however, concentrations that inhibit bacterial growth (Ruggiero et al., 2005) greatly exceed relevant environmental concentrations. Therefore, toxicity response pathways, which may affect Pu mobility, are likely most relevant at micro-scale processes in heterogeneous environments (i.e. mineral surface-associated biogeochemistry) or anthropogenic impacted environments where actinide levels can be quite high, such as nuclear waste disposal sites (e.g. Harley, 1980; Francis, 1985). The SRS lysimeter strains exhibited resistance to a variety of heavy metals that are: (i) more toxic than most actinides, including Pu; (ii) much more pervasive environmental contaminants than actinides; and (iii) whose toxicity is exerted at much lower, and thus relevant concentrations. Briefly, growth was uninterrupted by Cu^{2+} ranging from 250 to 1 mM. All strains were exceptionally sensitive to Ni with the exception of strains related to genus *Arthrobacter*, which tolerated 750 µM doses. Growth was measured, though severely retarded, by 1 mM Al; conversely, all isolates grew uninterrupted in the presence of 2 mM Cr^{6+}. Half of the strains exhibited growth rate and/or biomass stimulation in the presence of 1–250 µM Cu^{2+}. Specific concentrations of Al and Cr^{6+} stimulated growth rates of bacteria exhibiting the high levels of desiccation tolerance. Stress resistance implies that the majority of recovered strains could survive, sustain metabolic activity if only intermittently, and perhaps propagate in metal-contaminated environments or mixed waste sites.

Metabolites and organic ligands

Microbial metabolites and ligands can complex with and affect the environmental mobility, and potentially the bioavailability, of Pu in soil (John et al., 2001; Kauri et al., 2006; Roberts et al., 2008; Thompson et al., 2009). Four isolates reacted positively on CAS-agar plates prepared for colorimetric assay of iron chelation and nearly half of the strains produced sufficient organic acids to significantly decrease the pH of their growth environment (Table 19.2). These Fe-binding siderophores were not chemically or structurally defined; however, because of the similarity in charge-to-ion radius, specific siderophores and transport proteins may not be able to distinguish between Fe(III) and Pu(IV). Uptake and accumulation of Pu(IV)-desferrioxamine-B complexes has been demonstrated for the soil bacterium *Microbacterium flavescens* (John et al., 2001) and maize plants (Demirkanli et al., 2009; Thompson et al., 2009), though desferrioxamine siderophores are less effective at promoting dissolution of Pu(IV) hydroxide (solubilized 7 µmol g^{-1} aged PuO_2) compared to other organic chelators (including organic acids) in simple solution systems. Therefore, biotic systems for the acquisition of Fe may inadvertently mobilize soluble or complexed Pu, which could have contributed to the upward mobility of Pu already in soil solution in our lysimeter system. Siderophores appear less likely to control the dissolution of tightly sorbed or aged Pu from mineral complexes.

The spent medium from organic acid-producing strains was analysed by ion chromatography, the composition of those mixtures are shown in Table 19.2. Two product peaks were consistently detected from numerous acid-producing isolates but

Table 19.2. Phylogenic identification and physiological stress resistance of bacterial isolates collected from vadose soil lysimeters at the US DOE Savannah River Site.

Identity	Closest match (%ID)	% survival-radiation (4KGy)	UV-C (mw cm^{-2})	Dehydration (weeks)	Siderophore production	Acid Production (pH 6.84)	LA	AA	FA	PrA	PyA	UVA #1	UVA #2
GW1	Streptomyces thermotolerans S001573938 (79%)	100	>160	0	Fe Starvation	3.74	1.16	BDL	0.10	0.13	0.64		
GW2	Xanthomonas campestris S000639468 (97%)	0	120	4	ND	6.36							
GW3	Kitasatospora kifunense U93322 (98%)	100	>160	0	ND	4.14	1.54	BDL	0.06	0.08	0.65		
GW4	Dermacoccus nishinomiyaensis X87757 (99%)	0	>160	5	ND	4.66	2.83	BDL	0.68	BDL	BDL		
GW6	Dermacoccus nishinomiyaensis AM992178 (95%)	0	80	2	ND	4.83	BDL	0.47	BDL	0.08	0.38	+	
GW7	Dienococcus grandis AY424359 (85%)	100	>160	1	TSA	6.4							
GW9	Arthrobacter pascens S001248364 (68%)	0	30	3	ND	5.56	0.98	0.46	0.13	0.17	1.44	+	+
GW11	Arthrobacter sp. S001020196 (83%)	0	>160	1	TSA	4.84	BDL	BDL	0.83	0.05	0.06	+	+
GW12	Terrabacter sp. S000717117 (94%)	0	50	0	Fe Starvation	4.89	1.75	0.58	BDL	BDL	1.26	+	+
R2A2	Corynebacterium cyclohexanicum AB210282 (98%)	0	80	1	ND	5.05	BDL	1.18	0.41	0.08	BDL		+
R2A4	Amycolatopsis echigonensis AB248535 (99%)	0	50	2	Fe Starvation	7.66							
R2A5	Uncultured Bacillus sp. S001047854 (97%)	100	>160	8	TSA	4.44	BDL	0.84	0.36	0.02	0.01		+
R2A6	Micrococcus sp. S001020197 (94%)	0.01	50	5	TSA	5.42	BDL	2.58	BDL	0.03	1.43		+
R2A7	Bacillus cereus S001046766 (100%)	100	>160	>9	TSA	4.73	BDL	0.89	0.40	0.02	BDL		+
R2A8	Micrococcus luteus S001577226 (95%)	0	40	0	ND	7.87							
R2A9	Arthrobacter sp. S000368538 (98%)	100	75	>9	TSA	6.88							
R2A10	Anaplasma phagocytophilum S001589792 (98%)	0.01	50	8	ND	5.4	BDL	BDL	BDL	0.19	BDL	+	
R2A11	Arthrobacter sp. S001020203 (90%)	0	50	8	ND	7.43							
R2A12	Agromyces sp. S000489357 (88%)	0	75	8	TSA	4.86	2.48	1.25	BDL	0.07	0.02		+
R2A13	Arthrobacter sp. S001020203 (89%)	0	50	8	ND	7.06							
D. radiodurans		100	100	8									
E. coli K12		0	30	1									

Strain designations indicate primary enrichment on groundwater medium (GW) or solidified R2A; siderophore production assays were prepared by transferring cultures on to CAS agar plates following Fe(III) starvation or directly from TSA plating medium. ND indicates that no halos were observed. Acid production shows the end point pH of weakly buffered grown medium after 5 days of incubation. Conformed organic acids included lactic acid (LA), acetic acid (AA), formic acid (FA), propionic acid (PyA) and two unidentified compounds; BDL indicates peaks were below detection.

could not be confidently identified. We have been able to eliminate anions (Cl^-, SO_4^{2-}, NO_3^-, PO_4^-), amino acids, metabolic intermediates (glucose-6-phosphate, α-ketoglutarate) and the following organic acids: citric acid, isocitric acid, oxalic acid, succinic acid, fumaric acid, butyric acid, gluconic acid, glycolic acid, malic acid, caproic acid, valeric acid and carbonic acid (and bicarbonate). All of the acid-producing strains demonstrated the ability to withstand medium acidification during prolonged incubation, which could be ecologically relevant for vadose soils or a potentially effective strategy for mitigating cellular toxicity by increasing metal solubility. In two-staged experiments we have confirmed that bacterial growth and excretion of organic acids significantly contributes to dissolution of aged Pu(IV) soils, thus increasing aqueous phase concentration (0.6–0.9%), and that organic acid mixtures strongly complex soluble Pu(V) (98% mass balance), 2–50% of the total Pu complexed with each of the individual compounds. It is interesting to note that approximately 1% Pu solubilization was achieved here with active bacterial cultures in 5 days; the same percentage release was measured in open-top lysimeters after 11 years. Likewise, soil humic acids have been shown to be effective at remobilizing Pu from soils; as much as 1.2% Pu can be released by fulvic acids (Santschi et al., 2002). Viewed from this perspective, these results are not insignificant. The quantity of labile carbon used in these experiments was excessive (50 mg C l^-) compared to typical levels present in SRS vadose soils, but are not unrealistic for certain waste repositories where carbon bioavailability supports high levels of bacterial metabolism (Gillow et al., 2000; Francis, 2001).

Because of the long half-life associated with most isotopes of Pu, slow but persistent biogeochemical processes are critically important to the long term fate and transport of this actinide. The vadose zone is almost always of direct concern, because this is where most accidental releases occur, tremendous volumes of legacy materials are buried in shallow storage configurations, Pu can have a long residence time in the vadose and finally, this zone marks the transition between sensitive surface receptors and subsurface aquifers. These results are preliminary but are intended to help provide a clearer picture of how microbes that are able to persist in low nutrient, dry sedimentary conditions could potentially affect the long-term fate and stability of plutonium. These pathways need better definition and quantitative measurements are essential for accurate reactive transport models, which are relied upon for environmental management, remediation and long term stewardship of low-level burial sites.

Conclusion

The Cold War legacy persists in tanks, trenches, casks and the environment and future generations must continue to confront the challenges of safe, permanent storage and disposal. Today, nuclear power is the safest, cleanest and most cost effective source of energy, and increased production will likely become necessary for future generations throughout the developing world, but the reality is that nuclear waste will accumulate and future nuclear disasters and accidental releases are inevitable. At the time of this writing the Fukushima Daiichi nuclear plant disaster is unfolding in Japan following an 8.9 magnitude earthquake and damaging tsunami. We can look to the past for examples of the lasting effects of nuclear production. While it can be inherently difficult to quantify precisely the long term impacts of nuclear exposure on humans and other mammals, genetic and molecular adaptations caused directly by the Chernobyl disaster and chronic exposure to radioisotopes are evident (e.g. Kovalchuk *et al*., 2004; Vornam *et al*., 2004; Zhdanova *et al*., 2004). Heavily degraded lands and those impacted by long-lived pollutants (e.g. the exclusion zones around Chernobyl and very likely the Fukushima reactor, Japan) will be of little to no value for future generations. Stabilization, remediation and long term stewardship of these lands will not be a trivial task; inorganic pollutants are only affected, and thus controlled, by complex biogeochemistry, which dictates solubility and reactivity, and unlike organic pollutants, safe removal by decomposition pathways is not an option. Microbes do have an important role to play but these processes are difficult to parameterize; we lack the fundamental understanding of the complex and often slow reactions that can occur with many inorganics over the life cycle of these materials. Microorganisms will interact with and microbial catalysed processes will affect nuclear waste. Only through continued investigation will we be able to decipher the magnitude and outcome of these interactions, but through this effort we will also establish our ability to harness and control these activities in order to achieve a desirable end state.

References

Adamson, D.T., McDade, J.M. and Hughes, J.B. (2003) Inoculation of a DNAPL source zone to initiate reductive dechlorination of PCE. *Environmental Science and Technology* 37, 2525–2533.

Anderson R.T., Vrionis, H.A., Ortiz-Bernad, I., Resch, C.T., Long, P.E., Dayvault, R., Karp, K., Marutzky, S., Metzler, D.R., Peacock, A., White, D.C., Lowe, M., and Lovley, D.R. (2003) Stimulating the *in situ* activity of *Geobacter* species to remove uranium from the groundwater of a uranium-contaminated aquifer. *Applied and Environmental Microbiology* 69, 5884–5891.

Au, F.H.F. (1974) The role of microorganisms in the movement of plutonium. In: *The Dynamics of Plutonium in the Desert Environments*. Nevada Applied Ecology Group Progress Report, NVO-142.

Backman, A. and Jansson, J.K. (2004) Degradation of 4-chlorophenol at low temperature and during extreme temperature fluctuations by *Arthrobacter chlorophenolicus* A6. *Microbial Ecology* 48, 246–253.

Bagwell, C.E., Bhat, S., Hawkins, G.M., Smith, B.W., Biswas, T., Hoover, T.R., Saunders, E., Han, C.S., Tsodikov, O.V. and Shimkets, L.J. (2008a) Survival in nuclear waste, extreme resistance, and potential applications gleaned from the genome sequence of *Kineococcus radiotolerans* SRS30216. *PLoS ONE* 3, e3878

Bagwell, C.E., Milliken, C.E., Ghoshroy, S. and Blom, D.A. (2008b) Intracellular copper accumulation enhances the growth of *Kineococcus radiotolerans* during chronic irradiation. *Applied and Environmental Microbiology* 74, 1376–1384.

Bagwell, C.E., Hixson, K.K., Milliken, C.E., Lopez-Ferrer, D. and Weitz, K.K. (2010) Proteomic and physiological responses of *Kineococcus radiotolerans* to copper. *PLoS ONE* 5, e12427.

Battista, J.R. (1997) Against all odds: the survival strategies of *Deinococcus radiodurans*. *Annual Reviews of Microbiology* 51, 203–224.

Bond, D.R. and Lovley, D.R. (2003) Electricity production by *Geobacter sulfurreducens* attached to electrodes. *Applied and Environmental Microbiology* 69, 1548–1555.

Brim, H., McFarlan, S.C., Fredrickson, J.K., Minton, K.W., Zhai, M., Wackett, L.P. and Daly, M.J. (2000) Engineering *Deinococcus radiodurans* for metal remediation in radioactive mixed waste environments. *Nature Biotechnology* 18, 85–90.

Cox, M.M. and Battista, J.R. (2005) *Deinococcus radiodurans* – the consummate survivor. *Nature Reviews Microbiology* 3, 882–892.

Dai, M., Kelley, J.M. and Buesseler, K.O. (2002) Sources and migration of plutonium in groundwater at the Savannah River Site. *Environmental Science and Technology* 36, 3690–3699.

Daly, M.J. (2009) A new perspective on radiation resistance based on *Deinococcus radiodurans*. *Nature Reviews Microbiology* 7, 237–245.

Daly, M.J., Gaidamakova, E.K., Matrosova, V.Y., Vasilenko, A., Zhai, M., Venkateswaran, A., Hess, M., Ornelchenko, M.V., Kostandarithes, H.M., Makarova, K.S., Wackett, L.P., Fredrickson, J.K. and Ghosal, D. (2004) Accumulation of Mn(II) in *Deinococcus radiodurans* facilitates gamma-radiation resistance. *Science* 306, 1025–1028.

Daly, M.J., Gaidamakova, E.K., Matrosova, V.Y., Vasilenko, A., Zhai, M., Leapman, R.D., Lai, B., Ravel, B., Li, K., Kemner, M. and Fredrickson, J.K. (2007) Protein oxidation implicated as the primary determinant of bacterial radioresistance. *PLoS Biology* 5, 1–11.

Demirkanli, D.I., Molz, F.J., Kaplan, D.I., Fjeld, R.A. and Serkiz, S.M. (2007) Modeling long-term plutonium transport in the Savannah River Site vadose zone. *Vadose Zone Journal* 6, 344–353.

Demirkanli, D.I., Molz, F.J., Kaplan, D.I. and Fjeld, R.A. (2009) Soil-Root Interactions Controlling Upward Plutonium Transport in Variably Saturated Soils. *Vadose Zone Journal* 9, 574–585.

Diercks, M., Sand, W. and Bock, E. (1991) Microbial corrosion of concrete. *Cellular and Molecular Life Sciences* 47, 514–516.

Ernst, W.G. (1983) The early Earth and the Archean rock record. In: Schopf, J.W. (ed.) *Earth's Earliest Biosphere. Its Origin and Evolution*. Princeton University Press, Princeton, New Jersey, pp. 41–52.

Fioravanti, M. and Makhijani, A. (1997) *Containing the Cold War mess: Restructuring the environmental management of the US nuclear weapons complex*. Institute for Energy and Environmental Research. Technical Report, pp. 1–323.

Francis, A.J. (1985) Low-level radioactive wastes in subsurface soils. In: Tate, R.L. and Klein, D.L. (eds) *Soil Reclamation Processes: Microbiological Analyses and Applications*. Marcel Dekker, New York, pp. 279–331.

Francis, A.J. (1990) Microbial dissolution and stabilization of toxic metals and radionuclides in mixed waste. *Experientia* 46, 840–851.

Francis, A.J. (2001) Microbial transformations of plutonium and implications for its mobility. *Radioactivity in the Environment* 1, 201–219.

Francis, A.J., Dobbs, S. and Nine, B.J. (1980a) Microbial activity of trench leachates from shallow-land, low-level radioactive waste disposal sites. *Applied and Environmental Microbiology* 40, 108-113.

Francis, A.J., Iden, C.R., Nine, B.J. and Chang, C.K. (1980b) Characterization of organics in leachates from the low-level radioactive waste disposal sites. *Nuclear Technology* 50, 158–163.

Francis, A.J., Gillow, J.B., Dodge, C.J., Dunn, M., Mantione, K., Strietelmeier, B.A., Pansoy-Hjelvik, M.E. and Papenguth, H.W. (1998) Role of bacteria as biocolloids in the transport of actinides from a deep underground radioactive waste repository. *Radiochimica Acta* 82, 347–354.

Francis, A.J., Dodge, C.J., Gillow, J.B. and Papenguth, H.W. (2000) Biotransformation of uranium compounds in high ionic strength brine by a halophile bacterium under denitrifying conditions. *Environmental Science and Technology* 34, 2311–2317.

Fredrickson, J.K., Kostandarithes, H.M., Li, S.W., Plymale, A.E. and Daly, M.J. (2000) Reduction of Fe(III), Cr(VI), U(VI), and Tc(VII) by *Deinococcus radiodurans* R1. *Applied and Environmental Microbiology* 66, 2006–2011.

Fredrickson, J.K., Zachara, J.M., Balkwill, D.L., Kennedy, D., Li, S.M., Kostandarithes, H.M., Daly, M.J., Romine, M.F. and Brockman, F.J. (2004) Geomicrobiology of high-level nuclear waste-contaminated vadose sediments at the Hanford site, Washington state. *Applied and Environmental Microbiology* 70, 4230–4241.

Gillow, J.B., Dunn, M., Francis, A.J., Lucero, D.A. and Papenguth, H.W. (2000) The potential of subterranean microbes in facilitating actinide migration at the Grimsel Test Site and Waste Isolation Pilot Plant. *Radiochimica Acta* 88, 769–774.

Granger, A.C., Gaidamakova, E.K., Matrosova, V.Y., Daly, M.J. and Setlow, P. (2011) Effects of Mn and Fe levels on *Bacillus subtilis* spore resistance and effects of Mn^{2+}, other divalent cations, orthophosphate, and dipicolinic acid on protein resistance to ionizing radiation. *Applied and Environmental Microbiology* 77, 32–40.

Harley, J.H. (1980) Plutonium in the environment – a review. *Journal of Radiation Research* 21, 83–104.

Horn, J.M. and Meike, A. (1995) Microbial activity at Yucca Mountain. OSTI ID 177391, Legacy ID: DE96003672.

Imlay, J.A. (2003) Pathways of oxidative damage. *Annual Reviews of Microbiology* 57, 395–418.

John, S.G., Ruggiero, C.E., Hersman, L.E., Tung, C.-S. and Neu, M. (2001) Siderophore mediated plutonium accumulation by *Microbacterium flavescens* (JG-9). *Environmental Science and Technology* 35, 2942–2948.

Jonkers, H. (2008) Self healing concrete: A biological approach. *Self Healing Materials* 100, 195–204.

Kaplan, D.I., Powell, B.A., Demirkanli, D.I., Fjeld, R.A., Molz, F.J., Serkiz, S.M. and Coates, J.T. (2004) Influence of oxidation states on plutonium mobility during long-term transport through an unsaturated subsurface environment. *Environmental Science and Technology* 38, 5053–5058.

Kaplan, D.I., Demirkanli, D.I., Gumapas, L., Powell, B.A., Fjeld, R.A., Molz, F.J. and Serkiz, S.M. (2006) Eleven-year field study of Pu migration from Pu III, IV, and VI sources. *Environmental Science and Technology* 40, 443–448.

Kaplan, D.I., Demirkanli, D.I., Molz, F.J., Beals, D.M., Cadieux, J.R. and Halverson, J.E. (2010) Upward movement of plutonium to surface sediments during an 11-year field study. *Journal of Environmental Radioactivity* 101, 338–344.

Kashefi, K., Tor, J.M., Holmes, D.E., Gaw Van Praagh, C.V., Reysenbach, A.-L. and Lovley, D.R. (2002) *Geoglobus ahangari* gen. nov., sp. nov., a novel hyperthermophilic archaeon capable of oxidizing organic acids and growing autotrophically on hydrogen with Fe(III) serving as the sole electron acceptor. *International Journal of Systematic and Evolutionary Microbiology* 52, 719–728.

Kauri, T., Kauri, T., Santry, D.C., Kudo, A. and Kushner, D.J. (2006) Uptake and exclusion of plutonium by bacteria isolated from soil near Nagasaki Japan. *Environmental Toxicology and Water Quality* 6, 109–112.

Kersting, A.B., Efurd, D.W., Finnegan, D.L., Rokop, D.J., Smith, D.K. and Thompson, J.L. (1999) Migration of plutonium in ground water at the Nevada Test Site. *Nature* 397, 56–59.

Ketterer, M.E., Gulin, S.B., MacLellan, G.D. and Hartsock, W.J. (2010) Fluvial transport of Chernobyl plutonium (Pu) to the Black Sea: Evidence from 240Pu/239Pu atom ratios in Danube Delta sediments. *The Open Chemical and Biomedical Methods Journal* 3, 197–201.

Kovalchuk, I., Abramov, V., Pogribny, I. and Kovalchuk, O. (2004) Molecular aspects of plant adaptation to life in the Chernobyl zone. *Plant Physiology* 135, 357–363.

Lendvay, J.M., Löffler, F.E., Dollhopf, M., Aiello, M.R., Daniels, G., Fathepure, B.Z., Gebhard, M., Heine, R., Helton, R., Shi, J., Krajmalnik-Brown, R., Major, Jr, C.L., Barcelona, M.J., Petrovskis, E., Hickey, R., Tiedje, J.M. and Adriaens, P. (2003) Bioreactive Barriers: A Comparison of Bioaugmentation and Biostimulation for Chlorinated Solvent Remediation. *Environmental Science & Technology* 37, 1422–1431.

Letelier, M.E., Lepe, A.M., Faundez, M., Salazar, J., Marin, R., Aracena, P. and Speisky, H. (2005) Possible mechanisms underlying copper-induced damage in biological membranes leading to cellular toxicity. *Chemico-Biological Interactions* 151, 71–82.

Levin-Zaidman, S., Englander, J., Shimoni, E., Sharma, A.K., Minton, K.W. and Minsky, A. (2003) Ringlike structures of the *Deinococcus radiodurans* genome: a key to radioresistance? *Science* 299, 254–256.

Liu, C., Gorby, Y.A., Zachara, J.M., Fredrickson, J.K. and Brown, C.F. (2002) Reduction kinetics of Fe(III), Co(III), U(VI), Cr(VI), and Tc(VII) in cultures of dissimilatory metal-reducing bacteria. *Biotechnology and Bioengineering* 80, 637–649.

Lloyd, J.R. and Macaskie, L.E. (2000) Bioremediation of radionuclide-containing wastewaters. In: Lovley, D.R. (ed.) *Environmental Microbe-Metal Interactions*. ASM Press, Washington, DC, pp. 277–328.

Lovitt, R.W., Longin, R. and Zeikus, J.G. (1984) Ethanol production by thermophilic bacteria: physiological comparisons of solvent effects on parent and alcohol-tolerant strains of *Clostridium thermohydrosulfuricum*. *Applied Environmental Microbiology* 48, 171–177.

Lovley, D.R. and Nevin, K.P. (2011) A shift in the current: New applications and concepts for microbe-electrode electron exchange. *Current Opinion in Biotechnology* 22, 1–8.

Major, D.W., McMaster, M.L., Cox, E.E., Edwards, E.A., Dworatzek, S.M., Hendrickson, E.R., Starr, M.G., Payne, J.A. and Buonamici, L.W. (2002) Field demonstration of successful bioaugmentation to achieve dechlorination of tetrachloroethene to ethene. *Environmental Science and Technology* 36, 5106–5116.

Makarova, K.S., Aravind, L., Wolf, Y.I., Tatusov, R.L., Minton, K.W., Koonin, E.V. and Daly, M.J. (2001) Genome of the extremely radiation-resistant bacterium *Deinococcus radiodurans* viewed from the perspective of comparative genomics. *Microbiology and Molecular Biology Reviews* 65, 44–79.

Margesin, R. and Schinner, F. (1996) Bacterial heavy metal-tolerance – extreme resistance to nickel in *Arthrobacter* spp. strains. *Journal of Basic Microbiology* 36, 269–282.

Marnett, L.J. (2000) Oxyradicals and DNA damage. *Carcinogenesis* 21, 361–370.

Martinez, R.J., Beazley, M.J., Taillefert, M., Arakaki, A.K., Skolnick, J. and Sobecky, P.A. (2007) Aerobic uranium (VI) bioprecipitation by metal-resistant bacteria isolated from radionuclide- and metal-contaminated subsurface soils. *Environmental Microbiology* 9, 3122–3133.

Mattimore, V. and Battista, J.R. (1996) Radioresistance of *Deinococcus radiodurans*: Functions necessary to survive ionizing radiation are also necessary to survive prolonged desiccation. *Journal of Bacteriology* 178, 633–637.

Neu, M.P., Icopini, G.A. and Boukhalfa, H. (2005) Plutonium speciation affected by environmental bacteria. *Radiochimica Acta* 93, 705-714.

Novikov, A.P., Kalmykov, S.N., Utsunomiya, S., Ewing, R.C., Horreard, F., Merkulov, A., Clark, S.B., Tkachev, V.V. and Myasoedov, B.F. (2006) Colloid transport of plutonium in the far-field of the Mayak Production Association, Russia. *Science* 314, 638–641.

Padmanabhan, P., Padmanabhan, S., DeRito, C., Gray, A., Gannon, D., Snape, J.R., Tsai, C.S., Park, W., Jeon, C. and Madsen, E.L. (2003) Respiration of ^{13}C-labeled substrates added to soil in the field and subsequently 16S rRNA gene analysis of ^{13}C-labeled soil DNA. *Applied and Environmental Microbiology* 69, 1614–1622.

Pedersen, K. (1999) Subterranean microorganisms and radioactive waste disposal in Sweden. *Engineering Geology* 52, 163-176.

Phillips, R.W., Wiegel, J., Berry, C.J., Fliermans, C., Peacock, A.D., White, D.C. and Shimkets, L.J. (2002) *Kineococcus radiotolerans* sp. nov., a radiation-resistant, Gram-positive bacterium. *International Journal of Systematic and Evolutionary Microbiology* 52, 933–938.

Proctor, P.H. and Reynolds, E.S. (1984) Free radicals and disease in man. *Physiological Chemistry, Physics, and Medical NMR* 16, 175–195.

Rae, T.D., Schmidt, P.J., Pufahl, R.A., Culotta, V.C. and O'Halloran, T.V. (1999) Undetectable intracellular free copper: the requirement of a copper chaperone for superoxide dismutase. *Science* 284, 805–808.

Riley, R.G. and Zachara, J.M. (1992) *Chemical contaminants on DOE lands and selection of contaminant mixtures for subsurface science research*. Washington DC: Office of Energy Research, Subsurface Science Program, US Department of Energy, pp. 1–71.

Roberts, K.A., Santschi, P.H. and Honeyman, B.D. (2008) Pu(V) reduction and enhancement of particle-water portioning by exopolymeric substances. *Radiochimica Acata* 96, 739–745.

Roh, Y., Chon, C.-M. and Moon, J.-W. (2007) Metal reduction and biomineralization by an alkaliphilic metal-reducing bacterium, *Alkaliphilus metalliredigens* (QYMF). *Geosciences Journal* 11, 415–423.

Ruggiero, C.E., Boukhalfa, H., Forsythe, J.H., Lack, J.G., Hersman, L.E. and Neu, M.P. (2005) Actinide and metal toxicity to prospective bioremediation bacteria. *Environmental Microbiology* 7, 88–97.

Runde, W. (2000) The chemical interactions of actinides in the environment. *Los Alamos Science* 26, 392–411.

Santschi, P.H., Roberts, K.A. and Guo, L. (2002) Organic nature of colloidal actinides transported in surface water environments. *Environmental Science and Technology* 36, 3711–3719.

Sardessai, Y. and Bhosle, S. (2002) Tolerance of bacteria to organic solvents. *Research in Microbiology* 153, 263–268.

Shukla, M., Chaturvedi, R., Tamhane, D., Vyas, P., Archana, G., Apte, S., Bandekar, J. and Desai, A. (2007) Multiple-stress tolerance of ionizing radiation-resistant bacterial isolates obtained from various habitats: correlation between stresses. *Current Microbiology* 54, 142–148.

Thompson, S.W., Molz, F.J., Fjeld, R.A. and Kaplan, D.I. (2009) Plutonium velocity in *Zea mays* (corn) and implications for plant uptake of Pu in the root zone. *Journal of Radioanalytical and Nuclear Chemistry* 282, 439–442.

Udupa, K.S., O'Cain, P.A., Mattimore, V. and Battista, J.R (1994) Novel ionizing radiation-sensitive mutants of *Deinococcus radiodurans*. *Journal of Bacteriology* 176, 7439–7446.

Vornam, B., Kuchma, O., Kuchma, N., Arkhipov, A. and Finkeldey, R. (2004) SSR markers as tools to reveal mutation events in Scots pine (*Pinus sylvestris* L.) from Chernobyl. *European Journal of Forest Research* 123, 245–248.

Wang, W., Zhang, Z., Tang, Q., Mao, J., Wei D., Huang, Y., Liu, Z., Shi, Y. and Goodfellow, M. (2007) *Lechevalieria xinjiangensis* sp. nov., a novel actinomycete isolated from radiation-polluted soil in China. *International Journal of Systematic and Evolutionary Microbiology* 57, 2819–2822.

Whicker, F.W., Hinton, T.G., Orlandini, K.A. and Clark, S.B. (1999) Uptake of natural and anthropogenic actinides in vegetable crops grown on s contaminated lake bed. *Journal of Environmental Radioactivity* 45(1), 1–12.

Whitman, W.B., Coleman, D.C. and Wiebe, W.J. (1998) Prokaryotes: The unseen majority. *Proceedings of the National Academy of Science USA* 95, 6578–6583.

Wicks, G.G. and Bickford, D.F. (1989) Doing something about high-level nuclear waste. *Technology Review (USA)* 92, 50–59.

Wolfram, J.H., Rogers, R.D. and Gazso, L.G. (1996) *Microbial Degradation Processes in Radioactive Waste Repository and in Nuclear Fuel Storage Areas*. Kluwer Academic Publishers, Dordrecht, the Netherlands.

Xu, C., Santschi, P.H., Zhong, J.Y., Hatcher, P.G., Francis, A.J., Dodge, C.J., Roberts, K.A., Hung, C.-C. and Honeyman, B.D. (2008) Colloid cutin-like substances cross-linked to siderophore decomposition products mobilizing plutonium from contaminated soils. *Environmental Science and Technology* 42, 8211–8217.

Yokota, A., Tamura, T., Nishii, T. and Hasegawa, T. (1993) *Kineococcus aurantiacus* gen. nov., sp. nov., a new aerobic, Gram-positive, motile coccus with meso-diaminopimelic acid and arabinogalactan in the cell wall. *International Journal of Systematic and Evolutionary Microbiology* 43, 52–57.

Zhang, C., Lehman, R.M., Pfiffner, S.M., Scarborough, S.P., Palumbo, A.V., Phelps, T.J., Beauchamp, J.J. and Colwell, F.S. (1997) Spatial and temporal variations of microbial properties at different scales in shallow subsurface sediments. *Applied Biochemistry and Biotechnology* 63–65, 797–808.

Zhdanova, N.N., Tugay, T., Dighton, J., Zheltonozhsky, V. and McDermott, P. (2004) Ionizing radiation attracts soil fungi. *Mycological Research* 108, 1089–1096.

Zimmerman, J.M. and Battista, J.R. (2005) A ring-like nucleoid is not necessary for radioresistance in the Deinococcaceae. *BMC Microbiology* 5, 17.

INDEX

16S rRNA	384, 387, 404-405
2-furoic acid	166
3, 4 dichlorophenol (DCP)	326
ABE process	160
Acetobacterium woodii	31
acetogens	29-31
acidogenesis	206
Acidophiles	399
Acidothiobacillus ferrioxidans	27
Acintobacteria	20
acrA	163
AcrAB	166, 170
AcrAB-TolC	170
Actinobacillus succinogenes	51, 66, 69, 91, 178
Actinobacteria	5, 402, 404, 406
activated carbon nanofibre (ACNF)	6
adaptive mutations (ADAM)	167
Adsorption	28, 54, 91, 126, 127, 141, 145, 213, 253, 324, 343, 345, 359, 368-369, 387
Advanced Water Management Centre	108
aeration (harrowing)	291
Aerobacter hydrophila	8
Aeromonas hydrophilia	24
Agency for Toxic Substances and Disease Registry (ATSDR, 2010)	304
Agrivida	124
Alcanivorax borkumensis	166
Algae Venture Systems (AVS)	214
Algal Cultivation Systems	250
Alkaliphiles	399
Alphaproteobacteria,	20
alternative energy sources	3
Ameirus nebulosus	305
Ames Laboratory	143-144
amperometric enzyme electrodes	48
amphitrophic	212
Anabaena cylindrica	110
Anaerobic digestion	5, 47, 77, 205-208, 212, 213, 215, 217, 223-229, 249, 262
anaerobic sludge blanket (UASB)	206
anion exchange membrane (AEM)	53
anode biofilm	5
anolyte	63, 65, 190

anthraquinone-2, 6-disulfonate (AQDS)	27, 178
Aquatic Species Program	212, 263, 267
ASKA	170
asparagine	168-169
autoflocculation	266
avermectin	161
Azospirillum	213
Bacillus	
B. polymyxa	8
B. sphaericus	12
B. thermooleovorans	309
Bacteroidetes	5, 75
BALINIT	147
Benthic Unattended Generator (BUG)	194
benzene, toluene, ethylbenzene and xylenes (BTEX)	272
benzylviologen	2, 23, 49
Bermuda grass straw	86, 89
Betaproteobacteria	20, 178
Bioaccumulation	305, 344-345, 404
Bioaugmentation	286-287, 290, 307, 315, 370, 371, 380, 385, 398
biocathodes	5, 27-28, 50- 52, 74, 192, 194, 197
Biocrystallization	345
biodegradation	4, 84, 107, 272-275, 278, 280- 284, 288- 292, 307-309, 314- 315, 324-335, 361-362
Biodegradation index	324-326
Biodiesel	7-9, 124, 132-133, 142, 148, 166, 207, 209-210, 212-215, 244-250, 255-256, 261-268
bioelectricity	105-106
bioelectrochemical system (BES)	107
bioenergy	4-5, 50, 62-63, 71, 98, 105-107, 113, 120-125, 135, 140-143, 146, 149, 205, 206, 228, 243, 245, 249
Bioethanol	7-8, 129, 215, 244, 248-250, 256
Biofilm	5-6, 20, 24, 26, 28, 32, 52, 66, 70, 74, 77, 90-92, 94, 97, 107-108, 177, 181-183, 185-186, 191, 195-197, 345, 362, 364-368, 372-373, 381, 383, 386
Biofractionation	371
Biofuels	3, 7-9, 13, 27, 29, 38, 50, 98, 105, 112, 120-121, 123-125, 135, 141-143, 148, 162, 165-166, 195, 215, 243-245, 248-250, 261-262
Algal based	252-253, 255- 256
First generation	255
second generation	256
third generation	256

bio-hydrogen	84
Biological fuel cells (BioFCs)	43
Biological U(VI) Reduction	346
Biomineralization	345
Biophotolysis	224
direct	225
indirect	225
Biopiling	283, 288, 289
bioprecipitation	12, 344-345, 404
Bioprospecting	251, 254
Bioreactors	7, 228, 283, 289, 310-311
photobioreactors	9, 111, 175, 206, 211, 226, 238, 244, 250, 253, 254, 265-268, 324, 326, 359
Bioremediation	9-10, 12, 13, 18, 20, 22, 27-28, 47, 63, 77, 106, 176, 274, 280-283, 286, 289, 290, 292, 304, 307, 309, 311, 311, 315, 324, 335, 337
in situ	357, 369-371, 380-381, 383-385, 387,
ex situ	388-389, 390, 396-400
aerobic	11, 283, 284, 285, 292, 384, 398
anaerobic	283, 284, 286, 292, 308, 314
Biosensors	47-48, 55, 75-76, 106, 388-390
Biosorption	344-345, 358-359, 361, 382-383
Biosortive processes	359
Biostimulation	284-285, 287, 291, 307, 308, 315, 383-387, 390, 397
bipolar membrane (BPM)	53, 192
Bjerkandera	311
Botryococcus braunii	210, 248
Bradyrhizobium	5
Brevibacillus sp.	20
Burkholderia	166, 309-310
Burkholderia ambifaria AMMD	166
Caldicellulosiruptor saccharolyticus	229
Candida melibiosica 2491	69
carbon nanofibre (CNF)	6
carbon nanotube/polyaniline composite	52
carbon nanotubes (CNTs)	6-7, 54, 130, 137
carbon-based electrodes	6
Carnot cycle	193
Carrier-binding	127
catechol	278-279, 310-311, 332, 334
catechol 2, 3-dioxygenase (C23DO)	332
Cathodes	28, 52-54, 71, 73, 89, 192, 196
aerobic	50
bio	5, 27, 51-52, 74, 192, 194, 197
abiotic	74
two-chambered	6

open air carbon	74, 188
cation exchange membrane (CEM)	53, 188
Catlin	124
Cellulosic	
biofuels	3
biomass	23, 125- 126, 128
cellulosic biomass	23, 125-126, 128
Central Food Technological Research Institute, Mysore	264
centrifugal sedimentation	214
Chaim Weizmann	160
chemical free energy	62
chemisorption	144-145
chemotaxis	184
Chesapeake Bay	305
Chlamydomonas	
C. perigranulata	249
C. reinhardtii	210, 248-249
chloramphenicol	163, 167
Chlorella protothecoides	207, 210, 212, 263
sorokini-ana	309
Chlorobaculum	5
Chloroflex	5
chromium nitride	147
Chrysene	306, 311
circular dichroism	131
Citrobacter sp.	11-12, 345
Closed photo-bioreactors	250, 253-254, 265
Clostridia	20, 213, 228, 230-231
ljungdahlii	31
Clostridiaceae	110, 224
Clostridium	5, 24, 84, 110, 227, 229-232, 235 388
C. butyricum	69
C. acetobutylicum	8, 134, 160
C. ljungdahlii	31
C. kluyveri	31
C. pasteurianum	227
C. pascui	314
C. sphenoides	348
C. thermohydrosulfuricum	399
C. thermocellum	
co-polymerization	145
Cobalt tetra methoxy phenyl porphyrin (CoTMPP)	73
COD (chemical oxygen demand)	182
Coelastrum	210, 214
Cold War	395-397, 411
Composting	224, 283, 290, 307-308

compression-ignition engines	8
Conventional activated sludge (CAS)	47
Copiotrophic cultures	182
Coptotermes formosanus	305
Cornyebacterium	160
critical control points (CCP).	324
Crypthecodinium	210, 264
C. cohnii	210
cyanobacteria	5, 70, 109, 110-111, 214-215, 226, 246, 309
Cyclic voltammetry	5, 66
Cylindrotheca	264
Cystoseria	382-383
Cytochromes	23-26, 31-32, 45, 93, 179-181, 183, 192, 347, 349, 350, 382, 389
cytoplasmic nanoparticles	135
dark fermentation	109, 224, 226, 229, 231
Deepwater Horizon oil spill	2, 397
Defence Research and Development Organization (DRDO)	264
deformations, erosions, lesions and tumours (DELTs)	305
Deinococcus	
radiodurans	11, 167, 348, 399-400
geothermalis	11
denitrification	28, 74, 207
DENSE	111-112
Desulfobulbus propionicus	24, 178
Desulfotomaculumgeothermicum	231
Desulfovibrio	11, 12, 348, 351, 388, 389
D. desulfuricans	66, 69, 178, 347-348, 355, 389
D. vulgaris	11, 348, 355, 382, 389
diphenols	51
direct electron transfer (DET)	46, 179
dissimilatory metal-reducing bacteria (DMRB)	350
Dunaliella salina	210, 214, 248, 264
	72
EcoBot-II	194
EcoCyc v12.5	164
electricigens	17, 27, 77
Electricity	2-8, 13, 17-18, 21-22, 28, 32, 45-26, 50-51, 62-63, 66, 68, 70-71, 73, 75- 77, 84-89, 93, 97-98, 105-106, 108, 120, 123, 135, 136, 175-178, 181-182, 185, 193-194, 197, 223, 225, 244, 238, 338, 380

Electrochemically active bacteria (EAB)	179
electrocytes	62
electron shuttles	17-18, 22-23, 27-29, 32, 66, 69, 177
electron work function (EWF)	184
Electrophs	27
Electrospinning	6, 140
elephant grass (*Pennisetum purpureum*)	288
Encapsulation	53, 127, 340-341
Enterobacter aerogenes	109, 231
Enterococcus durans	227
Entrapment	46, 127, 213, 280,
environmental clean-up	9-10, 288
Epicoccum nigrum	216,
Erwinia dissolven	178
Escherichia coli	8, 50, 106, 109, 160-161, 164, 178, 180, 232, 356, 400
Ethanol-type fermentation	224
Euglena gracilis	208, 210
exoelectrogenic bacteria	177
exoelectrogens	65, 195
extracellular electron transport	5, 24
Exxon Valdez oil spill	284
fabricated Pyridylthio-modified multiwalled carbon nanotubes (pythio-MWNTs)	130
FeCrNiAl coatings	147
Ferribacterium	5
Ferruginibacter	5
field portable sensor (FPS)	389
Fields Brook	306
Firmicutes	5, 20, 179, 382-383, 404
Flavobacteriaceae	20
flocculation	214, 254, 266
flotation	214, 266
fosmids	166
fossil fuels	2-3, 25, 62, 109, 113, 120-121, 132, 135, 148, 223, 243, 255- 256, 261, 337
Freundlich isotherm	343
Froth flotation	266
Fukushima incident	9
Fukushima nuclear reactor accident	2
Fusarium oxysporum	8
FUTURA NANO	147
Gammaproteobacteria	20, 404
gaseous voltaic battery	42
gastrobots	47
gatY	163

GEMM riboswitch	24
gene knockout	111
Gene Ontology (GO) annotations	163
Genentech	160
Genomic Science Program	106
Geobacter	12, 22, 24-25, 68, 107-108, 181, 183-184, 197, 347, 384-385, 387, 399
metallireducens	22, 107, 177-178, 347-348, 382
sulfurreducens	11, 23, 25, 48, 50-51, 66, 90, 107, 177-178, 347-348, 382, 388
Geobacter sufurreducens	11
geopilins	66
Geopsychrobacter electrodiphilus	179
geothermal energy	3, 337
Geotrichum candidum	216
global warming	4, 109, 120, 160, 223, 261
Gluconobacter oxydans	178
Glucose oxidase (GOx)	49
Glycine max	287
glycine-betaine synthesis	163
goose grass (*Eleusine indica*)	288
granular-activated carbon	52
graphene	6-7
greenhouse gas	2, 7, 98, 124, 132, 206, 208, 223, 243, 249, 261-262
GSU1771	25
GSU3274	32
H2OIL	124
Haber Weiss reactions	401
Halomonas aromativorans	309
Halophiles	399
Hansenula anomala	69, 70
Harvesting, Dewatering and Drying (AVS-HDD)	214
Headwaters	124
heat-shock stress response	163
Henry's law	344
heterocysts	110
hexadecane	272, 278, 280
hexaferricyanide	28
High solid anaerobic digestion (HSAD)	223
High temperature gas-cooled reactors (HTGR)	339
High-level radioactive waste (HLW)	12, 396
High-level waste (HLW)	338
homoacetogenesis	206
Host Engineering	160, 166, 168
Humulin	160
hydraulic retention time (HRT)	209
Hydrogenases	110, 112, 237-238

hydrophobe/amphiphile (HAE1) family	166
Immobilization	46, 51, 53-55, 141, 149, 213, 381, 384
enzyme	123-126-128, 130, 131, 134
immunobiosensors	388-390
Immunogold localization	26
impedance spectroscopy	5
Indirect electron transfer system	23
Intermediate level waste (ILW)	338
International Energy Agency	3
International Energy Outlook (IEO) 2011	2-3
Ion exchange	344, 360
Irpex lacteus	311
IrrE	22, 167, 285, 313
Isochrysis	264
Jatropha	9, 256, 247, 262
Kan^R transposon	167, 169
KEIO	170
Kineococcus radiotolerans	11, 400-401, 403
Klebsiella oxytoca	8, 231
Klebsiella pneumoniae	50, 178-179, 353
Kluyveromyces marxians	8
KN400	24
Kudankulam nuclear plant	10
labelling techniques	110
Lachnospiraceae	224
Lactobacillus paracasei	227
Lactobacillus plantarum	178
Lactococcus lactis	107
lacZ	167
Lake Erie tributaries	305
Lake Taihu	306
Land-fill leachate	21-22
Landfarming	291, 307-308
Langmuir and Langmuir-Hinshelwood	343
Langmuir-Blodgett (LB) method	130
Lead dioxide	73, 192
lecithin	145
Leptothrix discophora	27, 51, 74
Lipid accumulation	207-209, 212, 215
liposomes	128
Long Beach soil	286
Low molecular weight (LMW)	304

Low-level waste (LLW)	338
lycopene	168
marCRAB	163
marCRAB,	163
Marinobacter aqueolei	166
mediated electron transfer (MET)	46, 179
meldolas blue (MelB)	177
Membrane processes	344
merA	11
mesoporous matrix	132
metagenomic DNA	166
Metagenomics	108
Metal hydrides	125, 145
Methanobacterium palustre	29
Methanococcus jannaschii,	11
methanogens	29, 195, 206, 224, 228, 237
methyl viologen	29, 180, 402
methyl-tert-butyl-ether (MTBE)	369
MFCs	
single chamber	95, 176, 188
Two-chamber	186-188
Micractinium pusillum	210
micro-array-based gene expression	107
microalgae	9, 205, 211-217, 224-225, 243, 249, 252- 255, 261-268
Microalgae harvesting	266
Microbacterium flavescens	410
MicrobesOnline	164
microbial biofuels	3
Microbial biorefineries	47
microbial consortia	89, 107-108
microbial desalination cells (MDCs)	44
Microbial electric systems	17, 32
microbial electrolysis cell (MEC)	44, 46, 195
Microbial electrosynthesis cell (MEC)	17
Microbial fuel cells (MFC's)	4, 5, 23- 25, 44, 47, 52, 62-63, 68, 70, 77, 98, 106- 107, 112, 120, 123, 135,
rumen	175
	84-97
Microbial nanowires	23-25
microbial solar cells (MSC)	44
Microbial Uranium (VI) Reduction Kinetics	355
Microcystis aeruginosa	208
microfiltration membrane (MFM)	53
molecular beam epitaxy	147
Moorella thermoaceticum	31

Mtr
    A   26, 349-350
    B   26, 349-350
    C   25, 183, 349-350
    F   183

Mucor sp   8
Multiplex Automated Genome Engineering (MAGE)   168
Muscodor vitigenus   305

Nafion   6, 54, 92-93, 97, 141, 187-188, 192, 196
Nano fibre   123
Nannochloris   264
Nannochloropsis   264
nano wires   66
nanocatalysts   123-125, 135
Nanocomposite membranes   6
Nanocomposites   123, 130-131, 147
Nanoemulsions   148
    additive-based   148
Nanoparticle nucleotide conjugates   134-135
Nanoparticles   7, 54, 123, 129-131, 139, 140, 143, 145, 147-148, 389-390

Nanophase   123
nanopowders   138, 147
Nanoribbon   123, 149,
Nanoscale   123, 131, 135, 139-140, 142

Nanoscience   123

Nanotechnology   6, 13, 120-126, 128-129, 134, 136, 138, 141, 143, 145-149, 380, 383, 389-390
Nanotechnology-enabled Coatings   147
nanotechnology-enabled information   149
intensive' extraction
National biofuel policy   9
National Renewable Energy Laboratory (NREL)   263
National Waste Minimization Program of the US   304
EPA (USEPA, 2009)
Neochloris   209, 264
Neochloris oleoabundans   209
Neurospora crassa   98
neutral red   17, 23, 27, 66, 72, 107, 177-179, 191
nicotinamide adenine dinucleotide   237
hydrogen (NADH)
NiFe hydrogenase   112
Nigeria's Niger Delta   272
Nitrogen deprivation   212
Nitrogenases   110
Nitzschia   264, 309

non-native regulators	167
nonsense mutations	168
nuclear magnetic resonance	131
Ochrobactrumanthropi YZ-1	178
Oder River	307
oleaginous microalgae	205, 211-215
open circuit voltage (OCV)	70
Open ponds	250, 253-254, 265
Operons	165
nuo	165
cyo	165
organic fraction of municipal solid waste (OFSWM)	224
Organization for Economic Cooperation and Development (OECD)	3
organometallics	148
Osmotic shock	267
Oxonica	124
P1vir lysate	167
Pacemakers	22, 47
painted turtles (*Chrysemys picta*)	305
partition coefficients (K_{ow})	304
PCP (pentachlorophenol)	325
Pearl River Delta	307
Pelobacter propionicus DSM 2379	166
Permeable reactive barriers	368-369, 372
Petroleum	2, 4, 8, 10, 21-22, 148, 261-262, 264, 267, 272-275, 280, 282-291, 305, 315, 403
based biodeisel	132
based chemicals	160
derived liquid fuels	246
Phacus	331
Phaeodactylum	210, 264
Phaeodactylum tricornutum	210
Phanearochaete chrysosporium	311
phase-transfer catalysis	132
Phenanthrene	272-273, 280, 288, 309-312, 315
PhoN	11-12
Phormidium bohneri	208
photo- and dark fermentation	109
photo-catalysis	133
photo-fermentation	109, 225-226, 229, 238
photosynthesis	27, 70, 109, 111, 212, 225, 248, 252, 268
photosynthetic active radiation (PAR)	70

photosynthetic algal microbial fuel cell (PAMFC 1)	70
photosynthetic cultures	5
Phycoremediation	252-254
physiorption	144-145
phytoremediation	285, 287, 289, 307, 313-315
Pichia anomala	178
Pichia stipitis	8
PilD	25
plant growth-promoting bacteria (PGPB)	213
platinum electrodes	42, 106
Pleurotus ostreatus	311
plexiglas	187, 189
polyacrylonitrile (PAN)	140, 148
Polyaniline (PANI)–Pt	72
polybenzimidazoles	139,
polychlorinated biphenyls (PCBs)	361
polycyclic aromatic hydrocarbons (PAHs)	272, 304
polydimethylsiloxane (PDMS)	148
polymer electrolyte membrane fuel cells (PEMFC)	136
Polypyrrole	5-6, 54
polyunsaturated fatty acids (PUFAs)	243
Porous chitosan	54
Porphyridium	264
post-translational modifications	111
pre-treated feedstock	125
Prepilins	244–253, 255–256, 259, 262, 264, 312
primary energy	2, 310, 337
propionate-type of fermentation	88
Proteobacteria	5
gammaproteobacteria	20
betaproteobacteria	20
proteogenomics	108, 388
Proteus	
mirabilis	66, 69, 91, 178, 190
vulgaris	66, 69
Protocatechuate	278, 334
proton exchange membrane (PEM)	92, 175
proton extrusion	161
PRP-Finder	111
Pseudomonas	20, 23, 66, 275, 286, 309-310, 328-329, 333, 352, 399
P. aeruginosa	
P. fluorescens	50, 69, 160, 163, 178, 231, 286, 290, 308, 345, 352
P. palustris	
P. putida	66, 69, 364
P. migulae	50
P. stutzeri	11, 310, 348, 364, 384, 387, 309 352-354, 357
pyrosequencing	5
Quantum dot	123

Raceways	211, 212, 265, 267, 268
random immobilization	53
reactive oxygen species (ROS)	12, 400
Redox enzymes	48, 54-55, 65, 77
Reed mannagrass	70
Remediation by enhanced natural attenuation (RENA)	291
Renewable energy source	17, 120, 206, 223, 256, 261
reticulated vitreous carbon	52, 72, 188, 191
Rhizobiales	20
rhizodeposits	70
Rhodobacter	5
R. sphaeroides	71
Rhodoferax	5, 24, 68, 90, 107, 177, 181
R. ferrireducens	178
Rhodopseudomonas	5
R. palustris	50
Rhodopseudomonas palustris DX-1	178
Rhodospirillum rubrum	226
Rhodotorula glutinis	217
RNA polymerase	161
RNAi	111
Rubrobacter xylanophilus	11
rumen	84-98
Rumuekpe oilfield	290
Saccharomyces cerevisiae	8, 50, 161
SARP-like regulator	25
scanning tunnelling microscopy	25
Scenedesmus obliquus	208, 210
Schizochytrium	210, 264
Schizochytrium sp.	210
Selenastrum capricornutum	309
SEM/EDS spectra	402
semivolatile organic compounds (SVOC)	304
sequencing batch reactor (SBR)	334
Shewanella	12, 23-26, 48, 66, 68-69, 181, 184, 197, 347-350, 382
S. putrefaciens	11, 48, 50, 66, 107, 177-178, 181, 347
S. oneidensis	11, 23, 50, 107, 178, 348-349, 371, 389, 400
Shewanella reductase(s)	347
Shine-Dalgarno canonical sequence	168
short chain fatty acids (SCFA)	88
Sigma factors	162
Silicone rubber membrane	231, 237
single-wall carbon nanotubes (SWNT)	136
single-walled carbon nanotubes	54
singleprostheca (stalk).	364

Skeletonema costatum	309
Sludge granulation	231, 237
sludge retention times (SRTs)	206
snapping turtles (*Chelydra serpentina*)	305
solar energy	1, 3, 225- 226, 245, 248, 252
solvent-resistant nanofiltration (SRNF)	148
Soxhlet apparatus	266
Spartina anglica	71
special conductive pili	177, 180
Sphingomonas yanoikuyae	309
Spirulina platensis	209-120
Sporolactobacillaceae	224,
Sporomusa	
ovata	30, 31
sphaeroides	31
silvacetica	31
sputtering	147
Stacked MFCs	189-190
Steam explosion	76
Streptococcaceae	224
Streptococcus fragilis	8
Streptomyces avermitillis	161
Stripping	344
Sulfurospirillum deleyianum	180
Surface attachment	127-128, 136
Surface loading	127
Synechocystis sp. PCC 6803	112
syngas	130, 230, 244, 248, 249
syringaldehyde	166-167
systems microbiology	98, 105, 107, 109-113
Terrabacter sp	310, 408,
Tetraselmis	264
thallophytes	243
Thermoanaerobacteriaceae	224
Thermoanerobacter ethanolicus	8
Thermococcus kodakaraensis KOD1	231
Thermotogales spp.	231
Thermus brockii	309
Thermus thermophilis	11
Thiocapsa roseopersicina	112
TiAlCrSiYn coatings	284, 340
Titania	133-136, 139
tnaA	163
Tohoku (Great East Japan) Earthquake	10
Toluene	11-12, 22, 166, 272, 306, 314-315
total heterotrophic bacteria (THB)	288, 291
total hydrocarbon content (THC)	385-386
total petroleum hydrocarbon (TPH)	283

Trametes versicolor	311
transesterification	132, 143, 215, 248-249, 263
transmission electron microscopy (TEM) imaging	350
Transuranic Waste	337-338
trichlorophenol (TCP)	326
Trichothecium roseum	216
Tungsten carbide anodes	72
Two-step Detoxication Technology	334
UDiamond	147
ultrafiltration membrane (UFM)	53, 192
UNH Biodiesel Group	263
United States Department of Energy (DOE)	105
US Environmental Protection Agency (USEPA, 2011)	304
US Geological Survey National Water Quality Assessment	306
vadose zone	396, 404-405, 410
vetiver grass (*Vetiveria zizanioides*)	287
Vibrio pelagius	315
Virtual Institute for Microbial Stress and Survival	164
volatile fatty acid (VFA)	206
Waste Isolation Pilot Plant (WIPP)	404
wastewater treatment	5, 7, 27-28, 42, 47, 52, 54, 63, 74-77, 106, 176-177, 181, 187, 192, 194, 196, 205, 207-211, 214, 230, 238, 250, 252, 264, 361-362
Water–gas shift reaction	225, 226, 229-230
white blood cells	22
white rot fungi (WRF)	311
Wood-Ljungdahl pathway	29
X-ray absorption near-edge structural analysis (XANES)	389
Xenobiotics	324-325, 332-334
ycjF	164-165
ycjX	165
Yeast	70, 176, 178, 205, 207, 211, 215, 216, 217, 274-274, 344, 380
Y. lipolytica	216
yhbJ	163
Zymomonas mobilis	8
β-oxidation	275-277